STUDENT SOLUTIONS MANUAL

Richard N. Aufmann
Palomar College

Vernon C. Barker
Palomar College

Joanne S. Lockwood
Plymouth State University

INTERMEDIATE ALGEBRA: AN APPLIED APPROACH

SEVENTH EDITION

Aufmann/Barker/Lockwood

HOUGHTON MIFFLIN COMPANY BOSTON NEW YORK

Senior Sponsoring Editor: Lynn Cox
Associate Editor: Melissa Parkin
Editorial Assistant: Noel Kamm
Senior Project Editor: Tamela Ambush
Editorial Assistant: Sage Anderson
Assistant Manufacturing Coordinator: Karmen Chong
Senior Marketing Manager: Ben Rivera

Printed in the U.S.A.

ISBN: 0-618-52037-6

4 5 6 7 8 9-CRS-09 08 07 06

Contents

Student Solutions Manual

Chapter 1: Review of Real Numbers

Prep Test

1. $\frac{5}{12}+\frac{7}{30}=\frac{25}{60}+\frac{14}{60}=\frac{25+14}{60}$

$=\frac{39}{60}=\frac{\overset{1}{\cancel{3}}\cdot 13}{\underset{1}{\cancel{3}}\cdot 2\cdot 2\cdot 5}=\frac{13}{20}$

2. $\frac{8}{15}-\frac{7}{20}=\frac{32}{60}-\frac{21}{60}=\frac{32-21}{60}$

$=\frac{11}{60}$

3. $\frac{5}{6}\cdot\frac{4}{15}=\frac{\overset{1}{\cancel{5}}\cdot\overset{1}{\cancel{2}}\cdot 2}{\underset{1}{\cancel{2}}\cdot 3\cdot 3\cdot\underset{1}{\cancel{5}}}=\frac{2}{9}$

4. $\frac{4}{15}\div\frac{2}{5}=\frac{4}{15}\cdot\frac{5}{2}=\frac{\overset{1}{\cancel{2}}\cdot 2\cdot\overset{1}{\cancel{5}}}{3\cdot\underset{1}{\cancel{5}}\cdot\underset{1}{\cancel{2}}}=\frac{2}{3}$

5.
$$\begin{array}{r} 8.000 \\ 29.340 \\ +\ 7.065 \\ \hline 44.405 \end{array}$$

6.
$$\begin{array}{r} 92.00 \\ -18.37 \\ \hline 73.63 \end{array}$$

7. $2.19(3.4) = 7.446$

8.
$$\begin{array}{r} 54.06 \\ 6\overline{)324.36} \\ \underline{30} \\ 24 \\ \underline{24} \\ 03 \\ \underline{0} \\ 36 \\ \underline{36} \\ 0 \end{array}$$

• $32.436 \div 0.6 = 324.36 \div 6$

9. a. $-6 > -8$, Yes
b. $-10 < -8$, No
c. $0 > -8$, Yes
d. $8 > -8$, Yes

10. a. $\frac{1}{2}= 0.5$; C
b. $\frac{7}{10}= 0.7$; D
c. $\frac{3}{4}= 0.75$; A
d. $\frac{89}{100}= 0.89$; B

Go Figure

If the areas of the known rectangles are 2, 3, and 6, the corresponding rectangles have to be sizes 2×1, 3×1, and 2×3 (a rectangle of size 6×1 would not meet the perpendicular line requirement). In order for the 4th rectangle to be attached to the three above-sized rectangles, it must be of size 1×1, 2×2, or 3×3. Therefore, the possible values of x, or the areas of these three rectangles, are 1, 4, and 9.
Answer: 1, 4, 9

Section 1.1

Objective A Exercises

1. **a.** Integers: 0, −3
b. Rational numbers: $-\frac{15}{2}, 0, -3, 2.\overline{33}$
c. Irrational numbers: π, 4.232232223..., $\frac{\sqrt{5}}{4}, \sqrt{7}$
d. Real numbers: all

3. −27

5. $-\frac{3}{4}$

7. 0

9. $\sqrt{33}$

11. 91

13. Replace y with each element in the set and determine whether the inequality is true.
$y > -4$
$-6 > -4$ False
$-4 > -4$ False
$7 > -4$ True

15. Replace w with each element in the set and determine whether the inequality is true.
$w \le -1$
$-2 \le -1$ True
$-1 \le -1$ True
$0 \le -1$ False
$1 \le -1$ False

17. Replace b with each element in the set and evaluate the expression.
$-b$
$-(-9) = 9$
$-(0) = 0$
$-(9) = -9$

19. Replace c with each element in the set and evaluate the expression.
$|c|$
$|-4| = 4$
$|0| = 0$
$|4| = 4$

Objective B Exercises

21. $\{-2, -1, 0, 1, 2, 3, 4\}$

23. $\{2, 4, 6, 8, 10, 12\}$

25. $\{3, 6, 9, 12, 15, 18, 21, 24, 27, 30\}$

27. $\{x|x > 4, x \in \text{integers}\}$

29. $\{x|x \geq -2\}$

31. $\{x|0 < x < 1\}$

33. $\{x|1 \leq x \leq 4\}$

Objective C Exercises

35. $A \cup B = \{1, 2, 4, 6, 9\}$

37. $A \cup B = \{2, 3, 5, 8, 9, 10\}$

39. $A \cup B = \{-4, -2, 0, 2, 4, 8\}$

41. $A \cup B = \{1, 2, 3, 4, 5\}$

43. $A \cap B = \{6\}$

45. $A \cap B = \{5, 10, 20\}$

47. $A \cap B = \varnothing$

49. $A \cap B = \{4, 6\}$

51. $\{x|x < 2\}$
−5 −4 −3 −2 −1 0 1 2 3 4 5

53. $\{x|x \geq 1\}$
−5 −4 −3 −2 −1 0 1 2 3 4 5

55. $\{x|-1 < x < 5\}$
−5 −4 −3 −2 −1 0 1 2 3 4 5

57. $\{x|0 \leq x \leq 3\}$
−5 −4 −3 −2 −1 0 1 2 3 4 5

59. $\{x|x > 1\} \cup \{x|x < -1\}$
−5 −4 −3 −2 −1 0 1 2 3 4 5

61. $\{x|x \leq 2\} \cap \{x|x \geq 0\}$
−5 −4 −3 −2 −1 0 1 2 3 4 5

63. $\{x|x > 1\} \cap \{x|x \geq -2\}$
−5 −4 −3 −2 −1 0 1 2 3 4 5

65. $\{x|x > 2\} \cup \{x|x > 1\}$
−5 −4 −3 −2 −1 0 1 2 3 4 5

67. $\{x|0 < x < 8\}$

69. $\{x|-5 \leq x \leq 7\}$

71. $\{x|-3 \leq x < 6\}$

73. $\{x|x \leq 4\}$

75. $\{x|x > 5\}$

77. $(-2, 4)$

79. $[-1, 5]$

81. $(-\infty, 1)$

83. $[-2, \infty)$

85. $(-2, 5)$
−5 −4 −3 −2 −1 0 1 2 3 4 5

87. $[-1, 2]$
−5 −4 −3 −2 −1 0 1 2 3 4 5

89. $(-\infty, 3]$
−5 −4 −3 −2 −1 0 1 2 3 4 5

91. $[3, \infty)$
−5 −4 −3 −2 −1 0 1 2 3 4 5

93. $[-5, 0) \cup (1, 4]$
−5 −4 −3 −2 −1 0 1 2 3 4 5

95. $[-3, 3] \cap [0, 5]$
−5 −4 −3 −2 −1 0 1 2 3 4 5

Applying the Concepts

97. $A \cup B = \{x|-1 \leq x \leq 1\} \cup \{x|0 \leq x \leq 1\}$
$= \{x|-1 \leq x \leq 1\}$
$= A$

99. $B \cap B$ is set B.

101. $A \cap R$ is $\{x|-1 \leq x \leq 1\}$,which is set A.

103. $B \cup R$ is the set of real numbers, R.

105. $R \cup R$ is the set of real numbers, R.

107. $B \cap C$ is $\{x|0 \leq x \leq 1\} \cap \{x|-1 \leq x \leq 0\}$, which contains only the number 0.

Section 1.2

Objective A Exercises

1. **a.** Students should paraphrase the following rule: Add the absolute values of the numbers; then attach the sign of the addends.
   b. Students should paraphrase the following rule: Find the absolute value of each number; subtract the smaller of the two numbers from the larger; then attach the sign of the number with the larger absolute value.

3. $-18 + (-12) = -30$

5. $5 - 22 = 5 + (-22) = -17$

7. $3 \cdot 4 \cdot (-8) = 12 \cdot (-8) = -96$

9. $18 \div (-3) = -6$

11. $-60 \div (-12) = 5$

13. $-20(35)(-16) = -700(-16) = 11{,}200$

15. $8 - (-12) = 8 + 12 = 20$

17. $|12(-8)| = |-96| = 96$

19. $|15-(-8)| = |15 + 8| = |23| = 23$

21. $|-56 \div 8| = |-7| = 7$

23. $|-153 \div (-9)| = |17| = 17$

25. $-|-8| + |-4| = -8 + 4 = -4$

27. $$\begin{aligned} -30 + (-16) - 14 - 2 &= -30 + (-16) + (-14) + (-2) \\ &= -46 + (-14) + (-2) \\ &= -60 + (-2) \\ &= -62 \end{aligned}$$

29. $$\begin{aligned} -2 + (-19) - 16 + 12 &= -2 + (-19) + (-16) + 12 \\ &= -21 + (-16) + 12 \\ &= -37 + 12 \\ &= -25 \end{aligned}$$

31. $$\begin{aligned} 13 - |6 - 12| &= 13 - |6 + (-12)| \\ &= 13 - |-6| \\ &= 13 - 6 \\ &= 13 + (-6) = 7 \end{aligned}$$

33. $$\begin{aligned} &738 - 46 + (-105) - 219 \\ &= 738 + (-46) + (-105) + (-219) \\ &= 692 + (-105) + (-219) \\ &= 587 + (-219) \\ &= 368 \end{aligned}$$

35. $-442 \div (-17) = 26$

37. $-4897 \div 59 = -83$

Objective B Exercises

39. **a.** The least common multiple of two numbers is the smallest number that is a multiple of each of those numbers.

b. The greatest common factor of two numbers is the largest integer that divides evenly into both numbers.

41. $\frac{7}{12} - \left(-\frac{5}{16}\right) = \frac{7}{12} + \frac{5}{16} = \frac{28}{48} + \frac{15}{48} = \frac{28 + 15}{48} = \frac{43}{48}$

43. $$\begin{aligned} -\frac{5}{9} - \frac{14}{15} &= -\frac{25}{45} - \frac{42}{45} \\ &= \frac{-25 - 42}{45} \\ &= -\frac{67}{45} \end{aligned}$$

45. $$\begin{aligned} -\frac{1}{3} + \frac{5}{9} - \frac{7}{12} &= -\frac{12}{36} + \frac{20}{36} - \frac{21}{36} \\ &= \frac{-12 + 20 - 21}{36} \\ &= -\frac{13}{36} \end{aligned}$$

47. $$\begin{aligned} \frac{2}{3} - \frac{5}{12} + \frac{5}{24} &= \frac{16}{24} - \frac{10}{24} + \frac{5}{24} \\ &= \frac{16 - 10 + 5}{24} \\ &= \frac{11}{24} \end{aligned}$$

49. $$\begin{aligned} \frac{5}{8} - \frac{7}{12} + \frac{1}{2} &= \frac{15}{24} - \frac{14}{24} + \frac{12}{24} \\ &= \frac{15 - 14 + 12}{24} \\ &= \frac{13}{24} \end{aligned}$$

51. $$\begin{aligned} \left(\frac{6}{35}\right)\left(-\frac{5}{16}\right) &= -\frac{6 \cdot 5}{35 \cdot 16} \\ &= -\frac{\overset{1}{\cancel{2}} \cdot 3 \cdot \overset{1}{\cancel{5}}}{\underset{1}{\cancel{5}} \cdot 7 \cdot \underset{1}{\cancel{2}} \cdot 2 \cdot 2 \cdot 2} = -\frac{3}{56} \end{aligned}$$

53. $$\begin{aligned} -\frac{8}{15} \div \frac{4}{5} &= -\frac{8}{15} \cdot \frac{5}{4} \\ &= -\frac{8 \cdot 5}{15 \cdot 4} \\ &= -\frac{\overset{1}{\cancel{2}} \cdot \overset{1}{\cancel{2}} \cdot 2 \cdot \overset{1}{\cancel{5}}}{3 \cdot \underset{1}{\cancel{5}} \cdot \underset{1}{\cancel{2}} \cdot \underset{1}{\cancel{2}}} = -\frac{2}{3} \end{aligned}$$

55. $$\begin{aligned} -\frac{11}{24} \div \frac{7}{12} &= -\frac{11}{24} \cdot \frac{12}{7} \\ &= -\frac{11 \cdot 12}{24 \cdot 7} \\ &= -\frac{11 \cdot \overset{1}{\cancel{2}} \cdot \overset{1}{\cancel{2}} \cdot \overset{1}{\cancel{3}}}{\underset{1}{\cancel{2}} \cdot \underset{1}{\cancel{2}} \cdot 2 \cdot \underset{1}{\cancel{3}} \cdot 7} = -\frac{11}{14} \end{aligned}$$

57. $$\begin{aligned} \left(-\frac{5}{12}\right)\left(\frac{4}{35}\right)\left(\frac{7}{8}\right) &= -\frac{5 \cdot 4 \cdot 7}{12 \cdot 35 \cdot 8} \\ &= -\frac{\overset{1}{\cancel{5}} \cdot \overset{1}{\cancel{2}} \cdot \overset{1}{\cancel{2}} \cdot \overset{1}{\cancel{7}}}{\underset{1}{\cancel{2}} \cdot \underset{1}{\cancel{2}} \cdot 3 \cdot \underset{1}{\cancel{5}} \cdot \underset{1}{\cancel{7}} \cdot 2 \cdot 2 \cdot 2} \\ &= -\frac{1}{24} \end{aligned}$$

59. $$\begin{array}{r} -14.270 \\ +\ 1.296 \\ \hline -12.974 \end{array}$$

61. $$\begin{aligned} 1.832 - 7.84 &= 1.832 + (-7.84) \\ &= -6.008 \end{aligned}$$

63. $(0.03)(10.5)(6.1) = (0.315)(6.1) = 1.9215$

65.
$$\begin{array}{r} 6.02 \\ 0.9\overline{)5.418} \\ \underline{-5\,4} \\ 0\,01 \\ \underline{-0} \\ 18 \\ \underline{-18} \\ 0 \end{array}$$

$5.418 \div (-0.9) = -6.02$

67.
$$\begin{array}{r} 6.7 \\ 0.065\overline{)0.4355} \\ \underline{-390} \\ 455 \\ \underline{-455} \\ 0 \end{array}$$

$-0.4355 \div 0.065 = -6.7$

69. $38.241 \div [-(-6.027)] - 7.453$
$= 38.241 \div 6.027 + (-7.453)$
≈ -1.11

71. $-287.3069 \div 0.1415 \approx -2030.44$

Objective C Exercises

73. $5^3 = 5 \cdot 5 \cdot 5 = 125$

75. $-2^3 = -(2 \cdot 2 \cdot 2) = -8$

77. $(-5)^3 = (-5)(-5)(-5) = -125$

79. $2^2 \cdot 3^4 = (2)(2) \cdot (3)(3)(3)(3)$
$= 4 \cdot 81$
$= 324$

81. $-2^2 \cdot 3^2 = -(2)(2) \cdot (3)(3)$
$= -4 \cdot 9$
$= -36$

83. $(-2)^3(-3)^2 = (-2)(-2)(-2) \cdot (-3)(-3)$
$= -8 \cdot 9$
$= -72$

85. $4 \cdot 2^3 \cdot 3^3 = 4 \cdot (2)(2)(2) \cdot (3)(3)(3)$
$= 4 \cdot 8 \cdot 27$
$= 32 \cdot 27$
$= 864$

87. $2^2(-10)(-2)^2 = 2 \cdot 2 \cdot (-10)(-2)(-2)$
$= 4 \cdot (-10)(4)$
$= -40(4)$
$= -160$

89. $\left(-\frac{2}{3}\right)^2 \cdot 3^3 = \frac{2 \cdot 2 \cdot \overset{1}{\cancel{3}} \cdot \overset{1}{\cancel{3}} \cdot 3}{\underset{1}{\cancel{3}} \cdot \underset{1}{\cancel{3}}} = 12$

91. $2^5(-3)^4 \cdot 4^5 = 32 \cdot (81) \cdot 1024$
$= 2592 \cdot 1024$
$= 2{,}654{,}208$

Objective D Exercises

93. We need an Order of Operations Agreement to ensure that there is only one way in which an expression can be correctly simplified.

95. $5 - 3(8 \div 4)^2 = 5 - 3(2)^2$
$= 5 - 3(4)$
$= 5 - 12 = -7$

97. $16 - \frac{2^2 - 5}{3^2 + 2} = 16 - \frac{4 - 5}{9 + 2}$
$= 16 - \frac{-1}{11}$
$= 16 + \frac{1}{11} = \frac{177}{11}$

99. $\frac{3 + \frac{2}{3}}{\frac{11}{16}} = \frac{\frac{11}{3}}{\frac{11}{16}} = \frac{11}{3} \cdot \frac{16}{11} = \frac{16}{3}$

101. $5[(2 - 4) \cdot 3 - 2] = 5[(-2) \cdot 3 - 2]$
$= 5[-6 - 2]$
$= 5[-8] = -40$

103. $16 - 4\left(\frac{8 - 2}{3 - 6}\right) \div \frac{1}{2} = 16 - 4\left(\frac{6}{-3}\right) \div \frac{1}{2}$
$= 16 - 4(-2) \div \frac{1}{2}$
$= 16 - (-8) \div \frac{1}{2}$
$= 16 - (-8) \cdot 2$
$= 16 - (-16) = 16 + 16 = 32$

105. $6[3 - (-4 + 2) \div 2] = 6[3 - (-2) \div 2]$
$= 6[3 - (-1)]$
$= 6[3 + 1] = 6[4] = 24$

107. $\frac{1}{2} - \left(\frac{2}{3} \div \frac{5}{9}\right) + \frac{5}{6} = \frac{1}{2} - \left(\frac{2}{3} \cdot \frac{9}{5}\right) + \frac{5}{6}$
$= \frac{1}{2} - \frac{6}{5} + \frac{5}{6}$
$= \frac{15}{30} - \frac{36}{30} + \frac{25}{30}$
$= \frac{15 - 36 + 25}{30}$
$= \frac{4}{30} = \frac{2}{15}$

109. $$\frac{1}{2}-\frac{\frac{17}{25}}{4-\frac{3}{5}}\div\frac{1}{5}=\frac{1}{2}-\frac{\frac{17}{25}}{\frac{17}{5}}\div\frac{1}{5}$$
$$=\frac{1}{2}-\left(\frac{17}{25}\cdot\frac{5}{17}\right)\div\frac{1}{5}$$
$$=\frac{1}{2}-\frac{1}{5}\div\frac{1}{5}$$
$$=\frac{1}{2}-\frac{1}{5}\cdot\frac{5}{1}$$
$$=\frac{1}{2}-1=-\frac{1}{2}$$

111. $$\frac{2}{3}-\left[\frac{3}{8}+\frac{5}{6}\right]\div\frac{3}{5}=\frac{2}{3}-\left[\frac{9}{24}+\frac{20}{24}\right]\div\frac{3}{5}$$
$$=\frac{2}{3}-\frac{29}{24}\div\frac{3}{5}$$
$$=\frac{2}{3}-\frac{29}{24}\cdot\frac{5}{3}$$
$$=\frac{2}{3}-\frac{145}{72}$$
$$=\frac{48}{72}-\frac{145}{72}=-\frac{97}{72}$$

113. $$0.4(1.2-2.3)^2+5.8=0.4(-1.1)^2+5.8$$
$$=0.4(1.21)+5.8$$
$$=0.484+5.8$$
$$=6.284$$

115. $$1.75\div 0.25-(1.25)^2=1.75\div 0.25-1.5625$$
$$=7-1.5625$$
$$=5.4375$$

117. $$25.76\div(6.96-3.27)^2=25.76\div(3.69)^2$$
$$=25.76\div 13.6161$$
$$=1.891878$$

Applying the Concepts

119. 0

121. No, the number zero has a multiplicative inverse that is undefined.

123. $7^{18}=1{,}628{,}413{,}597{,}910{,}449$
The ones digit is 9.

125. 5^{234} has over 150 digits. The last three are 625.

127. First find b^c; then find $a^{(b^c)}$.

Section 1.3

Objective A Exercises

1. $3\cdot 4=4\cdot 3$

3. $(3+4)+5=3+(4+5)$

5. $\frac{5}{0}$ is undefined.

7. $3(x+2)=3x+6$

9. $\frac{0}{-6}=0$

11. $\frac{1}{mn}(mn)=1$

13. $2(3x)=(2\cdot 3)\cdot x$

15. The Division Property of Zero

17. The Inverse Property of Multiplication

19. The Addition Property of Zero

21. The Division Property of Zero

23. The Distributive Property

25. The Associative Property of Multiplication

Objective B Exercises

27. $$ab+dc=(2)(3)+(-4)(-1)$$
$$=6+4$$
$$=10$$

29. $$4cd\div a^2=4(-1)(-4)\div(2)^2$$
$$=4(-1)(-4)\div 4$$
$$=(-4)(-4)\div 4$$
$$=16\div 4=4$$

31. $$(b-2a)^2+c=[3-2(2)]^2+(-1)$$
$$=[3-4]^2+(-1)$$
$$=[-1]^2+(-1)$$
$$=1+(-1)=0$$

33. $$(bc+a)^2\div(d-b)=[(3)(-1)+2]^2\div(-4-3)$$
$$=[-3+2]^2\div(-7)$$
$$=[-1]^2\div(-7)$$
$$=1\div(-7)=-\frac{1}{7}$$

35. $$\frac{1}{4}a^4-\frac{1}{6}bc=\frac{1}{4}(2)^4-\frac{1}{6}(3)(-1)$$
$$=\frac{1}{4}(16)-\frac{1}{6}(3)(-1)$$
$$=4-\frac{1}{6}(3)(-1)$$
$$=4-\frac{1}{2}(-1)$$
$$=4-\left(-\frac{1}{2}\right)$$
$$=4+\frac{1}{2}=\frac{9}{2}$$

37. $$\frac{3ac}{-4}-c^2=\frac{3(2)(-1)}{-4}-(-1)^2$$
$$=\frac{6(-1)}{-4}-(-1)^2$$
$$=\frac{-6}{-4}-(-1)^2$$
$$=\frac{3}{2}-(-1)^2$$
$$=\frac{3}{2}-1=\frac{1}{2}$$

39. $\frac{3b-5c}{3a-c}=\frac{3(3)-5(-1)}{3(2)-(-1)}$
$=\frac{9-(-5)}{6-(-1)}$
$=\frac{9+5}{6+1}$
$=\frac{14}{7}=2$

41. $\frac{a-d}{b+c}=\frac{2-(-4)}{3+(-1)}$
$=\frac{2+4}{3+(-1)}$
$=\frac{6}{2}=3$

43. $-a|a+2d|=-2|2+2(-4)|$
$=-2|2+(-8)|$
$=-2|-6|$
$=-2(6)=-12$

45. $\frac{2a-4d}{3b-c}=\frac{2(2)-4(-4)}{3(3)-(-1)}$
$=\frac{4-(-16)}{9-(-1)}$
$=\frac{4+16}{9+1}=\frac{20}{10}=2$

47. $-3d\div\left|\frac{ab-4c}{2b+c}\right|=-3(-4)\div\left|\frac{2(3)-4(-1)}{2(3)+(-1)}\right|$
$=-3(-4)\div\left|\frac{6-(-4)}{6+(-1)}\right|$
$=-3(-4)\div\left|\frac{6+4}{6+(-1)}\right|$
$=-3(-4)\div\left|\frac{10}{5}\right|$
$=-3(-4)\div|2|$
$=-3(-4)\div 2$
$=12\div 2=6$

49. $2(d-b)\div(3a-c)=2(-4-3)\div[3(2)-(-1)]$
$=2(-7)\div[6-(-1)]$
$=2(-7)\div[6+1]$
$=2(-7)\div 7$
$=-14\div 7=-2$

51. $-d^2-c^3a=-(-4)^2-(-1)^3 2$
$=-16-(-1)2$
$=-16+2=-14$

53. $-d^3+4ac=-(-4)^3+4(2)(-1)$
$=-(-64)+8(-1)$
$=64-8=56$

55. $4^{(a^2)}=4^{(2^2)}=4^4=256$

Objective C Exercises

57. $5x+7x=12x$

59. $-8ab-5ab=-13ab$

61. $3x-5x+9x=-2x+9x=7x$

63. $5b-8a-12b=-8a-7b$

65. $\frac{1}{3}(3y)=y$

67. $-5(x-9)=-5x+45$

69. $-(x+y)=-x-y$

71. $3(a-5)=3a-15$

73. $4x-3(2y-5)=4x-6y+15$

75. $3x-2(5x-7)=3x-10x+14$
$=-7x+14$

77. $3[a-5(5-3a)]=3[a-25+15a]$
$=3[16a-25]$
$=48a-75$

79. $3[x-2(x+2y)]=3[x-2x-4y]$
$=3[-x-4y]$
$=-3x-12y$

81. $-2(x-3y)+2(3y-5x)=-2x+6y+6y-10x$
$=-12x+12y$

83. $5(3a-2b)-3(-6a+5b)=15a-10b+18a-15b$
$=33a-25b$

85. $3x-2[y-2(x+3[2x+3y])]$
$=3x-2[y-2(x+6x+9y)]$
$=3x-2[y-2(7x+9y)]$
$=3x-2[y-14x-18y]$
$=3x-2[-17y-14x]$
$=3x+34y+28x$
$=31x+34y$

87. $4-2(7x-2y)-3(-2x+3y)$
$=4-14x+4y+6x-9y$
$=4-8x-5y$

89. $\frac{1}{3}[8x-2(x-12)+3]=\frac{1}{3}[8x-2x+24+3]$
$=\frac{1}{3}[6x+27]$
$=2x+9$

Applying the Concepts

91. $4(3y+1)=12y+4$
The statement is correct; it uses the Distributive Property.

93. $2+3x+(2+3)x=5x$
The statement is not correct; it incorrectly uses the Distributive Property. It is an irreducible statement. That is, the answer is $2+3x$.

95. $2(3y)=(2\cdot 3)(2y)=12y$
The statement is not correct; it incorrectly uses the Associative Property of Multiplication. The correct answer is $(2\cdot 3)y=6y$.

97. $-x^2 + y^2 = y^2 - x^2$
The statement is correct; it uses the Commutative Property of Addition.

Section 1.4

Objective A Exercises

1. The unknown number: n
Eight less than a number: $n - 8$

3. The unknown number: n
Four-fifths of a number: $\frac{4}{5}n$

5. The unknown number: n
The quotient of a number and fourteen: $\frac{n}{14}$

7. The unknown number: n
The sum of the number and two: $n + 2$
$$n - (n + 2) = n - n - 2$$
$$= -2$$

9. The unknown number: n
The product of eight and the number: $8n$
$5(8n) = 40n$

11. The unknown number: n
The product of seventeen and the number: $17n$
twice the number: $2n$
$17n - 2n = 15n$

13. The unknown number: n
The square of the number: n^2
The total of twelve and the square of the number: $12 + n^2$
$n^2 - (12 + n^2) = n^2 - 12 - n^2 = -12$

15. The unknown number: n
The sum of five times the number and 12: $5n + 12$
The product of the number and fifteen: $15n$
$$15n + (5n + 12) = 15n + 5n + 12$$
$$= 20n + 12$$

17. Let the smaller number be x.
The larger number is $15 - x$.
The sum of twice the smaller number and two more than the larger number.
$$(15 - x + 2) + 2x = (17 - x) + 2x$$
$$= x + 17$$

19. Let the larger number be x.
The smaller number is $34 - x$.
The quotient of five times the smaller number and the difference between the larger number and three.
$$\frac{5(34 - x)}{x - 3}$$

Objective B Exercises

21. Distance from Earth to moon: d
Distance from Earth to sun is 390 times the distance from Earth to the moon: $390d$

23. Amount of caramel in the mixture: c
Amount of milk chocolate in the mixture is 3 lb more than the amount of caramel: $c + 3$

25. The amount in the first account: x
The total amount is 10,000.
The amount in the second account: $10{,}000 - x$

27. The measure of angle B: x
The measure of angle A is twice the measure of B: $2x$
The measure of angle C is twice the measure of A: $4x$

Applying the Concepts

29. The sum of twice x and 3

31. Twice the sum of x and 3

33. a. One-half the acceleration due to gravity: $\frac{1}{2}g$
Time squared: t^2
The product: $\frac{1}{2}gt^2$
b. The product of m and a: ma
c. The product of A and v^2: Av^2

Chapter 1 Review Exercises

1. $\{-2, -1, 0, 1, 2, 3\}$

2. $A \cap B = \{2, 3\}$

3. $(-2, 4]$
−5 −4 −3 −2 −1 0 1 2 3 4 5

4. The Associative Property of Multiplication

5. $-4.07 + 2.3 - 1.07 = -1.77 - 1.07 = -2.84$

6. $$(a - 2b^2) \div ab = (4 - 2(-3)^2) \div [(4)(-3)]$$
$$= (4 - 2(9)) \div [(4)(-3)]$$
$$= (4 - 18) \div [(4)(-3)]$$
$$= -14 \div [(4)(-3)]$$
$$= -14 \div (-12) = \frac{-14}{-12} = \frac{7}{6}$$

7. $$-2 \cdot (4^2) \cdot (-3)^2 = -2 \cdot 16 \cdot 9$$
$$= -32 \cdot 9$$
$$= -288$$

8. $$4y - 3[x - 2(3 - 2x) - 4y]$$
$$= 4y - 3[x - 6 + 4x - 4y]$$
$$= 4y - 3[5x - 6 - 4y]$$
$$= 4y - 15x + 18 + 12y$$
$$= 16y - 15x + 18$$

9. $\frac{3}{4}; -\frac{3}{4}+\frac{3}{4}=0$

10. $\{x|x<-3\}$

11. $\{x|x<1\}$

-5 -4 -3 -2 -1 0 1 2 3 4 5

12. $-10-(-3)-8=-10+3+(-8)$
$=-7+(-8)$
$=-15$

13. $-\frac{2}{3}+\frac{3}{5}-\frac{1}{6}=-\frac{20}{30}+\frac{18}{30}-\frac{5}{30}$
$=\frac{-20+18-5}{30}$
$=\frac{-7}{30}=-\frac{7}{30}$

14. 4

15. $-\frac{3}{8}\div\frac{3}{5}=-\frac{3}{8}\cdot\frac{5}{3}$
$=-\frac{\cancel{3}\cdot 5}{8\cdot\cancel{3}}$
$=-\frac{5}{8}$

16. Replace x with the elements in the set and determine whether the inequality is true.
$x>-1$
$-4>-1$ False
$-2>-1$ False
$0>-1$ True
$2>-1$ True

17. $2a^2-\frac{3b}{a}=2(-3)^2-\frac{3(2)}{-3}$
$=2(-3)^2-\frac{6}{-3}$
$=2(-3)^2-(-2)$
$=2(9)-(-2)$
$=18+2=20$

18. $18-|-12+8|=18-|-4|$
$=18-4$
$=14$

19. $20\div\frac{3^2-2^2}{3^2+2^2}=20\div\frac{9-4}{9+4}$
$=20\div\frac{5}{13}$
$=20\cdot\frac{13}{5}=52$

20. $[-3,\infty)$

-5 -4 -3 -2 -1 0 1 2 3 4 5

21. $A\cup B=\{1,2,3,4,5,6,7,8\}$

22. $-204\div(-17)=12$

23. $\{x|-2\le x\le 3\}$

24. $\frac{3}{5}\left(-\frac{10}{21}\right)\left(-\frac{7}{15}\right)=\frac{3\cdot 10\cdot 7}{5\cdot 21\cdot 15}$
$=\frac{\cancel{3}\cdot 2\cdot\cancel{5}\cdot\cancel{7}}{\cancel{5}\cdot\cancel{3}\cdot\cancel{7}\cdot 3\cdot 5}$
$=\frac{2}{15}$

25. 3

26. $\{x|x\le -3\}\cup\{x|x>0\}$

-5 -4 -3 -2 -1 0 1 2 3 4 5

27. $-2(x-3)+4(2-x)=-2x+6+8-4x$
$=-6x+14$

28. $p\in\{-4,0,7\}$
$-|p|$
$-|-4|=-4$
$-|0|=0$
$-|7|=-7$

29. The Inverse Property of Addition

30. $-3.286\div(-1.06)=3.1$

31. The addition inverse of −87 is 87.

32. Replace y with the element in the set and determine whether the inequality is true.
$y>-2$
$-4>-2$ False
$-1>-2$ True
$4>-2$ True

33. The set of integers between −4 and 2 in roster notation is {−3, −2,−1, 0, 1}

34. The set of real numbers less than 7 in set-builder notation is $\{x|x<7\}$

35. A ∪ B = {−4, −2, 0 , 2, 4, 5, 10}

36. A and B have no elements in common, Therefore, A ∩ B = ∅

37. $\{x|x\le 3\}\cap\{x|x>-2\}$

-5 -4 -3 -2 -1 0 1 2 3 4 5

38. $(-3, 4)\cup[-1, 5]$

-5 -4 -3 -2 -1 0 1 2 3 4 5

39. $9-(-3)-7=9+3+(-7)$
$=12+(-7)$
$=5$

40. $-\frac{2}{3}+\left(-\frac{1}{4}\right)-\left(-\frac{5}{12}\right)=-\frac{2}{3}+\left(-\frac{1}{4}\right)+\frac{5}{12}$
$=\frac{-8}{12}+\left(-\frac{3}{12}\right)+\frac{5}{12}$
$=\frac{-8+(-3)+5}{12}$
$=\frac{-11+5}{12}$
$=-\frac{6}{12}$
$=-\frac{1}{2}$

41. $\left(\frac{2}{3}\right)^3(-3)^4=\frac{8}{27}\cdot 81$
$=24$

42. $(-3)^3-(2-6)^2\cdot 5=(-3)^3-(-4)^2\cdot 5$
$=-27-16.5$
$=-27-80$
$=-107$

43. $-8ac\div b^2=-8(-1)(-3)\div 2^2$
$=-8(-1)(-3)\div 4$
$=-24\div 4$
$=-6$

44. $-(3a+b)-2(-4a-5b)=-3a-b+8a+10b$
$=5a+9b$

45. The unknown number: x
The sum of a number and four: $x+4$
Four times the sum of a number and four: $4(x+4)$
$4(x+4)=4x+16$

46. Flying time between San Diego and New York: t
Total time: 13
Flying time between New York and San Diego: $13-t$

47. Number of calories burned by walking at 4 mph for one hour: c
Number of calories burned by cross-country skiing for one hour is 396 more than the number walking: $c+396$

48. The unknown number: x
The difference between the number and two: $x-2$
Twice this difference: $2(x-2)$
Eight more than $2(x-2)$: $2(x-2)+8$
$2(x-2)+8=2x-4+8$
$=2x+4$

49. First integer: x
Second integer: $4x+5$

50. The unknown number: x
The quotient of three more than the number and four: $\frac{x+3}{4}$
Twelve minus the quotient of three more than a number and four: $12-\frac{x+3}{4}$
$12-\frac{x+3}{4}=\frac{48}{4}-\frac{x+3}{4}$
$=\frac{48-(x+3)}{4}$
$=\frac{48-x-3}{4}$
$=\frac{-x+45}{4}$

51. Let x be the smaller of the numbers. Then the larger number is $40-x$.
$2x+(40-x)+5=x+45$

52. The width of the rectangle: w
The length is 3 feet less than $3w$.
The length is $3w-3$

Chapter 1 Test

1. $(-2)(-3)(-5)=(6)(-5)=-30$

2. $A\cap B=\{5, 7\}$

3. $(-2)^3(-3)^2=(-8)(9)=-72$

4. $(-\infty, 1]$

−5 −4 −3 −2 −1 0 1 2 3 4 5

5. $A\cap B=\{-1,0,1\}$

6. $(a-b)^2\div(2b+1)=(2-(-3))^2\div(2(-3)+1)$
$=(5)^2\div(-6+1)$
$=(5)^2\div(-5)$
$=25\div(-5)=-5$

7. $|-3-(-5)|=|-3+5|$
$=|2|$
$=2$

8. $2x-4[2-3(x+4y)-2]=2x-4[2-3x-12y-2]$
$=2x-4[-3x-12y]$
$=2x+12x+48y$
$=14x+48y$

9. 12

10. $-5^2\cdot 4=-25\cdot 4=-100$

11. $\{x|x<3\}\cap\{x|x>-2\}$

−5 −4 −3 −2 −1 0 1 2 3 4 5

12. $2-(-12)+3-5=2+12+3+(-5)$
$=14+3+(-5)$
$=17+(-5)$
$=12$

13. $$\frac{2}{3}-\frac{5}{12}+\frac{4}{9}=\frac{24}{36}-\frac{15}{36}+\frac{16}{36}$$
$$=\frac{24-15+16}{36}$$
$$=\frac{25}{36}$$

14. 4

15. $$\left(-\frac{2}{3}\right)\left(\frac{9}{15}\right)\left(\frac{10}{27}\right)=-\frac{2\cdot\cancel{3}\cdot\cancel{3}\cdot 2\cdot\cancel{5}}{\cancel{3}\cdot\cancel{3}\cdot\cancel{5}\cdot 3\cdot 3\cdot 3}=-\frac{4}{27}$$

16. Replace x with each element in the set and determine whether the inequality is true.
$x < -1$
$-5 < -1$ True
$3 < -1$ False
$7 < -1$ False

17. $$\frac{b^2-c^2}{a-2c}=\frac{(3)^2-(-1)^2}{2-2(-1)}$$
$$=\frac{9-1}{2-(-2)}$$
$$=\frac{8}{4}=2$$

18. $-180 \div 12 = -15$

19. $$12-4\left(\frac{5^2-1}{3}\right)\div 16=12-4\left(\frac{25-1}{3}\right)\div 16$$
$$=12-4\left(\frac{24}{3}\right)\div 16$$
$$=12-4(8)\div 16$$
$$=12-32\div 16$$
$$=12-2=10$$

20. $(3,\infty)$

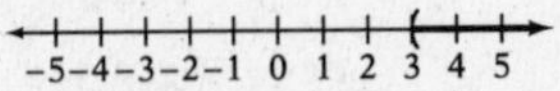

21. $A \cup B = \{1,2,3,4,5,7\}$

22. $$3x-2(x-y)-3(y-4x)=3x-2x+2y-3y+12x$$
$$=13x-y$$

23. $$8-4(2-3)^2\div 2=8-4(-1)^2\div 2$$
$$=8-4(1)\div 2$$
$$=8-4\div 2$$
$$=8-2=6$$

24. $$\frac{3}{5}\left(-\frac{10}{21}\right)\left(-\frac{7}{15}\right)=\frac{\cancel{3}\cdot 2\cdot\cancel{5}\cdot\cancel{7}}{5\cdot\cancel{3}\cdot\cancel{7}\cdot 3\cdot\cancel{5}}=\frac{2}{15}$$

25. The Distributive Property

26. $\{x|x \le 3\} \cup \{x|x < -2\}$

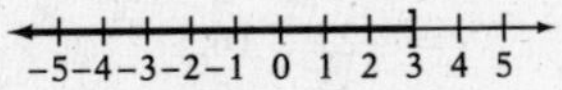

27. $4.27 - 6.98 + 1.3 = -2.71 + 1.3 = -1.41$

28. $A \cup B = \{-2, -1, 0, 1, 2, 3\}$

29. The larger number: x
The smaller number: $9-x$
The difference between one more than the larger number and twice the smaller number.
$$(x+1)-2(9-x)=x+1-18+2x$$
$$=3x-17$$

30. Amount of cocoa produced in Ghana: x
Amount produced in Ivory Coast: $3x$

Chapter 2: First-Degree Equations and Inequalities

Prep Test

1. $8 - 12 = -4$

2. $-9 + 3 = -6$

3. $\frac{-18}{-6} = 3$

4. $-\frac{3}{4}\left(-\frac{4}{3}\right) = -\frac{\overset{1}{\cancel{3}}}{\underset{1}{\cancel{2}} \cdot \underset{1}{\cancel{2}}}\left(-\frac{\overset{1}{\cancel{2}} \cdot \overset{1}{\cancel{2}}}{\underset{1}{\cancel{3}}}\right) = 1$

5. $-\frac{5}{8}\left(\frac{4}{5}\right) = -\frac{\overset{1}{\cancel{5}}}{\underset{1}{\cancel{2}} \cdot \underset{1}{\cancel{2}} \cdot 2}\left(\frac{\overset{1}{\cancel{2}} \cdot \overset{1}{\cancel{2}}}{\underset{1}{\cancel{5}}}\right) = -\frac{1}{2}$

6. $3x - 5 + 7x$
   $= (3x + 7x) - 5$
   $= 10x - 5$

7. $6(x - 2) + 3$
   $= 6x - 12 + 3$
   $= 6x - 9$

8. $n + (n + 2) + (n + 4)$
   $= (n + n + n) + (2 + 4)$
   $= 3n + 6$

9. $0.08x + 0.05(400 - x)$
   $= 0.08x + 20 - 0.05x$
   $= 0.03x + 20$

10. Ounces of nuts in mixture: n
    Ounces of snack mixture: 20
    Ounces of pretzels in mixture: $20 - n$

Go Figure

1. Let n be the total number of votes cast. Let $n - 1$ be the number of votes received by the candidate because the candidate cannot receive 100% of the votes. Then the percentage of votes can be expressed as
   $\frac{\text{number of votes received}}{\text{total number of votes}}$; $\frac{n-1}{n} > \frac{94}{100}$.

   $100(n - 1) > 94n$
   $100n - 100 > 94n$
   $6n > 100$
   $n > 16.6$

   Therefore, when rounding up, the least possible number of votes cast must be 17.
   Answer: 17 votes

Section 2.1

Objective A Exercises

1. An equation contains an equals sign; an expression does not.

3. The Addition Property of Equations states that the same quantity can be added to each side of an equation without changing the solution of the equation. This property is used to remove a term from one side of an equation by adding the opposite of that term to each side of the equation.

5. $7 - 3(1) = 4$
   $7 - 3 = 4$
   $4 = 4$; Yes, 1 is a solution.

7. $6(-2) - 1 = 7(-2) + 1$
   $-12 - 1 = -14 + 1$
   $-13 = -13$; Yes, -2 is a solution.

9. $x - 2 = 7$
   $x - 2 + 2 = 7 + 2$
   $x = 9$
   The solution is 9.

11. $a + 3 = -7$
    $a + 3 + (-3) = -7 + (-3)$
    $a = -10$
    The solution is -10.

13. $b - 3 = -5$
    $b - 3 + 3 = -5 + 3$
    $b = -2$
    The solution is -2.

15. $-7 = x + 8$
    $-7 + (-8) = x + 8 + (-8)$
    $-15 = x$
    The solution is -15.

17. $3x = 12$
    $\frac{1}{3}(3x) = \frac{1}{3}(12)$
    $x = 4$
    The solution is 4.

19. $-3x = 2$
    $-\frac{1}{3}(-3x) = -\frac{1}{3}(2)$
    $x = -\frac{2}{3}$
    The solution is $-\frac{2}{3}$.

21.
$$-\frac{3}{2}+x=\frac{4}{3}$$
$$-\frac{3}{2}+\frac{3}{2}+x=\frac{4}{3}+\frac{3}{2}$$
$$x=\frac{8}{6}+\frac{9}{6}$$
$$x=\frac{17}{6}$$
The solution is $\frac{17}{6}$.

23.
$$x+\frac{2}{3}=\frac{5}{6}$$
$$x+\frac{2}{3}+\left(-\frac{2}{3}\right)=\frac{5}{6}+\left(-\frac{2}{3}\right)$$
$$x=\frac{5}{6}+\left(-\frac{4}{6}\right)$$
$$x=\frac{1}{6}$$
The solution is $\frac{1}{6}$.

25.
$$\frac{2}{3}y=5$$
$$\frac{3}{2}\left(\frac{2}{3}\right)y=\frac{3}{2}(5)$$
$$y=\frac{15}{2}$$
The solution is $\frac{15}{2}$.

27.
$$-\frac{5}{8}x=\frac{4}{5}$$
$$-\frac{8}{5}\left(-\frac{5}{8}x\right)=-\frac{8}{5}\left(\frac{4}{5}\right)$$
$$x=-\frac{32}{25}$$
The solution is $-\frac{32}{25}$.

29.
$$\frac{4x}{7}=-12$$
$$\frac{7}{4}\left(\frac{4}{7}x\right)=\frac{7}{4}(-12)$$
$$x=-21$$
The solution is -21.

31.
$$-\frac{5y}{7}=\frac{10}{21}$$
$$\left(-\frac{7}{5}\right)\left(-\frac{5}{7}y\right)=-\frac{7}{5}\left(\frac{10}{21}\right)$$
$$y=-\frac{2}{3}$$
The solution is $-\frac{2}{3}$.

33.
$$-\frac{3b}{5}=-\frac{3}{5}$$
$$-\frac{5}{3}\left(-\frac{3}{5}b\right)=-\frac{5}{3}\left(-\frac{3}{5}\right)$$
$$b=1$$
The solution is 1.

35.
$$-\frac{2}{3}x=-\frac{5}{8}$$
$$-\frac{3}{2}\left(-\frac{2}{3}x\right)=-\frac{3}{2}\left(-\frac{5}{8}\right)$$
$$x=\frac{15}{16}$$
The solution is $\frac{15}{16}$.

37.
$$-\frac{5}{8}x=40$$
$$-\frac{8}{5}\left(-\frac{5}{8}x\right)=-\frac{8}{5}(40)$$
$$x=-64$$
The solution is -64.

39.
$$-\frac{5}{6}y=-\frac{25}{36}$$
$$-\frac{6}{5}\left(-\frac{5}{6}y\right)=-\frac{6}{5}\left(-\frac{25}{36}\right)$$
$$y=\frac{5}{6}$$
The solution is $\frac{5}{6}$.

41.
$$3x+5x=12$$
$$8x=12$$
$$\frac{1}{8}(8x)=\frac{1}{8}(12)$$
$$x=\frac{3}{2}$$
The solution is $\frac{3}{2}$.

43.
$$3y-5y=0$$
$$-2y=0$$
$$-\frac{1}{2}(-2y)=-\frac{1}{2}(0)$$
$$y=0$$
The solution is 0.

Objective B Exercises

45.
$$2x-4=12$$
$$2x-4+4=12+4$$
$$2x=16$$
$$\frac{1}{2}(2x)=\frac{1}{2}(16)$$
$$x=8$$
The solution is 8.

47.
$$\begin{aligned} 4x - 6 &= 3x \\ 4x + (-4x) - 6 &= 3x + (-4x) \\ -6 &= -x \\ (-1)(-6) &= (-1)(-x) \\ 6 &= x \end{aligned}$$
The solution is 6.

49.
$$\begin{aligned} 7x + 12 &= 9x \\ 7x + (-7x) + 12 &= 9x + (-7x) \\ 12 &= 2x \\ \frac{1}{2}(12) &= \frac{1}{2}(2x) \\ 6 &= x \end{aligned}$$
The solution is 6.

51.
$$\begin{aligned} 4x + 2 &= 4x \\ 4x + (-4x) + 2 &= 4x + (-4x) \\ 2 &= 0 \end{aligned}$$
The equation has no solution.

53.
$$\begin{aligned} 2x + 2 &= 3x + 5 \\ 2x + (-3x) + 2 &= 3x + (-3x) + 5 \\ -x + 2 &= 5 \\ -x + 2 + (-2) &= 5 + (-2) \\ -x &= 3 \\ (-1)(-x) &= (-1)(3) \\ x &= -3 \end{aligned}$$
The solution is −3.

55.
$$\begin{aligned} 2 - 3t &= 3t - 4 \\ 2 - 3t + (-3t) &= 3t + (-3t) - 4 \\ 2 - 6t &= -4 \\ 2 + (-2) - 6t &= -4 + (-2) \\ -6t &= -6 \\ -\frac{1}{6}(-6t) &= -\frac{1}{6}(-6) \\ t &= 1 \end{aligned}$$
The solution is 1.

57.
$$\begin{aligned} 3b - 2b &= 4 - 2b \\ b &= 4 - 2b \\ b + 2b &= 4 - 2b + 2b \\ 3b &= 4 \\ \frac{1}{3}(3b) &= \frac{1}{3}(4) \\ b &= \frac{4}{3} \end{aligned}$$
The solution is $\frac{4}{3}$.

59.
$$\begin{aligned} 3x + 7 &= 3 + 7x \\ 3x + (-7x) + 7 &= 3 + 7x + (-7x) \\ -4x + 7 &= 3 \\ -4x + 7 + (-7) &= 3 + (-7) \\ -4x &= -4 \\ -\frac{1}{4}(-4x) &= -\frac{1}{4}(-4) \\ x &= 1 \end{aligned}$$
The solution is 1.

61.
$$\begin{aligned} \frac{1}{3} - 2b &= 3 \\ \frac{1}{3} + \left(-\frac{1}{3}\right) - 2b &= 3 + \left(-\frac{1}{3}\right) \\ -2b &= \frac{8}{3} \\ -\frac{1}{2}(-2b) &= -\frac{1}{2}\left(\frac{8}{3}\right) \\ b &= -\frac{4}{3} \end{aligned}$$
The solution is $-\frac{4}{3}$.

63.
$$\begin{aligned} 5.3y + 0.35 &= 5.02y \\ 5.3y - 5.3y + 0.35 &= 5.02y - 5.3y \\ 0.35 &= -0.28y \\ \frac{0.35}{-0.28} &= \frac{-0.28y}{0.28} \\ -1.25 &= y \end{aligned}$$
The solution is −1.25.

65.
$$\begin{aligned} \frac{2x}{3} - \frac{1}{2} &= \frac{5}{6} \\ \frac{2x}{3} - \frac{1}{2} + \frac{1}{2} &= \frac{5}{6} + \frac{1}{2} \\ \frac{2x}{3} &= \frac{5}{6} + \frac{3}{6} \\ \frac{2x}{3} &= \frac{8}{6} \\ \frac{3}{2}\left(\frac{2}{3}x\right) &= \frac{3}{2}\left(\frac{8}{6}\right) \\ x &= 2 \end{aligned}$$
The solution is 2.

67.
$$\begin{aligned} \frac{3}{2} - \frac{10x}{9} &= -\frac{1}{6} \\ \frac{3}{2} - \frac{3}{2} - \frac{10x}{9} &= -\frac{1}{6} - \frac{3}{2} \\ -\frac{10x}{9} &= -\frac{1}{6} - \frac{9}{6} \\ -\frac{10x}{9} &= -\frac{10}{6} \\ \left(-\frac{9}{10}\right)\left(-\frac{10}{9}x\right) &= \left(-\frac{9}{10}\right)\left(-\frac{10}{6}\right) \\ x &= \frac{3}{2} \end{aligned}$$
The solution is $\frac{3}{2}$.

69.
$$
\begin{aligned}
2y - 6 &= 3y + 2\\
2y - 2y - 6 &= 3y - 2y + 2\\
-6 &= y + 2\\
-6 - 2 &= y + 2 - 2\\
-8 &= y
\end{aligned}
$$

Now, evaluate $7y + 1$ when $y = -8$.

$$
\begin{aligned}
7(-8) + 1 &= -56 + 1\\
&= -55
\end{aligned}
$$

The value of $7y + 1$ when $y = -8$ is -55.

Objective C Exercises

71.
$$
\begin{aligned}
2x + 3(x - 5) &= 15\\
2x + 3x - 15 &= 15\\
5x - 15 &= 15\\
5x &= 30\\
\frac{1}{5}(5x) &= \frac{1}{5}(30)\\
x &= 6
\end{aligned}
$$

The solution is 6.

73.
$$
\begin{aligned}
5(2 - b) &= -3(b - 3)\\
10 - 5b &= -3b + 9\\
10 - 2b &= 9\\
-2b &= -1\\
-\frac{1}{2}(-2b) &= -\frac{1}{2}(-1)\\
b &= \frac{1}{2}
\end{aligned}
$$

The solution is $\frac{1}{2}$.

75.
$$
\begin{aligned}
3(y - 5) - 5y &= 2y + 9\\
3y - 15 - 5y &= 2y + 9\\
-2y - 15 &= 2y + 9\\
-4y - 15 &= 9\\
-4y &= 24\\
-\frac{1}{4}(-4y) &= -\frac{1}{4}(24)\\
y &= -6
\end{aligned}
$$

The solution is -6.

77.
$$
\begin{aligned}
4 - 3x &= 7x - 2(3 - x)\\
4 - 3x &= 7x - 6 + 2x\\
4 - 3x &= 9x - 6\\
4 - 12x &= -6\\
-12x &= -10\\
-\frac{1}{12}(-12x) &= -\frac{1}{12}(-10)\\
x &= \frac{5}{6}
\end{aligned}
$$

The solution is $\frac{5}{6}$.

79.
$$
\begin{aligned}
-3x - 2(4 + 5x) &= 14 - 3(2x - 3)\\
-3x - 8 - 10x &= 14 - 6x + 9\\
-13x - 8 &= 23 - 6x\\
-13x + 6x - 8 &= 23 - 6x + 6x\\
-7x - 8 &= 23\\
-7x - 8 + 8 &= 23 + 8\\
-7x &= 31\\
-\frac{1}{7}(-7x) &= -\frac{1}{7}(31)\\
x &= -\frac{31}{7}
\end{aligned}
$$

The solution is $-\frac{31}{7}$.

81.
$$
\begin{aligned}
3y &= 2[5 - 3(2 - y)]\\
3y &= 2[5 - 6 + 3y]\\
3y &= 2[-1 + 3y]\\
3y &= -2 + 6y\\
-3y &= -2\\
-\frac{1}{3}(-3y) &= -\frac{1}{3}(-2)\\
y &= \frac{2}{3}
\end{aligned}
$$

The solution is $\frac{2}{3}$.

83.
$$
\begin{aligned}
2[4 + 2(5 - x) - 2x] &= 4x - 7\\
2[4 + 10 - 2x - 2x] &= 4x - 7\\
2[14 - 4x] &= 4x - 7\\
28 - 8x &= 4x - 7\\
28 - 12x &= -7\\
-12x &= -35\\
-\frac{1}{12}(-12x) &= -\frac{1}{12}(-35)\\
x &= \frac{35}{12}
\end{aligned}
$$

The solution is $\frac{35}{12}$.

85.
$$
\begin{aligned}
2[3 - 2(z + 4)] &= 3(4 - z)\\
2[3 - 2z - 8] &= 12 - 3z\\
2[-5 - 2z] &= 12 - 3z\\
-10 - 4z &= 12 - 3z\\
-z &= 22\\
z &= -22
\end{aligned}
$$

The solution is -22.

87.
$$
\begin{aligned}
3[x - (2 - x) - 2x] &= 3(4 - x)\\
3[x - 2 + x - 2x] &= 12 - 3x\\
3[-2] &= 12 - 3x\\
-6 &= 12 - 3x\\
-18 &= -3x\\
-\frac{1}{3}(-18) &= -\frac{1}{3}(-3x)\\
6 &= x
\end{aligned}
$$

The solution is 6.

89.
$$\frac{2x-5}{12}-\frac{3-x}{6}=\frac{11}{12}$$
$$12\left(\frac{2x-5}{12}-\frac{3-x}{6}\right)=12\cdot\frac{11}{12}$$
$$\frac{12(2x-5)}{12}-\frac{12(3-x)}{6}=11$$
$$2x-5-2(3-x)=11$$
$$2x-5-6+2x=11$$
$$4x-11=11$$
$$4x=22$$
$$\frac{1}{4}(4x)=\frac{1}{4}(22)$$
$$x=\frac{11}{2}$$
The solution is $\frac{11}{2}$.

91.
$$\frac{2x-1}{4}+\frac{3x+4}{8}=\frac{1-4x}{12}$$
$$24\left(\frac{2x-1}{4}+\frac{3x+4}{8}\right)=24\left(\frac{1-4x}{12}\right)$$
$$\frac{24(2x-1)}{4}+\frac{24(3x+4)}{8}=\frac{24(1-4x)}{12}$$
$$6(2x-1)+3(3x+4)=2(1-4x)$$
$$12x-6+9x+12=2-8x$$
$$21x+6=2-8x$$
$$29x+6=2$$
$$29x=-4$$
$$\frac{1}{29}(29x)=\frac{1}{29}(-4)$$
$$x=-\frac{4}{29}$$
The solution is $-\frac{4}{29}$.

93.
$$-1.6(b-2.35)=-11.28$$
$$-1.6b+3.76=-11.28$$
$$-1.6b+3.76-3.76=-11.28-3.76$$
$$-1.6b=-15.04$$
$$\frac{-1.6b}{-1.6}=\frac{-15.04}{-1.6}$$
$$b=9.4$$
The solution is 9.4.

95.
$$0.05(300-x)+0.07x=45$$
$$15-0.05x+0.07x=45$$
$$15+0.02x=45$$
$$15-15+0.02x=45-15$$
$$0.02x=30$$
$$x=1500$$
The solution is 1500.

97.
$$3(2x+1)=5-2(x-2)$$
$$6x+3=5-2x+4$$
$$6x+3=9-2x$$
$$8x+3=9$$
$$8x=6$$
$$\frac{1}{8}(8x)=\frac{1}{8}(6)$$
$$x=\frac{3}{4}$$
$$2x^2+1=2\left(\frac{3}{4}\right)^2+1$$
$$=2\left(\frac{9}{16}\right)+1$$
$$=\frac{18}{16}+1$$
$$=\frac{9}{8}+\frac{8}{8}$$
$$=\frac{17}{8}$$
The solution is $\frac{17}{8}$.

99.
$$5-2(4x-1)=3x+7$$
$$5-8x+2=3x+7$$
$$7-8x=3x+7$$
$$7-11x=7$$
$$-11x=0$$
$$-\frac{1}{11}(-11x)=-\frac{1}{11}(0)$$
$$x=0$$
$$x^4-x^2=0^4-0^2$$
$$=0-0$$
$$=0$$
The solution is 0.

Objective D Exercises

101.
$$C=2\pi r$$
$$\frac{C}{2\pi}=\frac{2\pi r}{2\pi}$$
$$r=\frac{C}{2\pi}$$

103.
$$A=\frac{1}{2}bh$$
$$2\cdot A=2\cdot\frac{1}{2}bh$$
$$2A=bh$$
$$\frac{2A}{b}=\frac{bh}{b}$$
$$h=\frac{2A}{b}$$

105.
$$I=\frac{100M}{C}$$
$$\frac{C}{100}\cdot I=\frac{C}{100}\cdot\frac{100}{C}M$$
$$M=\frac{IC}{100}$$

107.
$$A = P + Prt$$
$$A - P = Prt$$
$$\frac{A-P}{Pt} = \frac{Prt}{Pt}$$
$$r = \frac{A-P}{Pt}$$

109.
$$s = \frac{1}{2}(a+b+c)$$
$$2 \cdot s = \frac{2}{1} \cdot \frac{1}{2}(a+b+c)$$
$$2s = a+b+c$$
$$c = 2s - a - b$$

111.
$$S = 2\pi r^2 + 2\pi rh$$
$$S - 2\pi r^2 = 2\pi rh$$
$$\frac{S - 2\pi r^2}{2\pi r} = \frac{2\pi rh}{2\pi r}$$
$$h = \frac{S - 2\pi r^2}{2\pi r}$$

113.
$$P = \frac{R-C}{n}$$
$$P \cdot n = \frac{R-C}{n} \cdot n$$
$$Pn = R - C$$
$$R = Pn + C$$

115.
$$S = 2WH + 2WL + 2LH$$
$$S - 2WL = 2WH + 2LH$$
$$S - 2WL = H(2W + 2L)$$
$$\frac{S - 2WL}{2W + 2L} = \frac{H(2W+2L)}{2W+2L}$$
$$H = \frac{S - 2WL}{2W + 2L}$$

Applying the Concepts

117. When the Multiplication Property of Equations is used, the quantity that multiplies each side of the equation must not be zero because the result will be the identity $0 = 0$, which does not help solve the equation. Some students might notice that multiplying each side of an inequality by zero can result in an equality. For instance, $2 \neq 3$, but $2 \cdot 0 = 3 \cdot 0$.

Section 2.2

Objective A Exercises

1. **Strategy** The number added to the numerator: x

The fraction $\frac{3+x}{10}$ must equal $\frac{4}{5}$.

Solution
$$\frac{3+x}{10} = \frac{4}{5}$$
$$10\left(\frac{3+x}{10}\right) = 10\left(\frac{4}{5}\right)$$
$$3 + x = 8$$
$$x = 5$$

The number is 5.

3. **Strategy**
- The smaller integer: n
  The larger integer: $10 - n$
- Three times the larger integer is three less than eight times the smaller integer.

Solution
$$3(10 - n) = 8n - 3$$
$$30 - 3n = 8n - 3$$
$$-11n = -33$$
$$n = 3$$
$$10 - n = 10 - 3 = 7$$

The integers are 3 and 7.

5. **Strategy**
- The larger integer: n
  The smaller integer: $n - 8$
- The sum of the two integers is fifty.

Solution
$$n + (n - 8) = 50$$
$$2n - 8 = 50$$
$$2n = 58$$
$$n = 29$$
$$n - 8 = 29 - 8 = 21$$

The integers are 21 and 29.

7. **Strategy**
- The first number: n
  The second number: $2n + 2$
  The third number: $3n - 5$
- The sum of the three numbers is 123.

Solution
$$n + (2n + 2) + (3n - 5) = 123$$
$$6n - 3 = 123$$
$$6n = 126$$
$$n = 21$$
$$2n + 2 = 2(21) + 2 = 42 + 2 = 44$$
$$3n - 5 = 3(21) - 5 = 63 - 5 = 58$$

The numbers are 21, 44, and 58.

9. Strategy • The first integer: n
The second consecutive integer: $n+1$
The third consecutive integer: $n+2$
• The sum of the integers is -57.

Solution
$$n+(n+1)+(n+2)=-57$$
$$3n+3=-57$$
$$3n=-60$$
$$n=-20$$
$$n+1=-20+1=-19$$
$$n+2=-20+2=-18$$
The integers are -20, -19, and -18.

11. Strategy • The first odd integer: n
The second consecutive odd integer: $n+2$
The third consecutive odd integer: $n+4$
• Five times the smallest of the three integers is ten more than twice the largest.

Solution
$$5n=2(n+4)+10$$
$$5n=2n+8+10$$
$$5n=2n+18$$
$$3n=18$$
$$n=6$$
Since 6 is not an odd integer, there is no solution.

13. Strategy • The first odd integer: n
The second consecutive odd integer: $n+2$
The third consecutive odd integer: $n+4$
• Three times the middle is seven more than the sum of the first and third integers.

Solution
$$3(n+2)=[n+(n+4)]+7$$
$$3n+6=2n+11$$
$$n=5$$
$$n+2=5+2=7$$
$$n+4=5+4=9$$
The integers are 5, 7, and 9.

Objective B Exercises

15. Strategy • Number of nickels: x
Number of dimes: $53-x$

Coin	Number	Value	Total Value
Nickel	x	5	$5x$
Dime	$53-x$	10	$10(53-x)$

• The sum of the total values of each denomination of coin equals the total value of all the coins (370 cents).

Solution
$$5x+10(53-x)=370$$
$$5x+530-10x=370$$
$$-5x+530=370$$
$$-5x=-160$$
$$x=32$$
$$53-x=53-32=21$$
There are 21 dimes in the collection.

17. Strategy • Number of quarters: x
Number of dimes: $4x$
Number of nickels: $22-5x$

Coin	Number	Value	Total Value
Quarter	x	25	$25x$
Dime	$4x$	10	$10(4x)$
Nickel	$22-5x$	5	$5(22-5x)$

• The sum of the total values of each denomination of coin equals the total value of all the coins (230 cents).

Solution
$$25x+10(4x)+5(22-5x)=230$$
$$25x+40x+110-25x=230$$
$$40x+110=230$$
$$40x=120$$
$$x=3$$
$$4x=4\cdot 3=12$$
There are 12 dimes in the bank.

19. Strategy • Number of 20¢ stamps: x
Number of 15¢ stamps: $3x-8$

Stamp	Number	Value	Total Value
20¢	x	20	$20x$
15¢	$3x-8$	15	$15(3x-8)$

• The sum of the total values of each denomination of stamp equals the total value of all the stamps (400 cents).

Solution
$$20x+15(3x-8)=400$$
$$20x+45x-120=400$$
$$65x-120=400$$
$$65x=520$$
$$x=8$$
$$3x-8=3(8)-8=24-8=16$$
There are eight 20¢ stamps and sixteen 15¢ stamps.

21. Strategy • Number of 3¢ stamps: x
Number of 8¢ stamps: $2x - 3$
Number of 13¢ stamps: $2(2x - 3)$

Stamp	Number	Value	Total Value
3¢	x	3	$3x$
8¢	$2x - 3$	8	$8(2x - 3)$
13¢	$2(2x - 3)$	13	$13(2)(2x - 3)$

• The sum of the total values of each denomination of stamp equals the total value of all the stamps (253 cents).

Solution

$$3x + 8(2x - 3) + 26(2x - 3) = 253$$
$$3x + 16x - 24 + 52x - 78 = 253$$
$$71x - 102 = 253$$
$$71x = 355$$
$$x = 5$$

There are five 3¢ stamps in the collection.

23. Strategy • Number of 18¢ stamps: x
Number of 8¢ stamps: $2x$
Number of 13¢ stamps: $x + 3$

Stamp	Number	Value	Total Value
18¢	x	18	$18x$
8¢	$2x$	8	$8(2x)$
13¢	$x + 3$	13	$13(x + 3)$

• The sum of the total values of each denomination of stamp equals the total value of all the stamps (368 cents).

Solution

$$18x + 8(2x) + 13(x + 3) = 368$$
$$18x + 16x + 13x + 39 = 368$$
$$47x + 39 = 368$$
$$47x = 329$$
$$x = 7$$

There are seven 18¢ stamps in the collection.

Applying the Concepts

25. Strategy • The first odd integer: n
The second consecutive odd integer: $n + 2$
The third consecutive odd integer: $n + 4$
• The product of the second and third minus the product of the first and second is 42.

Solution

$$(n + 2)(n + 4) - n(n + 2) = 42$$
$$n^2 + 6n + 8 - n^2 - 2n = 42$$
$$4n + 8 = 42$$
$$4n = 34$$
$$n = \frac{17}{2}$$

There is no solution, because n is not an integer.

Section 2.3

Objective A Exercises

1. Strategy • Cost of the mixture: x

	Amount	Cost	Value
Cashews	40	5.60	40(5.60)
Peanuts	100	1.89	100(1.89)
Mixture	140	x	$140x$

• The sum of the values before mixing equals the value after mixing.

Solution

$$40(5.60) + 100(1.89) = 140x$$
$$224 + 189 = 140x$$
$$413 = 140x$$
$$2.95 = x$$
$$x = 2.95$$

The cost of the mixture is \$2.95 per pound.

3. Strategy • Number of adult tickets: x
Number of children's tickets: $460 - x$

	Amount	Cost	Value
Adult	x	10	$10x$
Child	$460 - x$	4	$4(460 - x)$

• The total value of the tickets sold is \$3760.

Solution

$$10x + 4(460 - x) = 3760$$
$$10x + 1840 - 4x = 3760$$
$$6x + 1840 = 3760$$
$$6x = 1920$$
$$x = 320$$

320 adult tickets were sold.

5. Strategy • Liters of imitation maple syrup: x

	Amount	Cost	Value
Imitation	x	4.00	$4x$
Maple	5	9.50	9.50(5)
Mixture	$5+x$	5.00	$5(5+x)$

• The sum of the values before mixing equals the value after mixing.

Solution
$$4x+9.50(5)=5(5+x)$$
$$4x+47.5=25+5x$$
$$-x+47.5=25$$
$$-x=-22.5$$
$$x=22.5$$
The mixture must contain 22.5 L of imitation maple syrup.

7. Strategy • Ounces of pure gold: x
Ounces of gold alloy: $50-x$

	Amount	Cost	Value
Pure gold	x	400	$400x$
Gold alloy	$50-x$	150	$150(50-x)$
Mixture	50	250	250(50)

• The sum of the values before mixing equals the value after mixing.

Solution
$$400x+150(50-x)=250(50)$$
$$400x+7500-150x=12{,}500$$
$$250x=5000$$
$$x=20$$
$$50-x=50-20=30$$
20 oz of pure gold and 30 oz of the alloy were used.

9. Strategy • Cost per pound of the mixture: x

	Amount	Cost	Value
\$5.40 tea	40	5.40	5.40(40)
\$3.25 tea	60	3.25	3.25(60)
Mixture	100	x	$100x$

• The sum of the values before mixing equals the value after mixing.

Solution
$$5.40(40)+3.25(60)=100x$$
$$216+195=100x$$
$$411=100x$$
$$4.11=x$$
The cost of the mixture is \$4.11 per pound.

11. Strategy • Gallons of cranberry juice: x

	Amount	Cost	Value
Cranberry	x	5.60	$5.60x$
Apple	50	4.24	50(4.24)
Mixture	$50+x$	5.00	$5(50+x)$

• The sum of the values before mixing equals the value after mixing.

Solution
$$5.60x+50(4.24)=5(50+x)$$
$$5.60x+212=250+5x$$
$$0.60x+212=250$$
$$0.60x=38$$
$$x=63.\overline{3}$$
The mixture must contain 63.3 gal of cranberry juice.

Objective B Exercises

13. Strategy • Pounds of 15% aluminum alloy: x

	Amount	Percent	Quantity
15%	x	0.15	$0.15x$
22%	500	0.22	0.22(500)
20%	$500+x$	0.20	$0.20(500+x)$

• The sum of the quantities before mixing is equal to the quantity after mixing.

Solution
$$0.15x+0.22(500)=0.20(500+x)$$
$$0.15x+110=100+0.20x$$
$$-0.05x+110=100$$
$$-0.05x=-10$$
$$x=200$$
200 lb of the 15% aluminum alloy must be used.

15. Strategy • Ounces of pure water: x
Ounces of 70% alcohol: $3.5-x$

	Amount	Percent	Quantity
Water	x	0	$0(x)$
70% alcohol	$3.5-x$	0.70	$0.70(3.5-x)$
45% mixture	3.5	0.45	0.45(3.5)

• The sum of the quantities before mixing is equal to the quantity after mixing.

Solution
$$0(x)+0.70(3.5-x)=0.45(3.5)$$
$$2.45-0.70x=1.575$$
$$-0.70x=-0.875$$
$$x=1.25$$
$$3.5-x=3.5-1.25-2.25$$
The solution should contain 2.25 oz of rubbing alcohol and 1.25 oz of water.

17. Strategy • Ounces of pure water: x

	Amount	Percent	Quantity
Pure water	x	0	0
8%	75	0.08	0.08(75)
5%	$75+x$	0.05	$0.05(75+x)$

• The sum of the quantities before mixing is equal to the quantity after mixing.

Solution

$$0+0.08(75)=0.05(75+x)$$
$$6=3.75+0.05x$$
$$2.25=0.05x$$
$$45=x$$

45 oz of pure water must be added.

19. Strategy • Milliliters of alcohol: x

	Amount	Percent	Quantity
Alcohol	x	0	0
25% iodine	200	0.25	0.25(200)
10% iodine	$200+x$	0.10	$0.10(200+x)$

• The sum of the quantities before mixing is equal to the quantity after mixing.

Solution

$$0+0.25(200)=0.10(200+x)$$
$$50=20+0.10x$$
$$30=0.10x$$
$$300=x$$

300 ml of alcohol must be added.

21. Strategy • Percent concentration of the resulting drink: x

	Amount	Percent	Quantity
5% fruit	12	0.05	0.05(12)
Water	2	0	0(2)
Mixture	10	x	$10x$

• The sum of the quantities before mixing is equal to the quantity after mixing.

Solution

$$0.05(12)+0(2)=10(x)$$
$$0.6=10x$$
$$0.06=x$$

The resulting drink is 6% fruit juice.

23. Strategy • Quarts of 40% antifreeze replaced: x
Quarts of pure antifreeze added: x

	Amount	Percent	Quantity
40% antifreeze	12	0.40	0.40(12)
40% antifreeze	x	0.40	$0.40x$
replaced by pure antifreeze	x	1.00	x
added 60% antifreeze	12	0.60	0.60(12)

• The quantity in the radiator minus the quantity replaced plus the quantity added equals the quantity in the resulting solution.

Solution

$$0.40(12)-0.40x+x=0.60(12)$$
$$4.8+0.60x=7.2$$
$$0.60x=2.4$$
$$x=4$$

4 qt will have to be replaced with pure antifreeze.

Objective C Exercises

25. Strategy • Time for the car: t
Time for the helicopter: $t-\frac{1}{2}$

	Rate	Time	Distance
Car	80	t	$80t$
Helicopter	130	$t-\frac{1}{2}$	$130(t-\frac{1}{2})$

• The car and the helicopter travel the same distance.

Solution

$$80t=130\left(t-\frac{1}{2}\right)$$
$$80t=130t-65$$
$$-50t=-65$$
$$t=1.3$$
$$d=rt=80(1.3)=104$$

The helicopter will overtake the car 104 mi from the starting point.

27. Strategy • Rate of the first car: r
Rate of the second car: $r + 8$

	Rate	Time	Distance
1st car	r	2.5	$2.5r$
2nd car	$r+8$	2.5	$2.5(r+8)$

• The total distance traveled by the two cars is 310 mi.

Solution
$$2.5r + 2.5(r+8) = 310$$
$$2.5r + 2.5r + 20 = 310$$
$$5r + 20 = 310$$
$$5r = 290$$
$$r = 58$$
$r + 8 = 58 + 8 = 66$
The speed of the first car is 58 mph.
The speed of the second car is 66 mph.

29. Strategy • Time flying to a city: t
Time returning to the international airport: $4 - t$

	Rate	Time	Distance
Going	250	t	$250t$
Returning	150	$4-t$	$150(4-t)$

• The distance to the city is the same as the distance returning to the international airport.

Solution
$$250t = 150(4-t)$$
$$250t = 600 - 150t$$
$$400t = 600$$
$$t = 1.5$$
$d = rt = 250(1.5) = 375$
The distance between the two airports is 375 mi.

31. Strategy • Rate of freight train: r
Rate of the passenger train: $r + 18$

	Rate	Time	Distance
Freight train	r	4	$4r$
Passenger train	$r+18$	2.5	$2.5(r+18)$

• The distance the freight train travels is equal to the distance the passenger train travels.

Solution
$$4r = 2.5(r+18)$$
$$4r = 2.5r + 45$$
$$1.5r = 45$$
$$r = 30$$
$r + 18 = 30 + 18 = 48$
The rate of the freight train is 30 mph.
The rate of the passenger train is 48 mph.

33. Strategy • Rate of the jogger: x
Rate of the cyclist: $4x$

	Rate	Time	Distance
Jogger	x	2	$2x$
Cyclist	$4x$	2	$2(4x)$

• In 2 hours the cyclist is 33 mi ahead of the jogger.

Solution
$$2(4x) - 2x = 33$$
$$8x - 2x = 33$$
$$6x = 33$$
$$x = 5.5$$
$4x = 4(5.5) = 22$
The rate of the cyclist is 22 mph.
$d = rt = 22(2) = 44$
The cyclist traveled 44 mi.

Applying the Concepts

35. Strategy • Amount of 12-karat gold: x

	Amount	Percent	Quantity
12-karat gold	x	$\frac{12}{24}$	$\frac{12}{24}x$
24-karat gold	3	$\frac{24}{24}$	$\frac{24}{24}(3)$
14-karat gold	$x+3$	$\frac{14}{24}$	$\frac{14}{24}(x+3)$

• The sum of the quantities before mixing is equal to the quantity after mixing.

Solution
$$\frac{12}{24}x + \frac{24}{24}(3) = \frac{14}{24}(x+3)$$
$$12x + 24(3) = 14(x+3)$$
$$12x + 72 = 14x + 42$$
$$-2x = -30$$
$$x = 15$$
15 oz of the 12-karat gold should be used.

37. The distance between them 2 minutes before impact is equal to the sum of the distances each one can travel during 2 minutes.

$2 \text{ minutes} \dfrac{1 \text{ hour}}{60 \text{ minutes}} = 0.03\overline{3} \text{ hour}$

Distance between cars = rate of first car $\cdot$ $0.03\overline{3}$ + rate of second car $\cdot$ $0.03\overline{3}$

Distance between cars $= 40 \cdot 0.03\overline{3} + 60 \cdot 0.03\overline{3} = 3.3\overline{3}$

The cars are $3.3\overline{3}$ (or $3\frac{1}{3}$ mi) apart 2 min before impact.

39. Strategy Distance from 40-ft tower to grass seed: x

Distance from top of 40-ft tower to seed (using Pythagorean theorem): $\sqrt{40^2 + x^2}$

Distance from top of 30-ft tower to seed: $\sqrt{30^2 + (50 - x)^2}$

Solution Let the rate of both birds be r, because they are flying at the same rate.

	Rate	Distance	Time
Bird 1	r	$\sqrt{40^2 + x^2}$	$\frac{\sqrt{40^2 + x^2}}{r}$
Bird 2	r	$\sqrt{30^2 + (50 - x)^2}$	$\frac{\sqrt{30^2 + (50 - x)^2}}{r}$

They arrive at the seed at the same time.

$$\frac{\sqrt{40^2 + x^2}}{r} = \frac{\sqrt{30^2 + (50 - x)^2}}{r}$$

Thus

$$\sqrt{40^2 + x^2} = \sqrt{30^2 + (50 - x)^2}$$
$$40^2 + x^2 = 30^2 + (50 - x)^2$$
$$1600 + x^2 = 900 + 2500 - 100x + x^2$$
$$-1800 = -100x$$
$$x = 18$$

The grass seed is 18 ft from the 40-ft tower.

Section 2.4

Objective A Exercises

1. The Addition Property of Inequalities states that the same number can be added to each side of an inequality without changing the solution set of the inequality. Examples will vary. For instance,

$$8 > 6$$
$$8 + 4 > 6 + 4$$
$$12 > 10$$

and

$$-5 < -1$$
$$-5 + (-7) < -1 + (-7)$$
$$-12 < -8$$

3. a. $-17 + 7 \le -3$; $-10 \le -3$; solution

b. $8 + 7 \le -3$; $15 \le -3$; not a solution

c. $-10 + 7 \le -3$; $-3 \le -3$; solution

d. $0 + 7 \le -3$; $7 \le -3$; not a solution

5. $x - 3 < 2$
$x < 5$
$\{x|x < 5\}$

-5 -4 -3 -2 -1 0 1 2 3 4 5

7. $4x \le 8$
$\frac{1}{4}(4x) \le \frac{1}{4}(8)$
$x \le 2$
$\{x|x \le 2\}$

-5 -4 -3 -2 -1 0 1 2 3 4 5

9. $-2x > 8$
$-\frac{1}{2}(-2x) < -\frac{1}{2}(8)$
$x < -4$
$\{x|x < -4\}$

-5 -4 -3 -2 -1 0 1 2 3 4 5

11. $3x - 1 > 2x + 2$
$x - 1 > 2$
$x > 3$
$\{x|x > 3\}$

13. $2x - 1 > 7$
$2x > 8$
$\frac{1}{2}(2x) > \frac{1}{2}(8)$
$x > 4$
$\{x|x > 4\}$

15. $5x - 2 \le 8$
$5x \le 10$
$\frac{1}{5}(5x) \le \frac{1}{5}(10)$
$x \le 2$
$\{x|x \le 2\}$

17. $6x + 3 > 4x - 1$
$2x + 3 > -1$
$2x > -4$
$\frac{1}{2}(2x) > \frac{1}{2}(-4)$
$x > -2$
$\{x|x > -2\}$

19. $8x + 1 \ge 2x + 13$
$6x + 1 \ge 13$
$6x \ge 12$
$\frac{1}{6}(6x) \ge \frac{1}{6}(12)$
$x \ge 2$
$\{x|x \ge 2\}$

21. $4 - 3x < 10$
$-3x < 6$
$-\frac{1}{3}(-3x) > -\frac{1}{3}(6)$
$x > -2$
$\{x|x > -2\}$

23.
$$\begin{aligned}7-2x &\ge 1\\ -2x &\ge -6\\ -\frac{1}{2}(-2x) &\le -\frac{1}{2}(-6)\\ x &\le 3\end{aligned}$$
$\{x|x \le 3\}$

25.
$$\begin{aligned}-3-4x &> -11\\ -4x &> -8\\ -\frac{1}{4}(-4x) &< -\frac{1}{4}(-8)\\ x &< 2\end{aligned}$$
$\{x|x < 2\}$

27.
$$\begin{aligned}4x-2 &< x-11\\ 3x-2 &< -11\\ 3x &< -9\\ \frac{1}{3}(3x) &< \frac{1}{3}(-9)\\ x &< -3\end{aligned}$$
$\{x|x < -3\}$

29.
$$\begin{aligned}x+7 &\ge 4x-8\\ -3x+7 &\ge -8\\ -3x &\ge -15\\ -\frac{1}{3}(-3x) &\le -\frac{1}{3}(-15)\\ x &\le 5\end{aligned}$$
$\{x|x \le 5\}$

31.
$$\begin{aligned}3x+2 &\le 7x+4\\ -4x+2 &\le 4\\ -4x &\le 2\\ -\frac{1}{4}(-4x) &\ge -\frac{1}{4}(2)\\ x &\ge -\frac{1}{2}\end{aligned}$$
$\left\{x|x \ge -\frac{1}{2}\right\}$

33.
$$\begin{aligned}\frac{3}{5}x-2 &< \frac{3}{10}-x\\ 10\left(\frac{3}{5}x-2\right) &< 10\left(\frac{3}{10}-x\right)\\ 6x-20 &< 3-10x\\ 16x-20 &< 3\\ 16x &< 23\\ \frac{1}{16}(16x) &< \frac{1}{16}(23)\\ x &< \frac{23}{16}\end{aligned}$$
$\left(-\infty, \frac{23}{16}\right)$

35.
$$\begin{aligned}\frac{2}{3}x-\frac{3}{2} &< \frac{7}{6}-\frac{1}{3}x\\ 6\left(\frac{2}{3}x-\frac{3}{2}\right) &< 6\left(\frac{7}{6}-\frac{1}{3}x\right)\\ 4x-9 &< 7-2x\\ 6x-9 &< 7\\ 6x &< 16\\ \frac{1}{6}(6x) &< \frac{1}{6}(16)\\ x &< \frac{8}{3}\end{aligned}$$
$\left(-\infty, \frac{8}{3}\right)$

37.
$$\begin{aligned}\frac{1}{2}x-\frac{3}{4} &< \frac{7}{4}x-2\\ 4\left(\frac{1}{2}x-\frac{3}{4}\right) &< 4\left(\frac{7}{4}x-2\right)\\ 2x-3 &< 7x-8\\ -5x-3 &< -8\\ -5x &< -5\\ -\frac{1}{5}(-5x) &> -\frac{1}{5}(-5)\\ x &> 1\end{aligned}$$
$(1, \infty)$

39.
$$\begin{aligned}4(2x-1) &> 3x-2(3x-5)\\ 8x-4 &> 3x-6x+10\\ 8x-4 &> -3x+10\\ 11x-4 &> 10\\ 11x &> 14\\ \frac{1}{11}(11x) &> \frac{1}{11}(14)\\ x &> \frac{14}{11}\end{aligned}$$
$\left(\frac{14}{11}, \infty\right)$

41.
$$\begin{aligned}2-5(x+1) &\ge 3(x-1)-8\\ 2-5x-5 &\ge 3x-3-8\\ -5x-3 &\ge 3x-11\\ -8x-3 &\ge -11\\ -8x &\ge -8\\ -\frac{1}{8}(-8x) &\le -\frac{1}{8}(-8)\\ x &\le 1\end{aligned}$$
$(-\infty, 1]$

43. $3+2(x+5) \geq x+5(x+1)+1$
$3+2x+10 \geq x+5x+5+1$
$2x+13 \geq 6x+6$
$-4x+13 \geq 6$
$-4x \geq -7$
$-\frac{1}{4}(-4x) \leq -\frac{1}{4}(-7)$
$x \leq \frac{7}{4}$
$\left(-\infty, \frac{7}{4}\right]$

45. $3-4(x+2) \leq 6+4(2x+1)$
$3-4x-8 \leq 6+8x+4$
$-4x-5 \leq 10+8x$
$-12x-5 \leq 10$
$-12x \leq 15$
$-\frac{1}{12}(-12x) \geq -\frac{1}{12}(15)$
$x \geq -\frac{5}{4}$
$\left[-\frac{5}{4}, \infty\right)$

47. $12-2(3x-2) \geq 5x-2(5-x)$
$12-6x+4 \geq 5x-10+2x$
$16-6x \geq 7x-10$
$16-13x \geq -10$
$-13x \geq -26$
$-\frac{1}{13}(-13x) \leq -\frac{1}{13}(-26)$
$x \leq 2$
$(-\infty, 2]$

Objective B Exercises

49. Writing $-3 > x > 4$ does not make sense because there is no number that is less than -3 *and* greater than 4.

51. $x-3 \leq 1$ and $2x \geq -4$
$x \leq 4$; $x \geq -2$
$\{x|x \leq 4\}$; $\{x|x \geq -2\}$
$\{x|x \leq 4\} \cap \{x|x \geq -2\} = [-2, 4]$

53. $2x < 6$ or $x-4 > 1$
$x < 3$; $x > 5$
$\{x|x < 3\}$; $\{x|x > 5\}$
$\{x|x < 3\} \cup \{x|x > 5\} = (-\infty, 3) \cup (5, \infty)$

55. $\frac{1}{2}x > -2$ and $5x < 10$
$x > -4$; $x < 2$
$\{x|x > -4\}$; $\{x|x < 2\}$
$\{x|x > -4\} \cap \{x|x < 2\} = (-4, 2)$

57. $\frac{2}{3}x > 4$ or $2x < -8$
$x > 6$; $x < -4$
$\{x|x > 6\}$; $\{x|x < -4\}$
$\{x|x > 6\} \cup \{x|x < -4\} = (-\infty, -4) \cup (6, \infty)$

59. $3x < -9$ and $x-2 < 2$
$x < -3$; $x < 4$
$\{x|x < -3\}$; $\{x|x < 4\}$
$\{x|x < -3\} \cap \{x|x < 4\} = (-\infty, -3)$

61. $2x-3 > 1$ and $3x-1 < 2$
$2x > 4$; $3x < 3$
$x > 2$; $x < 1$
$\{x|x > 2\}$; $\{x|x < 1\}$
$\{x|x > 2\} \cap \{x|x < 1\} = \varnothing$

63. $4x+1 < 5$ and $4x+7 > -1$
$4x < 4$; $4x > -8$
$x < 1$; $x > -2$
$\{x|x < 1\}$; $\{x|x > -2\}$
$\{x|x < 1\} \cap \{x|x > -2\} = (-2, 1)$

65. $6x-2 < -14$ or $5x+1 > 11$
$6x < -12$; $5x > 10$
$x < -2$; $x > 2$
$\{x|x < -2\}$; $\{x|x > 2\}$
$\{x|x < -2\} \cup \{x|x > 2\} = \{x|x < -2 \text{ or } x > 2\}$

67. $5 < 4x-3 < 21$
$5+3 < 4x-3+3 < 21+3$
$8 < 4x < 24$
$\frac{1}{4}(8) < \frac{1}{4}(4x) < \frac{1}{4}(24)$
$2 < x < 6$
$\{x|2 < x < 6\}$

69. $-2 < 3x+7 < 1$
$-2+(-7) < 3+7+(-7) < 1+(-7)$
$-9 < 3x < -6$
$\frac{1}{3}(-9) < \frac{1}{3}(3x) < \frac{1}{3}(-6)$
$-3 < x < -2$
$\{x|-3 < x < -2\}$

71. $3x-5 > 10$ or $3x-5 < -10$
$3x > 15$; $3x < -5$
$x > 5$; $x < -\frac{5}{3}$
$\{x|x > 5\}$; $\left\{x \middle| x < -\frac{5}{3}\right\}$
$\{x|x > 5\} \cup \left\{x \middle| x < -\frac{5}{3}\right\} = \left\{x \middle| x > 5 \text{ or } x < -\frac{5}{3}\right\}$

73. $8x+2 \leq -14$ and $4x-2 > 10$
$8x \leq -16$; $4x > 12$
$x \leq -2$; $x > 3$
$\{x|x \leq -2\}$; $\{x|x > 3\}$
$\{x|x \leq -2\} \cap \{x|x > 3\} = \varnothing$

75. $5x + 12 \geq 2$ or $7x - 1 \leq 13$
$5x \geq -10$ $\quad 7x \leq 14$
$x \geq -2$ $\quad x \leq 2$
$\{x|x \geq -2\}$ $\quad \{x|x \leq 2\}$
$\{x|x \geq -2\} \cup \{x|x \leq 2\}$ = the set of real numbers

77. $$3 \leq 7x - 14 \leq 31$$
$$3 + 14 \leq 7x - 14 + 14 \leq 31 + 14$$
$$17 \leq 7x \leq 45$$
$$\frac{1}{7}(17) \leq \frac{1}{7}(7x) \leq \frac{1}{7}(45)$$
$$\frac{17}{7} \leq x \leq \frac{45}{7}$$
$$\left\{x \middle| \frac{17}{7} \leq x \leq \frac{45}{7}\right\}$$

79. $1 - 3x < 16$ and $1 - 3x > -16$
$-3x < 15$ $\quad -3x > -17$
$x > -5$ $\quad x < \frac{17}{3}$
$\{x|x > -5\}$ $\quad \left\{x \middle| x < \frac{17}{3}\right\}$
$$\{x|x > -5\} \cap \left\{x \middle| x < \frac{17}{3}\right\} = \left\{x \middle| -5 < x < \frac{17}{3}\right\}$$

81. $6x + 5 < -1$ or $1 - 2x < 7$
$6x < -6$ $\quad -2x < 6$
$x < -1$ $\quad x > -3$
$\{x|x < -1\}$ $\quad \{x|x > -3\}$
$\{x|x < -1\} \cup \{x|x > -3\}$ = the set of real numbers

83. $9 - x \geq 7$ and $9 - 2x < 3$
$-x \geq -2$ $\quad -2x < -6$
$x \leq 2$ $\quad x > 3$
$\{x|x \leq 2\}$ $\quad \{x|x > 3\}$
$\{x|x \leq 2\} \cap \{x|x > 3\} = \emptyset$

Objective C Exercises

85. **Strategy** The unknown number: x
Two times the difference between the number and eight: $2(x - 8)$
Five times the sum of the number and four: $5(x + 4)$

two times the difference between a number and eight $\leq$ five times the sum the number and four

$$2(x - 8) \leq 5(x + 4)$$
$$2x - 16 \leq 5x + 20$$
$$-3x - 16 \leq 20$$
$$-3x \leq 36$$
$$x \geq -12$$

The smallest number is -12

87. **Strategy** The width of the rectangle: x
The length of the rectangle: $2x - 5$
To find the maximum width, substitute the given values in the inequality $2L + 2W < 60$ and solve.

Solution
$$2L + 2W < 60$$
$$2(2x - 5) + 2x < 60$$
$$4x - 10 + 2x < 60$$
$$6x - 10 < 60$$
$$6x < 70$$
$$x < \frac{70}{6}$$
$$x < 11\frac{2}{3}$$

The maximum width of the rectangle is 11 cm.

89. **Strategy** To find the number of pages, write and solve an inequality using P to represent the number of pages. Then $P - 400$ is the number of pages for which you are charged an extra fee per page.

Solution Cost of Top Page < Cost of competitor
$$6.95 + 0.10(P - 400) < 3.95 + 0.15(P - 400)$$
$$6.95 + 0.10P - 40 < 3.95 + 0.15P - 60$$
$$0.10P - 33.05 < 0.15P - 56.05$$
$$-0.05P < -23$$
$$P > 460$$

The Top Page plan is less expensive when service is for more than 460 pages.

91. **Strategy** To find the number of minutes, write and solve an inequality using x to represent the number of minutes.

Solution Cost of paying with coins < Cost of paying with calling card
$$0.70 + 0.15(x - 3) < 0.35 + 0.196(1) + 0.126(x - 1)$$
$$0.70 + 0.15x - 0.45 < 0.35 + 0.196 + 0.126x - 0.126$$
$$0.15x + 0.25 < 0.42 + 0.126x$$
$$0.024x < 0.17$$
$$x < 7.08$$

Paying with cash is less expensive when the call is 7 min or less.

93. **Strategy** To find the temperature range in Fahrenheit degrees, write and solve a compound inequality using F to represent Fahrenheit degrees.

Solution
$$0 < \frac{5(F - 32)}{9} < 30$$
$$\frac{9}{5}(0) < \frac{9}{5}\left(\frac{5(F - 32)}{9}\right) < \frac{9}{5}(30)$$
$$0 < F - 32 < 54$$
$$0 + 32 < F - 32 + 32 < 54 + 32$$
$$32° < F < 86°$$

95. Strategy To find the minimum amount of sales, write and solve an inequality using N to represent the amount of sales.

Solution
$$1000 + 0.05N \geq 3200$$
$$0.05N \geq 2200$$
$$N \geq 44{,}000$$
George's amount of sales must be \$44,000 or more.

97. Strategy To find the number of checks, write and solve an inequality using N to represent the number of checks.

Solution Cost of the first account < Cost of the second account
$$8 + 0.12(N - 100) < 5 + 0.15(N - 100)$$
$$8 + 0.12N - 12 < 5 + 0.15N - 15$$
$$0.12N - 4 < 0.15N - 10$$
$$-0.03N < -6$$
$$N > 200$$
The first account is less expensive if more than 200 checks are written.

99. Strategy To find the range of scores, write and solve an inequality using N to represent the score on the last test.

Solution
$$70 \leq \frac{56 + 91 + 83 + 62 + N}{5} \leq 79$$
$$70 \leq \frac{292 + N}{5} \leq 79$$
$$5 \cdot 70 \leq 5\left(\frac{292 + N}{5}\right) \leq 5 \cdot 79$$
$$350 \leq 292 + N \leq 395$$
$$350 - 292 \leq 292 + N - 292 \leq 395 - 292$$
$$58 \leq N \leq 103$$
Since 100 is a maximum score, the range of scores to receive a C grade is $58 \leq N \leq 100$.

101. Strategy To find the three consecutive even integers, write and solve a compound inequality using x to represent the first even integer.

Solution Lower limit of the sum < sum < Upper limit of the sum
$$30 < x + (x + 2) + (x + 4) < 51$$
$$30 < 3x + 6 < 51$$
$$30 - 6 < 3x + 6 - 6 < 51 - 6$$
$$24 < 3x < 45$$
$$\frac{1}{3}(24) < \frac{1}{3}(3x) < \frac{1}{3}(45)$$
$$8 < x < 15$$
The three even integers are 10, 12, and 14; or 12, 14, and 16; or 14, 16, and 18.

Applying the Concepts

103.
a. Always true.
b. Sometimes true.
c. Sometimes true (true when the integer is negative).
d. Sometimes true (true when $0 < a < 1$).
e. Always true.

Section 2.5

Objective A Exercises

1. $|2 - 8| = 6$
$|-6| = 6$
$6 = 6$
Yes, 2 is a solution.

3. $|3(-1) - 4| = 7$
$|-3 - 4| = 7$
$|-7| = 7$
$7 = 7$
Yes, -1 is a solution.

5. $|x| = 7$
$x = 7$ or $x = -7$
The solutions are 7 and -7.

7. $|b| = 4$
$b = 4$ or $b = -4$
The solutions are 4 and -4.

9. $|-y| = 6$
$-y = 6$ or $-y = -6$
$y = -6$ $\quad y = 6$
The solutions are -6 and 6.

11. $|-a| = 7$
$-a = 7$ or $-a = -7$
$a = -7$ $\quad a = 7$
The solutions are -7 and 7.

13. $|x| = -4$
There is no solution to this equation because the absolute value of a number must be nonnegative.

15. $|-t| = -3$
There is no solution to this equation because the absolute value of a number must be nonnegative.

17. $|x + 2| = 3$
$x + 2 = 3$ or $x + 2 = -3$
$x = 1$ $\quad x = -5$
The solutions are 1 and -5.

19. $|y - 5| = 3$
$y - 5 = 3$ or $y - 5 = -3$
$y = 8$ $\quad y = 2$
The solutions are 8 and 2.

21. $|a-2|=0$
$a-2=0$
$a=2$
The solution is 2.

23. $|x-2|=-4$
There is no solution to this equation because the absolute value of a number must be nonnegative.

25. $|3-4x|=9$
$3-4x=9$ or $3-4x=-9$
$-4x=6$ $\quad -4x=-12$
$x=-\frac{3}{2}$ $\quad x=3$
The solutions are $-\frac{3}{2}$ and 3.

27. $|2x-3|=0$
$2x-3=0$
$2x=3$
$x=\frac{3}{2}$
The solution is $\frac{3}{2}$.

29. $|3x-2|=-4$
There is no solution to this equation because the absolute value of a number must be nonnegative.

31. $|x-2|-2=3$
$|x-2|=5$
$x-2=5$ or $x-2=-5$
$x=7$ $\quad x=-3$
The solutions are 7 and –3.

33. $|3a+2|-4=4$
$|3a+2|=8$
$3a+2=8$ $\quad 3a+2=-8$
$3a=6$ $\quad 3a=-10$
$a=2$ $\quad a=-\frac{10}{3}$
The solutions are 2 and $-\frac{10}{3}$.

35. $|2-y|+3=4$
$|2-y|=1$
$2-y=1$ or $2-y=-1$
$-y=-1$ $\quad -y=-3$
$y=1$ $\quad y=3$
The solutions are 1 and 3.

37. $|2x-3|+3=3$
$|2x-3|=0$
$2x-3=0$
$2x=3$
$x=\frac{3}{2}$
The solution is $\frac{3}{2}$.

39. $|2x-3|+4=-4$
$|2x-3|=-8$
There is no solution to this equation because the absolute value of a number must be nonnegative.

41. $|6x+5|-2=4$
$|6x-5|=6$
$6x-5=6$ $\quad 6x-5=-6$
$6x=11$ $\quad 6x=-1$
$x=\frac{11}{6}$ $\quad x=-\frac{1}{6}$
The solutions are $\frac{11}{6}$ and $-\frac{1}{6}$.

43. $|3t+2|+3=4$
$|3t+2|=1$
$3t+2=1$ $\quad 3t+2=-1$
$3t=-1$ $\quad 3t=-3$
$t=-\frac{1}{3}$ $\quad t=-1$
The solutions are $-\frac{1}{3}$ and –1.

45. $3-|x-4|=5$
$-|x-4|=2$
$|x-4|=-2$
There is no solution to this equation because the absolute value of a number must be nonnegative.

47. $8-|2x-3|=5$
$-|2x-3|=-3$
$|2x-3|=3$
$2x-3=3$ $\quad 2x-3=-3$
$2x=6$ $\quad 2x=0$
$x=3$ $\quad x=0$
The solutions are 3 and 0.

49. $|2-3x|+7=2$
$|2-3x|=-5$
There is no solution to this equation because the absolute value of a number must be nonnegative.

51. $|8-3x|-3=2$
$|8-3x|=5$
$8-3x=5$ $\quad 8-3x=-5$
$-3x=-3$ $\quad -3x=-13$
$x=1$ $\quad x=\frac{13}{3}$
The solutions are 1 and $\frac{13}{3}$.

53. $|2x-8|+12=2$
$|2x-8|=-10$
There is no solution to this equation because the absolute value of a number must be nonnegative.

55. $2+|3x-4|=5$
$|3x-4|=3$
$3x-4=3$ $\quad$ $3x-4=-3$
$3x=7$ $\quad$ $3x=1$
$x=\frac{7}{3}$ $\quad$ $x=\frac{1}{3}$
The solutions are $\frac{7}{3}$ and $\frac{1}{3}$.

57. $5-|2x+1|=5$
$-|2x+1|=0$
$|2x+1|=0$
$2x+1=0$
$2x=-1$
$x=-\frac{1}{2}$
The solution is $-\frac{1}{2}$.

59. $6-|2x+4|=3$
$-|2x+4|=-3$
$|2x+4|=3$
$2x+4=3$ $\quad$ $2x+4=-3$
$2x=-1$ $\quad$ $2x=-7$
$x=-\frac{1}{2}$ $\quad$ $x=-\frac{7}{2}$
The solutions are $-\frac{1}{2}$ and $-\frac{7}{2}$.

61. $8-|1-3x|=-1$
$-|1-3x|=-9$
$|1-3x|=9$
$1-3x=9$ $\quad$ $1-3x=-9$
$-3x=8$ $\quad$ $-3x=-10$
$x=-\frac{8}{3}$ $\quad$ $x=\frac{10}{3}$
The solutions are $-\frac{8}{3}$ and $\frac{10}{3}$.

63. $5+|2-x|=3$
$|2-x|=-2$
There is no solution to this equation because the absolute value of a number must be nonnegative.

Objective B Exercises

65. $|x|>3$
$x>3$ or $x<-3$
$\{x|x>3\}$ $\quad$ $\{x|x<-3\}$
$\{x|x>3\}\cup\{x|x<-3\}=\{x|x>3 \text{ or } x<-3\}$

67. $|x+1|>2$
$x+1>2$ or $x+1<-2$
$x>1$ $\quad$ $x<-3$
$\{x|x>1\}$ $\quad$ $\{x|x<-3\}$
$\{x|x>1\}\cup\{x|x<-3\}=\{x|x>1 \text{ or } x<-3\}$

69. $|x-5|\le 1$
$-1\le x-5\le 1$
$-1+5\le x-5+5\le 1+5$
$4\le x\le 6$
$\{x|4\le x\le 6\}$

71. $|2-x|\ge 3$
$2-x\le -3$ or $2-x\ge 3$
$-x\le -5$ $\quad$ $-x\ge 1$
$x\ge 5$ $\quad$ $x\le -1$
$\{x|x\ge 5\}$ $\quad$ $\{x|x\le -1\}$
$\{x|x\ge 5\}\cup\{x|x\le -1\}=\{x|x\ge 5 \text{ or } x\le -1\}$

73. $|2x+1|<5$
$-5<2x+1<5$
$-5+(-1)<2x+1+(-1)<5+(-1)$
$-6<2x<4$
$\frac{1}{2}(-6)<\frac{1}{2}(2x)<\frac{1}{2}(4)$
$-3<x<2$
$\{x|-3<x<2\}$

75. $|5x+2|>12$
$5x+2>12$ or $5x+2<-12$
$5x>10$ $\quad$ $5x<-14$
$x>2$ $\quad$ $x<-\frac{14}{5}$
$\{x|x>2\}$ $\quad$ $\left\{x\middle|x<-\frac{14}{5}\right\}$
$\{x|x>2\}\cup\left\{x\middle|x<-\frac{14}{5}\right\}=\left\{x\middle|x>2 \text{ or } x<-\frac{14}{5}\right\}$

77. $|4x-3|\le -2$
The absolute value of a number must be nonnegative. The solution set is the empty set, $\varnothing$.

79. $|2x+7|>-5$
$2x+7>-5$ or $2x+7<5$
$2x>-12$ $\quad$ $2x<-2$
$x>-6$ $\quad$ $x<-1$
$\{x|x>-6\}$ $\quad$ $\{x|x<-1\}$
$\{x|x>-6\}\cup\{x|x<-1\}=$ The set of real numbers

81. $|4-3x|\ge 5$
$4-3x\ge 5$ or $4-3x\le -5$
$-3x\ge 1$ $\quad$ $-3x\le -9$
$x\le -\frac{1}{3}$ $\quad$ $x\ge 3$
$\left\{x\middle|x\le -\frac{1}{3}\right\}$ $\quad$ $\{x|x\ge 3\}$
$\left\{x\middle|x\le -\frac{1}{3}\right\}\cup\{x|x\ge 3\}=\left\{x\middle|x\le -\frac{1}{3} \text{ or } x\ge 3\right\}$

83. $|5-4x|\le 13$
$-13\le 5-4x\le 13$
$-13+(-5)\le 5+(-5)-4x\le 13+(-5)$
$-18\le -4x\le 8$
$-\frac{1}{4}(-18)\ge -\frac{1}{4}(-4x)\ge -\frac{1}{4}(8)$
$\frac{9}{2}\ge x\ge -2$
$\left\{x\middle|-2\le x\le \frac{9}{2}\right\}$

85. $|6-3x| \le 0$

$6-3x \le 0$ or $6-3x \ge 0$

$-3x \le -6$ $\quad -3x \ge -6$

$x \ge 2$ $\quad x \le 2$

$\{x|x \ge 2\}$ $\quad \{x|x \le 2\}$

$\{x|x \ge 2\} \cup \{x|x \le 2\} = \{x|x = 2\}$

87. $|2-9x| > 20$

$2-9x > 20$ or $2-9x < -20$

$-9x > 18$ $\quad -9x < -22$

$x < -2$ $\quad x > \frac{22}{9}$

$\{x|x < -2\}$ $\quad \left\{x \middle| x > \frac{22}{9}\right\}$

$\{x|x < -2\} \cup \left\{x \middle| x > \frac{22}{9}\right\} = \left\{x \middle| x < -2 \text{ or } x > \frac{22}{9}\right\}$

89. $|2x-3|+2 < 8$

$|2x-3| < 6$

$-6 < 2x-3 < 6$

$-6+3 < 2x-3+3 < 6+3$

$-3 < 2x < 9$

$\frac{1}{2}(-3) < \frac{1}{2}(2x) < \frac{1}{2}(9)$

$-\frac{3}{2} < x < \frac{9}{2}$

$\left\{x \middle| -\frac{3}{2} < x < \frac{9}{2}\right\}$

91. $|2-5x|-4 > -2$

$|2-5x| > 2$

$2-5x > 2$ or $2-5x < -2$

$-5x > 0$ $\quad -5x < -4$

$x < 0$ $\quad x > \frac{4}{5}$

$\{x|x < 0\}$ $\quad \left\{x \middle| x > \frac{4}{5}\right\}$

$\{x|x < 0\} \cup \left\{x \middle| x > \frac{4}{5}\right\} = \left\{x \middle| x < 0 \text{ or } x > \frac{4}{5}\right\}$

93. $8-|2x-5| < 3$

$-|2x-5| < -5$

$|2x-5| > 5$

$2x-5 > 5$ or $2x-5 < -5$

$2x > 10$ $\quad 2x < 0$

$x > 5$ $\quad x < 0$

$\{x|x > 5\}$ $\quad \{x|x < 0\}$

$\{x|x > 5\} \cup \{x|x < 0\} = \{x|x > 5 \text{ or } x < 0\}$

Objective C Exercises

95. Strategy Let d represent the diameter of the bushing, T the tolerance, and x the lower and upper limits of the diameter. Solve the absolute value inequality $|x-d| \le T$ for x.

Solution

$|x-d| \le T$

$|x-1.75| \le 0.008$

$-0.008 \le x-1.75 \le 0.008$

$-0.008+1.75 \le x-1.75+1.75 \le 0.008+1.75$

$1.742 \le x \le 1.758$

The lower and upper limits of the diameter of the bushing are 1.742 in. and 1.758 in.

97. Strategy Let V represent the amount of voltage, T the tolerance, and A the given amount of voltage. Solve the absolute value inequality $|V-A| \le T$ for V.

Solution

$|V-A| \le T$

$|V-220| \le 25$

$-25 \le V-220 \le 25$

$-25+220 \le V-220+220 \le 25+220$

$195 \le V \le 245$

The lower and upper limits of the voltage of the electric motor are 195 volts and 245 volts.

99. Strategy Let r represent the length of the piston rod, T the tolerance, and L the lower and upper limits of the length. Solve the absolute value inequality $|L-r| \le T$ for L.

Solution

$|L-r| \le T$

$\left|L-9\frac{5}{8}\right| \le \frac{1}{32}$

$-\frac{1}{32} \le L-9\frac{5}{8} \le \frac{1}{32}$

$-\frac{1}{32}+9\frac{5}{8} \le L-9\frac{5}{8}+9\frac{5}{8} \le \frac{1}{32}+9\frac{5}{8}$

$9\frac{19}{32} \le L \le 9\frac{21}{32}$

The lower and upper limits of the length of the piston rod are $9\frac{19}{32}$ in. and $9\frac{21}{32}$ in.

101. Strategy Let M represent the amount of ohms, T the tolerance, and r the given amount of the resistor. Find the tolerance and solve $|M - r| \le T$ for M.

Solution

$$\begin{aligned} T &= (.02)(29{,}000) \\ &= 580 \text{ ohms} \\ |M - r| &\le T \\ |M - 29{,}000| &\le 580 \\ -580 \le M - 29{,}000 &\le 580 \\ -580 + 29{,}000 \le M - 29{,}000 + 29{,}000 &\le 580 + 29{,}000 \\ 28{,}420 \le M &\le 29{,}580 \end{aligned}$$

The lower and upper limits of the resistor are 28,420 ohms and 29,580 ohms.

103. Strategy Let M represent the amount of ohms, T the tolerance, and r the given amount of the resistor. Find the tolerance and solve $|M - r| \le T$ for M.

Solution

$$\begin{aligned} T &= (0.05)(25{,}000) \\ &= 1250 \text{ ohms} \\ |M - r| &\le T \\ |M - 25{,}000| &\le 1250 \\ -1250 \le M - 25{,}000 &\le 1250 \\ -1250 + 25{,}000 \le M - 25{,}000 + 25{,}000 &\le 1250 + 25{,}000 \\ 23{,}750 \le M &\le 26{,}250 \end{aligned}$$

The lower and upper limits of the resistor are 23,750 ohms and 26,250 ohms.

Applying the Concepts

105. **a.** $|x + 3| = x + 3$

Any value of x that makes $x + 3$ negative will result in a false equation because the left side of the equation will be positive and the right side of the equation will be negative. Therefore, the equation is true if $x + 3$ is greater than or equal to zero.

$$\begin{aligned} x + 3 &\ge 0 \\ x &\ge -3 \end{aligned}$$

$\{x|x \ge -3\}$

b. $|a - 4| = 4 - a$

Any value of a that makes $4 - a$ negative will result in a false equation because the left side of the equation will be positive and the right side of the equation will be negative. Therefore, the equation is true if $4 - a$ is greater than or equal to zero.

$$\begin{aligned} 4 - a &\ge 0 \\ a &\le 4 \end{aligned}$$

$\{a|a \le 4\}$

107. **a.** $|x + y| \le |x| + |y|$

b. $|x - y| \ge |x| - |y|$

c. $||x| - |y|| \ge |x| - |y|$

d. $\left|\dfrac{x}{y}\right| = \dfrac{|x|}{|y|}, y \ne 0$

e. $|xy| = |x||y|$

Chapter 2 Review Exercises

1.
$$\begin{aligned} 3t - 3 + 2t &= 7t - 15 \\ 5t - 3 &= 7t - 15 \\ 5t - 3 - 7t &= 7t - 15 - 7t \\ -2t - 3 &= -15 \\ -2t - 3 + 3 &= -15 + 3 \\ -2t &= -15 + 3 \\ -2t &= -12 \\ -\frac{1}{2}(-2t) &= -\frac{1}{2}(-12) \\ t &= 6 \end{aligned}$$

2.
$$\begin{aligned} 3x - 7 &> -2 \\ 3x &> 5 \\ \frac{3x}{3} &> \frac{5}{3} \\ x &> \frac{5}{3} \end{aligned}$$

$\left(\frac{5}{3}, \infty\right)$

3.
$$\begin{aligned} P &= 2L + 2W \\ P - 2L &= 2W \\ -2L &= -P + 2W \\ 2L &= P - 2W \\ L &= \frac{P - 2W}{2} \end{aligned}$$

4.
$$\begin{aligned} x + 4 &= -5 \\ x + 4 - 4 &= -5 - 4 \\ x &= -9 \end{aligned}$$

5. $3x < 4$ and $x + 2 > -1$

$x < \frac{4}{3}$ $\qquad x > -3$

$\left\{x \middle| x < \frac{4}{3}\right\}$ $\qquad \{x|x > -3\}$

$\left\{x \middle| x < \frac{4}{3}\right\} \cap \{x|x > -3\} = \left\{x \middle| -3 < x < \frac{4}{3}\right\}$

6. $$\begin{aligned}\frac{3}{5}x - 3 &= 2x + 5\\ 5\left(\frac{3}{5}x - 3\right) &= 5(2x+5)\\ 3x - 15 &= 10x + 25\\ 3x - 15 - 10x &= 10x + 25 - 10x\\ -7x - 15 &= 25\\ -7x - 15 + 15 &= 25 + 15\\ -7x &= 40\\ -\frac{1}{7}(-7x) &= -\frac{1}{7}(40)\\ x &= -\frac{40}{7}\end{aligned}$$

7. $$\begin{aligned}-\frac{2}{3}x &= \frac{4}{9}\\ -\frac{3}{2}\left(-\frac{2}{3}x\right) &= -\frac{3}{2}\left(\frac{4}{9}\right)\\ x &= -\frac{2}{3}\end{aligned}$$

8. $|x-4| - 8 = -3$
$|x-4| = 5$
$x - 4 = 5$ or $x - 4 = -5$
$x = 9$ $\quad$ $x = -1$
The solutions are –1 and 9.

9. $$\begin{aligned}|2x-5| &< 3\\ -3 < 2x &- 5 < 3\\ -3 + 5 < 2x - 5 &+ 5 < 3 + 5\\ 2 < 2x &< 8\\ \frac{1}{2}(2) < \frac{1}{2}(2x) &< \frac{1}{2}(8)\\ 1 < x &< 4\\ \{x|1 < x &< 4\}\end{aligned}$$

10. $$\begin{aligned}\frac{2x-3}{3} + 2 &= \frac{2-3x}{5}\\ 15\left(\frac{2x-3}{3} + 2\right) &= 15\left(\frac{2-3x}{5}\right)\\ \frac{15(2x-3)}{3} + 15(2) &= \frac{15(2-3x)}{5}\\ 5(2x-3) + 30 &= 3(2-3x)\\ 10x - 15 + 30 &= 6 - 9x\\ 10x + 15 &= 6 - 9x\\ 10x + 15 + 9x &= 6 - 9x + 9x\\ 19x + 15 &= 6\\ 19x + 15 - 15 &= 6 - 15\\ 19x &= -9\\ \frac{1}{19}(19x) &= \frac{1}{19}(-9)\\ x &= -\frac{9}{19}\end{aligned}$$

11. $$\begin{aligned}2(a-3) &= 5(4-3a)\\ 2a - 6 &= 20 - 15a\\ 2a - 6 + 15a &= 20 - 15a + 15a\\ 17a - 6 &= 20\\ 17a - 6 + 6 &= 20 + 6\\ 17a &= 26\\ \frac{1}{17}(17a) &= \frac{1}{17}(26)\\ a &= \frac{26}{17}\end{aligned}$$

12. $5x - 2 > 8$ or $3x + 2 < -4$
$5x > 10$ $\quad$ $3x < -6$
$x > 2$ $\quad$ $x < -2$
$\{x|x > 2\}$ $\quad$ $\{x|x < -2\}$
$\{x|x > 2\} \cup \{x|x < -2\} = \{x|x > 2 \text{ or } x < -2\}$

13. $|4x-5| \ge 3$ $\quad$ or $\quad$ $4x - 5 \le -3$
$4x - 5 \ge 3$ $\quad$ $4x \le 2$
$4x \ge 8$ $\quad$ $x \le \frac{1}{2}$
$x \ge 2$
$\{x|x \ge 2\}$ $\quad$ $\left\{x \,\middle|\, x \le \frac{1}{2}\right\}$
$\{x|x \ge 2\} \cup \left\{x \,\middle|\, x \le \frac{1}{2}\right\} = \left\{x \,\middle|\, x \ge 2 \text{ or } x \le \frac{1}{2}\right\}$

14. $$\begin{aligned}P &= \frac{R-C}{n}\\ n \cdot P &= \frac{R-C}{n} \cdot n\\ Pn &= R - C\\ C + Pn &= R\\ C &= R - Pn\end{aligned}$$

15. $$\begin{aligned}\frac{1}{2}x - \frac{5}{8} &= \frac{3}{4}x + \frac{3}{2}\\ 8\left(\frac{1}{2}x - \frac{5}{8}\right) &= 8\left(\frac{3}{4}x + \frac{3}{2}\right)\\ 8\left(\frac{1}{2}x\right) - 8\left(\frac{5}{8}\right) &= 8\left(\frac{3}{4}x\right) + 8\left(\frac{3}{2}\right)\\ 4x - 5 &= 6x + 12\\ 4x - 5 - 6x &= 6x + 12 - 6x\\ -2x - 5 &= 12\\ -2x - 5 + 5 &= 12 + 5\\ -2x &= 17\\ -\frac{1}{2}(-2x) &= -\frac{1}{2}(17)\\ x &= -\frac{17}{2}\end{aligned}$$

16. $6 + |3x-3| = 2$
$|3x-3| = -4$
There is no solution to this equation because the absolute value of a number must be nonnegative.

17. $3x - 2 > x - 4$ or $7x - 5 < 3x + 3$

$$2x - 2 > -4 \qquad 4x - 5 < 3$$
$$2x > -2 \qquad 4x < 8$$
$$\frac{2x}{2} > \frac{-2}{2} \qquad \frac{4x}{4} < \frac{8}{4}$$
$$x > -1 \qquad x < 2$$
$$\{x|x > -1\} \qquad \{x|x < 2\}$$

$\{x|x > -1\} \cup \{x|x < 2\} = \{x|x \text{ is any real number}\}$

The solution set is $(-\infty, \infty)$.

18.
$$2x - (3 - 2x) = 4 - 3(4 - 2x)$$
$$2x - 3 + 2x = 4 - 12 + 6x$$
$$4x - 3 = -8 + 6x$$
$$4x - 3 - 6x = -8 + 6x - 6x$$
$$-2x - 3 = -8$$
$$-2x - 3 + 3 = -8 + 3$$
$$-2x = -5$$
$$-\frac{1}{2}(-2x) = -\frac{1}{2}(-5)$$
$$x = \frac{5}{2}$$

19.
$$x + 9 = -6$$
$$x + 9 - 9 = -6 - 9$$
$$x = -15$$

20.
$$\frac{2}{3} = x + \frac{3}{4}$$
$$\frac{2}{3} - \frac{3}{4} = x + \frac{3}{4} - \frac{3}{4}$$
$$\frac{8}{12} - \frac{9}{12} = x$$
$$-\frac{1}{12} = x$$

21.
$$-3x = -21$$
$$\frac{-3x}{-3} = \frac{-21}{-3}$$
$$x = 7$$

22.
$$\frac{2}{3}a = \frac{4}{9}$$
$$\frac{3}{2}\left(\frac{2}{3}a\right) = \frac{3}{2}\left(\frac{4}{9}\right)$$
$$a = \frac{2}{3}$$

23.
$$3y - 5 = 3 - 2y$$
$$3y + 2y - 5 = 3 - 2y + 2y$$
$$5y - 5 = 3$$
$$5y - 5 + 5 = 3 + 5$$
$$5y = 8$$
$$\frac{5y}{5} = \frac{8}{5}$$
$$y = \frac{8}{5}$$

24.
$$4x - 5 + x = 6x - 8$$
$$5x - 5 = 6x - 8$$
$$5x - 6x - 5 = 6x - 6x - 8$$
$$-x - 5 = -8$$
$$-x - 5 + 5 = -8 + 5$$
$$-x = -3$$
$$(-1)(-x) = (-1)(-3)$$
$$x = 3$$

25.
$$3(x - 4) = -5(6 - x)$$
$$3x - 12 = -30 + 5x$$
$$3x - 5x - 12 = -30 + 5x - 5x$$
$$-2x - 12 = -30$$
$$-2x - 12 + 12 = -30 + 12$$
$$-2x = -18$$
$$-\frac{1}{2}(-2x) = -\frac{1}{2}(-18)$$
$$x = 9$$

26.
$$\frac{3x - 2}{4} + 1 = \frac{2x - 3}{2}$$
$$4\left(\frac{3x - 2}{4} + 1\right) = 4\left(\frac{2x - 3}{2}\right)$$
$$4\left(\frac{3x - 2}{4}\right) + 4(1) = 4\left(\frac{2x - 3}{2}\right)$$
$$3x - 2 + 4 = 4x - 6$$
$$3x + 2 = 4x - 6$$
$$3x - 4x + 2 = 4x - 4x - 6$$
$$-x + 2 = -6$$
$$-x + 2 - 2 = -6 - 2$$
$$-x = -8$$
$$-1(-x) = -1(-8)$$
$$x = 8$$

27.
$$5x - 8 < -3$$
$$5x - 8 + 8 < -3 + 8$$
$$5x < 5$$
$$\frac{5x}{5} < \frac{5}{5}$$
$$x < 1$$
$$(-\infty, 1)$$

28.
$$2x - 9 \le 8x + 15$$
$$-6x \le 24$$
$$x \ge -4$$
$$[-4, \infty)$$

29.
$$\frac{2}{3}x - \frac{5}{8} \ge \frac{3}{4}x + 1$$
$$24\left(\frac{2}{3}x - \frac{5}{8}\right) \ge 24\left(\frac{3}{4}x + 1\right)$$
$$16x - 15 \ge 18x + 24$$
$$-2x \ge 39$$
$$x \le -\frac{39}{2}$$
$$\left\{x\middle|x \le -\frac{39}{2}\right\}$$

30. $2 - 3(2x - 4) \le 4x - 2(1 - 3x)$
$2 - 6x + 12 \le 4x - 2 + 6x$
$-6x + 14 \le 10x - 2$
$-16x \le -16$
$x \ge 1$
$\{x|x \ge 1\}$

31. $-5 < 4x - 1 < 7$
$-4 < 4x < 8$
$\frac{-4}{4} < \frac{4x}{4} < \frac{8}{4}$
$-1 < x < 2$
$(-1, 2)$

32. $|2x - 3| = 8$
$2x - 3 = 8$ $\quad$ $2x - 3 = -8$
$2x = 11$ $\quad$ $2x = -5$
$x = \frac{11}{2}$ $\quad$ $x = -\frac{5}{2}$
$-\frac{5}{2}, \frac{11}{2}$

33. $|5x + 8| = 0$
$5x + 8 = 0$
$5x = -8$
$x = -\frac{8}{5}$

34. $|5x - 4| < -2$
The absolute value of a number must be nonnegative.
The solution set is the empty set, $\varnothing$.

35. Strategy • Time to travel to the island: t
Time to return to dock: $2\frac{1}{3} - t$

	Rate	Time	Distance
To island	16	t	$16t$
Back to dock	12	$\frac{7}{3} - t$	$12\left(\frac{7}{3} - t\right)$

• The distance to the island equals the distance back to the dock.
• Determine t and then find the distance.

Solution $16t = 12\left(\frac{7}{3} - t\right)$
$16t = 28 - 12t$
$28t = 28$
$t = 1$
The distance is $16t = 16$.
The island is 16 min from the dock.

36. Strategy • Gallons of apple juice: x

	Amount	Cost	Value
Apple	x	4.20	$4.20x$
Cranberry	40	6.50	$40(6.50)$
Mixture	$40 + x$	5.20	$5.20(40 + x)$

• The sum of the values before mixing equals the value after mixing.

Solution $4.20x + 40(6.50) = 5.20(40 + x)$
$4.20x + 260 = 208 + 5.20x$
$-x = -52$
$x = 52$
The mixture must contain 52 gal of apple juice.

37. Strategy To find the minimum amount of sales, write and solve an inequality using N to represent the amount of sales.

Solution $800 + 0.04N \ge 3000$
$0.04N \ge 2200$
$N \ge 55{,}000$
The executive's amount of sales must be \$55,000 or more.

38. Strategy • Number of nickels: x
Number of dimes: $x + 3$
Number of quarters:
$30 - (2x + 3) = 27 - 2x$

Coin	Number	Value	Total Value
Nickel	x	5	$5x$
Dime	$x + 3$	10	$10(x + 3)$
Quarter	$27 - 2x$	25	$25(27 - 2x)$

• The sum of the total values of each denomination of coin equals the total value of all the coins (355 cents).

Solution $5x + 10(x + 3) + 25(27 - 2x) = 355$
$5x + 10x + 30 + 675 - 50x = 355$
$-35x + 705 = 355$
$-35x = -350$
$x = 10$
$27 - 2x = 27 - 2(10) = 27 - 20 = 7$
There are 7 quarters in the collection.

39. Strategy Let b represent the diameter of the bushing, T the tolerance, and d the lower and upper limits of the diameter. Solve the absolute value inequality $|d-b| \le T$ for d.

Solution

$$|d-b| \le T$$
$$|d-2.75| \le 0.003$$
$$-0.003 \le d-2.75 \le 0.003$$
$$-0.003+2.75 \le d-2.75+2.75 \le 0.003+2.75$$
$$2.747 \le d \le 2.753$$

The lower and upper limits of the diameter of the bushing are 2.747 in. and 2.753 in.

40. Strategy
- The smaller integer: n
  The larger integer: $20-n$
- Five times the smaller integer is two more than twice the larger integer.

Solution

$$5n = 2+2(20-n)$$
$$5n = 2+40-2n$$
$$5n = 42-2n$$
$$7n = 42$$
$$n = 6$$
$$20-n = 20-6 = 14$$

The integers are 6 and 14.

41. Strategy To find the range of scores, write and solve an inequality using N to represent the score on the last test.

Solution

$$80 \le \frac{92+66+72+88+N}{5} \le 90$$
$$80 \le \frac{318+N}{5} \le 90$$
$$5\cdot 80 \le 5\cdot\frac{318+N}{5} \le 5\cdot 90$$
$$400 \le 318+N \le 450$$
$$400-318 \le 318+N-318 \le 450-318$$
$$82 \le N \le 132$$

Since 100 is the maximum score, the range of scores to receive a C grade is $82 \le x \le 100$.

42. Strategy
- Rate of the first plane: r
  Rate of the second plane: $r+80$

	Rate	Time	Distance
2nd plane	r	1.75	$1.75r$
1st plane	$r+80$	1.75	$1.75(r+80)$

- The total distance traveled by the two planes is 1680 mi.

Solution

$$1.75r+1.75(r+80) = 1680$$
$$1.75r+1.75r+140 = 1680$$
$$3.5r+140 = 1680$$
$$3.5r = 1540$$
$$r = 440$$
$$r+80 = 440+80 = 520$$

The speed of the second plane is 440 mph.
The speed of the first plane is 520 mph.

43. Strategy
- Pounds of 30% tin: x
  Pounds of 70% tin: $500-x$

	Amount	Percent	Quantity
30%	x	0.30	$0.30x$
70%	$500-x$	0.70	$0.70(500-x)$
40%	500	0.40	$0.40(500)$

- The sum of the quantities before mixing is equal to the quantity after mixing.

Solution

$$0.30x+0.70(500-x) = 0.40(500)$$
$$0.30x+350-0.70x = 200$$
$$-0.40x+350 = 200$$
$$-0.40x = -150$$
$$x = 375$$
$$500-x = 500-375 = 125$$

375 lb of the 30% tin alloy and 125 lb of the 70% tin alloy were used.

44. Strategy Let r represent the length of the piston rod, T the tolerance, and L the lower and upper limits of the length. Solve the absolute value inequality $|L-r| \le T$ for L.

Solution

$$|L-r| \le T$$
$$|L-10\tfrac{3}{8}| \le \frac{1}{32}$$
$$-\frac{1}{32} \le L-10\tfrac{3}{8} \le \frac{1}{32}$$
$$-\frac{1}{32}+10\tfrac{3}{8} \le L-10\tfrac{3}{8}+10\tfrac{3}{8} \le \frac{1}{32}+10\tfrac{3}{8}$$
$$10\tfrac{11}{32} \le L \le 10\tfrac{13}{32}$$

The lower and upper limits of the length of the piston are $10\frac{11}{32}$ in. and $10\frac{13}{32}$ in.

Chapter 2 Test

1.

$$x-2 = -4$$
$$x-2+2 = -4+2$$
$$x = -2$$

2. $$\begin{aligned} b+\frac{3}{4} &= \frac{5}{8} \\ b+\frac{3}{4}-\frac{3}{4} &= \frac{5}{8}-\frac{3}{4} \\ b &= \frac{5}{8}-\frac{6}{8} \\ b &= -\frac{1}{8} \end{aligned}$$

3. $$\begin{aligned} -\frac{3}{4}y &= -\frac{5}{8} \\ -\frac{4}{3}\left(-\frac{3}{4}y\right) &= -\frac{4}{3}\left(-\frac{5}{8}\right) \\ y &= \frac{5}{6} \end{aligned}$$

4. $$\begin{aligned} 3x-5 &= 7 \\ 3x-5+5 &= 7+5 \\ 3x &= 12 \\ \frac{3x}{3} &= \frac{12}{3} \\ x &= 4 \end{aligned}$$

5. $$\begin{aligned} \frac{3}{4}y-2 &= 6 \\ \frac{3}{4}y-2+2 &= 6+2 \\ \frac{3}{4}y &= 8 \\ \frac{4}{3}\left(\frac{3}{4}y\right) &= \frac{4}{3}(8) \\ y &= \frac{32}{3} \end{aligned}$$

6. $$\begin{aligned} 2x-3-5x &= 8+2x-10 \\ -3x-3 &= -2+2x \\ -3x-3+3x &= -2+2x+3x \\ -3 &= -2+5x \\ -3+2 &= -2+5x+2 \\ -1 &= 5x \\ \frac{1}{5}(-1) &= \frac{1}{5}(5x) \\ -\frac{1}{5} &= x \\ x &= -\frac{1}{5} \end{aligned}$$

7. $$\begin{aligned} 2[a-(2-3a)-4] &= a-5 \\ 2[a-2+3a-4] &= a-5 \\ 2[4a-6] &= a-5 \\ 8a-12 &= a-5 \\ 8a-a-12 &= a-a-5 \\ 7a-12 &= -5 \\ 7a-12+12 &= -5+12 \\ 7a &= 7 \\ \frac{7a}{7} &= \frac{7}{7} \\ a &= 1 \end{aligned}$$

8. $$\begin{aligned} E &= IR+Ir \\ E-IR &= Ir \\ -IR &= -E+Ir \\ IR &= E-Ir \\ R &= \frac{E-Ir}{I} \end{aligned}$$

9. $$\begin{aligned} \frac{2x+1}{3}-\frac{3x+4}{6} &= \frac{5x-9}{9} \\ 18\left(\frac{2x+1}{3}-\frac{3x+4}{6}\right) &= 18\left(\frac{5x-9}{9}\right) \\ 18\left(\frac{2x+1}{3}\right)-18\left(\frac{3x+4}{6}\right) &= 18\left(\frac{5x-9}{9}\right) \\ 6(2x+1)-3(3x+4) &= 2(5x-9) \\ 12x+6-9x-12 &= 10x-18 \\ 3x-6 &= 10x-18 \\ 3x-6-10x &= 10x-18-10x \\ -7x-6 &= -18 \\ -7x-6+6 &= -18+6 \\ -7x &= -12 \\ \frac{-7x}{-7} &= \frac{-12}{-7} \\ x &= \frac{12}{7} \end{aligned}$$

10. $$\begin{aligned} 3x-2 &\ge 6x+7 \\ -3x &\ge 9 \\ \frac{-3x}{-3} &\le \frac{9}{-3} \\ x &\le -3 \end{aligned}$$
$\{x|x \le -3\}$

11. $$\begin{aligned} 4-3(x+2) &< 2(2x+3)-1 \\ 4-3x-6 &< 4x+6-1 \\ -2-3x &< 4x+5 \\ -7x &< 7 \\ \frac{-7x}{-7} &> \frac{7}{-7} \\ x &> -1 \end{aligned}$$
$(-1, \infty)$

12. $4x-1>5$ or $2-3x<8$

$$\begin{aligned} 4x &> 6 & -3x &< 6 \\ \frac{4x}{4} &> \frac{6}{4} & \frac{-3x}{-3} &> \frac{6}{-3} \\ x &> \frac{3}{2} & x &> -2 \end{aligned}$$
$\left\{x \middle| x > \frac{3}{2}\right\}$ $\{x|x>-2\}$

$\left\{x|x > \frac{3}{2}\right\} \cup \{x|x>-2\} = \{x|x>-2\}$

13. $4-3x \ge 7$ and $2x+3 \ge 7$

$$\begin{aligned} -3x &\ge 3 & 2x &\ge 4 \\ \frac{-3x}{-3} &\le \frac{3}{-3} & \frac{2x}{2} &\ge \frac{4}{2} \\ x &\le -1 & x &\ge 2 \end{aligned}$$
$\{x|x \le -1\}$ $\{x|x \ge 2\}$

$\{x|x \le -1\} \cap \{x|x \ge 2\} = \varnothing$

14. $|3 - 5x| = 12$

$$\begin{aligned} 3 - 5x &= 12 \\ -5x &= 9 \\ x &= -\frac{9}{5} \end{aligned} \qquad \begin{aligned} 3 - 5x &= -12 \\ -5x &= -15 \\ x &= 3 \end{aligned}$$

The solutions are $-\frac{9}{5}$ and 3.

15.
$$\begin{aligned} 2 - |2x - 5| &= -7 \\ -|2x - 5| &= -9 \\ |2x - 5| &= 9 \end{aligned}$$

$$\begin{aligned} 2x - 5 &= 9 \\ 2x &= 14 \\ x &= 7 \end{aligned} \qquad \begin{aligned} 2x - 5 &= -9 \\ 2x &= -4 \\ x &= -2 \end{aligned}$$

The solutions are 7 and −2.

16.
$$\begin{gathered} |3x - 5| \le 4 \\ -4 \le 3x - 5 \le 4 \\ -4 + 5 \le 3x - 5 + 5 \le 4 + 5 \\ 1 \le 3x \le 9 \\ \frac{1}{3} \le \frac{3x}{3} \le \frac{9}{3} \\ \frac{1}{3} \le x \le 3 \end{gathered}$$

$$\left\{x \,\middle|\, \frac{1}{3} \le x \le 3\right\}$$

17. $|4x - 3| > 5$

$$\begin{aligned} 4x - 3 &< -5 \\ 4x &< -2 \\ x &< -\frac{2}{4} \\ x &< -\frac{1}{2} \end{aligned} \qquad \text{or} \qquad \begin{aligned} 4x - 3 &> 5 \\ 4x &> 8 \\ x &> 2 \end{aligned}$$

$$\left\{x \,\middle|\, x < -\frac{1}{2}\right\} \qquad \{x|x > 2\}$$

$$\left\{x \,\middle|\, x < -\frac{1}{2}\right\} \cup \{x|x > 2\} =$$
$$\left\{x \,\middle|\, x < -\frac{1}{2} \text{ or } x > 2\right\}$$

18. Strategy To find the number of miles, write and solve an inequality using N to represent the number of miles.

Solution Cost of Gambelli car < Cost of McDougal car

$$\begin{aligned} 12 + 0.10N &< 24 \\ 0.10N &< 12 \\ N &< 120 \end{aligned}$$

It costs less to rent from Gambelli agency if the car is driven less than 120 mi.

19. Strategy Let b represent the diameter of the bushing, T the tolerance, and d the lower and upper limits of the diameter. Solve the absolute value inequality $|d - b| \le T$ for d.

Solution
$$\begin{gathered} |d - b| \le T \\ |d - 2.65| \le 0.002 \\ -0.002 \le d - 2.65 \le 0.002 \\ -0.002 + 2.65 \le d - 2.65 + 2.65 \le 0.002 + 2.65 \\ 2.648 \le d \le 2.652 \end{gathered}$$

The lower and upper limits of the diameter of the bushing are 2.648 in. and 2.652 in.

20. Strategy
- The smaller integer: n
  The larger integer: $15 - n$
- Eight times the smaller integer is one less than three times the larger integer.

Solution
$$\begin{aligned} 8n &= 3(15 - n) - 1 \\ 8n &= 45 - 3n - 1 \\ 8n &= 44 - 3n \\ 8n + 3n &= 44 - 3n + 3n \\ 11n &= 44 \\ n &= 4 \end{aligned}$$

$15 - n = 15 - 4 = 11$

The integers are 4 and 11.

21. Strategy
- Number of 15¢ stamps: x
  Number of 11¢ stamps: $2x$
  Number of 24¢ stamps: $30 - 3x$

Stamp	Number	Value	Total Value
15¢	x	15	$15x$
11¢	$2x$	11	$11(2x)$
24¢	$30 - 3x$	24	$24(30 - 3x)$

- The sum of the total values of each denomination of stamp equals the total value of all the stamps (440 cents).

Solution
$$\begin{aligned} 15x + 11(2x) + 24(30 - 3x) &= 440 \\ 15x + 22x + 720 - 72x &= 440 \\ -35x + 720 &= 440 \\ -35x &= -280 \\ x &= 8 \end{aligned}$$

$30 - 3x = 30 - 3(8) = 30 - 24 = 6$

There are six 24¢ stamps.

22. Strategy • Price of hamburger mixture: x

	Amount	Cost	Value
\$2.10 hamburger	100	2.10	2.10(100)
\$3.70 hamburger	60	3.70	3.70(60)
Mixture	160	x	$160x$

• The sum of the values before mixing equals the value after mixing.

Solution
$$\begin{aligned} 2.10(100) + 3.70(60) &= 160x \\ 210 + 222 &= 160x \\ 432 &= 160x \\ 2.70 &= x \end{aligned}$$
The price of the hamburger mixture is \$2.70 per pound.

23. Strategy • Time jogger runs a distance: t
Time jogger returns same distance: $1\frac{45}{60} - t$

	Rate	Time	Distance
Jogger runs a distance	8	8	$8t$
Jogger returns same distance	6	$\frac{7}{4} - t$	$6\left(\frac{7}{4} - t\right)$

• The jogger runs a distance and returns the same distance.

Solution
$$\begin{aligned} 8t &= 6\left(\frac{7}{4} - t\right) \\ 8t &= \frac{21}{2} - 6t \\ 14t &= \frac{21}{2} \\ \frac{1}{14}(14t) &= \frac{1}{14}\left(\frac{21}{2}\right) \\ t &= \frac{3}{4} \end{aligned}$$
The jogger ran for $\frac{3}{4}$ h on the way there.
$$8t = 8 \cdot \frac{3}{4} = 6$$
The jogger ran a distance of 6 mi. one way. The jogger ran a total distance of 12 mi.

24. Strategy • Rate of the slower train: r
Rate of the faster train: $r + 5$

	Rate	Time	Distance
Slower train	r	2	$2r$
Faster train	$r + 5$	2	$2(r + 5)$

• The total distance traveled by the two trains is 250 mi.

Solution
$$\begin{aligned} 2r + 2(r + 5) &= 250 \\ 2r + 2r + 10 &= 250 \\ 4r + 10 &= 250 \\ 4r &= 240 \\ r &= 60 \\ r + 5 = 60 + 5 &= 65 \end{aligned}$$
The rate of the slower train is 60 mph. The rate of the faster train is 65 mph.

25. Strategy • Ounces of pure water: x

	Amount	Percent	Quantity
Pure water	x	0	0
8% salt	60	0.08	0.08(60)
3% salt	$60 + x$	0.03	$0.03(60 + x)$

• The sum of the quantities before mixing is equal to the quantity after mixing.

Solution
$$\begin{aligned} 0 + 0.08(60) &= 0.03(60 + x) \\ 4.8 &= 1.8 + 0.03x \\ 3 &= 0.03x \\ 100 &= x \end{aligned}$$
It is necessary to add 100 oz of pure water.

Cumulative Review Exercises

1. $$\begin{aligned} -4 - (-3) - 8 + (-2) &= -4 + 3 + (-8) + (-2) \\ &= -1 + (-8) + (-2) \\ &= -9 + (-2) \\ &= -11 \end{aligned}$$

2. $$\begin{aligned} -2^2 \cdot 3^3 &= -(2 \cdot 2)(3 \cdot 3 \cdot 3) \\ &= -(4)(27) \\ &= -108 \end{aligned}$$

3. $$\begin{aligned} 4 - (2 - 5)^2 \div 3 + 2 &= 4 - (-3)^2 \div 3 + 2 \\ &= 4 - 9 \div 3 + 2 \\ &= 4 - 3 + 2 \\ &= 1 + 2 \\ &= 3 \end{aligned}$$

4. $$\begin{aligned} 4 \div \frac{\frac{3}{8} - 1}{5} \cdot 2 &= 4 \div \frac{-\frac{5}{8}}{5} \cdot 2 \\ &= 4 \div \left(-\frac{5}{8} \cdot \frac{1}{5}\right) \cdot 2 \\ &= 4 \div \left(-\frac{1}{8}\right) \cdot 2 \\ &= 4 \cdot (-8) \cdot 2 \\ &= -32 \cdot 2 \\ &= -64 \end{aligned}$$

5. $$\begin{aligned} 2a^2 - (b - c)^2 &= 2(2)^2 - (3 - (-1))^2 \\ &= 2 \cdot 4 - (3 + 1)^2 \\ &= 2 \cdot 4 - 4^2 \\ &= 2 \cdot 4 - 16 \\ &= 8 - 16 \\ &= -8 \end{aligned}$$

6. $$\begin{aligned} \frac{a - b^2}{b - c} &= \frac{2 - (-3)^2}{-3 - 4} \\ &= \frac{2 - 9}{-3 - 4} \\ &= \frac{-7}{-7} \\ &= 1 \end{aligned}$$

7. The Commutative Property of Addition

8. the unknown number: n
three times the number: $3n$
the sum of three times the number and six:
$3n + 6$
$3n + (3n + 6) = 6n + 6$

9. $$\begin{aligned} F &= \frac{evB}{c} \\ Fc &= \frac{evB}{c} \cdot c \\ Fc &= evB \\ \frac{Fc}{ev} &= \frac{evB}{ev} \\ B &= \frac{Fc}{ev} \end{aligned}$$

10. $$\begin{aligned} 5[y - 2(3 - 2y) + 6] &= 5[y - 6 + 4y + 6] \\ &= 5[5y] \\ &= 25y \end{aligned}$$

11. $\{-4, 0\}$

12. $\{x | x \le 3\} \cap \{x | x > -1\}$

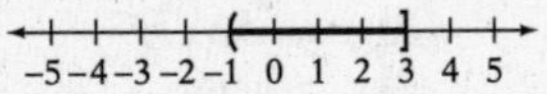

13. $$\begin{aligned} Ax + By + C &= 0 \\ By &= -C - Ax \\ \frac{By}{B} &= \frac{-C - Ax}{B} \\ y &= \frac{-Ax - C}{B} \end{aligned}$$

14. $$\begin{aligned} -\frac{5}{6}b &= -\frac{5}{12} \\ \left(-\frac{6}{5}\right)\left(-\frac{5}{6}\right)b &= \left(-\frac{6}{5}\right)\left(-\frac{5}{12}\right) \\ b &= \frac{1}{2} \end{aligned}$$

15. $$\begin{aligned} 2x + 5 &= 5x + 2 \\ 2x + 5 - 5x &= 5x + 2 - 5x \\ -3x + 5 &= 2 \\ -3x + 5 - 5 &= 2 - 5 \\ -3x &= -3 \\ \frac{-3x}{-3} &= \frac{-3}{-3} \\ x &= 1 \end{aligned}$$

16. $$\begin{aligned} \frac{5}{12}x - 3 &= 7 \\ \frac{5}{12}x - 3 + 3 &= 7 + 3 \\ \frac{5}{12}x &= 10 \\ \left(\frac{12}{5}\right)\left(\frac{5}{12}\right)x &= \left(\frac{12}{5}\right)10 \\ x &= 24 \end{aligned}$$

17. $$\begin{aligned} 2[3 - 2(3 - 2x)] &= 2(3 + x) \\ 2[3 - 6 + 4x] &= 6 + 2x \\ 2[-3 + 4x] &= 6 + 2x \\ -6 + 8x &= 6 + 2x \\ -6 + 8x - 2x &= 6 + 2x - 2x \\ -6 + 6x &= 6 \\ -6 + 6x + 6 &= 6 + 6 \\ 6x &= 12 \\ \frac{6x}{6} &= \frac{12}{6} \\ x &= 2 \end{aligned}$$

18. $$\begin{aligned} 3[2x - 3(4 - x)] &= 2(1 - 2x) \\ 3[2x - 12 + 3x] &= 2 - 4x \\ 3[5x - 12] &= 2 - 4x \\ 15x - 36 &= 2 - 4x \\ 15x - 36 + 4x &= 2 - 4x + 4x \\ 19x - 36 &= 2 \\ 19x - 36 + 36 &= 2 + 36 \\ 19x &= 38 \\ \frac{19x}{19} &= \frac{38}{19} \\ x &= 2 \end{aligned}$$

19. $$\begin{aligned} \frac{1}{2}y - \frac{2}{3}y + \frac{5}{12} &= \frac{3}{4}y - \frac{1}{2} \\ 12\left(\frac{1}{2}y - \frac{2}{3}y + \frac{5}{12}\right) &= 12\left(\frac{3}{4}y - \frac{1}{2}\right) \\ 6y - 8y + 5 &= 9y - 6 \\ -2y + 5 &= 9y - 6 \\ -2y + 5 - 9y &= 9y - 6 - 9y \\ -11y + 5 &= -6 \\ -11y + 5 - 5 &= -6 - 5 \\ -11y &= -11 \\ \frac{-11y}{-11} &= \frac{-11}{-11} \\ y &= 1 \end{aligned}$$

20.
$$\frac{3x-1}{4}-\frac{4x-1}{12}=\frac{3+5x}{8}$$
$$24\left(\frac{3x-1}{4}-\frac{4x-1}{12}\right)=24\left(\frac{3+5x}{8}\right)$$
$$\frac{24(3x-1)}{4}-\frac{24(4x-1)}{12}=\frac{24(3+5x)}{8}$$
$$6(3x-1)-2(4x-1)=3(3+5x)$$
$$18x-6-8x+2=9+15x$$
$$10x-4=9+15x$$
$$10x-4-15x=9+15x-15x$$
$$-5x-4=9$$
$$-5x-4+4=9+4$$
$$-5x=13$$
$$\frac{-5x}{-5}=\frac{13}{-5}$$
$$x=-\frac{13}{5}$$

21.
$$3-2(2x-1)\geq 3(2x-2)+1$$
$$3-4x+2\geq 6x-6+1$$
$$-4x+5\geq 6x-5$$
$$-4x+5-6x\geq 6x-5-6x$$
$$-10x+5\geq -5$$
$$-10x+5-5\geq -5-5$$
$$-10x\geq -10$$
$$\frac{-10x}{-10}\leq\frac{-10}{-10}$$
$$x\leq 1$$
$$(-\infty, 1]$$

22.
$$3x+2\leq 5 \quad\text{and}\quad x+5>1$$
$$3x+2-2\leq 5-2 \qquad x+5-5>1-5$$
$$3x\leq 3 \qquad x>-4$$
$$\frac{3x}{3}\leq\frac{3}{3}$$
$$x\leq 1$$
$$\{x|x\leq 1\} \qquad \{x|x>-4\}$$
$$\{x|x\leq 1\}\cap\{x|x>-4\}=\{x|-4<x\leq 1\}$$

23.
$$|3-2x|=5$$
$$3-2x=5 \quad\text{or}\quad 3-2x=-5$$
$$-2x=2 \qquad -2x=-8$$
$$x=-1 \qquad x=4$$

The solutions are −1 and 4.

24.
$$3-|2x-3|=-8$$
$$-|2x-3|=-11$$
$$|2x-3|=11$$
$$2x-3=11 \qquad 2x-3=-11$$
$$2x=14 \qquad 2x=-8$$
$$x=7 \qquad x=-4$$

The solutions are 7 and −4.

25.
$$|3x-1|>5$$
$$3x-1<-5 \quad\text{or}\quad 3x-1>5$$
$$3x<-4 \qquad 3x>6$$
$$x<-\frac{4}{3} \qquad x>2$$
$$\left\{x\,\middle|\,x<-\frac{4}{3}\right\} \qquad \{x|x>2\}$$
$$\left\{x\,\middle|\,x<-\frac{4}{3}\right\}\cup\{x|x>2\}=\left\{x\,\middle|\,x<-\frac{4}{3}\text{ or }x>2\right\}$$

26.
$$|2x-4|<8$$
$$-8<2x-4<8$$
$$-8+4<2x-4+4<8+4$$
$$-4<2x<12$$
$$\frac{-4}{2}<\frac{2x}{2}<\frac{12}{2}$$
$$-2<x<6$$
$$\{x|-2<x<6\}$$

27. Strategy To find the number of checks, write and solve an inequality using c to represent the number of checks.
$5.00+0.04c>2.00+0.10c$

Solution
$$5.00+0.04c>2.00+0.10c$$
$$5+0.04c-0.10c>2+0.10c-0.10c$$
$$5-0.06c>2$$
$$5-0.06c-5>2-5$$
$$-0.06c>-3$$
$$\frac{-0.06c}{-0.06}<\frac{-3}{-0.06}$$
$$c<50$$

The second account is cheaper if the customer writes fewer than 50 checks.

28. Strategy
- First odd integer: n
  Second odd integer: $n+2$
  Third odd integer: $n+4$
- Four times the sum of the first and third integers is one less than seven times the second.
  $4(n+n+4)=7(n+2)-1$

Solution
$$4(n+n+4)=7(n+2)-1$$
$$4(2n+4)=7n+14-1$$
$$8n+16=7n+13$$
$$8n+16-7n=7n+13-7n$$
$$n+16=13$$
$$n+16-16=13-16$$
$$n=-3$$

The first integer is −3.

29. Strategy • Number of quarters: x
Number of dimes: $2x - 5$

Coin	Number	Value	Total Value
Quarter	x	25	$25x$
Dime	$2x - 5$	10	$10(2x - 5)$

• The sum of the total values of each denomination of coin equals the total value of all the coins (400 cents).

Solution
$$\begin{aligned} 25x + 10(2x - 5) &= 400 \\ 25x + 20x - 50 &= 400 \\ 45x - 50 &= 400 \\ 45x &= 450 \\ x &= 10 \\ 2x - 5 = 2(10) - 5 &= 15 \end{aligned}$$
There are 15 dimes.

30. Strategy • Ounces of pure silver: x
Ounces of silver alloy: 100

	Amount	Cost	Value
Silver	x	8.50	$8.50x$
Alloy	100	4.00	100(4.00)
Mixture	$100 + x$	6.00	$(100 + x)6.00$

• The sum of the values before mixing equals the value after mixing.

Solution
$$\begin{aligned} 8.5x + 100(4) &= (100 + x)6 \\ 8.5x + 400 &= 600 + 6x \\ 2.5x + 400 &= 600 \\ 2.5x &= 200 \\ x &= 80 \end{aligned}$$
80 oz of pure silver were used in the mixture.

31. Strategy • Slower plane: x
Faster plane: $x + 120$

	Rate	Time	Distance
Slower plane	x	2.5	$2.5x$
Faster plane	$x + 120$	2.5	$2.5(x + 120)$

• The two planes travel a total distance of 1400 miles.

Solution
$$\begin{aligned} 2.5x + 2.5(x + 120) &= 1400 \\ 2.5x + 2.5x + 300 &= 1400 \\ 5x + 300 &= 1400 \\ 5x &= 1100 \\ x &= 220 \end{aligned}$$
The speed of the slower plane is 220 mph.

32. Strategy Let b represent the diameter of the bushing, T the tolerance, and d the lower and upper limits of the diameter. Solve the absolute value inequality $|d - b| \le T$ for d.

Solution
$$\begin{aligned} |d - b| &\le T \\ |d - 2.45| &\le 0.001 \\ -0.001 \le d - 2.45 &\le 0.001 \\ -0.001 + 2.45 \le d - 2.45 + 2.45 &\le 0.001 + 2.45 \\ 2.449 \le d &\le 2.451 \end{aligned}$$
The lower and upper limits of the diameter of the bushing are 2.449 in. and 2.451 in.

33. Strategy • Liters of 12% acid solution: x

	Amount	Percent	Quantity
12% solution	x	0.12	$0.12x$
5% solution	4	0.05	4(0.05)
8% solution	$x + 4$	0.08	$0.08(x + 4)$

• The sum of the quantities before mixing is equal to the quantity after mixing.

Solution
$$\begin{aligned} 0.12x + 4(0.05) &= 0.08(x + 4) \\ 0.12x + 0.2 &= 0.08x + 0.32 \\ 0.04x + 0.2 &= 0.32 \\ 0.04x &= 0.12 \\ x &= 3 \end{aligned}$$
3 L of 12% acid solution must be in the mixture.

Chapter 3: Linear Functions and Inequalities in Two Variables

Prep Test

1. $-4(x-3)$
   $= -4x + 12$

2. $\sqrt{(-6)^2 + (-8)^2}$
   $= \sqrt{36 + 64}$
   $= \sqrt{100} = 10$

3. $\dfrac{3-(-5)}{2-6}$
   $= \dfrac{8}{-4} = -2$

4. $-2(-3) + 5$
   $= 6 + 5 = 11$

5. $\dfrac{2(5)}{5-1}$
   $= \dfrac{10}{4} = \dfrac{5}{2} = 2.5$

6. $2(-1)^3 - 3(-1) + 4$
   $= -2 + 3 + 4$
   $= 5$

7. $\dfrac{7+(-5)}{2}$
   $= \dfrac{2}{2} = 1$

8. $3x - 4(0) = 12$
   $\left(\dfrac{1}{3}\right)3x = 12\left(\dfrac{1}{3}\right)$
   $x = 4$

9. $2x - y = 7$
   $-y = -2x + 7$
   $y = 2x - 7$

Go Figure

From $\boxed{5} = 4$, one determines that the value for a square is equal to one less than itself.
From ⑤ $= 6$, one determines that the value for a circle is equal to one more than itself.
From $y = x - 1$, one determines that y's value will always be one less than the value of x.

Thus, ⓧ $>$ ⓨ $>$ $\boxed{x}$ $>$ $\boxed{y}$

$(x+1)$ $\quad$ $y+1 = (x-1)+1 = x$ $\quad$ $x-1$ $\quad$ $y-1 = (x-1)-1 = x-2$

Answer: ⓧ

Section 3.1

Objective A Exercises

1.

3. $A(0, 3)$, $B(1, 1)$, $C(3, -4)$, $D(-4, 4)$

5.

7.

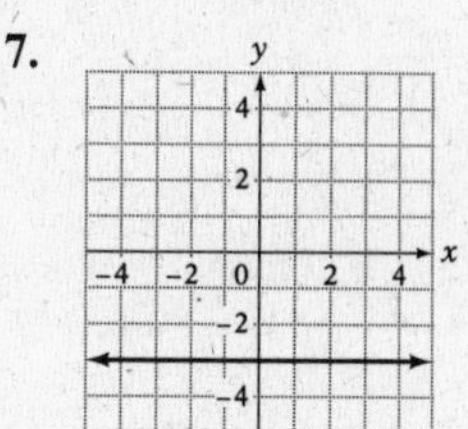

9.

$y = x^2$
Ordered pairs: $(-2, 4)$, $(-1, 1)$, $(0, 0)$, $(1, 1)$, $(2, 4)$

11.

$y = |x+1|$
Ordered pairs: $(-5, 4)$, $(-3, 2)$, $(0, 1)$, $(3, 4)$, $(5, 6)$

13.

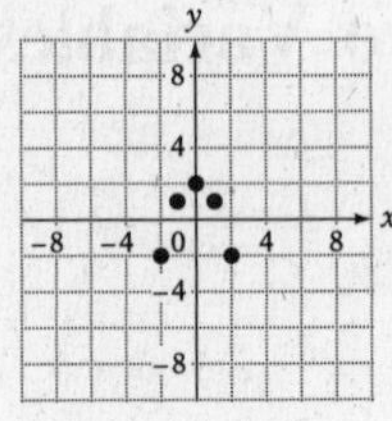

$y = -x^2 + 2$

Ordered pairs: (−2, −2), (−1, 1), (0, 2), (1, 1), (2, −2)

15.

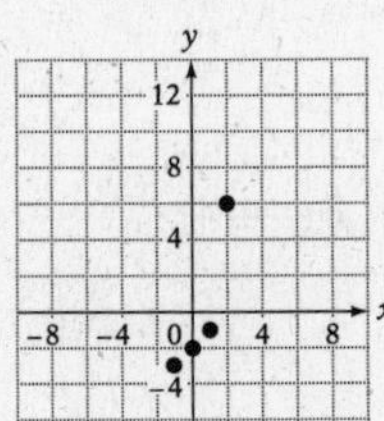

$y = x^3 - 2$

Ordered pairs: (−1, −3), (0, −2), (1, −1), (2, 6)

Objective B Exercises

17. $d = \sqrt{(3-5)^2 + (5-1)^2}$

$d = \sqrt{20} \approx 4.47$

$x_m = \dfrac{3+5}{2} = 4$

$y_m = \dfrac{5+1}{2} = 3$

Length: 4.47; midpoint: (4, 3)

19. $d = \sqrt{[0-(-2)]^2 + (3-4)^2}$

$d = \sqrt{5} \approx 2.24$

$x_m = \dfrac{0+(-2)}{2} = -1$

$y_m = \dfrac{3+4}{2} = \dfrac{7}{2}$

Length: 2.24; midpoint: $\left(-1, \dfrac{7}{2}\right)$

21. $d = \sqrt{(-3-2)^2 + [-5-(-4)]^2}$

$d = \sqrt{26} \approx 5.10$

$x_m = \dfrac{-3+2}{2} = -\dfrac{1}{2}$

$y_m = \dfrac{-5+(-4)}{2} = -\dfrac{9}{2}$

Length: 5.10; midpoint: $\left(-\dfrac{1}{2}, -\dfrac{9}{2}\right)$

23. $d = \sqrt{[5-(-2)]^2 + (-2-5)^2}$

$d = \sqrt{98} \approx 9.90$

$x_m = \dfrac{5+(-2)}{2} = \dfrac{3}{2}$

$y_m = \dfrac{-2+5}{2} = \dfrac{3}{2}$

Length: 9.90; midpoint: $\left(\dfrac{3}{2}, \dfrac{3}{2}\right)$

25. $d = \sqrt{(5-2)^2 + [-5-(-5)]^2}$

$d = 3$

$x_m = \dfrac{5+2}{2} = \dfrac{7}{2}$

$y_m = \dfrac{-5+(-5)}{2} = -5$

Length: 3; midpoint: $\left(\dfrac{7}{2}, -5\right)$

27. $d = \sqrt{\left[\dfrac{3}{2} - \left(-\dfrac{1}{2}\right)\right]^2 + \left(-\dfrac{4}{3} - \dfrac{7}{3}\right)^2}$

$d = \sqrt{(2)^2 + \left(-\dfrac{11}{3}\right)^2}$

$d = \sqrt{\dfrac{157}{9}} \approx 4.18$

$x_m = \dfrac{\dfrac{3}{2} + \left(-\dfrac{1}{2}\right)}{2} = \dfrac{1}{2}$

$y_m = \dfrac{-\dfrac{4}{3} + \dfrac{7}{3}}{2} = \dfrac{1}{2}$

Length: 4.18; midpoint: $\left(\dfrac{1}{2}, \dfrac{1}{2}\right)$

Objective C Exercises

29. **a.** After 20 min., the temperature is 280°F.

b. After 50 min., the temperature is 160°F.

31.

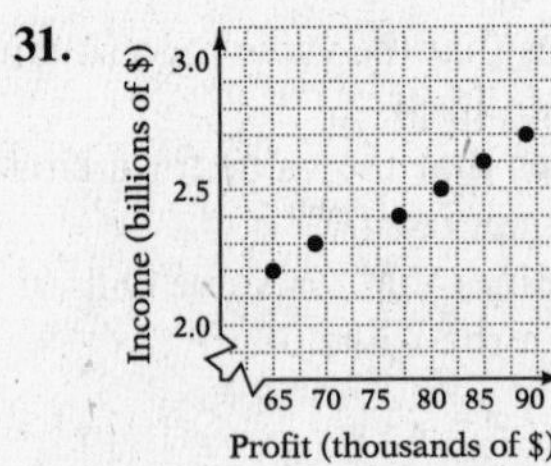

Applying the Concepts

33.

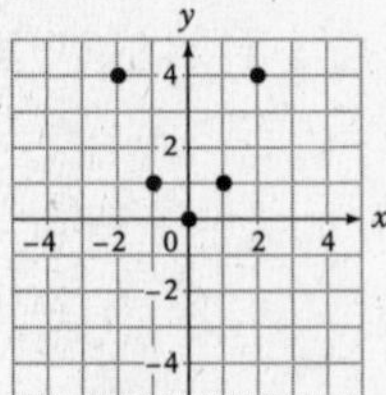

Ordered pairs: (−2, 4), (−1, 1), (0, 0), (1, 1), (2, 4)

35. The graph of all ordered pairs (x, y) that are 5 units from the origin is a circle of radius 5 that has its center at (0, 0).

37.

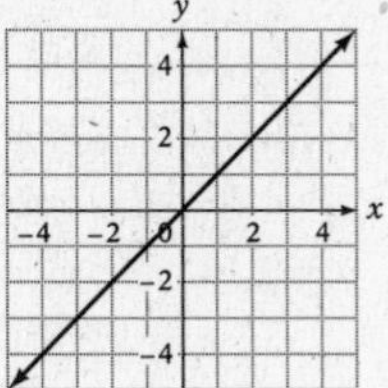

Section 3.2

Objective A Exercises

1. A function is a set of ordered pairs in which no two ordered pairs can have the same x-coordinate and different y-coordinates.

3&5. The diagram does represent a function because each number in the domain is paired with one number in the range.

7. No, the diagram does not represent a function. The 6 in the domain is paired with two different numbers in the range.

9. Function

11. Function

13. Function

15. Not a function

17. a. Yes, this table defines a function because there is only one cost for any given weight.

    b. If $x = 2.75$ lb, then $y = \$34.75$.

19. $f(3) = 11$

21. $f(0) = -4$

23. $G(0) = 4$

25. $G(-2) = 10$

27. $q(3) = 5$

29. $q(-2) = 0$

31. $F(4) = 24$

33. $F(-3) = -4$

35. $H(1) = 1$

37. $H(t) = \dfrac{3t}{t+2}$

39. $s(-1) = 6$

41. $s(a) = a^3 - 3a + 4$

43. $$\begin{aligned} P(-2+h) - P(-2) &= 4(-2+h) + 7 - [4(-2) + 7] \\ &= -8 + 4h + 7 + 8 - 7 \\ &= 4h \end{aligned}$$

45. a. \$4.75 per game

    b. \$4.00 per game

47. a. \$3000

    b. \$950

49. Domain = {1, 2, 3, 4, 5}
    Range = {1, 4, 7, 10, 13}

51. Domain = {0, 2, 4, 6}
    Range = {1, 2, 3, 4}

53. Domain = {1, 3, 5, 7, 9}
    Range = {0}

55. Domain = {−2, −1, 0, 1, 2}
    Range = {0, 1, 2}

57. Domain = {−2, −1, 0, 1, 2}
    Range = {−3, 3, 6, 7, 9}

59. $x = 1$

61. $x = -8$

63. None, no values are excluded.

65. None, no values are excluded.

67. $x = 0$

69. None, no values are excluded.

71. None, no values are excluded.

73. $x = -2$

75. None, no values are excluded.

77. Range = {−3, 1, 5, 9}

79. Range = {−23, −13, −8, −3, 7}

81. Range = {0, 1, 4}

83. Range = {2, 14, 26, 42}

85. Range = $\left\{-5, \dfrac{5}{3}, 5\right\}$

87. Range = $\left\{-1, -\dfrac{1}{2}, -\dfrac{1}{3}, 1\right\}$

89. Range = {−38, −8, 2}

Applying the Concepts

91. A relation and a function are similar in that both are sets of ordered pairs. A function is a specific type of relation. A function is a relation in which there are no two ordered pairs with the same first element. Alternatively, this can be stated as follows: A function is a relation in which no two ordered pairs with the same first coordinate(s) have different second coordinate(s).

93. Set of ordered pairs:
   a. $\{(-2, -8), (-1, -1), (0, 0), (1, 1), (2, 8)\}$.
   b. Yes, this set of ordered pairs defines a function because each member of the domain is assigned to exactly one member of the range.

95. Find the value of the function
$P = f(v) = 0.015v^3$ when $v = 15$.
$f(15) = 0.015(15)^3 = 50.625$
The power produced will be 50.625 watts.

97. a. 22.5 ft/s
   b. 30 ft/s

99. a. 68°F
   b. 52°F

Section 3.3

Objective A Exercises

1.

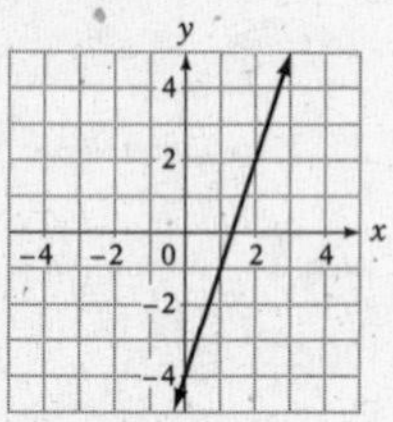

3.

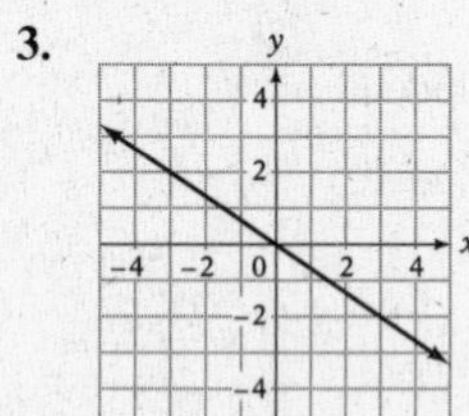

5.

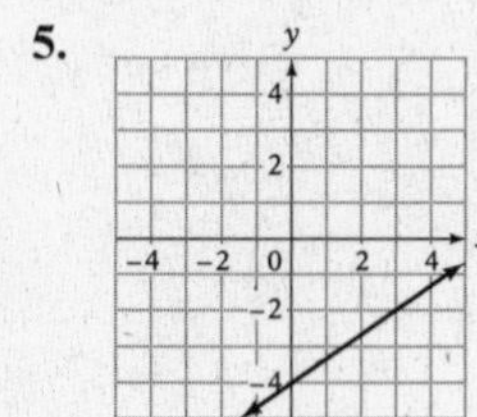

7.

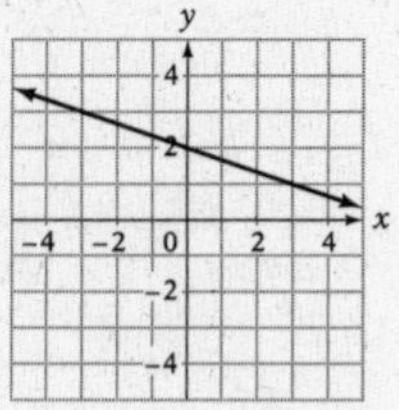

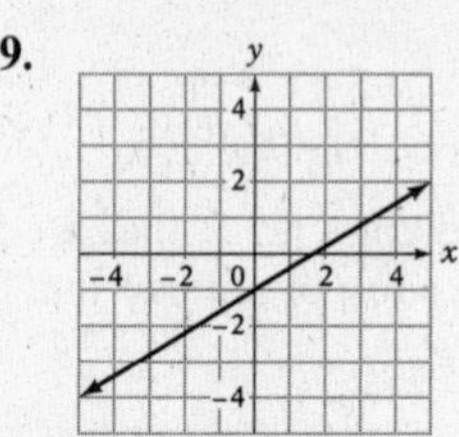

9.

Objective B Exercises

11. $2x + y = -3$
$y = -2x - 3$

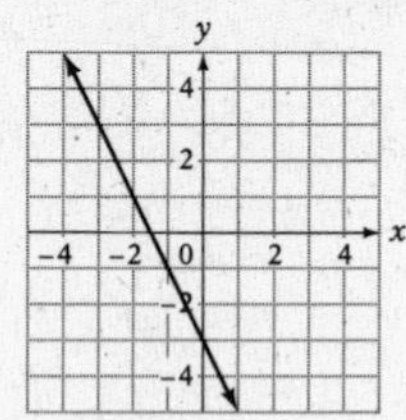

13. $x - 4y = 8$
$-4y = -x + 8$
$y = \frac{1}{4}x - 2$

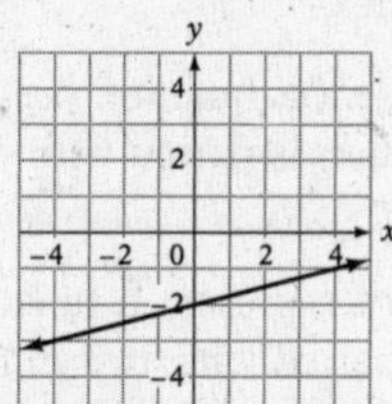

15. $y = \frac{1}{3}x$

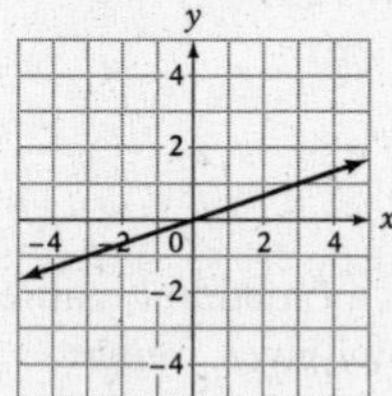

17. $3x - y = -2$
$-y = -3x - 2$
$y = 3x + 2$

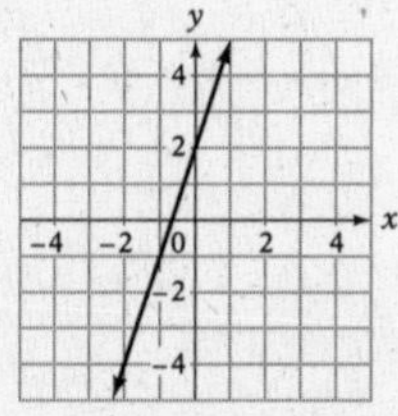

Objective C Exercises

19. x-intercept:
$$x - 2y = -4$$
$$x - 2(0) = -4$$
$$x = -4$$
$(-4, 0)$
y-intercept:
$$x - 2y = -4$$
$$0 - 2y = -4$$
$$-2y = -4$$
$$y = 2$$
$(0, 2)$

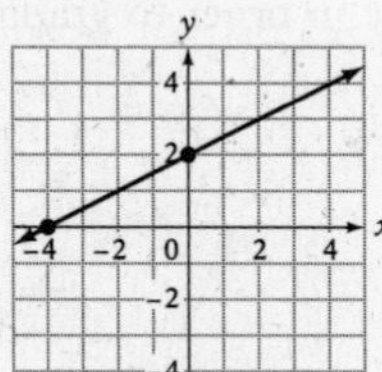

21. x-intercept:
$$2x - 3y = 9$$
$$2x - 3(0) = 9$$
$$2x = 9$$
$$x = \frac{9}{2}$$
$\left(\frac{9}{2}, 0\right)$
y-intercept:
$$2x - 3y = 9$$
$$2(0) - 3y = 9$$
$$-3y = 9$$
$$y = -3$$
$(0, -3)$

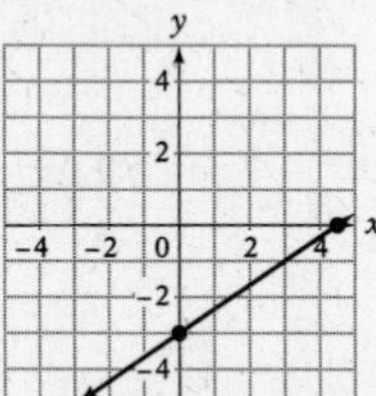

23. x-intercept:
$$2x + y = 3$$
$$2x + 0 = 3$$
$$2x = 3$$
$$x = \frac{3}{2}$$
$\left(\frac{3}{2}, 0\right)$
y-intercept:
$$2x + y = 3$$
$$2(0) + y = 3$$
$$y = 3$$
$(0, 3)$

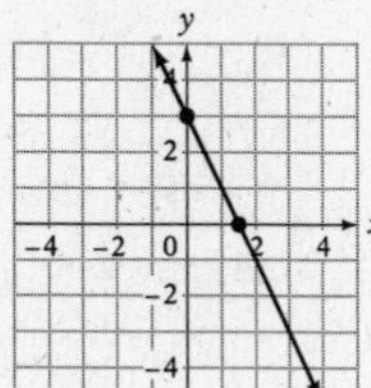

25. x-intercept:
$$3x + 2y = 4$$
$$3x + 2(0) = 4$$
$$3x = 4$$
$$x = \frac{4}{3}$$
$\left(\frac{4}{3}, 0\right)$
y-intercept:
$$3x + 2y = 4$$
$$3(0) + 2y = 4$$
$$2y = 4$$
$$y = 2$$
$(0, 2)$

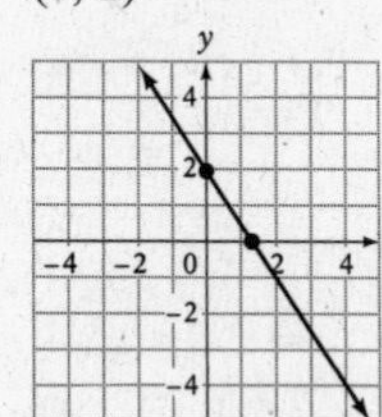

27. x-intercept:

$$\begin{aligned} 4x - 3y &= 6 \\ 4x - 3(0) &= 6 \\ 4x &= 6 \\ x &= \frac{3}{2} \end{aligned}$$

$\left(\frac{3}{2}, 0\right)$

y-intercept:

$$\begin{aligned} 4x - 3y &= 6 \\ 4(0) - 3y &= 6 \\ -3y &= 6 \\ y &= -2 \end{aligned}$$

$(0, -2)$

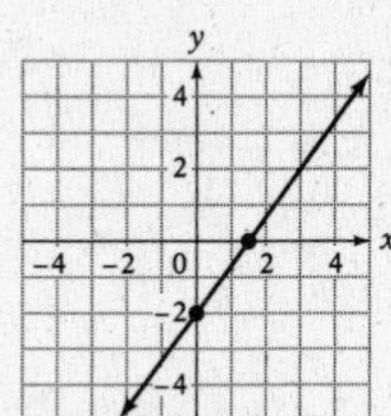

Objective D Exercises

29.

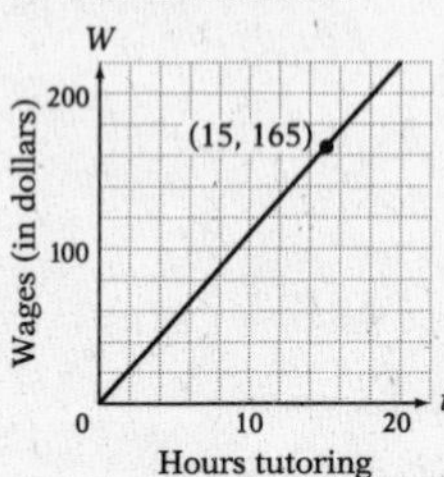

Marlys receives \$165 for tutoring 15 h.

31.

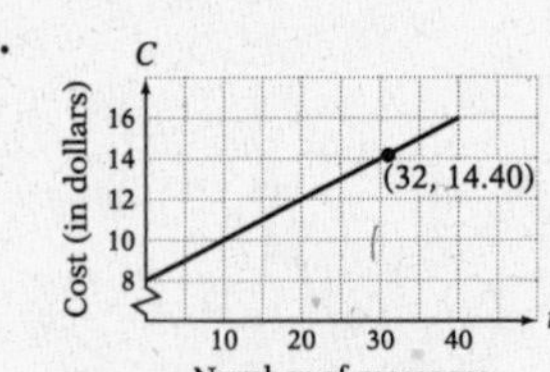

The cost of receiving 32 messages is \$14.40.

33.

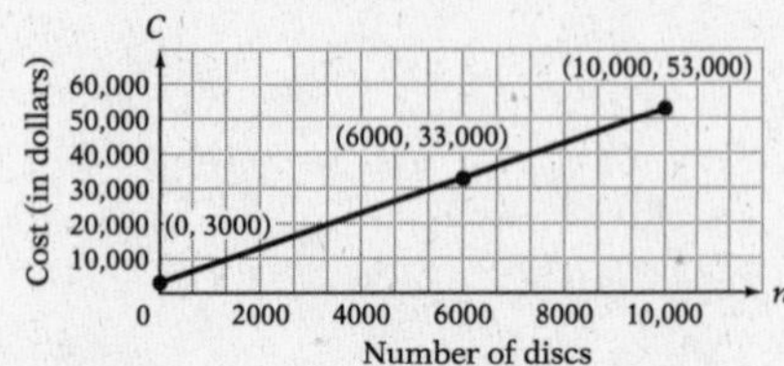

The cost of manufacturing 6000 compact discs is \$33,000.

Applying the Concepts

35. To graph the equation of a straight line by plotting points, find three ordered-pair solutions of the equation. Plot these ordered pairs in a rectangular coordinate system. Draw a straight line through the points.

37. The x-intercept of the graph of the equation $4x + 3y = 0$ is $(0, 0)$. The y-intercept of this graph is also $(0, 0)$. Therefore, the x-intercept and y-intercept are the same point. A straight line is determined by two points, so we must find another point on the line in order to graph the equation.

39.

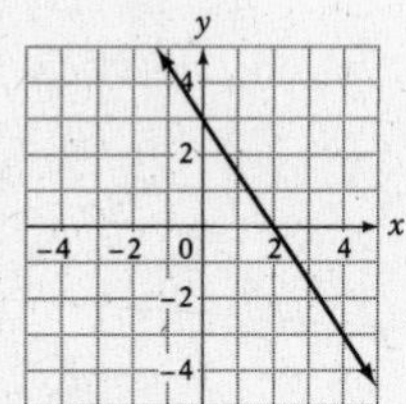

Section 3.4

Objective A Exercises

1. $P_1(1,3),\ P_2(3,1)$
$m = \frac{y_2 - y_1}{x_2 - x_1} = \frac{1-3}{3-1} = \frac{-2}{2} = -1$
The slope is -1.

3. $P_1(-1,4),\ P_2(2,5)$
$m = \frac{y_2 - y_1}{x_2 - x_1} = \frac{5-4}{2-(-1)} = \frac{1}{3}$
The slope is $\frac{1}{3}$.

5. $P_1(-1,3),\ P_2(-4,5)$
$m = \frac{y_2 - y_1}{x_2 - x_1} = \frac{5-3}{-4-(-1)} = \frac{2}{-3} = -\frac{2}{3}$
The slope is $-\frac{2}{3}$.

7. $P_1(0,3),\ P_2(4,0)$
$m = \frac{y_2 - y_1}{x_2 - x_1} = \frac{0-3}{4-0} = \frac{-3}{4} = -\frac{3}{4}$
The slope is $-\frac{3}{4}$.

9. $P_1(2,4),\ P_2(2,-2)$
$m = \frac{y_2 - y_1}{x_2 - x_1} = \frac{-2-4}{2-2} = \frac{-6}{0}$
The slope is undefined.

11. $P_1(2,5),\ P_2(-3,-2)$
$m = \frac{y_2 - y_1}{x_2 - x_1} = \frac{-2-5}{-3-2} = \frac{-7}{-5} = \frac{7}{5}$
The slope is $\frac{7}{5}$.

13. $P_1(2,3)$, $P_2(-1,3)$

$$m = \frac{y_2 - y_1}{x_2 - x_1} = \frac{3-3}{-1-2} = \frac{0}{-3} = 0$$

The line has zero slope.

15. $P_1(0,4)$, $P_2(-2,5)$

$$m = \frac{y_2 - y_1}{x_2 - x_1} = \frac{5-4}{-2-0} = \frac{1}{-2} = -\frac{1}{2}$$

The slope is $-\frac{1}{2}$.

17. $P_1(-3,-1)$, $P_2(-3,4)$

$$m = \frac{y_2 - y_1}{x_2 - x_1} = \frac{4-(-1)}{-3-(-3)} = \frac{5}{0}$$

The slope is undefined.

19. $m = \frac{240-80}{6-2}$

$m = 40$

The average speed of the motorist is 40 mph.

21. $m = \frac{275-125}{20-50} = \frac{150}{-30} = -5$

The temperature of the oven decreases 5°/min.

23. $m = \frac{13-6}{40-180} = -0.05$

For each mile the car is driven, approximately 0.05 gallon of fuel is used.

25. $m = \frac{5000}{14.54} \approx 343.9$

The average speed of the runner was 343.9 m/min.

27. $\frac{6\text{ in}}{5\text{ ft}} = \frac{6\text{ in}}{60\text{ in}} = \frac{1}{10} > \frac{1}{12}$; no, it does not meet the requirements for ANSI.

Objective B Exercises

	Equation	Value of m	Value of b	Slope	y-intercept
29.	$y = -3x + 5$	-3	5	-3	$(0,5)$
31.	$y = 4x$	4	0	4	$(0,0)$

33.

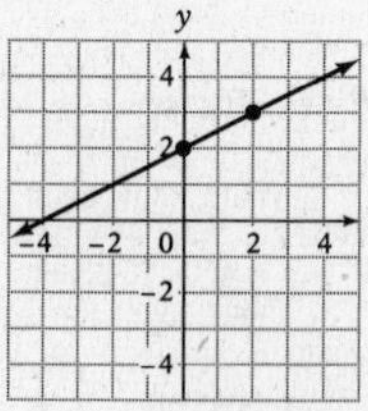

35.

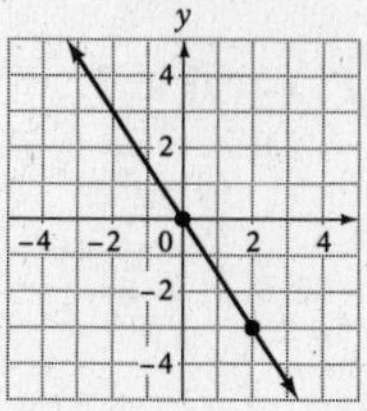

37. 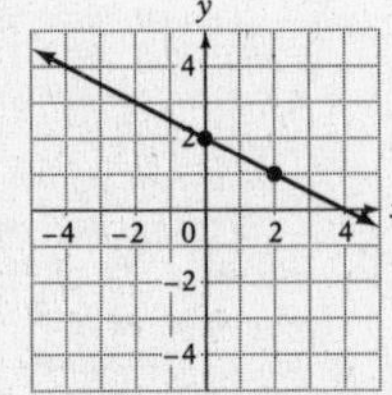

39. 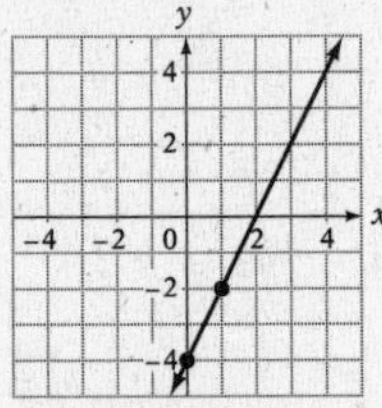

41. 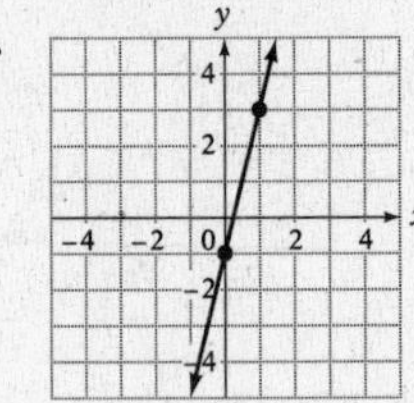

43. 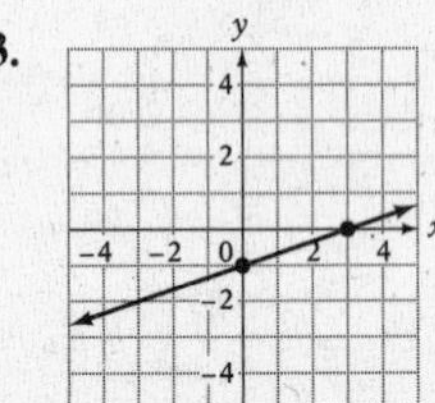

45. 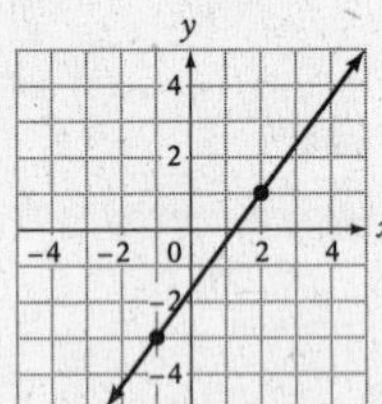

47. 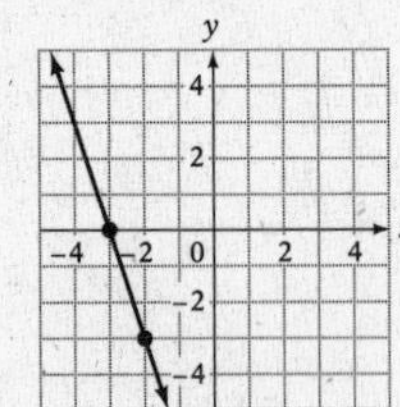

49. 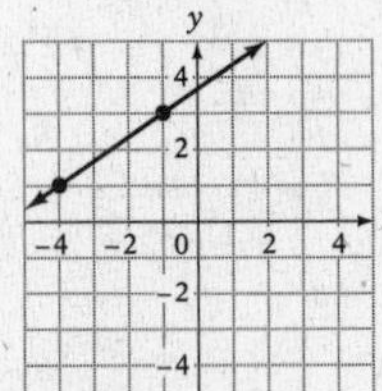

Applying the Concepts

51. increases by 2

53. increases by 2

55. decreases by $\frac{2}{3}$

57. i. D ii. C
iii. B iv. F
v. E vi. A

59. The slope between each set of points must be the same for the points to lie on the same line.

a. $P_1 = (2,5)$
$P_2 = (-1,-1)$
$P_3 = (3,7)$
P_1 to P_2: $m = \frac{-6}{-3} = 2$
P_2 to P_3: $m = \frac{8}{4} = 2$
P_1 to P_3: $m = \frac{2}{1} = 2$
Yes, the points lie on the same line.

b. $P_1 = (-1,5)$
$P_2 = (0,3)$
$P_3 = (-3,4)$
P_1 to P_2: $m = \frac{2}{-1} = -2$
P_2 to P_3: $m = \frac{-1}{3} = -\frac{1}{3}$
P_1 to P_3: $m = \frac{1}{2}$
No, the points do not lie on the same line.

61. $P_1 = (-2,3)$
$P_2 = (1,0)$
$P_3 = (k,2)$
P_1 to P_2: $m = -1$
The slope from P_1 to P_3 and that from P_2 to P_3 must also be -1. Set the slope from P_1 to P_3 equal to -1.

$$\frac{1}{-2-k} = -1$$
$$k+2 = 1$$
$$k = -1$$

This checks out against P_2 to P_3, so $k = -1$.

63. $P_1 = (4,-1)$
$P_2 = (3,-4)$
$P_3 = (k, k)$
P_1 to P_2: $m = 3$
The slope from P_1 to P_3 and that from P_1 to P_3 must also be 3.
Set the slope from P_1 to P_3 equal to 3.

$$\frac{-1-k}{4-k} = 3$$
$$-1-k = 3(4-k)$$
$$-1-k = 12-3k$$
$$2k = 13$$
$$k = \frac{13}{2}$$

This checks out against P_2 to P_3, so $k = \frac{13}{2}$.

Section 3.5

Objective A Exercises

1. When we know the slope and y-intercept, we can find the equation by using the slope-intercept form, $y = mx + b$. The slope can be substituted for m, and the y-coordinate of the y-intercept can be substituted for b.

3. $m = 2, b = 5$
$y = mx + b$
$y = 2x + 5$
The equation of the line is $y = 2x + 5$.

5. $m = \frac{1}{2}$ $(x_1, y_1) = (2, 3)$
$y - y_1 = m(x - x_1)$
$y - 3 = \frac{1}{2}(x - 2)$
$y - 3 = \frac{1}{2}x - 1$
$y = \frac{1}{2}x + 2$
The equation of the line is $y = \frac{1}{2}x + 2$.

7. $m = \frac{5}{4}$ $(x_1, y_1) = (-1, 4)$
$y - y_1 = m(x - x_1)$
$y - 4 = \frac{5}{4}[x - (-1)]$
$y - 4 = \frac{5}{4}(x + 1)$
$y - 4 = \frac{5}{4}x + \frac{5}{4}$
$y = \frac{5}{4}x + \frac{21}{4}$
The equation of the line is $y = \frac{5}{4}x + \frac{21}{4}$.

9. $m = -\frac{5}{3}$ $(x_1, y_1) = (3, 0)$
$y - y_1 = m(x - x_1)$
$y - 0 = -\frac{5}{3}(x - 3)$
$y = -\frac{5}{3}x + 5$
The equation of the line is $y = -\frac{5}{3}x + 5$.

11. $m = -3$ $(x_1, y_1) = (2, 3)$
$y - y_1 = m(x - x_1)$
$y - 3 = -3(x - 2)$
$y - 3 = -3x + 6$
$y = -3x + 9$
The equation of the line is $y = -3x + 9$.

13. $m = -3$ $(x_1, y_1) = (-1, 7)$
$y - y_1 = m(x - x_1)$
$y - 7 = -3[x - (-1)]$
$y - 7 = -3(x + 1)$
$y - 7 = -3x - 3$
$y = -3x + 4$
The equation of the line is $y = -3x + 4$.

15. $m = \frac{2}{3}$ $(x_1, y_1) = (-1, -3)$
$y - y_1 = m(x - x_1)$
$y - (-3) = \frac{2}{3}[x - (-1)]$
$y + 3 = \frac{2}{3}(x + 1)$
$y + 3 = \frac{2}{3}x + \frac{2}{3}$
$y = \frac{2}{3}x - \frac{7}{3}$
The equation of the line is $y = \frac{2}{3}x - \frac{7}{3}$.

17. $m = \frac{1}{2}$ $(x_1, y_1) = (0, 0)$
$y - y_1 = m(x - x_1)$
$y - 0 = \frac{1}{2}(x - 0)$
$y = \frac{1}{2}x$
The equation of the line is $y = \frac{1}{2}x$.

19. $m = 3$ $(x_1, y_1) = (2, -3)$
$y - y_1 = m(x - x_1)$
$y - (-3) = 3(x - 2)$
$y + 3 = 3x - 6$
$y = 3x - 9$
The equation of the line is $y = 3x - 9$.

21. $m = -\frac{2}{3}$ $(x_1, y_1) = (3, 5)$
$y - y_1 = m(x - x_1)$
$y - 5 = -\frac{2}{3}(x - 3)$
$y - 5 = -\frac{2}{3}x + 2$
$y = -\frac{2}{3}x + 7$
The equation of the line is $y = -\frac{2}{3}x + 7$.

23. $m = -1$ $(x_1, y_1) = (0, -3)$
$y - y_1 = m(x - x_1)$
$y - (-3) = -1(x - 0)$
$y + 3 = -x + 0$
$y = -x - 3$
The equation of the line is $y = -x - 3$.

25. $m = \frac{7}{5}$ $(x_1, y_1) = (1, -4)$

$$
\begin{aligned}
y - y_1 &= m(x - x_1) \\
y - (-4) &= \frac{7}{5}(x - 1) \\
y + 4 &= \frac{7}{5}x - \frac{7}{5} \\
y &= \frac{7}{5}x - \frac{27}{5}
\end{aligned}
$$

The equation of the line is $y = \frac{7}{5}x - \frac{27}{5}$.

27. $m = -\frac{2}{5}$ $(x_1, y_1) = (4, -1)$

$$
\begin{aligned}
y - y_1 &= m(x - x_1) \\
y - (-1) &= -\frac{2}{5}(x - 4) \\
y + 1 &= -\frac{2}{5}x + \frac{8}{5} \\
y &= -\frac{2}{5}x + \frac{3}{5}
\end{aligned}
$$

The equation of the line is $y = -\frac{2}{5}x + \frac{3}{5}$.

29. slope is undefined; $(x_1, y_1) = (3, -4)$
The line is a vertical line. All points on the line have an abscissa of 3. The equation of the line is $x = 3$.

31. $m = -\frac{5}{4}$ $(x_1, y_1) = (-2, -5)$

$$
\begin{aligned}
y - y_1 &= m(x - x_1) \\
y - (-5) &= -\frac{5}{4}[x - (-2)] \\
y + 5 &= -\frac{5}{4}(x + 2) \\
y + 5 &= -\frac{5}{4}x - \frac{5}{2} \\
y &= -\frac{5}{4}x - \frac{15}{2}
\end{aligned}
$$

The equation of the line is $y = -\frac{5}{4}x - \frac{15}{2}$.

33. $m = 0$ $(x_1, y_1) = (-2, -3)$

$$
\begin{aligned}
y - y_1 &= m(x - x_1) \\
y - (-3) &= 0[x - (-2)] \\
y + 3 &= 0 \\
y &= -3
\end{aligned}
$$

The equation of the line is $y = -3$.

35. $m = -2$ $(x_1, y_1) = (4, -5)$

$$
\begin{aligned}
y - y_1 &= m(x - x_1) \\
y - (-5) &= -2(x - 4) \\
y + 5 &= -2x + 8 \\
y &= -2x + 3
\end{aligned}
$$

The equation of the line is $y = -2x + 3$.

37. slope is undefined; $(x_1, y_1) = (-5, -1)$
The line is a vertical line. All points on the line have an abscissa of -5. The equation of the line is $x = -5$.

Objective B Exercises

39. $P_1(0, 2)$, $P_2(3, 5)$

$$
\begin{aligned}
m &= \frac{y_2 - y_1}{x_2 - x_1} = \frac{5 - 2}{3 - 0} = \frac{3}{3} = 1 \\
y - y_1 &= m(x - x_1) \\
y - 2 &= 1(x - 0) \\
y - 2 &= x \\
y &= x + 2
\end{aligned}
$$

The equation of the line is $y = x + 2$.

41. $P_1(0, -3)$, $P_2(-4, 5)$

$$
\begin{aligned}
m &= \frac{y_2 - y_1}{x_2 - x_1} = \frac{5 - (-3)}{-4 - 0} = \frac{8}{-4} = -2 \\
y - y_1 &= m(x - x_1) \\
y - (-3) &= -2(x - 0) \\
y + 3 &= -2x \\
y &= -2x - 3
\end{aligned}
$$

The equation of the line is $y = -2x - 3$.

43. $P_1(2, 3)$, $P_2(5, 5)$

$$
\begin{aligned}
m &= \frac{y_2 - y_1}{x_2 - x_1} = \frac{5 - 3}{5 - 2} = \frac{2}{3} \\
y - y_1 &= m(x - x_1) \\
y - 3 &= \frac{2}{3}(x - 2) \\
y - 3 &= \frac{2}{3}x - \frac{4}{3} \\
y &= \frac{2}{3}x + \frac{5}{3}
\end{aligned}
$$

The equation of the line is $y = \frac{2}{3}x + \frac{5}{3}$.

45. $P_1(-1, 3)$, $P_2(2, 4)$

$$
\begin{aligned}
m &= \frac{y_2 - y_1}{x_2 - x_1} = \frac{4 - 3}{2 - (-1)} = \frac{1}{3} \\
y - y_1 &= m(x - x_1) \\
y - 3 &= \frac{1}{3}[x - (-1)] \\
y - 3 &= \frac{1}{3}(x + 1) \\
y - 3 &= \frac{1}{3}x + \frac{1}{3} \\
y &= \frac{1}{3}x + \frac{10}{3}
\end{aligned}
$$

The equation of the line is $y = \frac{1}{3}x + \frac{10}{3}$.

47. $P_1(-1,-2),\ P_2(3,4)$

$$m=\frac{y_2-y_1}{x_2-x_1}=\frac{4-(-2)}{3-(-1)}=\frac{6}{4}=\frac{3}{2}$$
$$y-y_1=m(x-x_1)$$
$$y-(-2)=\frac{3}{2}[x-(-1)]$$
$$y+2=\frac{3}{2}(x+1)$$
$$y+2=\frac{3}{2}x+\frac{3}{2}$$
$$y=\frac{3}{2}x-\frac{1}{2}$$

The equation of the line is $y=\frac{3}{2}x-\frac{1}{2}$.

49. $P_1(0,3),\ P_2(2,0)$

$$m=\frac{y_2-y_1}{x_2-x_1}=\frac{0-3}{2-0}=\frac{-3}{2}=-\frac{3}{2}$$
$$y-y_1=m(x-x_1)$$
$$y-3=-\frac{3}{2}(x-0)$$
$$y-3=-\frac{3}{2}x$$
$$y=-\frac{3}{2}x+3$$

The equation of the line is $y=-\frac{3}{2}x+3$.

51. $P_1(-3,-1),\ P_2(2,-1)$

$$m=\frac{y_2-y_1}{x_2-x_1}=\frac{-1-(-1)}{2-(-3)}=\frac{0}{5}=0$$
$$y-y_1=m(x-x_1)$$
$$y-(-1)=0[x-(-3)]$$
$$y+1=0$$
$$y=-1$$

The equation of the line is $y=-1$.

53. $P_1(-2,-3),\ P_2(-1,-2)$

$$m=\frac{y_2-y_1}{x_2-x_1}=\frac{-2-(-3)}{-1-(-2)}=\frac{1}{1}=1$$
$$y-y_1=m(x-x_1)$$
$$y-(-3)=1[x-(-2)]$$
$$y+3=x+2$$
$$y=x-1$$

The equation of the line is $y=x-1$.

55. $P_1(-2,3),\ P_2(2,-1)$

$$m=\frac{y_2-y_1}{x_2-x_1}=\frac{-1-3}{2-(-2)}=\frac{-4}{4}=-1$$
$$y-y_1=m(x-x_1)$$
$$y-3=-1[x-(-2)]$$
$$y-3=-1(x+2)$$
$$y-3=-x-2$$
$$y=-x+1$$

The equation of the line is $y=-x+1$.

57. $P_1(2,3),\ P_2(5,-5)$

$$m=\frac{y_2-y_1}{x_2-x_1}=\frac{-5-3}{5-2}=-\frac{8}{3}$$
$$y-y_1=m(x-x_1)$$
$$y-3=-\frac{8}{3}(x-2)$$
$$y-3=-\frac{8}{3}x+\frac{16}{3}$$
$$y=-\frac{8}{3}x+\frac{25}{3}$$

The equation of the line is $y=-\frac{8}{3}x+\frac{25}{3}$.

59. $P_1(2,0),\ P_2(0,-1)$

$$m=\frac{y_2-y_1}{x_2-x_1}=\frac{-1-0}{0-2}=\frac{-1}{-2}=\frac{1}{2}$$
$$y-y_1=m(x-x_1)$$
$$y-0=\frac{1}{2}(x-2)$$
$$y=\frac{1}{2}x-1$$

The equation of the line is $y=\frac{1}{2}x-1$.

61. $P_1(3,-4),\ P_2(-2,-4)$

$$m=\frac{y_2-y_1}{x_2-x_1}=\frac{-4-(-4)}{-2-3}=\frac{0}{-5}=0$$
$$y-y_1=m(x-x_1)$$
$$y-(-4)=0(x-3)$$
$$y+4=0$$
$$y=-4$$

The equation of the line is $y=-4$.

63. $P_1(0,0),\ P_2(4,3)$

$$m=\frac{y_2-y_1}{x_2-x_1}=\frac{3-0}{4-0}=\frac{3}{4}$$
$$y-y_1=m(x-x_1)$$
$$y-0=\frac{3}{4}(x-0)$$
$$y=\frac{3}{4}x$$

The equation of the line is $y=\frac{3}{4}x$.

65. $P_1(2,-1),\ P_2(-1,3)$

$$m=\frac{y_2-y_1}{x_2-x_1}=\frac{3-(-1)}{-1-2}=\frac{4}{-3}=-\frac{4}{3}$$
$$y-y_1=m(x-x_1)$$
$$y-(-1)=-\frac{4}{3}(x-2)$$
$$y+1=-\frac{4}{3}x+\frac{8}{3}$$
$$y=-\frac{4}{3}x+\frac{5}{3}$$

The equation of the line is $y=-\frac{4}{3}x+\frac{5}{3}$.

67. $P_1(-2, 5), P_2(-2, -5)$

$$m = \frac{y_2 - y_1}{x_2 - x_1} = \frac{-5 - 5}{-2 - (-2)} = \frac{-10}{0}$$

The slope is undefined. The line is a vertical line. All points on the line have an abscissa of -2. The equation of the line is $x = -2$.

69. $P_1(2, 1), P_2(-2, -3)$

$$m = \frac{y_2 - y_1}{x_2 - x_1} = \frac{-3 - 1}{-2 - 2} = \frac{-4}{-4} = 1$$

$$y - y_1 = m(x - x_1)$$
$$y - 1 = 1(x - 2)$$
$$y - 1 = x - 2$$
$$y = x - 1$$

The equation of the line is $y = x - 1$.

71. $P_1(-4, -3), P_2(2, 5)$

$$m = \frac{y_2 - y_1}{x_2 - x_1} = \frac{5 - (-3)}{2 - (-4)} = \frac{8}{6} = \frac{4}{3}$$

$$y - y_1 = m(x - x_1)$$
$$y - (-3) = \frac{4}{3}[x - (-4)]$$
$$y + 3 = \frac{4}{3}(x + 4)$$
$$y + 3 = \frac{4}{3}x + \frac{16}{3}$$
$$y = \frac{4}{3}x + \frac{7}{3}$$

The equation of the line is $y = \frac{4}{3}x + \frac{7}{3}$.

73. $P_1(0, 3), P_2(3, 0)$

$$m = \frac{y_2 - y_1}{x_2 - x_1} = \frac{0 - 3}{3 - 0} = \frac{-3}{3} = -1$$

$$y - y_1 = m(x - x_1)$$
$$y - 3 = -1(x - 0)$$
$$y - 3 = -x$$
$$y = -x + 3$$

The equation of the line is $y = -x + 3$.

Objective C Exercises

75. Strategy
- Let x represent the number of minutes after takeoff.
- Let y represent the height of the plane in feet.
- Use the slope-intercept form of an equation to find the equation of the line.

Solution

a. The y-intercept is (0, 0). The slope is 1200.
$y = mx + b$
$y = 1200x + 0$
The linear equation is $y = 1200x$, $0 \le x \le 26\frac{2}{3}$.

b. Find the height of the plane 11 minutes after takeoff.
$y = 1200(11) = 13{,}200$
The height of the plane will be 13,200 ft 11 minutes after takeoff.

77. Strategy
- Let x represent the number of square feet of floor space.
- Let y represent the cost to build the house.
- Use the slope-intercept form of an equation to find the equation of the line.

Solution

a. The y-intercept is (0, 30,000). The slope is 85.
$y = 85x + 30{,}000$
The linear equation is $y = 85x + 30{,}000$.

b. Find the cost to build a house that has 1800 square feet of floor space.
$y = 85(1800) + 30{,}000$
$= 183{,}000$
It will cost \$183,000 to build an 1800-ft^2 house.

79. Strategy
- Let x represent the number of miles driven.
- Let y represent the number of gallons of gas in the tank.
- Use the slope-intercept form of an equation to find the equation of the line.

Solution

a. The y-intercept is (0, 16). The slope is -0.032.
$y = -0.032x + 16$.
Since $0 \le y \le 16$, we have
$0 \le -0.032x + 16 \le 16$
$-16 \le -0.032x \le 0$
$500 \ge x \ge 0$
The linear function is
$y = -0.032x + 16,\ 0 \le x \le 500$

b. Find the number of gallons of gas left in the tank after driving 150 mi.
$y = -0.032(150) + 16 = 11.2$
After driving 150 mi, 11.2 gal are left in the tank.

81. Strategy
- Let x represent the price of a car.
- Let y represent the number of cars sold.
- Use the two given points to find the slope.
- Use the point-slope form to determine the equation of the line.

Solution a. $(x_1, y_1) = (9000, 50{,}000)$

$(x_2, y_2) = (8750, 55{,}000)$

$$m = \frac{y_2 - y_1}{x_2 - x_1} = \frac{55{,}000 - 50{,}000}{8750 - 9000} = -20$$

$$\begin{aligned} y - y_1 &= m(x - x_1) \\ y - 50{,}000 &= -20(x - 9000) \\ y - 50{,}000 &= -20x + 180{,}000 \\ y &= -20x + 230{,}000 \end{aligned}$$

The linear function is $y = -20x + 230{,}000$.

b. Find the number of cars sold when the price is \$8500.

$y = -20(8500) + 230{,}000 = 60{,}000$

60,000 cars would be sold at \$8500 each.

83. Strategy
- Let x represent the number of ounces of lean hamburger.
- Let y represent the number of Calories.
- Use the two given points to find the slope.
- Use the point-slope form to determine the equation of the line.

Solution a. $(x_1, y_1) = (2, 126)$

$(x_2, y_2) = (3, 189)$

$$m = \frac{y_2 - y_1}{x_2 - x_1} = \frac{189 - 126}{3 - 2} = 63$$

$$\begin{aligned} y - y_1 &= m(x - x_1) \\ y - 126 &= 63(x - 2) \\ y - 126 &= 63x - 126 \\ y &= 63x \end{aligned}$$

The linear function is $y = 63x$.

b. Find the number of Calories in a 5-ounce serving.

$y = 63(5) = 315$

There are 315 Calories in a 5-ounce serving.

85. Strategy
- Use the given information to determine two ordered pairs.
- Use the ordered pairs to determine the slope.
- Use the point-slope form to determine the equation.

Solution $(x_1, y_1) = (2, 5)$

$(x_2, y_2) = (0, 3)$

$$m = \frac{y_2 - y_1}{x_2 - x_1} = \frac{3 - 5}{0 - 2} = 1$$

$$\begin{aligned} y - y_2 &= m(x - x_2) \\ y - 3 &= 1(x - 0) \\ y &= x + 3 \\ f(x) &= x + 3 \end{aligned}$$

87. Strategy
- Use the given information to determine the slope.
- Use the point-slope form to determine the equation.
- Write the equation using functional notation and evaluate for $x = 4$.

Solution $(x_1, y_1) = (1, 3)$

$(x_2, y_2) = (-1, 5)$

$$m = \frac{y_2 - y_1}{x_2 - x_1} = \frac{5 - 3}{-1 - 1} = -1$$

$$\begin{aligned} y - y_1 &= m(x - x_1) \\ y - 3 &= -1(x - 1) \\ y - 3 &= -x + 1 \\ y &= -x + 4 \\ f(x) &= -x + 4 \\ f(4) &= -4 + 4 = 0 \end{aligned}$$

89. Strategy
- Use the point-slope form to determine the equation.
- Evaluate for the given values.

Solution a. $y - y_1 = m(x - x_1)$

$$\begin{aligned} y - 2 &= \frac{4}{3}(x - 3) \\ y - 2 &= \frac{4}{3}x - 4 \\ y &= \frac{4}{3}x - 2 \end{aligned}$$

For $x = -6$ we have

$y = \frac{4}{3}(-6) - 2 = -8 - 2 = -10.$

b. For $y = 6$ we have

$$\begin{aligned} 6 &= \frac{4}{3}x - 2 \\ 8 &= \frac{4}{3}x \\ \frac{3}{4} \cdot 8 &= \frac{3}{4} \cdot \frac{4}{3}x \\ 6 &= x \\ x &= 6 \end{aligned}$$

Applying the Concepts

91. The slope-intercept form of a straight line, $y = mx + b$, is the general form in which the equation of a straight line is written. Given the slope of a line and its y-intercept, we can write the equation of the line by substituting the slope for m and the y-coordinate of the y-intercept for b in the equation $y = mx + b$.

The point-slope formula, $y - y_1 = m(x - x_1)$, is used to find the equation of a line when the slope of the line and any point on the line other than the y-intercept are known. We substitute the slope and the coordinates of the known point into the point-slope formula and then rewrite the equation in the slope-intercept form of a straight line.

93. **a.** The slope of the line $y = -0.01x + 2.15$ is -0.01. The slope is the rate at which the number of CD players sold changes per dollar increase in price. For every \$1 increase in price, the number of CD players sold decreases by 0.01 million, or 10,000.

b. The y-intercept is (0, 2.15). This means that 2.15 million CD players would be given away if the price per CD player were \$0.

c. The x-intercept is (215, 0). This means that if the price were \$215 per CD player, no one would buy one.

95. Find the equation of the line.

$$m = \frac{0-6}{6-(-3)} = \frac{-6}{9} = -\frac{2}{3}$$

$$y - 6 = -\frac{2}{3}(x - (-3))$$

$$y - 6 = -\frac{2}{3}(x + 3)$$

$$y - 6 = -\frac{2}{3}x - 2$$

$$y = -\frac{2}{3}x + 4$$

$$x = 0,\ y = -\frac{2}{3}(0) + 4 = 4$$

$$x = 3,\ y = -\frac{2}{3}(3) + 4 = 2$$

$$x = 9,\ y = -\frac{2}{3}(9) + 4 = -2$$

Answers will vary. The possible answers include (0, 4), (3, 2), and (9, −2).

97. Find the midpoint of the line segment.

$$x_m = \frac{2+(-4)}{2} = \frac{-2}{2} = -1$$

$$y_m = \frac{5+1}{2} = \frac{6}{2} = 3$$

The midpoint is (−1, 3).

Use the point-slope formula to find the equation of the line.

$$y - y_1 = m(x - x_1)$$
$$y - 3 = -2(x - (-1))$$
$$y - 3 = -2(x + 1)$$
$$y - 3 = -2x - 2$$
$$y = -2x + 1$$

The equation of the line is $y = -2x + 1$.

Section 3.6

Objective A Exercises

1. The student should note that the slope of each graphed line must be determined. Look for the idea that two lines are parallel if they have the same slope and different y-intercepts.

3. $m = -5$

5.
$$m_1 \cdot m_2 = -1$$
$$4 \cdot m_2 = -1$$
$$m_2 = -\frac{1}{4}$$

7. $x = -2$ is a vertical line.
$y = 3$ is a horizontal line.
Yes, the lines are perpendicular.

9. $x = -3$ is a vertical line.
$y = \frac{1}{3}$ is a horizontal line.
No, the lines are not parallel.

11.
$$y = \frac{2}{3}x - 4 \quad m_1 = \frac{2}{3}$$
$$y = -\frac{3}{2}x - 4 \quad m_2 = -\frac{3}{2}$$
$$m_1 \neq m_2$$
No, the lines are not parallel.

13.
$$y = \frac{4}{3}x - 2 \quad m_1 = \frac{4}{3}$$
$$y = -\frac{3}{4}x + 2 \quad m_2 = -\frac{3}{4}$$
$$m_1 \cdot m_2 = \frac{4}{3}\left(-\frac{3}{4}\right) = -1$$
Yes, the lines are perpendicular.

15. $2x+3y=2$
$3y=-2x+2$
$y=-\frac{2}{3}x+\frac{2}{3}$
$m_1=-\frac{2}{3}$
$2x+3y=-4$
$3y=-2x-4$
$y=-\frac{2}{3}x-\frac{4}{3}$
$m_2=-\frac{2}{3}$
$m_1=m_2=-\frac{2}{3}$
Yes, the lines are parallel.

17. $x-4y=2$
$-4y=-x+2$
$y=\frac{1}{4}x-\frac{1}{2}$
$m_1=\frac{1}{4}$
$4x+y=8$
$y=-4x+8$
$m_2=-4$
$m_1\cdot m_2=\frac{1}{4}(-4)=-1$
Yes, the lines are perpendicular.

19. $m_1=\frac{6-2}{1-3}=\frac{4}{-2}=-2$
$m_2=\frac{-1-3}{-1-(-1)}=\frac{-4}{0}$.
$m_1\neq m_2$
No, the lines are not parallel.

21. $m_1=\frac{-1-2}{4-(-3)}=\frac{-3}{7}=-\frac{3}{7}$
$m_2=\frac{-4-3}{-2-1}=\frac{-7}{-3}=\frac{7}{3}$
$m_1\cdot m_2=-\frac{3}{7}\left(\frac{7}{3}\right)=-1$
Yes, the lines are perpendicular.

23. $2x-3y=2$
$-3y=-2x+2$
$y=\frac{2}{3}x-\frac{2}{3}$
$m=\frac{2}{3}$
$y-y_1=m(x-x_1)$
$y-(-4)=\frac{2}{3}[x-(-2)]$
$y+4=\frac{2}{3}(x+2)$
$y+4=\frac{2}{3}x+\frac{4}{3}$
$y=\frac{2}{3}x-\frac{8}{3}$
The equation of the line is $y=\frac{2}{3}x-\frac{8}{3}$.

25. $y=-3x+4$
$m_1=-3$
$m_1\cdot m_2=-1$
$-3\cdot m_2=-1$
$m_2=\frac{1}{3}$
$y-y_1=m(x-x_1)$
$y-1=\frac{1}{3}(x-4)$
$y-1=\frac{1}{3}x-\frac{4}{3}$
$y=\frac{1}{3}x-\frac{1}{3}$
The equation of the line is $y=\frac{1}{3}x-\frac{1}{3}$.

27. $3x-5y=2$
$-5y=-3x+2$
$y=\frac{3}{5}x-\frac{2}{5}$
$m_1=\frac{3}{5}$
$m_1\cdot m_2=-1$
$\frac{3}{5}\cdot m_2=-1$
$m_2=-\frac{5}{3}$
$y-y_1=m(x-x_1)$
$y-(-3)=-\frac{5}{3}[x-(-1)]$
$y+3=-\frac{5}{3}(x+1)$
$y+3=-\frac{5}{3}x-\frac{5}{3}$
$y=-\frac{5}{3}x-\frac{14}{3}$
The equation of the line is $y=-\frac{5}{3}x-\frac{14}{3}$.

Applying the Concepts

29. Using points $O(0,0)$ and $P(6,3)$:
$m_1=\frac{3-0}{6-0}=\frac{3}{6}=\frac{1}{2}$
$m_1\cdot m_2=-1$
$\frac{1}{2}\cdot m_2=-1$
$m_2=-2$
using $P(6,3)$,
$y-3=-2(x-6)$
$y-3=-2x+12$
$y=-2x+15$ is the equation of the line.

31. Write the equations of the lines in slope-intercept form.

(1) $A_1x + B_1y = C_1$

$$B_1y = C_1 - A_1x$$

$$y = \frac{C_1}{B_1} - \frac{A_1}{B_1}x$$

(2) $A_2x + B_2y = C_2$

$$B_2y = C_2 - A_2x$$

$$y = \frac{C_2}{B_2} - \frac{A_2}{B_2}x$$

The slopes of the two lines must be the same for the lines to be parallel. $-\frac{A_1}{B_1} = -\frac{A_2}{B_2}$, so $\frac{A_1}{B_1} = \frac{A_2}{B_2}$.

33. Strategy
- Use one of the lines to find the slope of the line perpendicular to it.
- Use the slope-intercept form to define the line that makes a right triangle with the other two lines.

Solution For instance, choose the line $y = -\frac{1}{2}x + 2$.

The slope is $-\frac{1}{2}$, so a line perpendicular to this line has a slope of 2. The y-intercept can be the same, or $b = 2$.

$y = mx + b$

$y = 2x + 2$

An equation of the line is $y = 2x + 2$.

The answer is any equation of the form $y = 2x + b$ where $b \neq -13$ or of the form $y = -\frac{3}{2}x + c$, where $c \neq 8$.

Section 3.7

Objective A Exercises

1. A half-plane is the set of points on one side of a line in the plane.

3. $0 > 2(0) - 7$, $0 > -7$; Yes, (0, 0) is a solution.

5. $0 \leq -\frac{2}{3}(0) - 8$, $0 \leq -8$; No, (0, 0) is not a solution.

7. $3x - 2y \geq 6$

$$-2y \geq -3x + 6$$

$$y \leq \frac{3}{2}x - 3$$

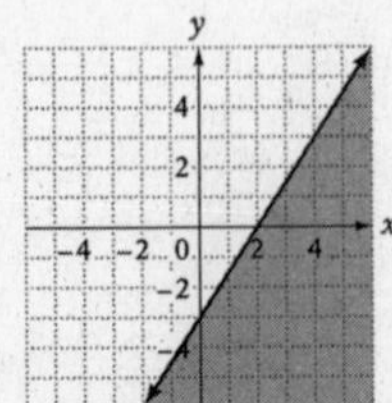

9. $x + 3y < 4$

$$3y < -x + 4$$

$$y < -\frac{1}{3}x + \frac{4}{3}$$

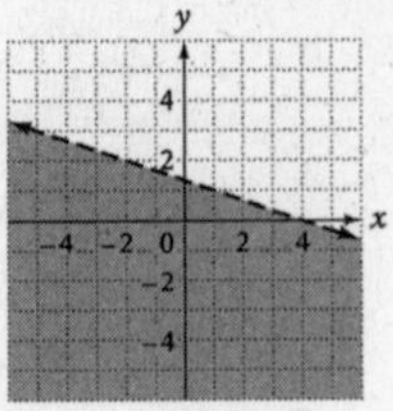

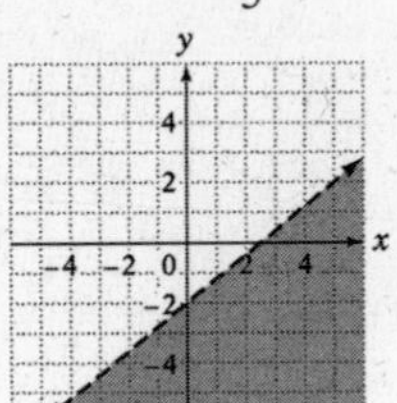

11. $4x - 5y > 10$

$$-5y > -4x + 10$$

$$y < \frac{4}{5}x - 2$$

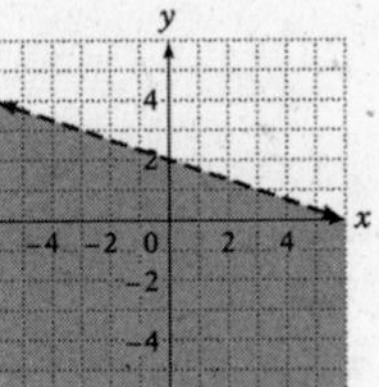

13. $x + 3y < 6$

$$3y < -x + 6$$

$$y < -\frac{1}{3}x + 2$$

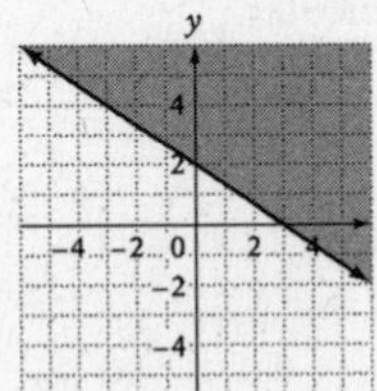

15. $2x + 3y \geq 6$

$$3y \geq -2x + 6$$

$$y \geq -\frac{2}{3}x + 2$$

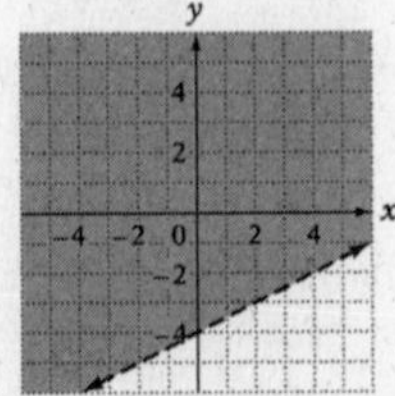

17. $-x + 2y > -8$

$$2y > x - 8$$

$$y > \frac{1}{2}x - 4$$

19. $y - 4 < 0$
$y < 4$

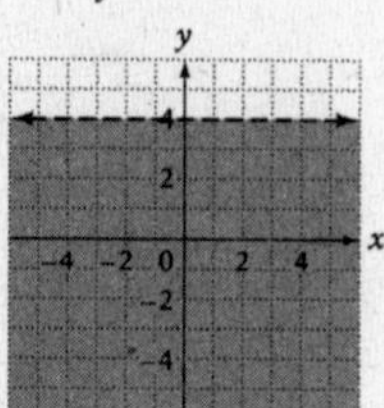

21. $6x + 5y < 15$
$5y < -6x + 15$
$y < -\frac{6}{5}x + 3$

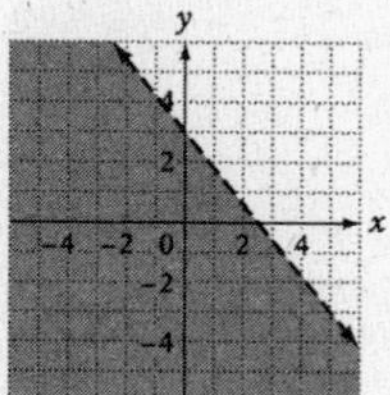

23. $-5x + 3y \geq -12$
$3y \geq 5x - 12$
$y \geq \frac{5}{3}x - 4$

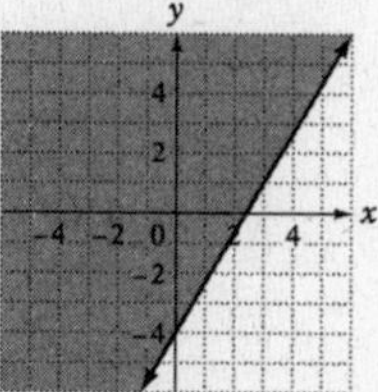

Applying the Concepts

25. The inequality $y < 3x - 1$ is not a function because given a value of x, there is more than one corresponding value of y. For example, both (3, 2) and (3, −1) are ordered pairs that satisfy the inequality. This contradicts the definition of a function because there are two ordered pairs with the same first coordinate and different second coordinates.

27. There are no points whose coordinates satisfy both $y \leq x - 1$ and $y \geq x + 2$. The solution set of $y \leq x - 1$ is all points on and below the line $y = x - 1$. The solution set of $y \geq x + 2$ is all points on and above the line $y = x + 2$. Since the lines $y = x - 1$ and $y = x + 2$ are parallel lines, and $y = x + 2$ is above the line $y = x - 1$, there are no points that lie both below $y = x - 1$ and above $y = x + 2$.

Chapter 3 Review Exercises

1. $y = \frac{x}{x-2}$
$y = \frac{4}{4-2}$
$y = 2$
The ordered pair is (4, 2).

2. $P(-2) = -2$
$P(a) = 3a + 4$

3. $y = 2x^2 - 5$
Ordered pairs: (−2, 3), (−1, −3), (0, −5), (1, −3), (2, 3)

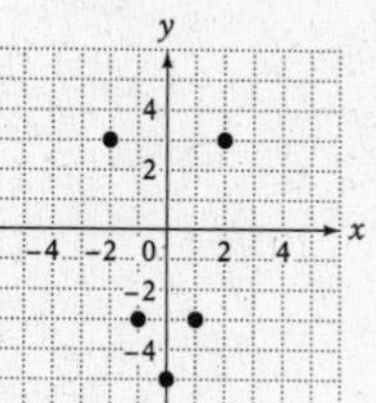

4.

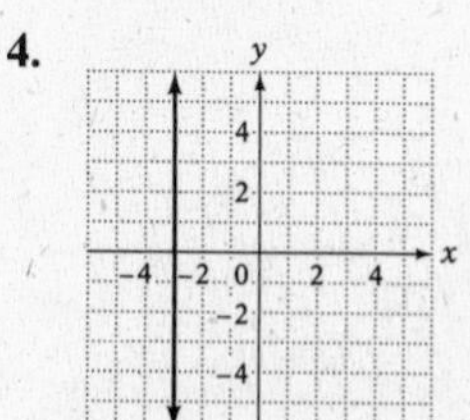

5. Evaluate the function at each element of the domain.
$f(x) = x^2 + x - 1$
$f(-2) = (-2)^2 + (-2) - 1 = 1$
$f(-1) = (-1)^2 + (-1) - 1 = -1$
$f(0) = 0^2 + 0 - 1 = -1$
$f(1) = 1^2 + 1 - 1 = 1$
$f(2) = 2^2 + 2 - 1 = 5$
Range = $\{-1, 1, 5\}$

6. Domain = $\{-1, 0, 1, 5\}$
Range = $\{0, 2, 4\}$

7. $y_m = \frac{y_1 + y_2}{2} = \frac{4+5}{2} = \frac{9}{2}$
$x_m = \frac{x_1 + x_2}{2} = \frac{-2+3}{2} = \frac{1}{2}$
Length $= \sqrt{(x_1 - x_2)^2 + (y_1 - y_2)^2}$
$= \sqrt{(-2-3)^2 + (4-5)^2}$
$= \sqrt{26} \approx 5.10$

Midpoint $\left(\frac{1}{2}, \frac{9}{2}\right)$; length 5.10

8. $f(x) = \dfrac{x}{x+4}$

The function is not defined for zero in the denominator.

$x + 4 = 0$

$x = -4$

$f(x)$ is not defined for $x = -4$.

9. x-intercept:

$y = -\frac{2}{3}x - 2$

$0 = -\frac{2}{3}x - 2$

$x = -3$

$(-3, 0)$

y-intercept:

$y = -\frac{2}{3}x - 2$

$y = -2$

$(0, -2)$

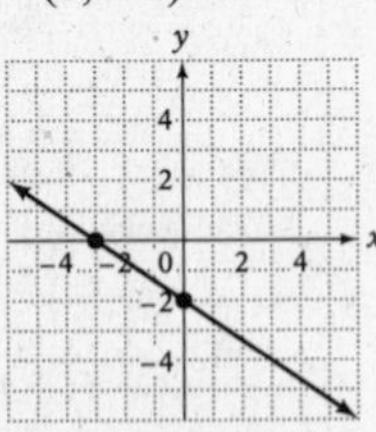

10. x-intercept:

$3x + 2y = -6$

$3x + 0 = -6$

$x = -2$

$(-2, 0)$

y-intercept:

$3x + 2y = -6$

$0 + 2y = -6$

$y = -3$

$(0, -3)$

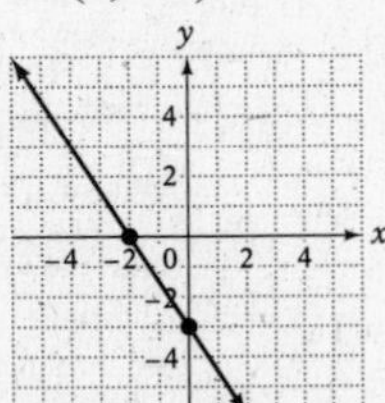

11.

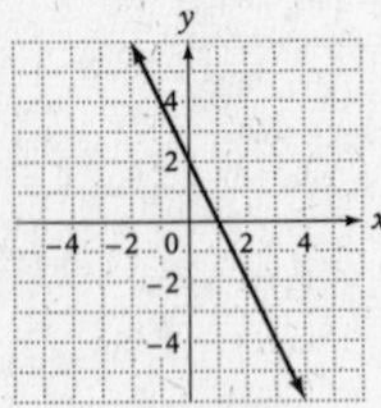

12.

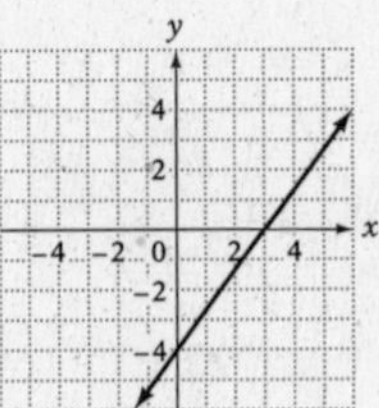

13. $m = \dfrac{y_2 - y_1}{x_2 - x_1}$

$m = \dfrac{2 - (-2)}{-1 - 3} = \dfrac{4}{-4} = -1$

14. Use the point-slope form to find the equation of the line.

$y - y_1 = m(x - x_1)$

$y - 4 = \frac{5}{2}[x - (-3)]$

$y - 4 = \frac{5}{2}x + \frac{15}{2}$

$y = \frac{5}{2}x + \frac{23}{2}$

The equation of the line is $y = \frac{5}{2}x + \frac{23}{2}$.

15.

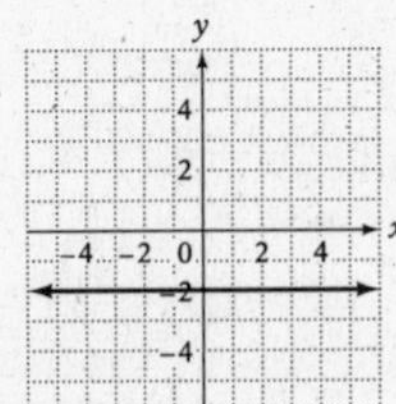

16. $(x_1, y_1) = (-2, 3)$

$m = -\frac{1}{4} = \frac{-1}{4}$

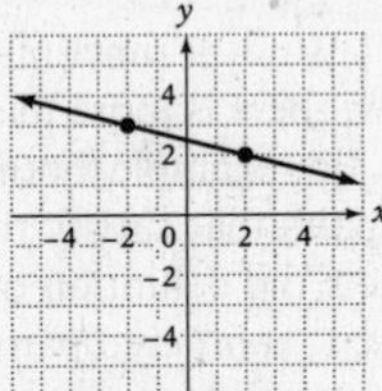

17. $f(x) = x^2 - 2$

$f(-2) = (-2)^2 - 2 = 4 - 2 = 2$

$f(-1) = (-1)^2 - 2 = 1 - 2 = -1$

$f(0) = 0^2 - 2 = -2$

$f(1) = 1^2 - 2 = 1 - 2 = -1$

$f(2) = 2^2 - 2 = 4 - 2 = 2$

The range is $\{-2, -1, 2\}$.

18. Strategy
- Let x represent the room rate.
- Let y represent the number of rooms occupied.
- Use the given information to determine the slope.
- Use the point-slope form to determine the equation of the line.

Solution **a.** $(x_1, y_2) = (95, 200)$
$(x_2, y_2) = (105, 190)$
$$m = \frac{y_2 - y_1}{x_2 - x_1} = \frac{190 - 200}{105 - 95} = -1$$
$$y - y_1 = m(x - x_1)$$
$$y - 200 = -1(x - 95)$$
$$y - 200 = -x + 95$$
$$y = -x + 295$$
The linear equation is
$y = -x + 295,\ 0 \le x \le 295.$

b. Find the number of rooms occupied when the rate is \$120.
$y = -120 + 295 = 175$
When the rate is \$120, 175 rooms will be occupied.

19. The slope for the parallel line is $m = -4$.
$$y - y_1 = m(x - x_1)$$
$$y - 3 = -4[x - (-2)] \cdot (x_1, y_1) = (-2, 3)$$
$$y - 3 = -4x - 8$$
$$y = -4x - 5$$

20. The slope for the perpendicular line is $m = \frac{5}{2}$.
$$y - y_1 = \frac{5}{2}(x - x_1)$$
$$y - 3 = \frac{5}{2}[x - (-2)] \cdot (x_1, y_1) = (-2, 3)$$
$$y - 3 = \frac{5}{2}x + 5$$
$$y = \frac{5}{2}x + 8$$

21.

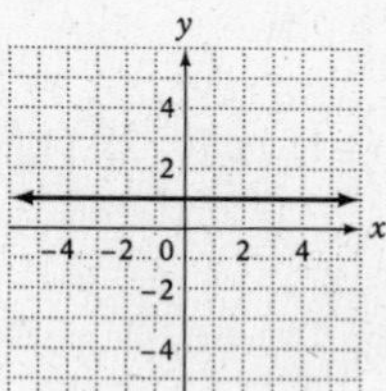

22.

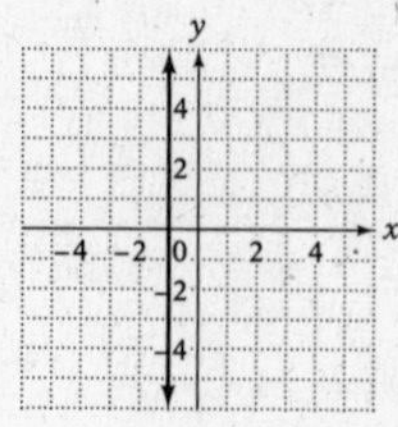

23. Use the point-slope formula with $m = -\frac{2}{3}$ and $(x_1, y_1) = (-3, 3)$.
$$y - y_1 = m(x - x_1)$$
$$y - 3 = -\frac{2}{3}(x + 3)$$
$$y - 3 = -\frac{2}{3}x - 2$$
$$y = -\frac{2}{3}x + 1$$

24. Use the two given points to determine the slope and then use the point-slope form to find the equation.
$(x_1, y_1) = (-8, 2)$
$(x_2, y_2) = (4, 5)$
$$m = \frac{y_2 - y_1}{x_2 - x_1} = \frac{5 - 2}{4 - (-8)} = \frac{3}{12} = \frac{1}{4}$$
$$y - y_1 = m(x - x_1)$$
$$y - 2 = \frac{1}{4}(x + 8)$$
$$y - 2 = \frac{1}{4}x + 2$$
$$y = \frac{1}{4}x + 4$$

25.
$$d = \sqrt{(x_1 - x_2)^2 + (y_1 - y_2)^2}$$
$$= \sqrt{[4 - (-2)]^2 + (-5 - 3)^2}$$
$$= \sqrt{6^2 + (-8)^2}$$
$$= \sqrt{100}$$
$$= 10$$

26. The coordinates of the midpoint are
$$\left(\frac{x_1 + x_2}{2}, \frac{y_1 + y_2}{2}\right) = \left(\frac{-3 + 5}{2}, \frac{8 - 2}{2}\right) = (1, 3).$$

27.

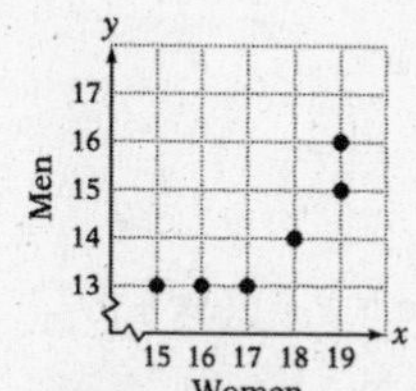

28. $y \ge 2x - 3$

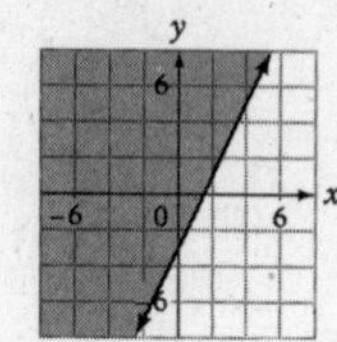

29. $3x - 2y < 6$
$$-2y < -3x + 6$$
$$y > \frac{3}{2}x - 3$$

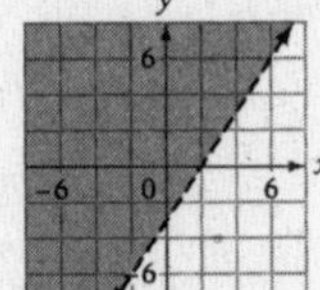

30. $P_1(-2, 4), P_2(4, -3)$
$$m = \frac{y_2 - y_1}{x_2 - x_1} = \frac{-3-4}{4-(-2)} = \frac{-7}{6} = -\frac{7}{6}$$
$$y - y_1 = m(x - x_1)$$
$$y - 4 = -\frac{7}{6}[x - (-2)]$$
$$y - 4 = -\frac{7}{6}(x + 2)$$
$$y - 4 = -\frac{7}{6}x - \frac{7}{3}$$
$$y = -\frac{7}{6}x + \frac{5}{3}$$
The equation of the line is $y = -\frac{7}{6}x + \frac{5}{3}$.

31. $4x - 2y = 7$
$$-2y = -4x + 7$$
$$y = 2x - \frac{7}{2}$$
The slope for a parallel line is $m = 2$.
$$y - y_1 = m(x - x_1)$$
$$y - (-4) = 2[x - (-2)]$$
$$y + 4 = 2x + 4$$
$$y = 2x$$
The equation of the parallel line is $y = 2x$.

32. $y = -3x + 4$
$$m = -3$$
$$y - y_1 = m(x - x_1)$$
$$y - (-2) = -3(x - 3)$$
$$y + 2 = -3x + 9$$
$$y = -3x + 7$$
The equation of the line is $y = -3x + 7$.

33. $y = -\frac{2}{3}x + 6$
$$m_1 = -\frac{2}{3}$$
$$m_1 \cdot m_2 = -1$$
$$-\frac{2}{3}m_2 = -1$$
$$m_2 = \frac{3}{2}$$
$$y - y_1 = m(x - x_1)$$
$$y - 5 = \frac{3}{2}(x - 2)$$
$$y - 5 = \frac{3}{2}x - 3$$
$$y = \frac{3}{2}x + 2$$
The equation of the line is $y = \frac{3}{2}x + 2$.

34. $(x_1, y_1) = (-1, 4)$
$$m = -\frac{1}{3} = \frac{-1}{3}$$

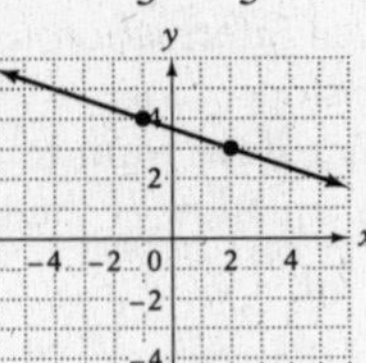

35.

d
300
200
100
(4, 220)
0 1 2 3 4 5 6
t
Distance (in miles)
Time (in hours)

After 4 h the car will have traveled 220 mi.

36. $m = \frac{y_2 - y_1}{x_2 - x_1} = \frac{12{,}000 - 6000}{500 - 200}$
$$= \frac{6000}{300} = 20$$
The slope is 20. The manufacturing cost is \$20 per calculator.

37. a. The y-intercept is (0, 25,000).
The slope is 80.
$$y = mx + b$$
$$y = 80x + 25{,}000$$
The linear function is $y = 80x + 25{,}000$.

b. Predict the cost of building a house with 2000 square feet.
$$y = 80(2000) + 25{,}000$$
$$y = 185{,}000$$
The house will cost \$185,000 to build.

Chapter 3 Test

1. $P(x) = 2 - x^2$
ordered pairs: $(-2, -2), (-1, 1), (0, 2), (1, 1), (2, -2)$

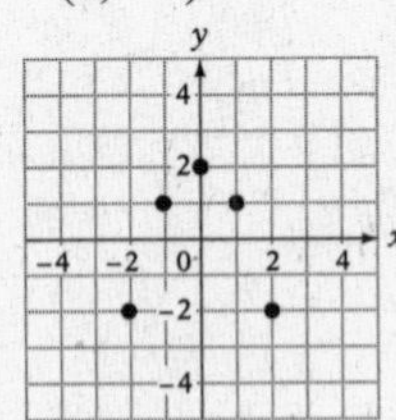

2. $y = 2x + 6$
$$y = 2(-3) + 6$$
$$y = -6 + 6$$
$$y = 0$$
The ordered pair is $(-3, 0)$.

3.

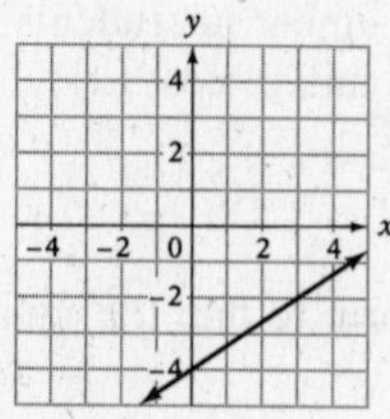

4. $2x + 3y = -3$

$3y = -2x - 3$

$y = -\frac{2}{3}x - 1$

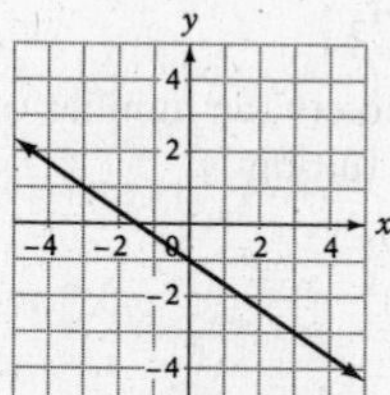

5. The equation of the vertical line that contains $(-2, 3)$ is $x = -2$.

6. Length $= \sqrt{(x_1 - x_2)^2 + (y_1 - y_2)^2}$

$= \sqrt{[4 - (-5)]^2 + (2 - 8)^2}$

$= \sqrt{81 + 36}$

$= \sqrt{117} \approx 10.82$

$x_m = \frac{x_1 + x_2}{2} = \frac{4 + (-5)}{2} = -\frac{1}{2}$

$y_m = \frac{y_1 + y_2}{2} = \frac{2 + 8}{2} = 5$

Length: 10.82; midpoint: $\left(-\frac{1}{2}, 5\right)$

7. $P_1(-2, 3)$, $P_2(4, 2)$

$m = \frac{y_2 - y_1}{x_2 - x_1} = \frac{2 - 3}{4 - (-2)} = -\frac{1}{6}$

The slope of the line is $-\frac{1}{6}$.

8. $P(x) = 3x^2 - 2x + 1$

$P(2) = 3(2)^2 - 2(2) + 1$

$P(2) = 9$

9. x-intercept:

$2x - 3(0) = 6$

$x = 3$

$(3, 0)$

y-intercept:

$2x - 3y = 6$

$2(0) - 3y = 6$

$-3y = 6$

$y = -2$

$(0, -2)$

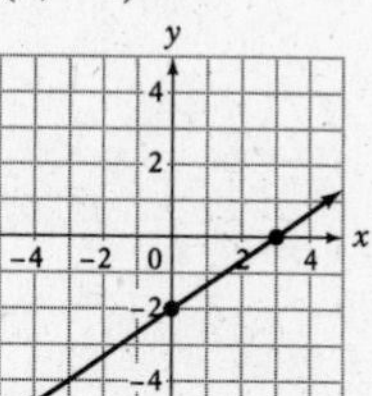

10.

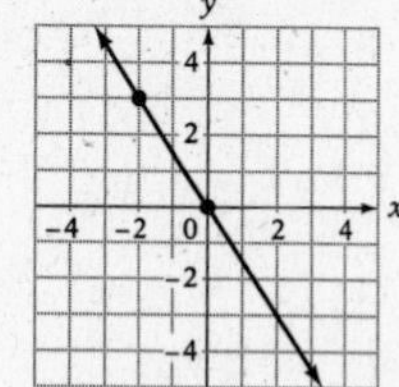

11. $m = \frac{2}{5}$ $(x_1, y_1) = (-5, 2)$

$y - 2 = \frac{2}{5}[x - (-5)]$

$y - 2 = \frac{2}{5}(x + 5)$

$y - 2 = \frac{2}{5}x + 2$

$y = \frac{2}{5}x + 4$

The equation of the line is $y = \frac{2}{5}x + 4$.

12. $x = 0$

13. $P_1(3, -4)$, $P_2(-2, 3)$

$m = \frac{y_2 - y_1}{x_2 - x_1} = \frac{3 - (-4)}{-2 - 3} = \frac{3 + 4}{-5} = -\frac{7}{5}$

$y - y_1 = m(x - x_1)$

$y - (-4) = -\frac{7}{5}(x - 3)$

$y + 4 = -\frac{7}{5}x + \frac{21}{5}$

$y = -\frac{7}{5}x + \frac{1}{5}$

The equation of the line is $y = -\frac{7}{5}x + \frac{1}{5}$.

14. A horizontal line has a slope of 0.

$y - y_1 = m(x - x_1)$

$y - (-3) = 0(x - 4)$

$y + 3 = 0$

$y = -3$

The equation of the line is $y = -3$.

15. Domain $= \{-4, -2, 0, 3\}$
Range $= \{0, 2, 5\}$

16. $y = -\frac{3}{2}x - 6$
$m = -\frac{3}{2}$
$y - y_1 = m(x - x_1)$
$y - 2 = -\frac{3}{2}(x - 1)$
$y - 2 = -\frac{3}{2}x + \frac{3}{2}$
$y = -\frac{3}{2}x + \frac{7}{2}$
The equation of the line is $y = -\frac{3}{2}x + \frac{7}{2}$.

17. $y = -\frac{1}{2}x - 3$
$m_1 = -\frac{1}{2}$
$m_1 \cdot m_2 = -1$
$-\frac{1}{2}m_2 = -1$
$m_2 = 2$
$y - y_1 = m(x - x_1)$
$y - (-3) = 2[x - (-2)]$
$y + 3 = 2(x + 2)$
$y + 3 = 2x + 4$
$y = 2x + 1$
The equation of the line is $y = 2x + 1$.

18. $3x - 4y > 8$
$-4y > -3x + 8$
$y < \frac{3}{4}x - 2$

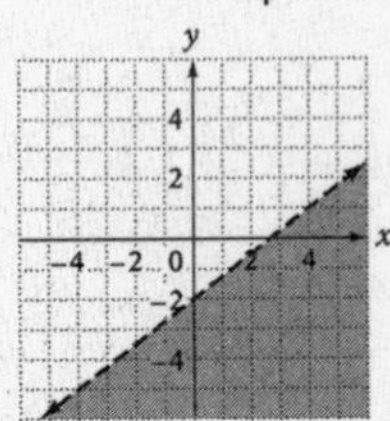

19. **Strategy** Use two points on the graph to find the slope of the line.

Solution $(x_1, y_1) = (3, 120{,}000)$
$(x_2, y_2) = (12, 30{,}000)$
$m = \frac{y_2 - y_1}{x_2 - x_1}$
$m = \frac{30{,}000 - 120{,}000}{12 - 3}$
$m = -10{,}000$
The value of the house decreases by \$10,000 per year.

20. a. Dependent variable: number of students (y)
Independent variable: tuition cost (x)
$m = \frac{\text{change of } y}{\text{change of } x} = \frac{-6}{20} = -\frac{3}{10}$
$P_1(250, 100)$
Use the point-slope form to find the equation.
$y - y_1 = m(x - x_1)$
$y - 100 = -\frac{3}{10}(x - 250)$
$y - 100 = -\frac{3}{10}x + 75$
$y = -\frac{3}{10}x + 175$
The equation that predicts the number of students for a certain tuition is
$y = -\frac{3}{10}x + 175$.

b. $y = -\frac{3}{10}x + 175$
$y = -\frac{3}{10}(300) + 175$
$y = 85$
When the tuition is \$300, 85 students will enroll.

Cumulative Review Exercises

1. The Commutative Property of Multiplication

2. $3 - \frac{x}{2} = \frac{3}{4}$
$4\left(3 - \frac{x}{2}\right) = \frac{3}{4}(4)$
$12 - 2x = 3$
$12 - 2x - 12 = 3 - 12$
$-2x = -9$
$x = \frac{9}{2}$

The solution is $\frac{9}{2}$.

3. $2[y - 2(3 - y) + 4] = 4 - 3y$
$2(y - 6 + 2y + 4) = 4 - 3y$
$2(3y - 2) = 4 - 3y$
$6y - 4 = 4 - 3y$
$9y - 4 = 4$
$9y = 8$
$y = \frac{8}{9}$
The solution is $\frac{8}{9}$.

4. $$\frac{1-3x}{2}+\frac{7x-2}{6}=\frac{4x+2}{9}$$
$$18\left(\frac{1-3x}{2}+\frac{7x-2}{6}\right)=18\left(\frac{4x+2}{9}\right)$$
$$9(1-3x)+3(7x-2)=2(4x+2)$$
$$9-27x+21x-6=8x+4$$
$$-6x+3=8x+4$$
$$-14x=1$$
$$x=-\frac{1}{14}$$
The solution is $-\frac{1}{14}$.

5. $x-3<-4$ or $2x+2>3$
$x<-1$ $\quad 2x>1$
$x>\frac{1}{2}$
$\{x|x<-1\}$ $\quad \left\{x\middle|x>\frac{1}{2}\right\}$
$\{x|x<-1\}$ or $\left\{x\middle|x>\frac{1}{2}\right\}$
$\left\{x\middle|x<-1 \text{ or } x>\frac{1}{2}\right\}$

6. $8-|2x-1|=4$
$-|2x-1|=-4$
$|2x-1|=4$
$2x-1=4$ or $2x-1=-4$
$2x=5$ $\quad 2x=-3$
$x=\frac{5}{2}$ $\quad x=-\frac{3}{2}$
The solutions are $\frac{5}{2}$ and $-\frac{3}{2}$.

7. $|3x-5|<5$
$$-5<3x-5<5$$
$$-5+5<3x-5+5<5+5$$
$$0<3x<10$$
$$\frac{1}{3}(0)<\frac{1}{3}(3x)<10\left(\frac{1}{3}\right)$$
$$0<x<\frac{10}{3}$$
$\left\{x\middle|0<x<\frac{10}{3}\right\}$

8. $4-2(4-5)^3+2=4-2(-1)^3+2$
$=4+2+2=8$

9. $(a-b)^2\div(ab)$ for $a=4$ and $b=-2$
$[4-(-2)]^2\div[4(-2)]=6^2\div(-8)$
$=36\div(-8)$
$=-4.5$

10. $\{x|x<-2\}\cup\{x|x>0\}$

−5 −4 −3 −2 −1 0 1 2 3 4 5

11. $$P=\frac{R-C}{n}$$
$$P\cdot n=\frac{R-C}{n}\cdot n$$
$$P\cdot n=R-C$$
$$C=R-Pn$$

12. $$2x+3y=6$$
$$2x=6-3y$$
$$x=3-\frac{3}{2}y$$

13. Solve each inequality.
$3x-1<4$ $\quad x-2>2$
$3x<5$ $\quad x>4$
$x<\frac{5}{3}$
$\left\{x\middle|x<\frac{5}{3}\right\}\cap\{x|x>4\}=\varnothing$

14. $$P(x)=x^2+5$$
$$P(-3)=(-3)^2+5$$
$$P(-3)=14$$

15. $$y=-\frac{5}{4}x+3$$
$$y=-\frac{5}{4}(-8)+3$$
$$y=10+3$$
$$y=13$$
The ordered-pair solution is $(-8, 13)$.

16. $P_1(-1,3)$, $P_2(3,-4)$
$$m=\frac{y_2-y_1}{x_2-x_1}=\frac{-4-3}{3-(-1)}=\frac{-7}{4}=-\frac{7}{4}$$

17. $m=\frac{3}{2}$, $(x_1, y_1)=(-1,5)$
$$y-y_1=m(x-x_1)$$
$$y-5=\frac{3}{2}[x-(-1)]$$
$$y-5=\frac{3}{2}(x+1)$$
$$y-5=\frac{3}{2}x+\frac{3}{2}$$
$$y=\frac{3}{2}x+\frac{13}{2}$$
The equation of the line is $y=\frac{3}{2}x+\frac{13}{2}$.

18. $(x_1, y_1) = (4, -2), (x_2, y_2) = (0, 3)$

$$m = \frac{y_2 - y_1}{x_2 - x_1} = \frac{3-(-2)}{0-4} = \frac{3+2}{-4} = -\frac{5}{4}$$

$$y - y_1 = m(x - x_1)$$
$$y - (-2) = -\frac{5}{4}(x-4)$$
$$y + 2 = -\frac{5}{4}(x-4)$$
$$y + 2 = -\frac{5}{4}x + 5$$
$$y = -\frac{5}{4}x + 3$$

The equation of the line is $y = -\frac{5}{4}x + 3$.

19. $y = -\frac{3}{2}x + 2, m = -\frac{3}{2}$

$$y - y_1 = m(x - x_1)$$
$$y - 4 = -\frac{3}{2}(x-2)$$
$$y - 4 = -\frac{3}{2}x + 3$$
$$y = -\frac{3}{2}x + 7$$

The equation of the line is $y = -\frac{3}{2}x + 7$.

20. $3x - 2y = 5$

$$y = \frac{3}{2}x - \frac{5}{2}$$
$$m_1 = \frac{3}{2}$$
$$m_1 \cdot m_2 = -1$$
$$\frac{3}{2}m_2 = -1$$
$$m_2 = -\frac{2}{3}$$
$$y - y_1 = m(x - x_1)$$
$$y - 0 = -\frac{2}{3}(x-4)$$
$$y = -\frac{2}{3}x + \frac{8}{3}$$

The equation of the line is $y = -\frac{2}{3}x + \frac{8}{3}$.

21. Strategy
- Number of quarters: x
  Number of nickels: $4x$
  Number of dimes: $17 - 5x$

Coin	Number	Value	Total Value
Quarters	x	25	$25x$
Nickels	$4x$	5	$4x(5)$
Dimes	$17-5x$	10	$10(17-5x)$

- The sum of the total values of each denomination of coin equals the total value of all the coins (160 cents).

Solution

$$10(17-5x) + 4x(5) + 25x = 160$$
$$170 - 50x + 20x + 25x = 160$$
$$170 - 5x = 160$$
$$-5x = -10$$
$$x = 2$$

$17 - 5x = 17 - 5(2) = 7$

There are 7 dimes in the coin purse.

22. Strategy
- Rate of first plane: x
  Rate of second plane: $2x$

	Rate	Time	Distance
1st Plane	x	3	$3x$
2nd Plane	$2x$	3	$3(2x)$

- The total distance traveled by the two planes is 1800 mi.

Solution

$$3x + 3(2x) = 1800$$
$$3x + 6x = 1800$$
$$9x = 1800$$
$$x = 200$$
$$2x = 2(200) = 400$$

The first plane is traveling at 200 mph and the second plane is traveling at 400 mph.

23. Strategy
- Pounds of coffee costing \$9.00: x
  Pounds of coffee costing \$6.00: $60 - x$

	Amount	Cost	Value
\$9 coffee	x	9	$9x$
\$6 coffee	$60-x$	6	$6(60-x)$
Mixture	60	8	$8(60)$

- The sum of the values before mixing equals the value after mixing.

Solution

$$9x + 6(60-x) = 8(60)$$
$$9x + 360 - 6x = 480$$
$$3x + 360 = 480$$
$$3x = 120$$
$$x = 40$$

$60 - x = 60 - 40 = 20$

The mixture consists of 40 lb of \$9.00 coffee and 20 lb of \$6.00 coffee.

24. x-intercept:

$$3x - 5(0) = 15$$
$$3x = 15$$
$$x = 5$$

$(5, 0)$

y-intercept:

$$3(0) - 5y = 15$$
$$-5y = 15$$
$$y = -3$$

$(0, -3)$

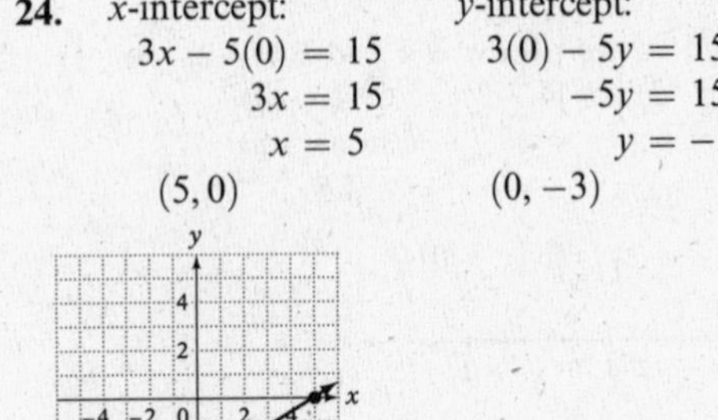

25.

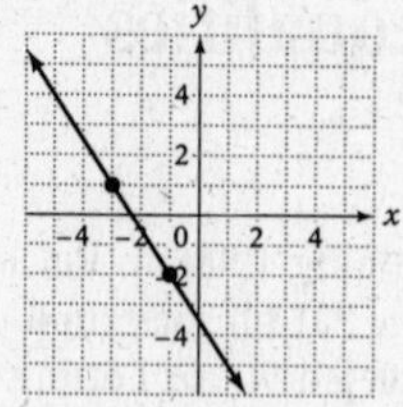

26. $3x - 2y \geq 6$

$-2y \geq -3x + 6$

$y \leq \frac{3}{2}x - 3$

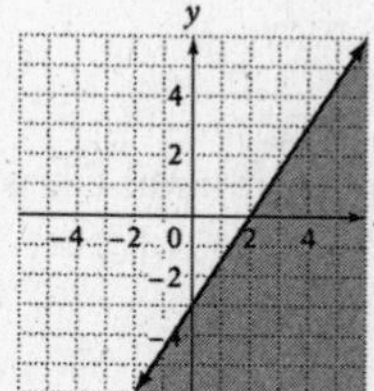

27. **Strategy** To write the equation:

- Use points on the graph to find the slope of the line.
- Locate the y-intercept of the line on the graph.
- Use the slope-intercept form of an equation to write the equation of the line.

Solution **a.** $(x_1, y_1) = (0, 30{,}000)$,

$(x_2, y_2) = (6, 0)$

$m = \frac{y_2 - y_1}{x_2 - x_1} = \frac{0 - 30{,}000}{6 - 0} = -5000$

The y-intercept is (0, 30,000).

$y = mx + b$

$y = -5000x + 30{,}000$

b. The value of the truck decreases by $5000 per year.

Chapter 4: Systems of Linear Equations and Inequalities

Prep Test

1. $\frac{10}{1}(\frac{3}{5}x) + \frac{10}{1}(\frac{1}{2}y) = \frac{30}{5}x + \frac{10}{2}y$
   $= 6x + 5y$

2. $3(-1) + 2(4) - (-2)$
   $= -3 + 8 + 2$
   $= 7$

3. $3x - 2(-2) = 4$
   $3x + 4 = 4$
   $\left(\frac{1}{3}\right)3x = 0\left(\frac{1}{3}\right)$
   $x = 0$

4. $3x + 4(-2x - 5) = -5$
   $3x - 8x - 20 = -5$
   $-5x - 20 = -5$
   $-5x = 15$
   $\frac{-5x}{-5} = \frac{15}{-5}$
   $x = -3$

5. $0.45x + 0.06(-x + 4000) = 630$
   $0.45x - 0.06x + 240 = 630$
   $0.39x + 240 = 630$
   $0.39x = 390$
   $x = 1000$

6.

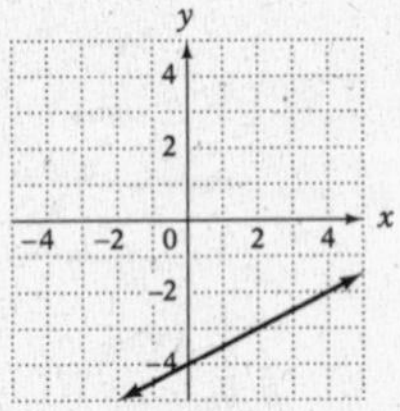

7.

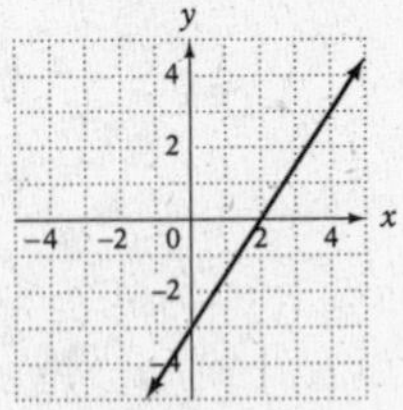

8.

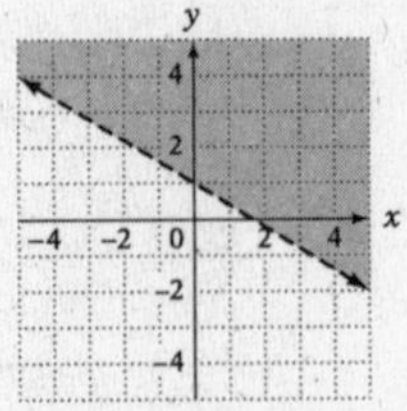

Go Figure

When Chris finishes running 1000 m, Pat has completed 900 m. When Pat finishes running 100 m, Leslie has completed 90 m. Letting x represent how many meters Leslie can run until Chris finishes in a 1000-meter race, we can set up a proportion:

$\frac{\text{Pat}}{\text{Leslie}} = \frac{100}{90} = \frac{900}{x}$
$100x = 81{,}000$
$x = 810$

Therefore, if Leslie can run 810 m until Chris finishes in a 1000-meter race, Chris will beat Leslie by 190 m.

$1000 - 810 = 190$

Answer: 190 m

Section 4.1

Objective A Exercises

1. $3(0) - 2(-1) = 2$
   $0 + 2 = 2$
   $2 = 2$
   $0 + 2(-1) = 6$
   $0 - 2 = 6$
   $-2 = 6$
   No, ordered pair is not a solution.

3. $(-3) + (-5) = -8$
   $-8 = -8$
   $2(-3) + 5(-5) = -31$
   $-6 + (-25) = -31$
   $-31 = -31$
   Yes, ordered pair is a solution.

5. Independent

7. Inconsistent

9. The solution is (3, −1).

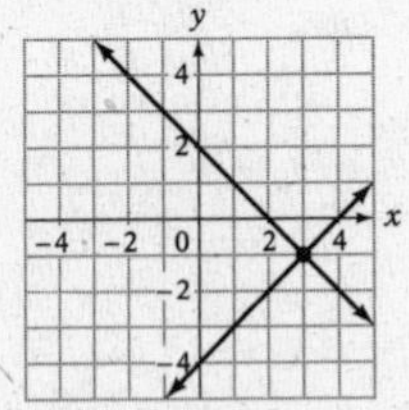

11.

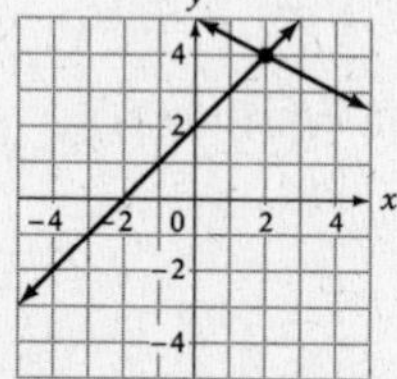

The solution is (2, 4).

13.

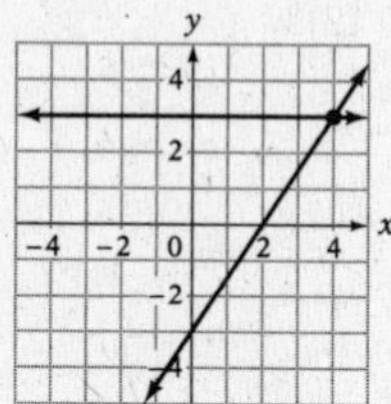

The solution is (4, 3).

15.

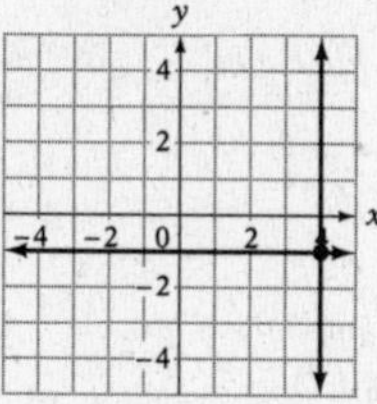

The solution is (4, −1).

17.

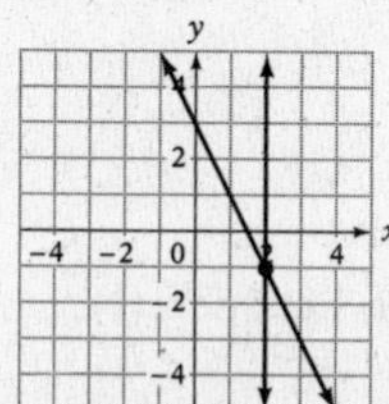

The solution is (2, −1).

19.

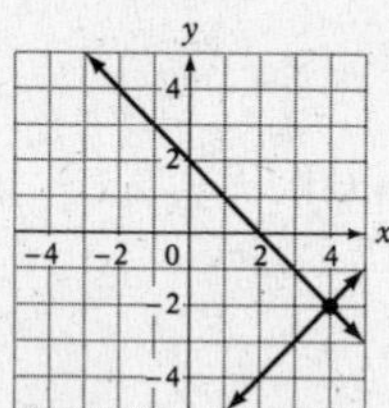

(4, −2)

21.

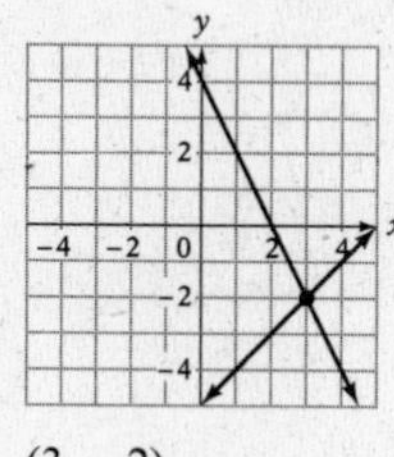

(3, −2)

23.

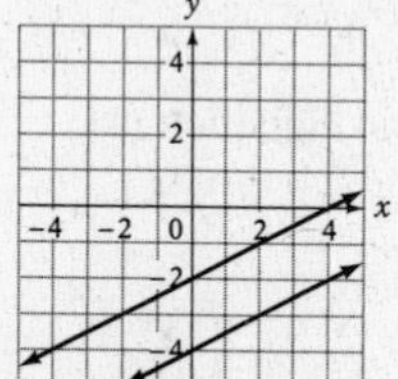

No solution.
The lines are parallel and therefore do not intersect.

25.

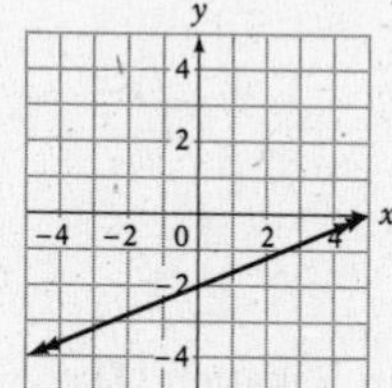

The two equations represent the same line.

$\left(x, \frac{2}{5}x - 2\right)$

Objective B Exercises

27. (1) $y = -x + 1$
(2) $2x - y = 5$
Substitute $-x + 1$ for y in Equation (2).

$$\begin{aligned} 2x - y &= 5 \\ 2x - (-x + 1) &= 5 \\ 2x + x - 1 &= 5 \\ 3x - 1 &= 5 \\ 3x &= 6 \\ x &= 2 \end{aligned}$$

Substitute into Equation (1).
$y = -x + 1$
$y = -2 + 1$
$y = -1$
The solution is (2, −1).

29. (1) $x = 2y - 3$
(2) $3x + y = 5$
Substitute $2y - 3$ for x in Equation (2).

$$\begin{aligned} 3x + y &= 5 \\ 3(2y - 3) + y &= 5 \\ 6y - 9 + y &= 5 \\ 7y - 9 &= 5 \\ 7y &= 14 \\ y &= 2 \end{aligned}$$

Substitute into Equation (1).
$x = 2y - 3$
$x = 2(2) - 3$
$x = 4 - 3$
$x = 1$
The solution is (1, 2).

31. (1) $3x + 5y = -1$
(2) $y = 2x - 8$
Substitute $2x - 8$ for y in Equation (1).
$3x + 5y = -1$
$3x + 5(2x - 8) = -1$
$3x + 10x - 40 = -1$
$13x - 40 = -1$
$13x = 39$
$x = 3$
Substitute into Equation (2).
$y = 2x - 8$
$y = 2(3) - 8$
$y = 6 - 8$
$y = -2$
The solution is (3, −2).

33. (1) $4x - 3y = 2$
(2) $y = 2x + 1$
Substitute $2x + 1$ for y in Equation (1).
$4x - 3(2x + 1) = 2$
$4x - 6x - 3 = 2$
$-2x - 3 = 2$
$-2x = 5$
$x = -\frac{5}{2}$
Substitute into Equation 2.
$y = 2\left(-\frac{5}{2}\right) + 1$
$y = -5 + 1$
$y = -4$
The solution is $\left(-\frac{5}{2}, -4\right)$.

35. (1) $3x - 2y = -11$
(2) $x = 2y - 9$
Substitute $2y - 9$ for x in Equation (1).
$3x - 2y = -11$
$3(2y - 9) - 2y = -11$
$6y - 27 - 2y = -11$
$4y - 27 = -11$
$4y = 16$
$y = 4$
Substitute into Equation (2).
$x = 2y - 9$
$x = 2(4) - 9$
$x = 8 - 9$
$x = -1$
The solution is (−1, 4).

37. (1) $3x + 2y = 4$
(2) $y = 1 - 2x$
Substitute $1 - 2x$ for y in Equation (1).
$3x + 2y = 4$
$3x + 2(1 - 2x) = 4$
$3x + 2 - 4x = 4$
$-x + 2 = 4$
$-x = 2$
$x = -2$
Substitute into Equation (2).
$y = 1 - 2x$
$y = 1 - 2(-2)$
$y = 1 + 4$
$y = 5$
The solution is (−2, 5).

39. (1) $5x + 2y = 15$
(2) $x = 6 - y$
Substitute $6 - y$ for x in Equation (1).
$5x + 2y = 15$
$5(6 - y) + 2y = 15$
$30 - 5y + 2y = 15$
$30 - 3y = 15$
$-3y = -15$
$y = 5$
Substitute into Equation (1).
$x = 6 - y$
$x = 6 - 5$
$x = 1$
The solution is (1, 5).

41. (1) $3x - 4y = 6$
(2) $x = 3y + 2$
Substitute $3y + 2$ for x in Equation (1).
$3x - 4y = 6$
$3(3y + 2) - 4y = 6$
$9y + 6 - 4y = 6$
$5y + 6 = 6$
$5y = 0$
$y = 0$
Substitute into Equation (2).
$x = 3y + 2$
$x = 3(0) + 2$
$x = 2$
The solution is (2, 0).

43. (1) $3x + 7y = -5$
(2) $y = 6x - 5$
Substitute $6x - 5$ for y in Equation (1).
$3x + 7(6x - 5) = -5$
$3x + 42x - 35 = -5$
$45x - 35 = -5$
$45x = 30$
$x = \frac{30}{45} = \frac{2}{3}$
Substitute into Equation (2).
$y = 6x - 5$
$y = 6\left(\frac{2}{3}\right) - 5$
$y = 4 - 5$
$y = -1$
The solution is $\left(\frac{2}{3}, -1\right)$.

45. (1) $3x - y = 10$
(2) $6x - 2y = 5$
Solve Equation (1) for y.
$3x - y = 10$
$-y = -3x + 10$
$y = 3x - 10$
Substitute into Equation (2).
$6x - 2(3x - 10) = 5$
$6x - 6x + 20 = 5$
$20 = 5$
No solution. This is not a true equation. The lines are parallel and the system is inconsistent.

47. (1) $3x + 4y = 14$
(2) $2x + y = 1$
Solve Equation (2) for y.
$2x + y = 1$
$y = -2x + 1$
Substitute into Equation (1).
$3x + 4y = 14$
$3x + 4(-2x + 1) = 14$
$3x - 8x + 4 = 14$
$-5x + 4 = 14$
$-5x = 10$
$x = -2$
Substitute into Equation (2).
$2x + y = 1$
$2(-2) + y = 1$
$-4 + y = 1$
$y = 5$
The solution is $(-2, 5)$.

49. (1) $3x + 5y = 0$
(2) $x - 4y = 0$
Solve Equation (2) for x.
$x - 4y = 0$
$x = 4y$
Substitute into Equation (1).
$3x + 5y = 0$
$3(4y) + 5y = 0$
$12y + 5y = 0$
$17y = 0$
$y = 0$
Substitute into Equation (2).
$x - 4y = 0$
$x - 4(0) = 0$
$x = 0$
The solution is $(0, 0)$.

51. (1) $2x - 4y = 16$
(2) $-x + 2y = -8$
Solve Equation (2) for x.
$-x = -2y - 8$
$x = 2y + 8$
Substitute into Equation (1).
$2x - 4y = 16$
$2(2y + 8) - 4y = 16$
$4y + 16 - 4y = 16$
$16 = 16$
This is a true equation. The equations are dependent. The solutions are the ordered pairs $\left(x, \frac{1}{2}x - 4\right)$.

53. (1) $y = 3x + 2$
(2) $y = 2x + 3$
Substitute $2x + 3$ for y in Equation (1).
$y = 3x + 2$
$2x + 3 = 3x + 2$
$2x = 3x - 1$
$-x = -1$
$x = 1$
Substitute into Equation (2).
$y = 2x + 3$
$y = 2(1) + 3$
$y = 2 + 3$
$y = 5$
The solution is $(1, 5)$.

55. (1) $y = 3x + 1$
(2) $y = 6x - 1$
Substitute $6x - 1$ for y in Equation (1).
$y = 3x + 1$
$6x - 1 = 3x + 1$
$6x = 3x + 2$
$3x = 2$
$x = \frac{2}{3}$
Substitute into Equation (2).
$y = 6x - 1$
$y = 6\left(\frac{2}{3}\right) - 1$
$y = 4 - 1$
$y = 3$
The solution is $\left(\frac{2}{3}, 3\right)$.

Objective C Exercises

57. **Strategy** • Amount invested at 3.5%: 2800
Amount invested at 4.2%: x

	Principal	Rate	Interest
Amount at 3.5%	2800	0.035	0.035(2800)
Amount at 4.2%	x	0.042	$0.042x$

• The sum of the interest earned is \$329.

Solution $0.035(2800) + 0.042x = 329$
$98 + 0.042x = 329$
$0.042x = 231$
$x = 5500$
The amount invested at 4.2% is \$5500.

59. **Strategy** • Amount invested at 4%: 6000
Amount invested at 6.5%: x
Total amount invested at 5%: y

	Principal	Rate	Interest
Amount at 4%	6000	0.04	0.04(6000)
Amount at 6.5%	x	0.065	$0.065x$
Total invested	y	0.05	$0.05y$

• The total amount invested is y.
$y = 6000 + x$
The total interest earned is equal to 5% of the total investment.

Solution (1) $y = 6000 + x$
(2) $0.04(6000) + 0.065x = 0.05y$
Substitute $6000 + x$ for y in Equation (2).
$0.04(6000) + 0.065x = 0.05(6000 + x)$
$240 + 0.065x = 300 + 0.05x$
$240 + 0.015x = 300$
$0.015x = 60$
$x = 4000$
The amount invested at 6.5% must be \$4000.

61. **Strategy** • Amount invested at 3.5%: x
Amount invested at 4.5%: y

	Principal	Rate	Interest
Amount at 3.5%	x	0.035	$0.035x$
Amount at 4.5%	y	0.045	$0.045y$

• The total amount invested is \$42,000.
$x + y = 42{,}000$
The interest earned from the 3.5% investment is equal to the interest earned from the 4.5% investment.

Solution (1) $x + y = 42{,}000$
(2) $0.035x = 0.045y$
Solve for y in Equation (1) and substitute for y in Equation (2).
$y = 42{,}000 - x$
$0.035x = 0.045(42{,}000 - x)$
$0.035x = 1890 - 0.045x$
$0.080x = 1890$
$x = 23{,}625$
$42{,}000 - x = 42{,}000 - 23{,}625 = 18{,}375$
The amount invested at 3.5% is \$23,625. The amount invested at 4.5% is \$18,375.

63. **Strategy** • Amount invested at 4.5%: x
Amount invested at 8%: y

	Principal	Rate	Interest
Amount at 4.5%	x	0.045	$0.045x$
Amount at 8%	y	0.08	$0.08y$

• The total amount invested is \$16,000.
$x + y = 16{,}000$
The total interest earned is \$1070.

Solution (1) $x + y = 16{,}000$
(2) $0.045x + 0.08y = 1070$
Solve Equation (1) for y and substitute for y in Equation (2).
$y = 16{,}000 - x$
$0.045x + 0.08(16{,}000 - x) = 1070$
$0.045x + 1280 - 0.08x = 1070$
$-0.035x + 1280 = 1070$
$-0.035x = -210$
$x = 6000$
There was \$6000 invested in the mutual bond fund.

Applying the Concepts

65.

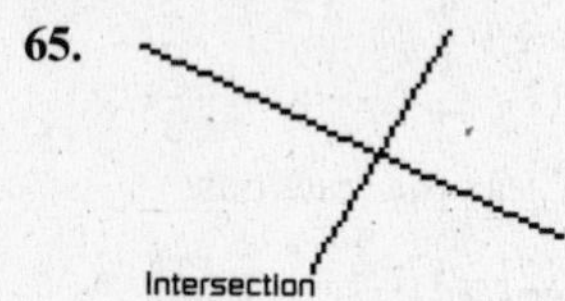

The solution is (1.20, 1.40).

67.

Intersection
X=.54024217 Y=1.0305542

The solution is (0.54, 1.03).

Section 4.2

Objective A Exercises

1. (1) $x - y = 5$
(2) $x + y = 7$
Eliminate y. Add the equations.
$2x = 12$
$x = 6$
Replace x in Equation (1).
$x - y = 5$
$6 - y = 5$
$-y = -1$
$y = 1$
The solution is (6, 1).

3. (1) $3x + y = 4$
(2) $x + y = 2$
Eliminate y.
$3x + y = 4$
$-1(x + y) = -1(2)$

$3x + y = 4$
$-x - y = -2$
Add the equations.
$2x = 2$
$x = 1$
Replace x in Equation (2).
$x + y = 2$
$1 + y = 2$
$y = 1$
The solution is (1, 1).

5. (1) $3x + y = 7$
(2) $x + 2y = 4$
Eliminate y.
$-2(3x + y) = -2(7)$
$x + 2y = 4$

$-6x - 2y = -14$
$x + 2y = 4$
Add the equations.
$-5x = -10$
$x = 2$
Replace x in Equation (2).
$x + 2y = 4$
$2 + 2y = 4$
$2y = 2$
$y = 1$
The solution is (2, 1).

7. (1) $2x + 3y = -1$
(2) $x + 5y = 3$
Eliminate x.
$2x + 3y = -1$
$-2(x + 5y) = -2(3)$

$2x + 3y = -1$
$-2x - 10y = -6$
Add the equations.
$-7y = -7$
$y = 1$
Replace y in Equation (2).
$x + 5y = 3$
$x + 5(1) = 3$
$x + 5 = 3$
$x = -2$
The solution is (−2, 1).

9. (1) $3x - y = 4$
(2) $6x - 2y = 8$
Eliminate y.
$-2(3x - y) = -2(4)$
$6x - 2y = 8$

$-6x + 2y = -8$
$6x - 2y = 8$
Add the equations.
$0 = 0$
This is a true equation. The equations are dependent. The solutions are the ordered pairs $(x, 3x - 4)$.

11. (1) $2x + 5y = 9$
(2) $4x - 7y = -16$
Eliminate x.
$-2(2x + 5y) = -2(9)$
$4x - 7y = -16$

$-4x - 10y = -18$
$4x - 7y = -16$
Add the equations.
$-17y = -34$
$y = 2$
Replace y in Equation (1).
$2x + 5y = 9$
$2x + 5(2) = 9$
$2x + 10 = 9$
$2x = -1$
$x = -\frac{1}{2}$
The solution is $\left(-\frac{1}{2}, 2\right)$.

13. (1) $4x - 6y = 5$
(2) $2x - 3y = 7$
Eliminate y.
$4x - 6y = 5$
$-2(2x - 3y) = -2(7)$

$4x - 6y = 5$
$-4x + 6y = -14$
Add the equations.
$0 = -9$
This is not a true equation. The system of equations is inconsistent and therefore has no solution.

15. (1) $3x - 5y = 7$
(2) $x - 2y = 3$
Eliminate x.
$3x - 5y = 7$
$-3(x - 2y) = -3(3)$

$3x - 5y = 7$
$-3x + 6y = -9$
Add the equations.
$y = -2$
Replace y in Equation (2).
$x - 2y = 3$
$x - 2(-2) = 3$
$x + 4 = 3$
$x = -1$
The solution is (−1, −2).

17. (1) $x + 3y = 7$
(2) $-2x + 3y = 22$
Eliminate y.
$-1(x + 3y) = -1(7)$
$-2x + 3y = 22$

$-x - 3y = -7$
$-2x + 3y = 22$
Add the equations.
$-3x = 15$
$x = -5$
Replace x in Equation (1).
$x + 3y = 7$
$-5 + 3y = 7$
$3y = 12$
$y = 4$
The solution is $(-5, 4)$.

19. (1) $3x + 2y = 16$
(2) $2x - 3y = -11$
Eliminate y.
$3(3x + 2y) = 3(16)$
$2(2x - 3y) = 2(-11)$

$9x + 6y = 48$
$4x - 6y = -22$
Add the equations.
$13x = 26$
$x = 2$
Replace x in Equation (1).
$3x + 2y = 16$
$3(2) + 2y = 16$
$6 + 2y = 16$
$2y = 10$
$y = 5$
The solution is $(2, 5)$.

21. (1) $4x + 4y = 5$
(2) $2x - 8y = -5$
Eliminate y.
$2(4x + 4y) = 2(5)$
$2x - 8y = -5$

$8x + 8y = 10$
$2x - 8y = -5$
Add the equations.
$10x = 5$
$x = \frac{1}{2}$
Replace x in Equation (1).
$4x + 4y = 5$
$4\left(\frac{1}{2}\right) + 4y = 5$
$2 + 4y = 5$
$4y = 3$
$y = \frac{3}{4}$
The solution is $\left(\frac{1}{2}, \frac{3}{4}\right)$.

23. (1) $5x + 4y = 0$
(2) $3x + 7y = 0$
Eliminate x.
$-3(5x + 4y) = -3(0)$
$5(3x + 7y) = 5(0)$

$-15x - 12y = 0$
$15x + 35y = 0$
Add the equations.
$23y = 0$
$y = 0$
Replace y in Equation (1).
$5x + 4y = 0$
$5x + 4(0) = 0$
$5x = 0$
$x = 0$
The solution is $(0, 0)$.

25. (1) $5x + 2y = 1$
(2) $2x + 3y = 7$
Eliminate y.
$-3(5x + 2y) = -3(1)$
$2(2x + 3y) = 2(7)$

$-15x - 6y = -3$
$4x + 6y = 14$
Add the equations.
$-11x = 11$
$x = -1$
Replace x in Equation (1).
$5x + 2y = 1$
$5(-1) + 2y = 1$
$-5 + 2y = 1$
$2y = 6$
$y = 3$
The solution is $(-1, 3)$.

27. (1) $3x - 6y = 6$
(2) $9x - 3y = 8$
Eliminate y.
$3x - 6y = 6$
$-2(9x - 3y) = -2(8)$

$3x - 6y = 6$
$-18x + 6y = -16$
Add the equations.
$-15x = -10$
$x = \frac{2}{3}$
Replace x in Equation (1).
$3x - 6y = 6$
$3\left(\frac{2}{3}\right) - 6y = 6$
$2 - 6y = 6$
$-6y = 4$
$y = -\frac{2}{3}$
The solution is $\left(\frac{2}{3}, -\frac{2}{3}\right)$.

29. (1) $\frac{3}{4}x + \frac{1}{3}y = -\frac{1}{2}$
(2) $\frac{1}{2}x - \frac{5}{6}y = -\frac{7}{2}$
Clear the fractions.
$12\left(\frac{3}{4}x + \frac{1}{3}y\right) = 12\left(-\frac{1}{2}\right)$
$6\left(\frac{1}{2}x - \frac{5}{6}y\right) = 6\left(-\frac{7}{2}\right)$

$9x + 4y = -6$
$3x - 5y = -21$
Eliminate x.
$9x + 4y = -6$
$-3(3x - 5y) = -3(-21)$

$9x + 4y = -6$
$-9x + 15y = 63$
Add the equations.
$19y = 57$
$y = 3$
Replace y in Equation (1).
$\frac{3}{4}x + \frac{1}{3}y = -\frac{1}{2}$
$\frac{3}{4}x + \frac{1}{3}(3) = -\frac{1}{2}$
$\frac{3}{4}x + 1 = -\frac{1}{2}$
$\frac{3}{4}x = -\frac{3}{2}$
$x = -2$
The solution is $(-2, 3)$.

31. (1) $\frac{5x}{6}+\frac{y}{3}=\frac{4}{3}$

(2) $\frac{2x}{3}-\frac{y}{2}=\frac{11}{6}$

Clear the fractions.

$6\left(\frac{5x}{6}+\frac{y}{3}\right)=6\left(\frac{4}{3}\right)$

$6\left(\frac{2x}{3}-\frac{y}{2}\right)=6\left(\frac{11}{6}\right)$

$5x+2y=8$
$4x-3y=11$

Eliminate y.

$3(5x+2y)=3(8)$
$2(4x-3y)=2(11)$

$15x+6y=24$
$8x-6y=22$

Add the equations.

$23x=46$
$x=2$

Replace x in Equation (1).

$\frac{5}{6}x+\frac{1}{3}y=\frac{4}{3}$

$\frac{5}{6}(2)+\frac{1}{3}y=\frac{4}{3}$

$\frac{5}{3}+\frac{1}{3}y=\frac{4}{3}$

$\frac{1}{3}y=-\frac{1}{3}$

$y=-1$

The solution is (2, −1).

33. (1) $\frac{2x}{5}-\frac{y}{2}=\frac{13}{2}$

(2) $\frac{3x}{4}-\frac{y}{5}=\frac{17}{2}$

Clear the fractions.

$10\left(\frac{2x}{5}-\frac{y}{2}\right)=10\left(\frac{13}{2}\right)$

$20\left(\frac{3x}{4}-\frac{y}{5}\right)=20\left(\frac{17}{2}\right)$

$4x-5y=65$
$15x-4y=170$

Eliminate y.

$-4(4x-5y)=-4(65)$
$5(15x-4y)=5(170)$

$-16x+20y=-260$
$75x-20y=850$

Add the equations.

$59x=590$
$x=10$

Replace x in Equation (1).

$\frac{2}{5}x-\frac{1}{2}y=\frac{13}{2}$

$\frac{2}{5}(10)-\frac{1}{2}y=\frac{13}{2}$

$4-\frac{1}{2}y=\frac{13}{2}$

$-\frac{1}{2}y=\frac{5}{2}$

$y=-5$

The solution is (10, −5).

35. (1) $\frac{3x}{2}-\frac{y}{4}=-\frac{11}{12}$

(2) $\frac{x}{3}-y=-\frac{5}{6}$

Clear the fractions.

$12\left(\frac{3x}{2}-\frac{y}{4}\right)=12\left(-\frac{11}{12}\right)$

$6\left(\frac{x}{3}-y\right)=6\left(-\frac{5}{6}\right)$

$18x-3y=-11$
$2x-6y=-5$

Eliminate y.

$-2(18x-3y)=-2(-11)$
$2x-6y=-5$

$-36x+6y=22$
$2x-6y=-5$

Add the equations.

$-34x=17$

$x=-\frac{1}{2}$

Replace x in Equation (2).

$\frac{x}{3}-y=-\frac{5}{6}$

$\frac{1}{3}\left(-\frac{1}{2}\right)-y=-\frac{5}{6}$

$-\frac{1}{6}-y=-\frac{5}{6}$

$-y=-\frac{4}{6}$

$y=\frac{2}{3}$

The solution is $\left(-\frac{1}{2},\frac{2}{3}\right)$.

37. (1) $4x - 5y = 3y + 4$
(2) $2x + 3y = 2x + 1$
Write the equations in the form $Ax + By = C$.
$4x - 5y = 3y + 4$
$4x - 8y = 4$

$2x + 3y = 2x + 1$
$3y = 1$
Solve the system.
$4x - 8y = 4$
$3y = 1$
Solve the equation $3y = 1$ for y.
$3y = 1$
$y = \frac{1}{3}$
Replace y in the equation $4x - 8y = 4$.
$4x - 8y = 4$
$4x - 8\left(\frac{1}{3}\right) = 4$
$4x - \frac{8}{3} = 4$
$4x = \frac{20}{3}$
$x = \frac{5}{3}$
The solution is $\left(\frac{5}{3}, \frac{1}{3}\right)$.

39. (1) $2x + 5y = 5x + 1$
(2) $3x - 2y = 3y + 3$
Write the equations in the form $Ax + By = C$.
$2x + 5y = 5x + 1$
$-3x + 5y = 1$

$3x - 2y = 3y + 3$
$3x - 5y = 3$
Solve the system.
$-3x + 5y = 1$
$3x - 5y = 3$
Add the equations.
$0 = 4$
This is not a true equation. The system of equations is inconsistent and therefore has no solution.

41. (1) $5x + 2y = 2x + 1$
(2) $2x - 3y = 3x + 2$
Write the equations in the form $Ax + By = C$.
$5x + 2y = 2x + 1$
$3x + 2y = 1$

$2x - 3y = 3x + 2$
$-x - 3y = 2$
Solve the system.
$3x + 2y = 1$
$-x - 3y = 2$
Eliminate x.
$3x + 2y = 1$
$3(-x - 3y) = 3(2)$

$3x + 2y = 1$
$-3x - 9y = 6$
Add the equations.
$-7y = 7$
$y = -1$
Replace y in the equation $-x - 3y = 2$.
$-x - 3y = 2$
$-x - 3(-1) = 2$
$-x + 3 = 2$
$-x = -1$
$x = 1$
The solution is $(1, -1)$.

Objective B Exercises

43. (1) $x + 2y - z = 1$
(2) $2x - y + z = 6$
(3) $x + 3y - z = 2$

Eliminate z. Add Equations (1) and (2).

$x + 2y - z = 1$
$2x - y + z = 6$

(4) $3x + y = 7$

Add Equations (2) and (3).

$2x - y + z = 6$
$x + 3y - z = 2$

(5) $3x + 2y = 8$

Multiply Equation (4) by -1 and add to Equation (5).

$-1(3x + y) = -1(7)$
$3x + 2y = 8$

$-3x - y = -7$
$3x + 2y = 8$
$y = 1$

Replace y by 1 in Equation (4).

$3x + y = 7$
$3x + 1 = 7$
$3x = 6$
$x = 2$

Replace x by 2 and y by 1 in Equation (1).

$x + 2y - z = 1$
$2 + 2(1) - z = 1$
$2 + 2 - z = 1$
$4 - z = 1$
$-z = -3$
$z = 3$

The solution is (2, 1, 3).

45. (1) $2x - y + 2z = 7$
(2) $x + y + z = 2$
(3) $3x - y + z = 6$

Eliminate y. Add Equations (1) and (2).

$2x - y + 2z = 7$
$x + y + z = 2$

$3x + 3z = 9$

Multiply both sides of the equation by $\frac{1}{3}$.

(4) $x + z = 3$

Add Equations (2) and (3).

$x + y + z = 2$
$3x - y + z = 6$

$4x + 2z = 8$

Multiply both sides of the equation by $\frac{1}{2}$.

(5) $2x + z = 4$

Multiply Equation (4) by -1 and add to Equation (5).

$-1(x + z) = -1(3)$
$2x + z = 4$

$-x - z = -3$
$2x + z = 4$
$x = 1$

Replace x by 1 in Equation (4).

$x + z = 3$
$1 + z = 3$
$z = 2$

Replace x by 1 and z by 2 in Equation (2).

$x + y + z = 2$
$1 + y + 2 = 2$
$3 + y = 2$
$y = -1$

The solution is (1, −1, 2).

47. (1) $3x + y = 5$
(2) $3y - z = 2$
(3) $x + z = 5$
Eliminate z. Add Equations (2) and (3).
$3y - z = 2$
$x + z = 5$

(4) $x + 3y = 7$
Multiply Equation (4) by -3 and add to Equation (1).
$-3(x + 3y) = -3(7)$
$3x + y = 5$

$-3x - 9y = -21$
$3x + y = 5$

$-8y = -16$
$y = 2$
Replace y by 2 in Equation (1).
$3x + y = 5$
$3x + 2 = 5$
$3x = 3$
$x = 1$
Replace x by 1 in Equation (3).
$x + z = 5$
$1 + z = 5$
$z = 4$
The solution is (1, 2, 4).

49. (1) $x - y + z = 1$
(2) $2x + 3y - z = 3$
(3) $-x + 2y - 4z = 4$
Eliminate z. Add Equations (1) and (2).
$x - y + z = 1$
$2x + 3y - z = 3$

(4) $3x + 2y = 4$
Multiply Equation (1) by 4 and add to Equation (3).
$4(x - y + z) = 4(1)$
$-x + 2y - 4z = 4$

$4x - 4y + 4z = 4$
$-x + 2y - 4z = 4$

(5) $3x - 2y = 8$
Multiply Equation (4) by -1 and add to Equation (5).
$-1(3x + 2y) = -1(4)$
$3x - 2y = 8$

$-3x - 2y = -4$
$3x - 2y = 8$

$-4y = 4$
$y = -1$
Replace y by -1 in Equation (5).
$3x - 2y = 8$
$3x - 2(-1) = 8$
$3x + 2 = 8$
$3x = 6$
$x = 2$
Replace x by 2 and y by -1 in Equation (1).
$x - y + z = 1$
$2 - (-1) + z = 1$
$3 + z = 1$
$z = -2$
The solution is $(2, -1, -2)$.

51. (1) $2x + 3z = 5$
(2) $3y + 2z = 3$
(3) $3x + 4y = -10$
Eliminate z. Multiply Equation (1) by -2 and Equation (2) by 3.
Then add the equations.

$$-2(2x + 3z) = -2(5)$$
$$3(3y + 2z) = 3(3)$$

$$-4x - 6z = -10$$
$$9y + 6z = 9$$

(4) $-4x + 9y = -1$
Multiply Equation (3) by 4 and Equation (4) by 3.
Then add the equations.

$$4(3x + 4y) = 4(-10)$$
$$3(-4x + 9y) = 3(-1)$$

$$12x + 16y = -40$$
$$-12x + 27y = -3$$

$$43y = -43$$
$$y = -1$$

Replace y by -1 in Equation (3).

$$3x + 4y = -10$$
$$3x + 4(-1) = -10$$
$$3x - 4 = -10$$
$$3x = -6$$
$$x = -2$$

Replace x by -2 in Equation (1).

$$2x + 3z = 5$$
$$2(-2) + 3z = 5$$
$$-4 + 3z = 5$$
$$3z = 9$$
$$z = 3$$

The solution is $(-2, -1, 3)$.

53. (1) $2x + 4y - 2z = 3$
(2) $x + 3y + 4z = 1$
(3) $x + 2y - z = 4$
Eliminate x. Multiply Equation (2) by -2 and add to Equation (1).

$$2x + 4y - 2z = 3$$
$$-2(x + 3y + 4z) = -2(1)$$

$$2x + 4y - 2z = 3$$
$$-2x - 6y - 8z = -2$$

(4) $-2y - 10z = 1$
Multiply Equation (2) by -1 and add to Equation (3).

$$-1(x + 3y + 4z) = -1(1)$$
$$x + 2y - z = 4$$

$$-x - 3y - 4z = -1$$
$$x + 2y - z = 4$$

(5) $-y - 5z = 3$
Multiply Equation (5) by -2 and add to Equation (4).

$$-2(-y - 5z) = -2(3)$$
$$-2y - 10z = 1$$

$$2y + 10z = -6$$
$$-2y - 10z = 1$$

$$0 = -5$$

This is not a true equation. The system of equations is inconsistent and therefore has no solution.

55. (1) $2x+y-z=5$
(2) $x+3y+z=14$
(3) $3x-y+2z=1$
Eliminate z. Add Equations (1) and (2).
$2x+y-z=5$
$x+3y+z=14$

(4) $3x+4y=19$
Multiply Equation (1) by 2 and add to Equation (3).
$2(2x+y-z)=2(5)$
$3x-y+2z=1$

$4x+2y-2z=10$
$3x-y+2z=1$

(5) $7x+y=11$
Multiply Equation (5) by -4 and add to Equation (4).
$-4(7x+y)=-4(11)$
$3x+4y=19$

$-28x-4y=-44$
$3x+4y=19$

$-25x=-25$
$x=1$
Replace x by 1 in Equation (5).
$7x+y=11$
$7(1)+y=11$
$7+y=11$
$y=4$
Replace x by 1 and y by 4 in Equation (1).
$2x+y-z=5$
$2(1)+4-z=5$
$2+4-z=5$
$6-z=5$
$-z=-1$
$z=1$
The solution is (1, 4, 1).

57. (1) $3x+y-2z=2$
(2) $x+2y+3z=13$
(3) $2x-2y+5z=6$
Eliminate y. Multiply Equation (1) by -2 and add to Equation (2).
$-2(3x+y-2z)=-2(2)$
$x+2y+3z=13$

$-6x-2y+4z=-4$
$x+2y+3z=13$

(4) $-5x+7z=9$
Add Equations (2) and (3).
$x+2y+3z=13$
$2x-2y+5z=6$

(5) $3x+8z=19$
Multiply Equation (4) by 3 and Equation (5) by 5. Then add the equations.
$3(-5x+7z)=3(9)$
$5(3x+8z)=5(19)$

$-15x+21z=27$
$15x+40z=95$

$61z=122$
$z=2$
Replace z by 2 in Equation (5).
$3x+8z=19$
$3x+8(2)=19$
$3x+16=19$
$3x=3$
$x=1$
Replace x by 1 and z by 2 in Equation (1).
$3x+y-2z=2$
$3(1)+y-2(2)=2$
$3+y-4=2$
$y-1=2$
$y=3$
The solution is (1, 3, 2).

59. (1) $2x - y + z = 6$
(2) $3x + 2y + z = 4$
(3) $x - 2y + 3z = 12$
Eliminate y. Multiply Equation (1) by 2 and add to Equation (2).
$2(2x - y + z) = 2(6)$
$3x + 2y + z = 4$

$4x - 2y + 2z = 12$
$3x + 2y + z = 4$

(4) $7x + 3z = 16$
Add Equations (2) and (3).
$3x + 2y + z = 4$
$x - 2y + 3z = 12$

$4x + 4z = 16$
Multiply each side of the equation by $\frac{1}{4}$.
(5) $x + z = 4$
Multiply Equation (5) by -3 and add to Equation (4).
$7x + 3z = 16$
$-3(x + z) = -3(4)$

$7x + 3z = 16$
$-3x - 3z = -12$

$4x = 4$
$x = 1$
Replace x by 1 in Equation (5).
$x + z = 4$
$1 + z = 4$
$z = 3$
Replace x by 1 and z by 3 in Equation (1).
$2x - y + z = 6$
$2(1) - y + 3 = 6$
$2 - y + 3 = 6$
$5 - y = 6$
$-y = 1$
$y = -1$
The solution is (1, −1, 3).

61. (1) $3x - 2y + 3z = -4$
(2) $2x + y - 3z = 2$
(3) $3x + 4y + 5z = 8$
Eliminate y. Multiply Equation (2) by 2 and add to Equation (1).
$3x - 2y + 3z = -4$
$2(2x + y - 3z) = 2(2)$

$3x - 2y + 3z = -4$
$4x + 2y - 6z = 4$

(4) $7x - 3z = 0$
Multiply Equation (2) by -4 and add to Equation (3).
$-4(2x + y - 3z) = -4(2)$
$3x + 4y + 5z = 8$

$-8x - 4y + 12z = -8$
$3x + 4y + 5z = 8$

(5) $-5x + 17z = 0$
Multiply Equation (4) by 5 and Equation (5) by 7.
$5(7x - 3z) = 5(0)$
$7(-5x + 17z) = 7(0)$

$35x - 15z = 0$
$-35x + 119z = 0$

$104z = 0$
$z = 0$
Replace z by 0 in Equation (4).
$7x - 3z = 0$
$7x - 3(0) = 0$
$7x = 0$
$x = 0$
Replace x by 0 and z by 0 in Equation (2).
$2x + y - 3z = 2$
$2(0) + y - 3(0) = 2$
$y = 2$
The solution is (0, 2, 0).

63. (1) $3x - y + 2z = 2$
(2) $4x + 2y - 7z = 0$
(3) $2x + 3y - 5z = 7$
Eliminate y. Multiply Equation (1) by 2 and add to Equation (2).
$2(3x - y + 2z) = 2(2)$
$4x + 2y - 7z = 0$

$6x - 2y + 4z = 4$
$4x + 2y - 7z = 0$

(4) $10x - 3z = 4$
Multiply Equation (1) by 3 and add to Equation (3).
$3(3x - y + 2z) = 3(2)$
$2x + 3y - 5z = 7$

$9x - 3y + 6z = 6$
$2x + 3y - 5z = 7$

(5) $11x + z = 13$
Multiply Equation (5) by 3 and add to Equation (4).
$10x - 3z = 4$
$3(11x + z) = 3(13)$

$10x - 3z = 4$
$33x + 3z = 39$

$43x = 43$
$x = 1$
Replace x by 1 in Equation (4).
$10x - 3z = 4$
$10(1) - 3z = 4$
$10 - 3z = 4$
$-3z = -6$
$z = 2$
Replace x by 1 and z by 2 in Equation (1).
$3x - y + 2z = 2$
$3(1) - y + 2(2) = 2$
$3 - y + 4 = 2$
$7 - y = 2$
$-y = -5$
$y = 5$
The solution is (1, 5, 2).

65. (1) $2x - 3y + 7z = 0$
(2) $x + 4y - 4z = -2$
(3) $3x + 2y + 5z = 1$
Eliminate x. Multiply Equation (2) by -2 and add to Equation (1).
$2x - 3y + 7z = 0$
$-2(x + 4y - 4z) = -2(-2)$

$2x - 3y + 7z = 0$
$-2x - 8y + 8z = 4$

(4) $-11y + 15z = 4$
Multiply Equation (2) by -3 and add to Equation (3).
$-3(x + 4y - 4z) = -3(-2)$
$3x + 2y + 5z = 1$

$-3x - 12y + 12z = 6$
$3x + 2y + 5z = 1$

(5) $-10y + 17z = 7$
Multiply Equation (4) by -10 and Equation (5) by 11. Then add the equations.
$-10(-11y + 15z) = -10(4)$
$11(-10y + 17z) = 11(7)$

$110y - 150z = -40$
$-110y + 187z = 77$

$37z = 37$
$z = 1$
Replace z by 1 in Equation (4).
$-11y + 15z = 4$
$-11y + 15(1) = 4$
$-11y + 15 = 4$
$-11y = -11$
$y = 1$
Replace y by 1 and z by 1 in Equation (1).
$2x - 3y + 7z = 0$
$2x - 3(1) + 7(1) = 0$
$2x - 3 + 7 = 0$
$2x + 4 = 0$
$2x = -4$
$x = -2$
The solution is $(-2, 1, 1)$.

Applying the Concepts

67a. The system of equations has no solutions; it is inconsistent.

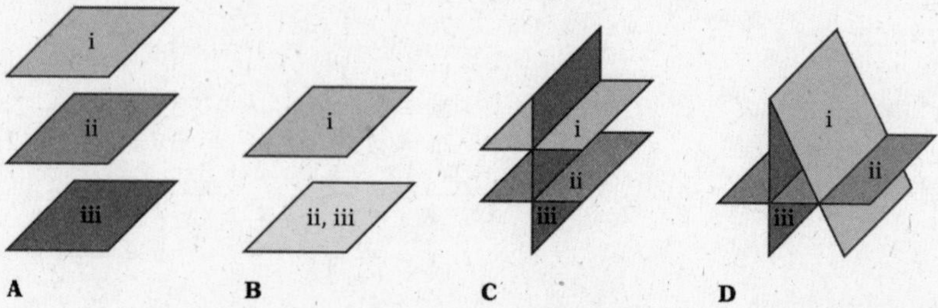

Graphs of Inconsistent Systems of Equations

b. The system has exactly one solution. This is an independent system whose solution is a point in space.

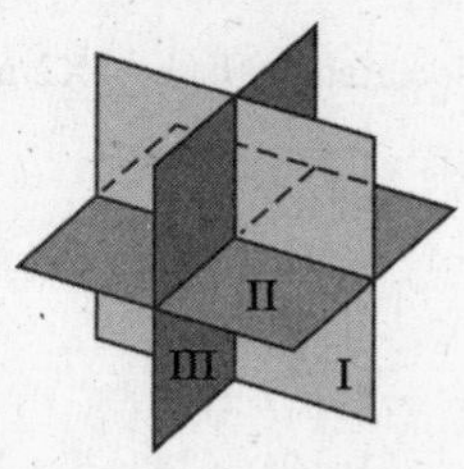

An Independent System of Equations

c. The system has infinitely many solutions. It is a dependent system.

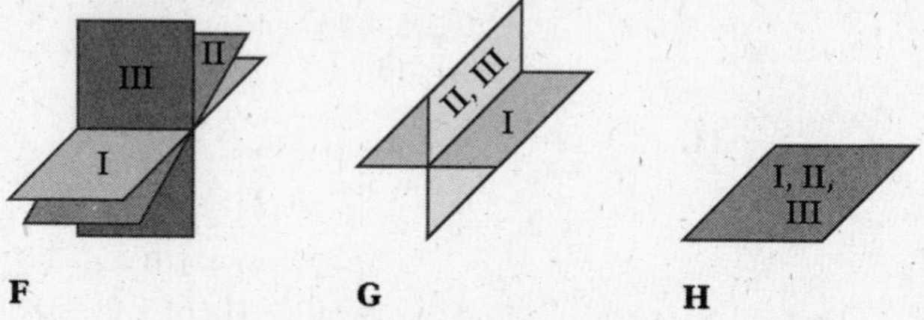

Dependent Systems of Equations

69. (1) $\frac{1}{x}-\frac{2}{y}=3$

(2) $\frac{2}{x}+\frac{3}{y}=-1$

Clear the fractions.

$$xy\left(\frac{1}{x}-\frac{2}{y}\right)=xy\cdot 3$$

$$xy\left(\frac{2}{x}+\frac{3}{y}\right)=xy\cdot(-1)$$

$$y-2x=3xy$$
$$2y+3x=-xy$$

Eliminate y. Multiply $y-2x=3xy$ by -2.

$$-2y+4x=-6xy$$
$$2y+3x=-xy$$
$$7x=-7xy$$
$$y=-1$$

Substitute into Equation (2).

$$\frac{2}{x}+\frac{3}{y}=-1$$
$$\frac{2}{x}+\frac{3}{-1}=-1$$
$$\frac{2}{x}=2$$
$$x=1$$

The solution is $(1, -1)$.

71. (1) $\frac{3}{x}+\frac{2}{y}=1$

(2) $\frac{2}{x}+\frac{4}{y}=-2$

Clear fractions.

$$xy\left(\frac{3}{x}+\frac{2}{y}\right)=xy\cdot 1$$

$$xy\left(\frac{2}{x}+\frac{4}{y}\right)=xy\cdot(-2)$$

$$3y+2x=xy$$
$$2y+4x=-2xy$$

Eliminate x. Multiply $3y+2x=xy$ by -2.

$$-6y-4x=-2xy$$
$$2y+4x=-2xy$$
$$-4y=-4xy$$
$$x=1$$

Substitute x into Equation (2).

$$\frac{2}{x}+\frac{4}{y}=-2$$
$$2+\frac{4}{y}=-2$$
$$\frac{4}{y}=-4$$
$$y=-1$$

The solution is $(1, -1)$.

73. Solve each equation for y.

$$\frac{1}{x}-\frac{2}{y}=3 \qquad \frac{2}{x}+\frac{3}{y}=-1$$
$$-\frac{2}{y}=3-\frac{1}{x} \qquad \frac{3}{y}=-1-\frac{2}{x}$$
$$-2x=3xy-y \qquad 3x=-xy-2y$$
$$-2x=y(3x-1) \qquad 3x=y(-x-2)$$
$$\frac{-2x}{3x-1}=y \qquad \frac{3x}{-x-2}=y$$
$$y=\frac{2x}{1-3x} \qquad y=\frac{-3x}{x+2}$$

Section 4.3

Objective A Exercises

1. The determinant associated with the 2X2 matrix $\begin{bmatrix} a & b \\ c & d \end{bmatrix}$ is $\begin{vmatrix} a & b \\ c & d \end{vmatrix}$. Its value is $ad-bc$.

3. $\begin{vmatrix} 2 & -1 \\ 3 & 4 \end{vmatrix}=2(4)-3(-1)=8+3=11$

5. $\begin{vmatrix} 6 & -2 \\ -3 & 4 \end{vmatrix}=6(4)-(-3)(-2)=24-6=18$

7. $\begin{vmatrix} 3 & 6 \\ 2 & 4 \end{vmatrix}=3(4)-2(6)=12-12=0$

9.
$$\begin{vmatrix} 1 & -1 & 2 \\ 3 & 2 & 1 \\ 1 & 0 & 4 \end{vmatrix}=1\begin{vmatrix} 2 & 1 \\ 0 & 4 \end{vmatrix}+1\begin{vmatrix} 3 & 1 \\ 1 & 4 \end{vmatrix}+2\begin{vmatrix} 3 & 2 \\ 1 & 0 \end{vmatrix}$$
$$=1(8-0)+1(12-1)+2(0-2)$$
$$=8+11-4$$
$$=15$$

11.
$$\begin{vmatrix} 3 & -1 & 2 \\ 0 & 1 & 2 \\ 3 & 2 & -2 \end{vmatrix}=3\begin{vmatrix} 1 & 2 \\ 2 & -2 \end{vmatrix}+1\begin{vmatrix} 0 & 2 \\ 3 & -2 \end{vmatrix}+2\begin{vmatrix} 0 & 1 \\ 3 & 2 \end{vmatrix}$$
$$=3(-2-4)+1(0-6)+2(0-3)$$
$$=3(-6)+1(-6)+2(-3)$$
$$=-18-6-6$$
$$=-30$$

13.
$$\begin{vmatrix} 4 & 2 & 6 \\ -2 & 1 & 1 \\ 2 & 1 & 3 \end{vmatrix}=4\begin{vmatrix} 1 & 1 \\ 1 & 3 \end{vmatrix}-2\begin{vmatrix} -2 & 1 \\ 2 & 3 \end{vmatrix}+6\begin{vmatrix} -2 & 1 \\ 2 & 1 \end{vmatrix}$$
$$=4(3-1)-2(-6-2)+6(-2-2)$$
$$=4(2)-2(-8)+6(-4)$$
$$=8+16-24$$
$$=0$$

Objective B Exercises

15. $2x - 5y = 26$
$5x + 3y = 3$

$D = \begin{vmatrix} 2 & -5 \\ 5 & 3 \end{vmatrix} = 31,\ D_x = \begin{vmatrix} 26 & -5 \\ 3 & 3 \end{vmatrix} = 93,$

$D_y = \begin{vmatrix} 2 & 26 \\ 5 & 3 \end{vmatrix} = -124$

$x = \frac{D_x}{D} = \frac{93}{31} = 3,\ y = \frac{D_y}{D} = \frac{-124}{31} = -4$

The solution is (3, −4).

17. $x - 4y = 8$
$3x + 7y = 5$

$D = \begin{vmatrix} 1 & -4 \\ 3 & 7 \end{vmatrix} = 19,\ D_x = \begin{vmatrix} 8 & -4 \\ 5 & 7 \end{vmatrix} = 76,$

$D_y = \begin{vmatrix} 1 & 8 \\ 3 & 5 \end{vmatrix} = -19$

$x = \frac{D_x}{D} = \frac{76}{19} = 4,\ y = \frac{D_y}{D} = \frac{-19}{19} = -1$

The solution is (4, −1).

19. $2x + 3y = 4$
$6x - 12y = -5$

$D = \begin{vmatrix} 2 & 3 \\ 6 & -12 \end{vmatrix} = -42,\ D_x = \begin{vmatrix} 4 & 3 \\ -5 & -12 \end{vmatrix} = -33,$

$D_y = \begin{vmatrix} 2 & 4 \\ 6 & -5 \end{vmatrix} = -34$

$x = \frac{D_x}{D} = \frac{-33}{-42} = \frac{11}{14},\ y = \frac{D_y}{D} = \frac{-34}{-42} = \frac{17}{21}$

The solution is $\left(\frac{11}{14}, \frac{17}{21}\right)$.

21. $2x + 5y = 6$
$6x - 2y = 1$

$D = \begin{vmatrix} 2 & 5 \\ 6 & -2 \end{vmatrix} = -34,\ D_x = \begin{vmatrix} 6 & 5 \\ 1 & -2 \end{vmatrix} = -17,$

$D_y = \begin{vmatrix} 2 & 6 \\ 6 & 1 \end{vmatrix} = -34$

$x = \frac{D_x}{D} = \frac{-17}{-34} = \frac{1}{2},\ y = \frac{D_y}{D} = \frac{-34}{-34} = 1$

The solution is $\left(\frac{1}{2}, 1\right)$.

23. $-2x + 3y = 7$
$4x - 6y = 9$

$D = \begin{vmatrix} -2 & 3 \\ 4 & -6 \end{vmatrix} = 0$

Since $D = 0$, $\frac{D_x}{D}$ is undefined. Therefore, the system of equations does not have a unique solution. The equations are not independent.

25. $2x - 5y = -2$
$3x - 7y = -3$

$D = \begin{vmatrix} 2 & -5 \\ 3 & -7 \end{vmatrix} = 1,\ D_x = \begin{vmatrix} -2 & -5 \\ -3 & -7 \end{vmatrix} = -1,$

$D_y = \begin{vmatrix} 2 & -2 \\ 3 & -3 \end{vmatrix} = 0$

$x = \frac{D_x}{D} = \frac{-1}{1} = -1,\ y = \frac{D_y}{D} = \frac{0}{1} = 0$

The solution is (−1, 0).

27. $2x - y + 3z = 9$
$x + 4y + 4z = 5$
$3x + 2y + 2z = 5$

$D = \begin{vmatrix} 2 & -1 & 3 \\ 1 & 4 & 4 \\ 3 & 2 & 2 \end{vmatrix} = -40,\ D_x = \begin{vmatrix} 9 & -1 & 3 \\ 5 & 4 & 4 \\ 5 & 2 & 2 \end{vmatrix} = -40,$

$D_y = \begin{vmatrix} 2 & 9 & 3 \\ 1 & 5 & 4 \\ 3 & 5 & 2 \end{vmatrix} = 40,\ D_z = \begin{vmatrix} 2 & -1 & 9 \\ 1 & 4 & 5 \\ 3 & 2 & 5 \end{vmatrix} = -80$

$x = \frac{D_x}{D} = \frac{-40}{-40} = 1,\ y = \frac{D_y}{D} = \frac{40}{-40} = -1$

$z = \frac{D_z}{D} = \frac{-80}{-40} = 2$

The solution is (1, −1, 2).

29. $3x - y + z = 11$
$x + 4y - 2z = -12$
$2x + 2y - z = -3$

$D = \begin{vmatrix} 3 & -1 & 1 \\ 1 & 4 & -2 \\ 2 & 2 & -1 \end{vmatrix} = -3,\ D_x = \begin{vmatrix} 11 & -1 & 1 \\ -12 & 4 & -2 \\ -3 & 2 & -1 \end{vmatrix} = -6,$

$D_y = \begin{vmatrix} 3 & 11 & 1 \\ 1 & -12 & -2 \\ 2 & -3 & -1 \end{vmatrix} = 6,\ D_z = \begin{vmatrix} 3 & -1 & 11 \\ 1 & 4 & -12 \\ 2 & 2 & -3 \end{vmatrix} = -9$

$x = \frac{D_x}{D} = \frac{-6}{-3} = 2,\ y = \frac{D_y}{D} = \frac{6}{-3} = -2$

$z = \frac{D_z}{D} = \frac{-9}{-3} = 3$

The solution is (2, −2, 3).

31. $4x - 2y + 6z = 1$
$3x + 4y + 2z = 1$
$2x - y + 3z = 2$

$$D = \begin{vmatrix} 4 & -2 & 6 \\ 3 & 4 & 2 \\ 2 & -1 & 3 \end{vmatrix} = 0$$

Since $D = 0$, $\dfrac{D_x}{D}$ is undefined. Therefore, the system of equations does not have a unique solution. The equations are not independent.

Applying the Concepts

33a. Sometimes true.

b. Always true.

c. Sometimes true.

35.
$$A = \frac{1}{2}\left\{\begin{vmatrix} 9 & 26 \\ -3 & 6 \end{vmatrix} + \begin{vmatrix} 26 & 18 \\ 6 & 21 \end{vmatrix} + \begin{vmatrix} 18 & 16 \\ 21 & 10 \end{vmatrix} + \begin{vmatrix} 16 & 1 \\ 10 & 11 \end{vmatrix} + \begin{vmatrix} 1 & 9 \\ 11 & -3 \end{vmatrix}\right\}$$
$$= \frac{1}{2}\{[9(6) - (-3)(26)] + [26(21) - 6(18)] + [18(10) - 21(16)] + [16(11) - 10(1)] + [1(-3) - 11(9)]\}$$
$$= \frac{1}{2}(132 + 438 - 156 + 166 - 102)$$
$$= \frac{1}{2}(478)$$
$$= 239 \text{ ft}^2$$

Section 4.4

Objective A Exercises

1. Strategy • Rate of the motorboat in calm water: x
Rate of the current: y

	Rate	Time	Distance
With current	$x+y$	2	$2(x+y)$
Against current	$x-y$	3	$3(x-y)$

• The distance traveled with the current is 36 mi.
The distance traveled against the current is 36 mi.
$2(x+y) = 36$
$3(x-y) = 36$

Solution $2(x+y) = 36$
$3(x-y) = 36$

$\frac{1}{2} \cdot 2(x+y) = \frac{1}{2} \cdot 36$
$\frac{1}{3} \cdot 3(x-y) = \frac{1}{3} \cdot 36$

$x + y = 18$
$x - y = 12$

$2x = 30$
$x = 15$

$x + y = 18$
$15 + y = 18$
$y = 3$

The rate of the motorboat in calm water is 15 mph. The rate of the current is 3 mph.

3. Strategy • Rate of the plane is calm air: p
Rate of the wind: w

	Rate	Time	Distance
With wind	$p+w$	4	$4(p+w)$
Against wind	$p-w$	4	$4(p-w)$

• The distance traveled with the wind is 2200 mi.
The distance traveled against the wind is 1820 mi.
$4(p+w)=2200$
$4(p-w)=1820$

Solution $4(p+w)=2200$
$4(p-w)=1820$

$$\frac{1}{4}\cdot 4(p+w)=\frac{1}{4}\cdot 2200$$
$$\frac{1}{4}\cdot 4(p-w)=\frac{1}{4}\cdot 1820$$

$p+w=550$
$p-w=455$

$2p=1005$
$p=502.5$

$p+w=550$
$502.5+w=550$
$w=47.5$

The rate of the plane in calm air is 502.5 mph. The rate of the wind is 47.5 mph.

5. Strategy • The rate of the team in calm water: x
The rate of the current: y

	Rate	Time	Distance
With current	$x+y$	2	$2(x+y)$
Against current	$x-y$	2	$2(x-y)$

• The distance traveled with the current is 20 km.
The distance traveled against the current is 12 km.
$2(x+y)=20$
$2(x-y)=12$

Solution $2(x+y)=20$
$2(x-y)=12$

$$\frac{1}{2}\cdot 2(x+y)=\frac{1}{2}\cdot 20$$
$$\frac{1}{2}\cdot 2(x-y)=\frac{1}{2}\cdot 12$$

$x+y=10$
$x-y=6$

$2x=16$
$x=8$

$x+y=10$
$8+y=10$
$y=2$

The rate of the team in calm water is 8 km/h. The rate of the current is 2 km/h.

7. Strategy • Rate of the plane in calm air: x
Rate of the wind: y

	Rate	Time	Distance
With wind	$x+y$	4	$4(x+y)$
Against wind	$x-y$	5	$5(x-y)$

• The distance traveled with the wind is 800 mi.
The distance traveled against the wind is 800 mi.
$4(x+y)=800$
$5(x-y)=800$

Solution $4(x+y)=800$
$5(x-y)=800$

$$\frac{1}{4}\cdot 4(x+y)=\frac{1}{4}\cdot 800$$
$$\frac{1}{5}\cdot 5(x-y)=\frac{1}{5}\cdot 800$$

$x+y=200$
$x-y=160$

$2x=360$
$x=180$

$x+y=200$
$180+y=200$
$y=20$

The rate of the plane in calm air is 180 mph. The rate of the wind is 20 mph.

9. Strategy • Rate of the plane in calm air: x
Rate of the wind: y

	Rate	Time	Distance
With wind	$x+y$	5	$5(x+y)$
Against wind	$x-y$	6	$6(x-y)$

• The distance traveled with the wind is 600 mi.
The distance traveled against the wind is 600 mi.
$5(x+y)=600$
$6(x-y)=600$

Solution

$$5(x+y)=600$$
$$6(x-y)=600$$

$$\frac{1}{5}\cdot 5(x+y)=\frac{1}{5}\cdot 600$$
$$\frac{1}{6}\cdot 6(x-y)=\frac{1}{6}\cdot 600$$

$$x+y=120$$
$$x-y=100$$

$$2x=220$$
$$x=110$$

$$x+y=120$$
$$110+y=120$$
$$y=10$$

The rate of the plane in calm air is 110 mph. The rate of the wind is 10 mph.

Objective B Exercises

11. Strategy • Number of nickels in the bank: n
Number of dimes in the bank: d
Coins in the bank now:

Coin	Number	Value	Total Value
Nickel	n	5	$5n$
Dime	d	10	$10d$

Coins in the bank if the nickels were dimes and the dimes were nickels:

Coin	Number	Value	Total Value
Nickel	d	5	$5d$
Dime	n	10	$10n$

• The value of the nickels and dimes in the bank is \$2.50. The value of the nickels and dimes in the bank would be \$3.50.
$5n+10d=250$
$10n+5d=350$

Solution

$$5n+10d=250$$
$$10n+5d=350$$

$$-1(5n+10d)=-1(250)$$
$$2(10n+5d)=2(350)$$

$$-5n-10d=-250$$
$$20n+10d=700$$

$$15n=450$$
$$n=30$$

There are 30 nickels in the bank.

13. Strategy • Cost of redwood: x
Cost of pine: y
First purchase:

	Amount	Rate	Total Value
Redwood	60	x	$60x$
Pine	80	y	$80y$

Second purchase:

	Amount	Rate	Total Value
Redwood	100	x	$100x$
Pine	60	y	$60y$

• The first purchase cost \$286.
The second purchase cost \$396.
$60x+80y=286$
$100x+60y=396$

Solution

$$60x+80y=286$$
$$100x+60y=396$$

$$3(60x+80y)=3(286)$$
$$-4(100x+60y)=-4(396)$$

$$180x+240y=858$$
$$-400x-240y=-1584$$

$$-220x=-726$$
$$x=3.3$$

$$60x+80y=286$$
$$60(3.3)+80y=286$$
$$198+80y=286$$
$$80y=88$$
$$y=1.1$$

The cost of the pine is \$1.10/ft.
The cost of the redwood is \$3.30/ft.

15. Strategy • Cost per yard of nylon carpet: x
Cost per yard of wool carpet: y
First purchase:

	Amount	Rate	Total Cost
Nylon	16	x	$16x$
Wool	20	y	$20y$

Second purchase:

	Amount	Rate	Total Cost
Nylon	18	x	$18x$
Wool	25	y	$25y$

• The first purchase cost \$1840.
The second purchase cost \$2200.

Solution
$$16x + 20y = 1840$$
$$18x + 25y = 2200$$

$$5(16x + 20y) = 5(1840)$$
$$-4(18x + 25y) = -4(2200)$$

$$80x + 100y = 9200$$
$$-72x - 100y = -8800$$

$$8x = 400$$
$$x = 50$$

$$16x + 20y = 1840$$
$$16(50) + 20y = 1840$$
$$800 + 20y = 1840$$
$$20y = 1040$$
$$y = 52$$

The cost of the wool carpet is \$52/yd.

17. Strategy • Number of mountain bikes to be manufactured: m
Number of trail bikes to be manufactured: t
Cost of materials:

Type	Number	Cost	Total Cost
Mountain	m	70	$70m$
Trail	t	50	$50t$

Cost of labor:

Type	Number	Cost	Total Cost
Mountain	m	80	$80m$
Trail	t	40	$40t$

• The company has budgeted \$2500 for materials. The company has budgeted \$2600 for labor.

$$70m + 50t = 2500$$
$$80m + 40t = 2600$$

Solve for m.

Solution
$$70m + 50t = 2500$$
$$80m + 40t = 2600$$

$$4(70m + 50t) = 4(2500)$$
$$-5(80m + 40t) = -5(2600)$$

$$280m + 200t = 10{,}000$$
$$-400m - 200t = -13{,}000$$

$$-120m = -3000$$
$$m = 25$$

The company plans to manufacture 25 mountain bikes during the week.

19. Strategy • Amount of first alloy to be used: x
Amount of second alloy to be used: y
Gold:

	Amount	Percent	Quantity
1st alloy	x	0.10	$0.10x$
2nd alloy	y	0.30	$0.30y$

Lead:

	Amount	Percent	Quantity
1st alloy	x	0.15	$0.15x$
2nd alloy	y	0.40	$0.40y$

• The resulting alloy contains 60 g of gold. The resulting alloy contains 88 g of lead.
$$0.10x + 0.30y = 60$$
$$0.15x + 0.40y = 88$$

Solution
$$0.10x + 0.30y = 60$$
$$0.15x + 0.40y = 88$$

$$3(0.10x + 0.30y) = 3(60)$$
$$-2(0.15x + 0.40y) = -2(88)$$

$$0.30x + 0.90y = 180$$
$$-0.30x - 0.80y = -176$$

$$0.10y = 4$$
$$y = 40$$

$$0.10x + 0.30y = 60$$
$$0.10x + 0.30(40) = 60$$
$$0.10x + 12 = 60$$
$$0.10x = 48$$
$$x = 480$$

The chemist should use 480 g of the first alloy and 40 g of the second alloy.

21. Strategy • Cost of the Model II computer: x
Cost of the Model VI computer: y
Cost of the Model IX computer: z

First shipment:

	Number	Unit Cost	Value
Model II	4	x	$4x$
Model VI	6	y	$6y$
Model IX	10	z	$10z$

Second shipment:

	Number	Unit Cost	Value
Model II	8	x	$8x$
Model VI	3	y	$3y$
Model IX	5	z	$5z$

Third shipment:

	Number	Unit Cost	Value
Model II	2	x	$2x$
Model VI	9	y	$9y$
Model IX	5	z	$5z$

• The value of the first shipment is $114,000. The value of the second shipment is $72,000. The value of the third shipment is $81,000.

Solution
(1) $4x + 6y + 10z = 114{,}000$
(2) $8x + 3y + 5z = 72{,}000$
(3) $2x + 9y + 5z = 81{,}000$

Equation (1) –2 times Equation (2)

$$4x + 6y + 10z = 114{,}000$$
$$-16x - 6y - 10z = -144{,}000$$
(4) $$-12x = -30{,}000$$
$$x = 2500$$

Equation (2) –1 times Equation (3)

$$8x + 3y + 5z = 72{,}000$$
$$-2x - 9y - 5z = -81{,}000$$
(5) $$6x - 6y = -9000$$

Substitute 2500 for x in Equation (5).

$$6(2500) - 6y = -9000$$
$$-6y = -24{,}000$$
$$y = 4000$$

The model VI computer costs $4000.

23. Strategy • Number of regular admission tickets: x
Number of member tickets: y
Number of student tickets: z

	Number	Unit Cost	Amount
Regular	x	10	$10x$
Member	y	7	$7y$
Student	z	5	$5z$

• 20 more student tickets than regular tickets were sold. The total number of tickets sold was 750. The total receipts for Saturday were $5400.

Solution
(1) $z = x + 20$
(2) $x + y + z = 750$
(3) $10x + 7y + 5z = 5400$

Multiply Equation (2) by –7 and add to Equation (3).

$$-7x - 7y - 7z = -5250$$
$$10x + 7y + 5z = 5400$$
(4) $$3x - 2z = 150$$

Substitute $x + 20$ for z in Equation (4).

$$3x - 2(x + 20) = 150$$
$$3x - 2x - 40 = 150$$
$$x = 190$$

Substitute 190 for x in Equation (1).

$$z = 190 + 20$$
$$z = 210$$

Substitute 190 for x and 210 for z in Equation (2).

$$190 + y + 210 = 750$$
$$y = 350$$

There were 190 regular tickets, 350 member tickets, and 210 student tickets sold for the Saturday performance.

25. Strategy • Amount deposited in the 8% account: x
Amount deposited in the 6% account: y
Amount deposited in the 4% account: z

	Principal	Rate	Interest
8%	x	0.08	$0.08x$
6%	y	0.06	$0.06y$
4%	z	0.04	$0.04z$

• The amount deposited in the 8% account is twice the amount deposited in the 6% account. The total amount invested is \$25,000. The total interest earned is \$1520.

Solution (1) $x = 2y$
(2) $x + y + z = 25{,}000$
(3) $0.08x + 0.06y + 0.04z = 1520$
Substitute $2y$ for x in Equation (2).
$2y + y + z = 25{,}000$
(4) $3y + z = 25{,}000$
Substitute $2y$ for x in Equation (3).
$0.08(2y) + 0.06y + 0.04z = 1520$
(5) $0.22y + 0.04z = 1520$
Solve Equation (4) for z and substitute for z in Equation (5).
$z = 25{,}000 - 3y$

$0.22y + 0.04(25{,}000 - 3y) = 1520$
$0.22y + 1000 - 0.12y = 1520$
$0.10y = 520$
$y = 5200$
Substitute 5200 for y in Equation (4).
$3(5200) + z = 25{,}000$
$z = 9400$
Substitute 5200 for y in Equation (1) and solve for x.
$x = 2(5200)$
$x = 10{,}400$
The investor placed \$10,400 in the 8% account, \$5200 in the 6% account, and \$9400 in the 4% account.

Applying the Concepts

27. Strategy • Measure of the smaller angle: n
Measure of the larger angle: m
First relationship: $m + n = 180$
Second relationship: $m = 3n + 40$

Solution Solve for m by substitution.
$n = 180 - m$
$m = 3(180 - m) + 40$
$m = 540 - 3m + 40$
$4m = 580$
$m = 145$
Substitute m into the equation $m + n = 180$ and solve for n.
$145 + n = 180$
$n = 35$
The measures of the two angles are 35° and 145°.

29. Strategy • Age of oil painting: x
Age of watercolor: y
First relationship: $x - y = 35$
Second relationship:
$x + 5 = 2(y - 5)$
$x + 5 = 2y - 10$
$x = 2y - 15$

Solution Solve for y by substitution.
$x - y = 35$
$2y - 15 - y = 35$
$y - 15 = 35$
$y = 50$

Substitute y into $x - y = 35$ and solve for x.
$x - 50 = 35$
$x = 85$
The age of the oil painting is 85 years, and the age of the watercolor is 50 years.

31. Strategy • Amount deposited in the 8% account: x
Amount deposited in the 6% account: y
Amount deposited in the 4% account: z

	Amount	Rate	Interest
8% account	x	0.08	$0.08x$
6% account	y	0.06	$0.06y$
4% account	z	0.04	$0.04z$

• The amount in the 8% account is \$1000 more than the amount in the 4% account. The total of the three investments is \$25,000. The total interest earned from the three investments is \$1520.

Solution
(1) $x = 1000 + z$
(2) $x + y + z = 25{,}000$
(3) $0.08x + 0.06y + 0.04z = 1520$

Substitute $1000 + z$ for x in Equation (2) and in Equation (3).

$1000 + z + y + z = 25{,}000$
(4) $y + 2z = 24{,}000$

$0.08(1000 + z) + 0.06y + 0.04z = 1520$
$80 + 0.08z + 0.06y + 0.04z = 1520$
(5) $0.06y + 0.12z = 1440$

Solve Equation (4) for y and substitute for y in Equation (5).

$y = 24{,}000 - 2z$

$0.06(24{,}000 - 2z) + 0.12z = 1440$
$1440 - 0.12z + 0.12z = 1440$
$1440 = 1440$

• In Exercise 25 above, the amount deposited in the 8% account is twice the amount in the 6% account. The total of the three investments is still \$25,000. The total interest earned from the three investments is still \$1520.
(1) $x = 2y$
(2) $x + y + z = 25{,}000$
Substitute $2y$ for x in Equation (2).
$2y + y + z = 25{,}000$
(3) $z = 25{,}000 - 3y$

(4) $0.08x + 0.06y + 0.04z = 1520$
Substitute $2y$ for x and $25{,}000 - 3y$ for z in Equation (4).
$0.08(2y) + 0.06y + 0.04(25{,}000 - 3y) = 1520$
$0.16y + 0.06y + 1000 - 0.12y = 1520$
$0.1y = 520$
$y = 5200$

Substitute 5200 for y in Equation (1).
$x = 2y = 2(5200) = 10{,}400$

Substitute 5200 for y in Equation (3).
$z = 25{,}000 - 3(5200) = 9400$

The system of equations is dependent. There is not a unique solution. The solution can be written $(z + 1000, 24{,}000 - 2z, z)$, which indicates that there is more than one way the investor can allocate money to each of the accounts to reach the goal. In Exercise 25, the system of equations is independent and there is a unique solution of (10,400, 5200, 9400); therefore, there is only one way to allocate the \$25,000.

Section 4.5

Objective A Exercises

1. Solve each inequality.
$x - y \geq 3$
$-y \geq 3 - x$
$y \leq -3 + x$

$x + y \leq 5$
$y \leq 5 - x$

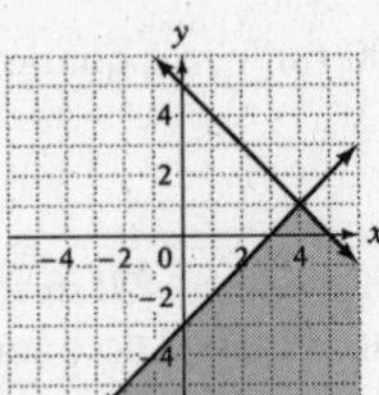

3. Solve each inequality.

$3x - y < 3$
$-y < 3 - 3x$
$y > -3 + 3x$

$2x + y \geq 2$
$y \geq 2 - 2x$

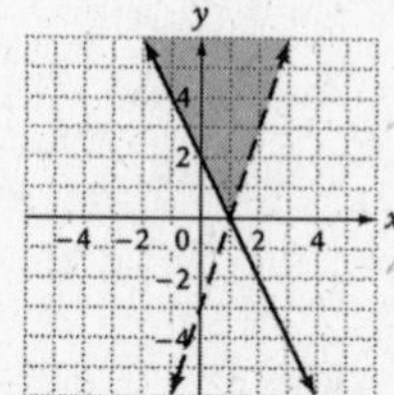

5. Solve each inequality.

$2x + y \geq -2$
$y \geq -2 - 2x$

$6x + 3y \leq 6$
$3y \leq 6 - 6x$
$y \leq 2 - 2x$

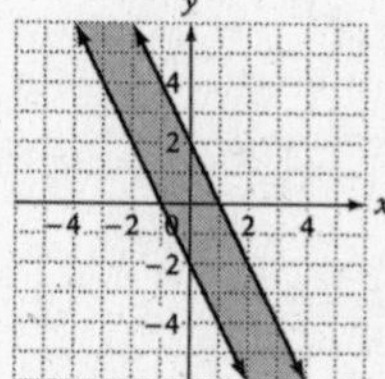

7. Solve each inequality.

$3x - 2y < 6$
$-2y < 6 - 3x$
$y > -3 + \frac{3}{2}x$

$y \leq 3$

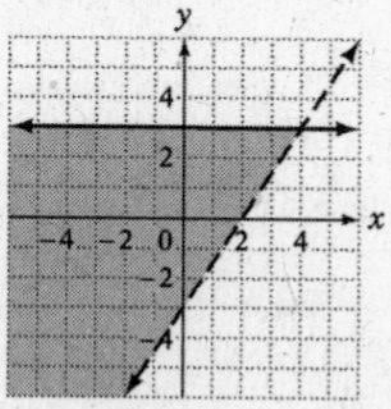

9. Solve each inequality.

$y > 2x - 6$

$x + y < 0$
$y < -x$

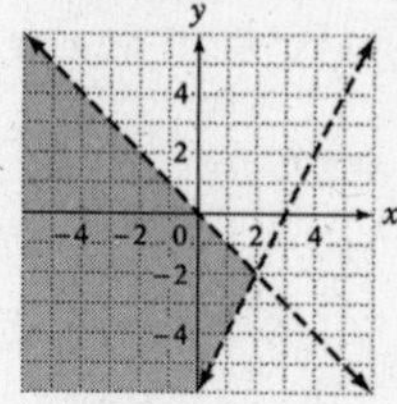

11. Solve each inequality.

$x + 1 \geq 0$
$x \geq -1$

$y - 3 \leq 0$
$y \leq 3$

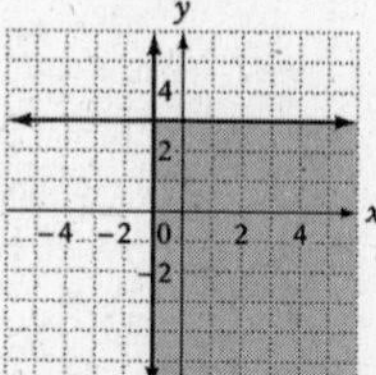

13. Solve each inequality.

$2x + y \geq 4$
$y \geq -2x + 4$

$3x - 2y < 6$
$-2y < -3x + 6$
$y > \frac{3}{2}x - 3$

$y \geq -2x + 4$
$y > \frac{3}{2}x - 3$

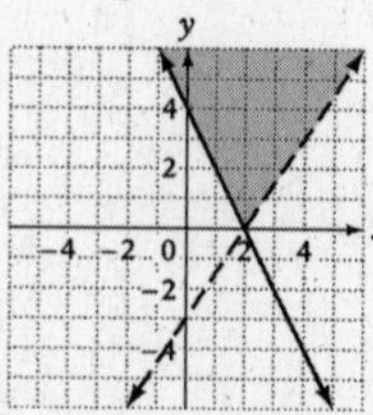

15. Solve each inequality.

$x - 2y \leq 6$
$-2y \leq -x + 6$
$y \geq \frac{1}{2}x - 3$

$2x + 3y \leq 6$
$3y \leq -2x + 6$
$y \leq -\frac{2}{3}x + 2$

$y \geq \frac{1}{2}x - 3$
$y \leq -\frac{2}{3}x + 2$

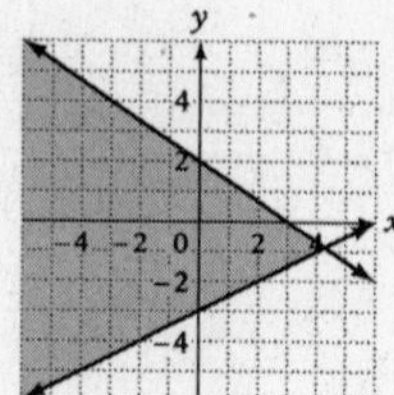

17. Solve each inequality.

$x - 2y \leq 4$
$-2y \leq -x + 4$
$y \geq \frac{1}{2}x - 2$

$3x + 2y \leq 8$
$2y \leq -3x + 8$
$y \leq -\frac{3}{2}x + 4$

$y \geq \frac{1}{2}x - 2$
$y \leq -\frac{3}{2}x + 4$

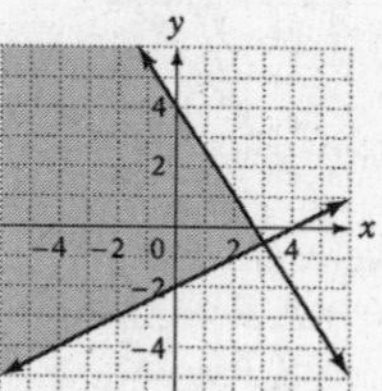

Applying the Concepts

19. Solve each inequality.

$2x + 3y \leq 15$
$3y \leq -2x + 15$
$y \leq -\frac{2}{3}x + 5$

$3x - y \leq 6$
$-y \leq -3x + 6$
$y \geq 3x - 6$

$y \leq -\frac{2}{3}x + 5$
$y \geq 3x - 6$
$y \geq 0$

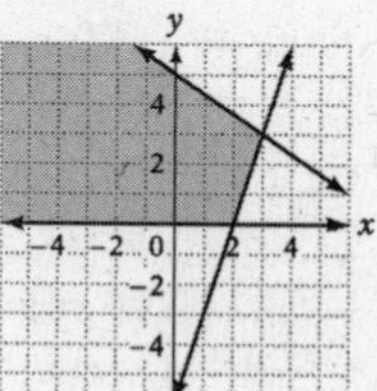

21. Solve each inequality.

$x - y \le 5$
$-y \le 5 - x$
$y \ge -5 + x$

$2x - y \ge 6$
$-y \ge 6 - 2x$
$y \le -6 + 2x$

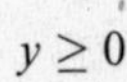
$y \ge 0$

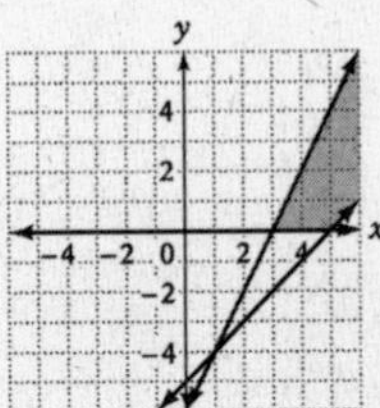

23. Solve each inequality.

$2x - y \le 4$
$-y \le 4 - 2x$
$y \ge -4 + 2x$

$3x + y < 1$
$y < 1 - 3x$

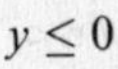
$y \le 0$

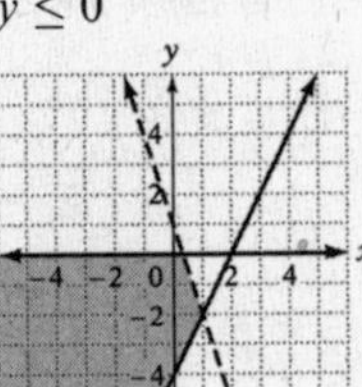

Chapter 4 Review Exercises

1. (1) $2x - 6y = 15$
(2) $x = 4y + 8$

Substitute $4y + 8$ for x in Equation (1).

$2(4y + 8) - 6y = 15$
$8y + 16 - 6y = 15$
$2y = -1$
$y = -\frac{1}{2}$

Substitute $-\frac{1}{2}$ for y in Equation (2).

$x = 4\left(-\frac{1}{2}\right) + 8$
$x = -2 + 8 = 6$

The solution is $\left(6, -\frac{1}{2}\right)$.

2. (1) $3x + 2y = 2$
(2) $x + y = 3$

Eliminate y. Multiply Equation (2) by -2 and add to Equation (1).

$3x + 2y = 2$
$-2(x + y) = 3(-2)$
$3x + 2y = 2$
$-2x - 2y = -6$

Add the equations.

$x = -4$

Replace x in Equation (2).

$x + y = 3$
$-4 + y = 3$
$y = 7$

The solution is $(-4, 7)$.

3.

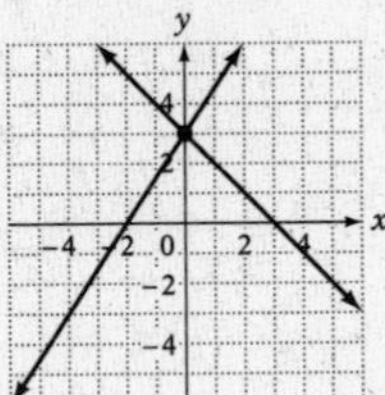

The solution is $(0, 3)$.

4.

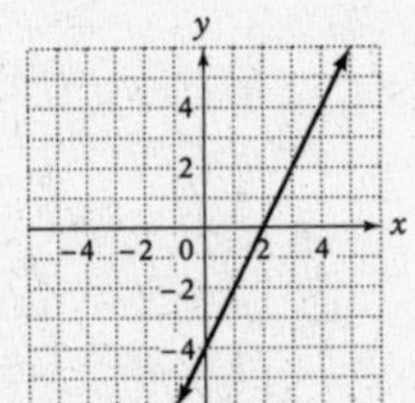

The two equations represent the same line.
$(x, 2x - 4)$

5. (1) $3x + 12y = 18$
(2) $x + 4y = 6$

Solve Equation (2) for x.

$x + 4y = 6$
$x = -4y + 6$

Substitute into Equation (1).

$3x + 12y = 18$
$3(-4y + 6) + 12y = 18$
$-12y + 18 + 12y = 18$
$18 = 18$

This is a true equation. The equations are dependent. The solutions are the ordered pairs $\left(x, -\frac{1}{4}x + \frac{3}{2}\right)$.

6. (1) $5x - 15y = 30$
(2) $x - 3y = 6$
Eliminate x. Multiply Equation (2) by -5 and add to Equation (1).
$5x - 15y = 30$
$-5(x - 3y) = 6(-5)$
$5x - 15y = 30$
$-5x + 15y = -30$
Add the equations.
$0 = 0$
This is a true equation. The equations are dependent. The solutions are the ordered pairs $\left(x, \frac{1}{3}x - 2\right)$.

7. (1) $3x - 4y - 2z = 17$
(2) $4x - 3y + 5z = 5$
(3) $5x - 5y + 3z = 14$
Eliminate z. Multiply Equation (1) by 3 and Equation (3) by 2. Then add the equations.
$3(3x - 4y - 2z) = 3(17)$
$2(5x - 5y + 3z) = 2(14)$

$9x - 12y - 6z = 51$
$10x - 10y + 6z = 28$

(4) $19x - 22y = 79$
Multiply Equation (1) by 5 and Equation (2) by 2. Then add the equations.
$5(3x - 4y - 2z) = 5(17)$
$2(4x - 3y + 5z) = 2(5)$

$15x - 20y - 10z = 85$
$8x - 6y + 10z = 10$

(5) $23x - 26y = 95$
Multiply Equation (4) by 23 and Equation (5) by -19. Then add the equations.
$23(19x - 22y) = 23(79)$
$-19(23x - 26y) = -19(95)$

$437x - 506y = 1817$
$-437x + 494y = -1805$

$-12y = 12$
$y = -1$
Replace y by -1 in Equation (4).
$19x - 22y = 79$
$19x - 22(-1) = 79$

$19x = 57$
$x = 3$
Replace x by 3 and y by -1 in Equation (1).
$3x - 4y - 2z = 17$
$3(3) - 4(-1) - 2z = 17$
$9 + 4 - 2z = 17$
$-2z = 4$
$z = -2$
The solution is (3,–1, –2).

8. (1) $3x + y = 13$
(2) $2y + 3z = 5$
(3) $x + 2z = 11$
Eliminate y. Multiply Equation (1) by -2, then add to Equation (2).
$-2(3x + y) = -2(13)$
$2y + 3z = 5$

$-6x - 2y = -26$
$2y + 3z = 5$

(4) $-6x + 3z = -21$
Multiply Equation (3) by 6. Then add to Equation (4).
$6(x + 2z) = 6(11)$
$-6x + 3z = -21$

$6x + 12z = 66$
$-6x + 3z = -21$

$15z = 45$
$z = 3$
Replace z by 3 in Equation (3).
$x + 2z = 11$
$x + 2(3) = 11$
$x = 5$
Replace x by 5 in Equation (1).
$3x + y = 13$
$3(5) + y = 13$
$y = -2$
The solution is (5,–2, 3).

9. $\begin{vmatrix} 6 & 1 \\ 2 & 5 \end{vmatrix} = 6(5) - 2(1) = 30 - 2 = 28$

10. $\begin{vmatrix} 1 & 5 & -2 \\ -2 & 1 & 4 \\ 4 & 3 & -8 \end{vmatrix} = 1\begin{vmatrix} 1 & 4 \\ 3 & -8 \end{vmatrix} - 5\begin{vmatrix} -2 & 4 \\ 4 & -8 \end{vmatrix} - 2\begin{vmatrix} -2 & 1 \\ 4 & 3 \end{vmatrix}$
$= 1(-8 - 12) - 5(16 - 16) - 2(-6 - 4)$
$= 1(-20) - 5(0) - 2(-10)$
$= -20 - 0 + 20$
$= 0$

11. $2x - y = 7$
$3x + 2y = 7$
$D = \begin{vmatrix} 2 & -1 \\ 3 & 2 \end{vmatrix} = 7$
$D_x = \begin{vmatrix} 7 & -1 \\ 7 & 2 \end{vmatrix} = 21$
$D_y = \begin{vmatrix} 2 & 7 \\ 3 & 7 \end{vmatrix} = -7$
$x = \frac{D_x}{D} = \frac{21}{7} = 3$
$y = \frac{D_y}{D} = \frac{-7}{7} = -1$
The solution is (3, –1).

12. $3x - 4y = 10$
$2x + 5y = 15$

$$D = \begin{vmatrix} 3 & -4 \\ 2 & 5 \end{vmatrix} = 23$$

$$D_x = \begin{vmatrix} 10 & -4 \\ 15 & 5 \end{vmatrix} = 110$$

$$D_y = \begin{vmatrix} 3 & 10 \\ 2 & 15 \end{vmatrix} = 25$$

$$x = \frac{D_x}{D} = \frac{110}{23}$$

$$y = \frac{D_y}{D} = \frac{25}{23}$$

The solution is $\left(\frac{110}{23}, \frac{25}{23}\right)$.

13. $x + y + z = 0$
$x + 2y + 3z = 5$
$2x + y + 2z = 3$

$$D = \begin{vmatrix} 1 & 1 & 1 \\ 1 & 2 & 3 \\ 2 & 1 & 2 \end{vmatrix} = 2$$

$$D_x = \begin{vmatrix} 0 & 1 & 1 \\ 5 & 2 & 3 \\ 3 & 1 & 2 \end{vmatrix} = -2$$

$$D_y = \begin{vmatrix} 1 & 0 & 1 \\ 1 & 5 & 3 \\ 2 & 3 & 2 \end{vmatrix} = -6$$

$$D_z = \begin{vmatrix} 1 & 1 & 0 \\ 1 & 2 & 5 \\ 2 & 1 & 3 \end{vmatrix} = 8$$

$$x = \frac{D_x}{D} = \frac{-2}{2} = -1$$

$$y = \frac{D_y}{D} = \frac{-6}{2} = -3$$

$$z = \frac{D_z}{D} = \frac{8}{2} = 4$$

The solution is $(-1, -3, 4)$.

14. $x + 3y + z = 6$
$2x + y - z = 12$
$x + 2y - z = 13$

$$D = \begin{vmatrix} 1 & 3 & 1 \\ 2 & 1 & -1 \\ 1 & 2 & -1 \end{vmatrix} = 7$$

$$D_x = \begin{vmatrix} 6 & 3 & 1 \\ 12 & 1 & -1 \\ 13 & 2 & -1 \end{vmatrix} = 14$$

$$D_y = \begin{vmatrix} 1 & 6 & 1 \\ 2 & 12 & -1 \\ 1 & 13 & -1 \end{vmatrix} = 21$$

$$D_z = \begin{vmatrix} 1 & 3 & 6 \\ 2 & 1 & 12 \\ 1 & 2 & 13 \end{vmatrix} = -35$$

$$x = \frac{D_x}{D} = \frac{14}{7} = 2$$

$$y = \frac{D_y}{D} = \frac{21}{7} = 3$$

$$z = \frac{D_z}{D} = \frac{-35}{7} = -5$$

The solution is $(2, 3, -5)$.

15. Solve each inequality.

$$x + 3y \le 6$$
$$3y \le 6 - x$$
$$y \le 2 - \frac{1}{3}x$$

$$2x - y \ge 4$$
$$-y \ge 4 - 2x$$
$$y \le -4 + 2x$$

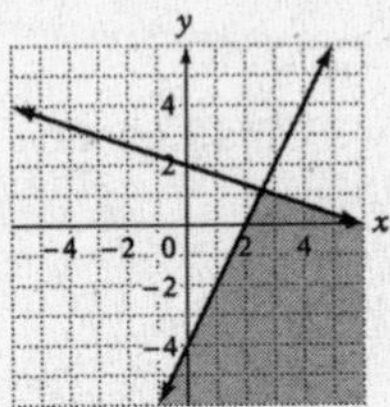

16. Solve each inequality.

$$2x + 4y \ge 8$$
$$4y \ge 8 - 2x$$
$$y \ge 2 - \frac{1}{2}x$$

$$x + y \le 3$$
$$y \le 3 - x$$

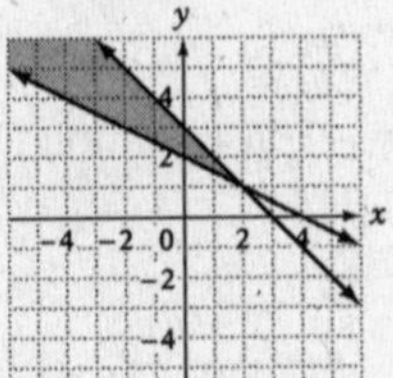

17. Strategy • Rate of cabin cruiser in calm water: x
Rate of the current: y

	Rate	Time	Distance
With current	$x+y$	3	$3(x+y)$
Against current	$x-y$	5	$5(x-y)$

• The distance traveled with the current is 60 mi. The distance traveled against the current is 60 mi.

$3(x+y)=60$
$5(x-y)=60$

Solution

$3(x+y)=60$
$5(x-y)=60$

$\frac{1}{3}\cdot 3(x+y)=\frac{1}{3}(60)$
$\frac{1}{5}\cdot 5(x-y)=\frac{1}{5}(60)$

$x+y=20$
$x-y=12$

$2x=32$
$x=16$

$x+y=20$
$16+y=20$
$y=4$

The rate of the cabin cruiser in calm water is 16 mph. The rate of the current is 4 mph.

18. Strategy • Rate of the plane in calm air: p
Rate of the wind: w

	Rate	Time	Distance
With wind	$p+w$	3	$3(p+w)$
Against wind	$p-w$	4	$4(p-w)$

• The distance traveled with the wind is 600 mi. The distance traveled against the wind is 600 mi.

$3(p+w)=600$
$4(p-w)=600$

Solution

$3(p+w)=600$
$4(p-w)=600$

$\frac{1}{3}\cdot 3(p+w)=\frac{1}{3}(600)$
$\frac{1}{4}\cdot 4(p-w)=\frac{1}{4}(600)$

$p+w=200$
$p-w=150$

$2p=350$
$p=175$

$p+w=200$
$175+w=200$
$w=25$

The rate of the plane in calm air is 175 mph. The rate of the wind is 25 mph.

19. Strategy • Number of children's tickets sold Friday: x
Number of adult's tickets sold Friday: y

	Amount	Rate	Quantity
Children	x	5	$5x$
Adults	y	8	$8y$

Saturday:

	Amount	Rate	Quantity
Children	$3x$	5	$5(3x)$
Adults	$\frac{1}{2}y$	8	$8(\frac{1}{2})y$

• The total receipts for Friday were \$2500. The total receipts for Saturday were \$2500.

$5x+8y=2500$
$5(3x)+8\left(\frac{1}{2}y\right)=2500$

Solution

$5x+8y=2500$
$15x+4y=2500$

$5x+8y=2500$
$-2(15x+4y)=-2(2500)$

$5x+8y=2500$
$-30x-8y=-5000$

$-25x=-2500$
$x=100$

On Friday, 100 children attended.

20. Strategy • The amount invested at 3%: x
The amount invested at 7%: y

	Principal	Rate	Interest
Amount at 3%	x	0.03	$0.03x$
Amount at 7%	y	0.07	$0.07y$

• The total amount invested is \$20,000. The total annual interest earned is \$1200.

Solution (1) $x + y = 20{,}000$
(2) $0.03x + 0.07y = 1200$
Multiply Equation (1) by -0.07 and add to Equation (2).

$$-0.07x - 0.07y = -1400$$
$$0.03x + 0.07y = 1200$$
$$-0.04x = -200$$
$$x = 5000$$

Substitute 5000 for x in Equation (1).

$$5000 + y = 20{,}000$$
$$y = 15{,}000$$

The amount invested at 3% is \$5000.
The amount invested at 7% is \$15,000.

Chapter 4 Test

1. 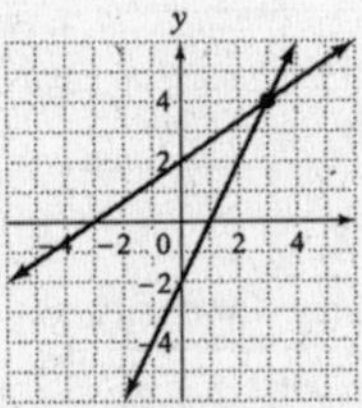

(3, 4).

2. 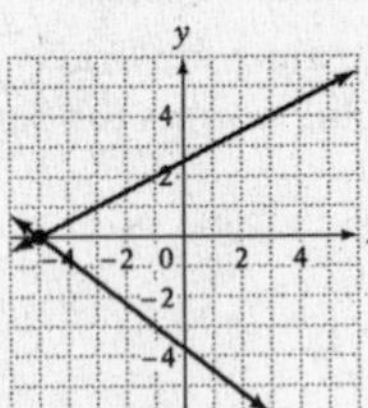

(−5, 0).

3. Solve each inequality.

$$2x - y < 3$$
$$-y < 3 - 2x$$
$$y > 2x - 3$$

$$4x + 3y < 11$$
$$3y < -4x + 11$$
$$y < -\frac{4}{3}x + \frac{11}{3}$$

$$y > 2x - 3$$
$$y < -\frac{4}{3}x + \frac{11}{3}$$

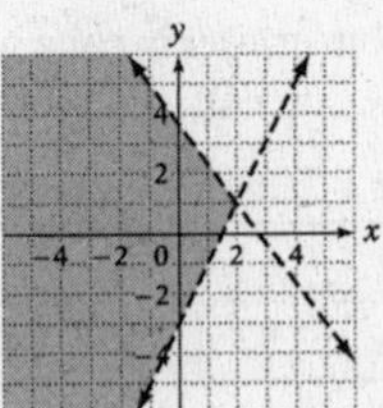

4. Solve each inequality.

$$x + y > 2$$
$$y > 2 - x$$

$$2x - y < -1$$
$$-y < -1 - 2x$$
$$y > 1 + 2x$$

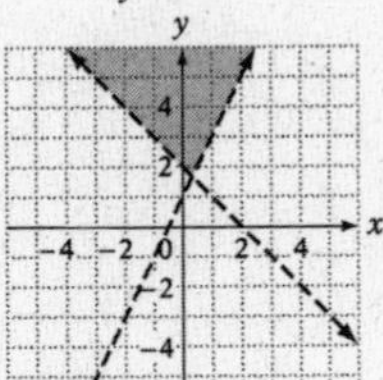

5. (1) $3x + 2y = 4$
(2) $x = 2y - 1$
Substitute $2y - 1$ for x in Equation (1).

$$3(2y - 1) + 2y = 4$$
$$6y - 3 + 2y = 4$$
$$8y = 7$$
$$y = \frac{7}{8}$$

Substitute into Equation (2).

$$x = 2y - 1$$
$$x = 2\left(\frac{7}{8}\right) - 1$$
$$x = \frac{7}{4} - 1 = \frac{3}{4}$$

The solution is $\left(\frac{3}{4}, \frac{7}{8}\right)$.

6. (1) $5x + 2y = -23$
(2) $2x + y = -10$
Solve Equation (2) for y.

$$2x + y = -10$$
$$y = -2x - 10$$

Substitute $-2x - 10$ for y in Equation (1).

$$5x + 2y = -23$$
$$5x + 2(-2x - 10) = -23$$
$$5x - 4x - 20 = -23$$
$$x - 20 = -23$$
$$x = -3$$

Substitute into Equation (2).

$$2x + y = -10$$
$$2(-3) + y = -10$$
$$-6 + y = -10$$
$$y = -4$$

The solution is (−3, −4).

7. (1) $y = 3x - 7$
(2) $y = -2x + 3$
Substitute Equation (2) into Equation (1).
$-2x + 3 = 3x - 7$
$-5x + 3 = -7$
$-5x = -10$
$x = 2$
Substitute into Equation (1).
$y = 3x - 7$
$y = 3(2) - 7 = 6 - 7 = -1$
The solution is (2, −1).

8. (1) $3x + 4y = -2$
(2) $2x + 5y = 1$
Multiply Equation (1) by −2 and Equation (2) by 3. Add the new equations.
$-2(3x + 4y) = -2(-2)$
$3(2x + 5y) = 3(1)$

$-6x - 8y = 4$
$6x + 15y = 3$

$7y = 7$
$y = 1$
Substitute into Equation (2).
$2x + 5y = 1$
$2x + 5(1) = 1$
$2x + 5 = 1$
$2x = -4$
$x = -2$
The solution is (−2, 1).

9. (1) $4x - 6y = 5$
(2) $6x - 9y = 4$
Multiply Equation (1) by −3. Multiply Equation (2) by 2. Add the new equations.
$-3(4x - 6y) = -3(5)$
$2(6x - 9y) = 2(4)$

$-12x + 18y = -15$
$12x - 18y = 8$

$0 = -7$

This is not a true equation. The system of equations is inconsistent and therefore has no solution.

10. (1) $3x - y = 2x + y - 1$
(2) $5x + 2y = y + 6$
Write the equation in the form $Ax + By = C$.
(3) $x - 2y = -1$
(4) $5x + y = 6$
Multiply Equation (4) by 2 and add to Equation (3).
$x - 2y = -1$
$2(5x + y) = 2(6)$

$x - 2y = -1$
$10x + 2y = 12$

$11x = 11$
$x = 1$
Substitute into Equation (4).
$5x + y = 6$
$5(1) + y = 6$
$y = 1$
The solution is (1, 1).

11. (1) $2x + 4y - z = 3$
(2) $x + 2y + z = 5$
(3) $4x + 8y - 2z = 7$
Eliminate z. Add Equations (1) and (2).
$2x + 4y - z = 3$
$x + 2y + z = 5$

(4) $3x + 6y = 8$
Multiply Equation (2) by 2 and add to Equation (3).
$2(x + 2y + z) = 2(5)$
$4x + 8y - 2z = 7$

$2x + 4y + 2z = 10$
$4x + 8y - 2z = 7$

(5) $6x + 12y = 17$
Multiply Equation (4) by −2 and add to Equation (5).
$-2(3x + 6y) = -2(8)$
$6x + 12y = 17$

$-6x - 12y = -16$
$6x + 12y = 17$
$0 = 1$
This is not a true equation. The system of equations is inconsistent and therefore has no solution.

12. (1) $x - y - z = 5$
(2) $2x + z = 2$
(3) $3y - 2z = 1$
Multiply Equation (1) by 3 and add to Equation (3).
$3(x - y - z) = 3(5)$
$3y - 2z = 1$

$3x - 3y - 3z = 15$
$3y - 2z = 1$

(4) $3x - 5z = 16$
Multiply Equation (2) by 5 and add to Equation (4).
$5(2x + z) = (2)5$
$3x - 5z = 16$

$10x + 5z = 10$
$3x - 5z = 16$

$13x = 26$
$x = 2$
Substitute into Equation (4).
$3x - 5z = 16$
$3(2) - 5z = 16$
$6 - 5z = 16$
$-5z = 10$
$z = -2$
Substitute into Equation (3).
$3y - 2z = 1$
$3y - 2(-2) = 1$
$3y + 4 = 1$
$3y = -3$
$y = -1$
The solution is $(2, -1, -2)$.

13. $\begin{vmatrix} 3 & -1 \\ -2 & 4 \end{vmatrix} = 3(4) - (-2)(-1) = 12 - 2 = 10$

14. $\begin{vmatrix} 1 & -2 & 3 \\ 3 & 1 & 1 \\ 2 & -1 & -2 \end{vmatrix} = 1\begin{vmatrix} 1 & 1 \\ -1 & -2 \end{vmatrix} - (-2)\begin{vmatrix} 3 & 1 \\ 2 & -2 \end{vmatrix} + 3\begin{vmatrix} 3 & 1 \\ 2 & -1 \end{vmatrix}$
$= 1(-2 - (-1)) + 2(-6 - 2) + 3(-3 - 2)$
$= 1(-2 + 1) + 2(-8) + 3(-5)$
$= -1 - 16 - 15$
$= -32$

15. $x - y = 3$
$2x + y = -4$

$D = \begin{vmatrix} 1 & -1 \\ 2 & 1 \end{vmatrix} = 3$

$D_x = \begin{vmatrix} 3 & -1 \\ -4 & 1 \end{vmatrix} = -1$

$D_y = \begin{vmatrix} 1 & 3 \\ 2 & -4 \end{vmatrix} = -10$

$x = \frac{D_x}{D} = -\frac{1}{3}$

$y = \frac{D_y}{D} = \frac{-10}{3}$

The solution is $\left(-\frac{1}{3}, -\frac{10}{3}\right)$.

16. $5x + 2y = 9$
$3x + 5y = -7$

$D = \begin{vmatrix} 5 & 2 \\ 3 & 5 \end{vmatrix} = 19$

$D_x = \begin{vmatrix} 9 & 2 \\ -7 & 5 \end{vmatrix} = 59$

$D_y = \begin{vmatrix} 5 & 9 \\ 3 & -7 \end{vmatrix} = -62$

$x = \frac{D_x}{D} = \frac{59}{19}$

$y = \frac{D_y}{D} = \frac{-62}{19}$

The solution is $\left(\frac{59}{19}, -\frac{62}{19}\right)$.

17. $x - y + z = 2$
$2x - y - z = 1$
$x + 2y - 3z = -4$

$D = \begin{vmatrix} 1 & -1 & 1 \\ 2 & -1 & -1 \\ 1 & 2 & -3 \end{vmatrix} = 5$

$D_x = \begin{vmatrix} 2 & -1 & 1 \\ 1 & -1 & -1 \\ -4 & 2 & -3 \end{vmatrix} = 1$

$D_y = \begin{vmatrix} 1 & 2 & 1 \\ 2 & 1 & -1 \\ 1 & -4 & -3 \end{vmatrix} = -6$

$D_z = \begin{vmatrix} 1 & -1 & 2 \\ 2 & -1 & 1 \\ 1 & 2 & -4 \end{vmatrix} = 3$

$x = \frac{D_x}{D} = \frac{1}{5}$

$y = \frac{D_y}{D} = -\frac{6}{5}$

$z = \frac{D_z}{D} = \frac{3}{5}$

The solution is $\left(\frac{1}{5}, -\frac{6}{5}, \frac{3}{5}\right)$.

18. Strategy • Rate of plane in calm air: x
Rate of wind: y

	Rate	Time	Distance
With wind	$x+y$	2	$2(x+y)$
Against wind	$x-y$	2.8	$2.8(x-y)$

• The distance traveled with the wind is 350 mi. The distance traveled against the wind is 350 mi.

$$2(x+y)=350$$
$$2.8(x-y)=350$$

Solution

$$2(x+y)=350$$
$$2.8(x-y)=350$$

$$\frac{1}{2}\cdot 2(x+y)=\frac{1}{2}\cdot 350$$
$$\frac{1}{2.8}\cdot 2.8(x-y)=\frac{1}{2.8}\cdot 350$$

$$x+y=175$$
$$x-y=125$$

$$2x=300$$
$$x=150$$

$$x+y=175$$
$$150+y=175$$
$$y=25$$

The rate of the plane in calm air is 150 mph. The rate of the wind is 25 mph.

19. Strategy • Cost per yard of cotton: x
Cost per yard of wool: y
First purchase:

	Amount	Rate	Total Value
Cotton	60	x	$60x$
Wool	90	y	$90y$

Second purchase:

	Amount	Rate	Total Value
Cotton	80	x	$80x$
Wool	20	y	$20y$

• The total cost of the first purchase was $1800. The total cost of the second purchase was $1000.

$$60x+90y=1800$$
$$80x+20y=1000$$

Solution

$$-4(60x+90y)=-4(1800)$$
$$3(80x+20y)=3(1000)$$

$$-240x-360y=-7200$$
$$240x+60y=3000$$

$$-300y=-4200$$
$$y=14$$

$$60x+90(14)=1800$$
$$60x+1260=1800$$
$$60x=540$$
$$x=9$$

The cost of cotton is $9.00/yd.
The cost of wool is $14.00/yd.

20. Strategy • Amount invested at 2.7%: x
Amount invested at 5.1%: y

	Principal	Rate	Interest
Amount at 2.7%	x	0.027	$0.027x$
Amount at 5.1%	y	0.051	$0.051y$

• The total amount invested is $15,000. The total annual interest earned is $549.

Solution (1) $x+y=15{,}000$
(2) $0.027x+0.051y=549$
Multiply Equation (1) by -0.051 and add to Equation (2).

$$-0.051x-0.051y=-765$$
$$0.027x+0.051y=549$$
$$-0.024x=-216$$
$$x=9000$$

Substitute 9000 for x in Equation (1) and solve for y.

$$9000+y=15{,}000$$
$$y=6000$$

The amount invested at 2.7% is $9000.
The amount invested at 5.1% is $6000.

Cumulative Review Exercises

1.

$$\frac{3}{2}x-\frac{3}{8}+\frac{1}{4}x=\frac{7}{12}x-\frac{5}{6}$$
$$24\left(\frac{3}{2}x-\frac{3}{8}+\frac{1}{4}x\right)=24\left(\frac{7}{12}x-\frac{5}{6}\right)$$
$$36x-9+6x=14x-20$$
$$42x-9=14x-20$$
$$28x-9=-20$$
$$28x=-11$$
$$x=-\frac{11}{28}$$

The solution is $-\frac{11}{28}$.

2. $(x_1, y_1) = (2, -1), (x_2, y_2) = (3, 4)$

$$m = \frac{y_2 - y_1}{x_2 - x_1} = \frac{4-(-1)}{3-2} = \frac{5}{1} = 5$$
$$y - y_1 = m(x - x_1)$$
$$y - (-1) = 5(x-2)$$
$$y + 1 = 5x - 10$$
$$y = 5x - 11$$
The equation of the line is $y = 5x - 11$.

3. $3[x - 2(5-2x) - 4x] + 6$
$= 3(x - 10 + 4x - 4x) + 6$
$= 3(x - 10) + 6$
$= 3x - 30 + 6$
$= 3x - 24$

4. $a + bc \div 2; a = 4, b = 8, c = -2$
$4 + 8(-2) \div 2 = 4 - 16 \div 2$
$= 4 - 8$
$= -4$

5. $2x - 3 < 9$ or $5x - 1 < 4$
Solve each inequality.
$2x - 3 < 9$ $\quad$ $5x - 1 < 4$
$2x < 12$ $\quad$ $5x < 5$
$x < 6$ $\quad$ or $\quad$ $x < 1$
$(-\infty, 6) \cup (-\infty, 1) = (-\infty, 6)$

6. $|x-2| - 4 < 2$
$|x-2| < 6$
$-6 < x - 2 < 6$
$-6 + 2 < x - 2 + 2 < 6 + 2$
$-4 < x < 8$
$\{x | -4 < x < 8\}$

7. $|2x - 3| > 5$
Solve each inequality.
$2x - 3 < -5$ or $2x - 3 > 5$
$2x < -2$ or $2x > 8$
$x < -1$ or $x > 4$
This is the set $\{x | x < -1\} \cup \{x | x > 4\}$ or $\{x | x < -1 \text{ or } x > 4\}$

8. $f(x) = 3x^3 - 2x^2 + 1$
$f(-3) = 3(-3)^3 - 2(-3)^2 + 1$
$f(-3) = 3(-27) - 2(9) + 1$
$f(-3) = -98$

9. The range is the set of numbers found by plugging in the set of numbers in the domain.
$f(-2) = 3(-2)^2 - 2(-2) = 16$
$f(-1) = 3(-1)^2 - 2(-1) = 5$
$f(0) = 3(0)^2 - 2(0) = 0$
$f(1) = 3(1)^2 - 2(1) = 1$
$f(2) = 3(2)^2 - 2(2) = 8$
The range is $\{0, 1, 5, 8, 16\}$

10. $F(x) = x^2 - 3$
$F(2) = 2^2 - 3 = 1$

11. $f(x) = 3x - 4$
$f(2+h) = 3(2+h) - 4$
$= 6 + 3h - 4$
$= 2 + 3h$
$f(2) = 3(2) - 4 = 2$
$f(2+h) - f(2) = 2 + 3h - 2$
$= 3h$

12. $\{x | x \le 2\} \cap \{x | x > -3\}$

−5 −4 −3 −2 −1 0 1 2 3 4 5

13. Slope $= -\frac{2}{3}$; Point $= (-2, 3)$.
$$y - 3 = \frac{-2}{3}[x - (-2)]$$
$$y - 3 = \frac{-2}{3}(x+2)$$
$$y = \frac{-2}{3}x - \frac{4}{3} + 3$$
$$y = -\frac{2}{3}x + \frac{5}{3}$$

14. The slope of the line $2x - 3y = 7$ is found by rearranging the equation as follows:
$-3y = 7 - 2x$
$$y = \frac{-7}{3} + \frac{2}{3}x$$
Slope $= \frac{2}{3}$; perpendicular slope: $-\frac{3}{2}$
The line is found using
$$y - 2 = \frac{3}{2}[x - (-1)]$$
$$y = -\frac{3}{2}(x+1) + 2$$
$$y = -\frac{3}{2}x - \frac{3}{2} + 2$$
$$y = -\frac{3}{2}x + \frac{1}{2}$$

15. The distance between points is $\sqrt{(x_1 - x_2)^2 + (y_1 - y_2)^2}$.
$$\sqrt{[2-(-4)]^2 + (0-2)^2} = \sqrt{6^2 + (-2)^2}$$
$$= \sqrt{36 + 4}$$
$$= \sqrt{40} = 2\sqrt{10} \approx 6.32$$

16. The midpoint is found using $\left(\frac{x_1 + x_2}{2}, \frac{y_1 + y_2}{2}\right)$.
$$\text{Midpoint} = \left(\frac{-4+3}{2}, \frac{3+5}{2}\right)$$
$$= \left(-\frac{1}{2}, 4\right)$$

17. $2x - 5y = 10$
$-5y = 10 - 2x$
$y = -2 + \frac{2}{5}x$
The y-intercept is -2.

The slope is $\frac{2}{5}$.

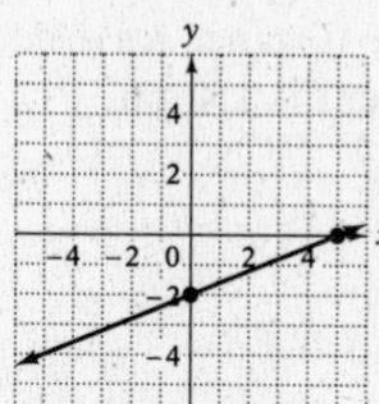

18. $3x - 4y \geq 8$
$-4y \geq 8 - 3x$
$y \leq -2 + \frac{3}{4}x$
The y-intercept is -2.

The slope is $\frac{3}{4}$.

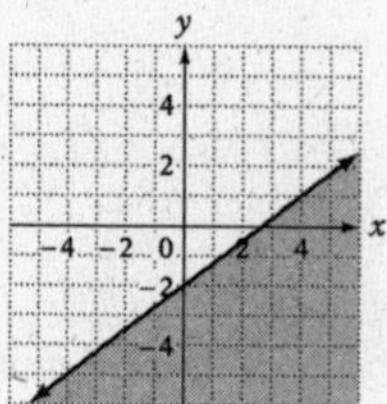

19.

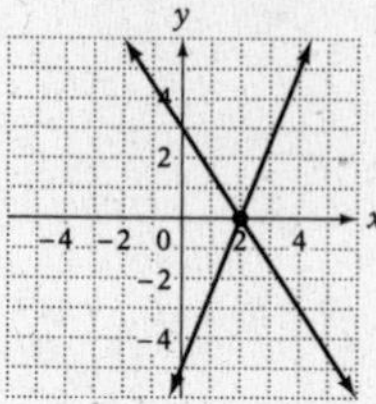

(2, 0).

20. Solve each inequality for y.
$3x - 2y \geq 4$
$-2y \geq 4 - 3x$
$y \leq -2 + \frac{3}{2}x$

$x + y < 3$
$y < 3 - x$

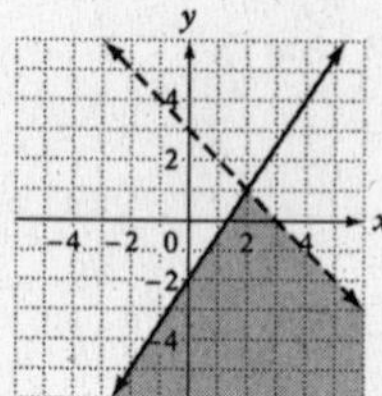

21. (1) $3x + 2z = 1$
(2) $2y - z = 1$
(3) $x + 2y = 1$
Multiply Equation (2) by -1 and add to Equation (3).
$-1(2y - z) = -1(1)$
$x + 2y = 1$

$-2y + z = -1$
$x + 2y = 1$

(4) $x + z = 0$
Multiply Equation (4) by -2 and add to Equation (1).
$(-2)(x + z) = -2(0)$
$3x + 2z = 1$

$-2x - 2z = 0$
$3x + 2z = 1$

$x = 1$
Substitute 1 for x in Equation (4).
$x + z = 0$
$1 + z = 0$
$z = -1$
Substitute 1 for x in Equation (3).
$x + 2y = 1$
$1 + 2y = 1$
$2y = 0$
$y = 0$
The solution is (1, 0, -1).

22. $$\begin{vmatrix} 2 & -5 & 1 \\ 3 & 1 & 2 \\ 6 & -1 & 4 \end{vmatrix} = 2\begin{vmatrix} 1 & 2 \\ -1 & 4 \end{vmatrix} - 3\begin{vmatrix} -5 & 1 \\ -1 & 4 \end{vmatrix} + 6\begin{vmatrix} -5 & 1 \\ 1 & 2 \end{vmatrix}$$
$= 2(4 + 2) - 3(-20 + 1) + 6(-10 - 1)$
$= 2(6) - 3(-19) + 6(-11)$
$= 12 + 57 - 66$
$= 3$

23. $4x - 3y = 17$
$3x - 2y = 12$
$$D = \begin{vmatrix} 4 & -3 \\ 3 & -2 \end{vmatrix} = 4(-2) - 3(-3) = 1$$
$$D_x = \begin{vmatrix} 17 & -3 \\ 12 & -2 \end{vmatrix} = 17(-2) - 12(-3) = 2$$
$$D_y = \begin{vmatrix} 4 & 17 \\ 3 & 12 \end{vmatrix} = 4(12) - 3(17) = -3$$
$$x = \frac{D_x}{D} = \frac{2}{1} = 2$$
$$y = \frac{D_y}{D} = \frac{-3}{1} = -3$$
The solution is (2, -3).

24. (1) $3x - 2y = 7$
(2) $y = 2x - 1$
Solve by the substitution method.
$3x - 2(2x - 1) = 7$
$3x - 4x + 2 = 7$
$-x + 2 = 7$
$-x = 5$
$x = -5$
Substitute -5 for x in Equation (2).
$y = 2x - 1$
$y = 2(-5) - 1 = -10 - 1 = -11$
The solution is $(-5, -11)$.

25. Strategy • The unknown number of quarters: x
The unknown number of dimes: $3x$
The unknown number of nickels:
$40 - (x + 3x)$

	Amount	Rate	Total Value
Quarters	x	25	$25x$
Dimes	$3x$	10	$10(3x)$
Nickels	$40-4x$	5	$5(40-4x)$

The sum of the total values of the denominations is \$4.10 (410 cents).
$25x + 10(3x) + 5(40 - 4x) = 410$

Solution $25x + 10(3x) + 5(40 - 4x) = 410$
$25x + 30x + 200 - 20x = 410$
$35x + 200 = 410$
$35x = 210$
$x = 6$
$40 - 4x = 40 - 24 = 16$
There are 16 nickels in the purse.

26. Strategy • The unknown amount of pure water: x

	Amount	Percent	Quantity
Water	x	0	$0 \cdot x$
4%	100	0.04	$100(0.04)$
2.5%	$100 + x$	0.025	$(100 + x)(0.025)$

• The sum of the quantities before mixing equals the quantity after mixing.
$0 \cdot x + 100(0.04) = (100 + x)0.025$

Solution $0 \cdot x + 100(0.04) = (100 + x)0.025$
$0 + 4 = 2.5 + 0.025x$
$1.5 = 0.025x$
$60 = x$
The amount of water that should be added is 60 ml.

27. Strategy • The Rate of the plane in calm air: x
The Rate of the wind: y

	Rate	Time	Distance
With wind	$x + y$	2	$2(x + y)$
Against wind	$x - y$	3	$3(x - y)$

• The distance traveled with the wind is 150 mi. The distance traveled against the wind is 150 mi.
$2(x + y) = 150$
$3(x - y) = 150$

Solution $2(x + y) = 150$
$3(x - y) = 150$

$\frac{1}{2} \cdot 2(x + y) = \frac{1}{2} \cdot 150$
$\frac{1}{3} \cdot 3(x - y) = \frac{1}{3} \cdot 150$

$x + y = 75$
$x - y = 50$

$2x = 125$
$x = 62.5$

$x + y = 75$
$62.5 + y = 75$
$y = 12.5$
The rate of the wind is 12.5 mph.

28. Strategy • Cost per pound of hamburger: x
Cost per pound of steak: y
First purchase:

	Amount	Percent	Quantity
Hamburger	100	x	$100x$
Steak	50	y	$50y$

Second purchase:

	Amount	Percent	Quantity
Hamburger	150	x	$150x$
Steak	100	y	$100y$

• The total cost of the first purchase is \$540. The total cost of the second purchase is \$960.

$$100x + 50y = 540$$
$$150x + 100y = 960$$

Solution

$$100x + 50y = 540$$
$$150x + 100y = 960$$

$$-2(100x + 50y) = -2(540)$$
$$150x + 100y = 960$$

$$-200x - 100y = -1080$$
$$150x + 100y = 960$$

$$-50x = -120$$
$$x = 2.4$$

$$100(2.4) + 50y = 540$$
$$240 + 50y = 540$$
$$50y = 300$$
$$y = 6$$

The cost of steak is \$6/lb.

29. Strategy • Let M be the number of ohms, T the tolerance, and r the given amount of the resistor. Find the tolerance and solve $|M - r| \leq T$ for M.

Solution $T = 0.15 \cdot 12{,}000 = 1800$ ohms

$$|M - 12{,}000| \leq 1800$$
$$-1800 \leq M - 12{,}000 \leq 1800$$
$$-1800 + 12{,}000 \leq M - 12{,}000 + 12{,}000 \leq 1800 + 12{,}000$$
$$10{,}200 \leq M \leq 13{,}800$$

The lower and upper limits of the resistor are 10,200 ohms and 13,800 ohms.

30. The slope of the line is $\dfrac{5000 - 1000}{100 - 0} = \dfrac{4000}{100} = 40$.
The commission rate of the executive is \$40 for every \$1000 worth of sales.

Chapter 5: Polynomials

Prep Test

1. $-4(3y)$
$= -12y$

2. $(-2)^3$
$= -2 \cdot -2 \cdot -2$
$= -8$

3. $(-4a + 7a) - 8b$
$= 3a - 8b$

4. $3x - 2[y - 4(x + 1) + 5]$
$= 3x - 2(y - 4x - 4 + 5)$
$= 3x - 2(y - 4x + 1)$
$= 3x - 2y + 8x - 2$
$= 11x - 2y - 2$

5. $-(x - y)$
$= -x + y$

6. $40 = 2 \cdot 20$
$= 2 \cdot 2 \cdot 10$
$= 2 \cdot 2 \cdot 2 \cdot 5$

7. $16 = \underline{2 \cdot 2} \cdot 2 \cdot 2$
$20 = \underline{2 \cdot 2} \cdot 5$
$24 = \underline{2 \cdot 2} \cdot 2 \cdot 3$
GCF is $2 \cdot 2$, which is 4.

8. $(-2)^3 - 2(-2)^2 + (-2) + 5$
$= -8 - 2(4) - 2 + 5$
$= -8 - 8 - 2 + 5$
$= -13$

9. $3x + 1 = 0$
$\left(\frac{1}{3}\right)3x = -1\left(\frac{1}{3}\right)$
$x = -\frac{1}{3}$

Go Figure

Let x represent the hours it takes to be "on time."

$57\left(x + \frac{1}{3}\right) = 64\left(x - \frac{1}{4}\right)$
$57x + 19 = 64x - 16$
$\left(\frac{1}{7}\right)35 = 7x\left(\frac{1}{7}\right)$
$x = 5$

If 5 hours is "on time," 20 minutes late is $5\frac{1}{3}$ hours. At 57 mph, the total distance traveled will be 304 miles.

$57\left(5\frac{1}{3}\right) = 304$

To travel 304 miles in 5 hours to be "on time" one would have to travel at a speed of 60.8 mph.

$\frac{304}{5} = 60.8$ Answer: 60.8 mph

Section 5.1

Objective A Exercises

1. $(ab^3)(a^3b) = a^4b^4$

3. $(9xy^2)(-2x^2y^2) = -18x^3y^4$

5. $(x^2y^4)^4 = x^8y^{16}$

7. $(-3x^2y^3)^4 = (-3)^4x^8y^{12} = 81x^8y^{12}$

9. $(3^3a^5b^3)^2 = 3^6a^{10}b^6 = 729a^{10}b^6$

11. $(x^2y^2)(xy^3)^3 = (x^2y^2)(x^3y^9) = x^5y^{11}$

13. $[(3x)^3]^2 = (3x)^6 = 3^6x^6 = 729x^6$

15. $[(ab)^3]^6 = (ab)^{18} = a^{18}b^{18}$

17. $[(2xy)^3]^4 = (2xy)^{12} = 2^{12}x^{12}y^{12} = 4096x^{12}y^{12}$

19. $[(2a^4b^3)^3]^2 = (2a^4b^3)^6 = 2^6a^{24}b^{18} = 64a^{24}b^{18}$

21. $x^n \cdot x^{n+1} = x^{n+n+1} = x^{2n+1}$

23. $y^{3n} \cdot y^{3n-2} = y^{3n+3n-2} = y^{6n-2}$

25. $(a^{n-3})^{2n} = a^{(n-3)2n} = a^{2n^2-6n}$

27. $(x^{3n+2})^5 = x^{(3n+2)5} = x^{15n+10}$

29. $(2xy)(-3x^2yz)(x^2y^3z^3) = -6x^5y^5z^4$

31. $(3b^5)(2ab^2)(-2ab^2c^2) = -12a^2b^9c^2$

33. $(-2x^2y^3z)(3x^2yz^4) = -6x^4y^4z^5$

35. $(-3ab^3)^3(-2^2a^2b)^2 = [(-3)^3a^3b^9][(-2^2)^2a^4b^2]$
$= (-27a^3b^9)(16a^4b^2)$
$= -432a^7b^{11}$

37. $(-2ab^2)(-3a^4b^5)^3 = (-2ab^2)[(-3)^3a^{12}b^{15}]$
$= (-2ab^2)(-27a^{12}b^{15})$
$= 54a^{13}b^{17}$

Objective B Exercises

39. $\frac{1}{3^{-5}} = 3^5 = 243$

41. $\frac{1}{y^{-3}} = y^3$

43. $\frac{a^3}{4b^{-2}} = \frac{a^3b^2}{4}$

45. $xy^{-4} = \frac{x}{y^4}$

47. $\frac{1}{2x^0} = \frac{1}{2}$

49. $\frac{-3^{-2}}{(2y)^0} = \frac{-1}{3^2} = -\frac{1}{9}$

51. $\frac{y^{-2}}{y^6} = y^{-2-6} = y^{-8} = \frac{1}{y^8}$

53. $(x^3y^5)^{-2} = x^{-6}y^{-10} = \frac{1}{x^6y^{10}}$

55. $\frac{x^4y^3}{x^{-1}y^{-2}} = x^5y^5$

57. $\frac{a^6b^{-4}}{a^{-2}b^5} = a^8b^{-9} = \frac{a^8}{b^9}$

59. $$\begin{aligned}(3a)^{-3}(9a^{-1})^{-2} &= (3a)^{-3}(3^2a^{-1})^{-2}\\ &= (3^{-3}a^{-3})(3^{-4}a^2)\\ &= 3^{-7}a^{-1}\\ &= \frac{1}{3^7a} = \frac{1}{2187a}\end{aligned}$$

61. $$\begin{aligned}(x^{-1}y^2)^{-3}(x^2y^{-4})^{-3} &= (x^3y^{-6})(x^{-6}y^{12})\\ &= x^{-3}y^6 = \frac{y^6}{x^3}\end{aligned}$$

63. $\frac{x^3y^6}{x^6y^2} = \frac{y^4}{x^3}$

65. $\frac{-6x^2y}{12x^4y} = -\frac{1}{2x^2}$

67. $\frac{-3ab^2}{(9a^2b^4)^3} = \frac{-3ab^2}{9^3a^6b^{12}} = \frac{-3ab^2}{729a^6b^{12}} = -\frac{1}{243a^5b^{10}}$

69. $\left(\frac{12x^3y^2z}{18xy^3z^4}\right)^4 = \left(\frac{2x^2}{3yz^3}\right)^4 = \frac{2^4x^8}{3^4y^4z^{12}} = \frac{16x^8}{81y^4z^{12}}$

71. $\frac{(3a^2b)^3}{(-6ab^3)^2} = \frac{3^3a^6b^3}{(-6)^2a^2b^6} = \frac{27a^6b^3}{36a^2b^6} = \frac{3a^4}{4b^3}$

73. $\frac{(-3a^2b^3)^2}{(-2ab^4)^3} = \frac{(-3)^2a^4b^6}{(-2)^3a^3b^{12}} = \frac{9a^4b^6}{-8a^3b^{12}} = -\frac{9a}{8b^6}$

75. $\frac{(-8x^2y^2)^4}{(16x^3y^7)^2} = \frac{(-8)^4x^8y^8}{16^2x^6y^{14}} = \frac{4096x^8y^8}{256x^6y^{14}} = \frac{16x^2}{y^6}$

77. $\frac{b^{6n}}{b^{10n}} = \frac{1}{b^{10n-6n}} = \frac{1}{b^{4n}}$

79. $\frac{y^{2n}}{-y^{8n}} = -\frac{1}{y^{8n-2n}} = -\frac{1}{y^{6n}}$

81. $\frac{y^{3n+2}}{y^{2n+4}} = y^{3n+2-(2n+4)} = y^{3n+2-2n-4} = y^{n-2}$

83. $\frac{x^ny^{3n}}{x^ny^{5n}} = \frac{1}{y^{5n-3n}} = \frac{1}{y^{2n}}$

85. $$\begin{aligned}\frac{x^{2n-1}y^{n-3}}{x^{n+4}y^{n+3}} &= x^{2n-1-(n+4)}y^{n-3-(n+3)}\\ &= x^{2n-1-n-4}y^{n-3-n-3}\\ &= x^{n-5}y^{-6}\\ &= \frac{x^{n-5}}{y^6}\end{aligned}$$

87. $$\begin{aligned}&\left(\frac{9ab^{-2}}{8a^{-2}b}\right)^{-2}\left(\frac{3a^{-2}b}{2a^2b^{-2}}\right)^3\\ &= \left(\frac{9a^3b^{-3}}{8}\right)^{-2}\left(\frac{3a^{-4}b^3}{2}\right)^3\\ &= \frac{9^{-2}a^{-6}b^6}{8^{-2}}\cdot\frac{3^3a^{-12}b^9}{2^3}\\ &= \frac{8^2\cdot 3^3a^{-18}b^{15}}{9^2\cdot 2^3}\\ &= \frac{64\cdot 27b^{15}}{81\cdot 8a^{18}}\\ &= \frac{8b^{15}}{3a^{18}}\end{aligned}$$

Objective C Exercises

89. $0.00000467 = 4.67 \times 10^{-6}$

91. $0.00000000017 = 1.7 \times 10^{-10}$

93. $200{,}000{,}000{,}000 = 2 \times 10^{11}$

95. $1.23 \times 10^{-7} = 0.000000123$

97. $8.2 \times 10^{15} = 8{,}200{,}000{,}000{,}000{,}000$

99. $3.9 \times 10^{-2} = 0.039$

101. $$\begin{aligned}(3 \times 10^{-12})(5 \times 10^{16}) &= (3)(5) \times 10^{-12+16}\\ &= 15 \times 10^4\\ &= 150{,}000\end{aligned}$$

103. $$\begin{aligned}&(0.0000065)(3{,}200{,}000{,}000{,}000)\\ &= (6.5 \times 10^{-6})(3.2 \times 10^{12})\\ &= (6.5)(3.2) \times 10^{-6+12}\\ &= 20.8 \times 10^6 = 20{,}800{,}000\end{aligned}$$

105. $$\begin{aligned}\frac{9 \times 10^{-3}}{6 \times 10^5} &= 1.5 \times 10^{-3-5}\\ &= 1.5 \times 10^{-8}\\ &\doteq 0.000000015\end{aligned}$$

107. $$\begin{aligned}\frac{0.0089}{500{,}000{,}000} &= \frac{8.9 \times 10^{-3}}{5 \times 10^8}\\ &= 1.78 \times 10^{-3-8}\\ &= 1.78 \times 10^{-11}\\ &= 0.0000000000178\end{aligned}$$

109. $$\begin{aligned}\frac{0.00056}{0.000000000004} &= \frac{5.6 \times 10^{-4}}{4 \times 10^{-12}}\\ &= 1.4 \times 10^{-4-(-12)}\\ &= 1.4 \times 10^8\\ &= 140{,}000{,}000\end{aligned}$$

111. $$\begin{aligned}&\frac{(3.2 \times 10^{-11})(2.9 \times 10^{15})}{8.1 \times 10^{-3}}\\ &= \frac{(3.2)(2.9) \times 10^{-11+15-(-3)}}{8.1}\\ &\approx 1.145679 \times 10^7\\ &= 11{,}456{,}790\end{aligned}$$

113. $\dfrac{(0.00000004)(84{,}000)}{(0.0003)(1{,}400{,}000)}$

$= \dfrac{4 \times 10^{-8} \times 8.4 \times 10^{4}}{3 \times 10^{-4} \times 1.4 \times 10^{6}}$

$= \dfrac{(4)(8.4) \times 10^{-8+4-(-4)-6}}{3(1.4)}$

$= 8 \times 10^{-6} = 0.000008$

Objective D Exercises

115. Strategy To find the time needed to cross the galaxy, divide the width of the galaxy (5.6×10^{19} mi) by the rate of the space ship (2.5×10^{4} mph)

Solution $\dfrac{5.6 \times 10^{19}}{2.5 \times 10^{4}} = 2.24 \times 10^{15}$

It would take a space ship 2.24×10^{15} h to cross the galaxy.

117. Strategy To find the time it takes the satellite to reach Saturn, divide the distance to Saturn (8.86×10^{8} mi) by the rate the satellite travels (1×10^{5} mph).

Solution $\dfrac{8.86 \times 10^{8}}{1 \times 10^{5}} = 8.86 \times 10^{3}$

It takes the satellite 8.86×10^{3} h to reach Saturn.

119. Strategy To find the number of times heavier the proton is, divide the mass of a proton (1.673×10^{-27} kg) by the mass of an electron (9.109×10^{-31} kg).

Solution $\dfrac{1.673 \times 10^{-27}}{9.109 \times 10^{-31}} \approx 1.83664508 \times 10^{3}$

A proton is 1.83664508×10^{3} times heavier than an electron.

121. Strategy To find the number of meters light travels in 8 h:

- Multiply the speed of light (3×10^{8} m/s) by the number of seconds per hour (3.6×10^{3} s/h).
- Multiply by 8 h.

Solution $(3 \times 10^{8}) \cdot (3.6 \times 10^{3}) \cdot 8 = 8.64 \times 10^{12}$

Light travels 8.64×10^{12} m in 8 h.

123. Strategy To find the rate of the signals:

- Write the distance (119 million mi) in scientific notation.
- Divide the distance by the time (1.1×10 min).

Solution $\dfrac{1.19 \times 10^{8}}{1.1 \times 10} = 1.0\overline{81} \times 10^{7}$

The signals from Earth to Mars traveled $1.0\overline{81} \times 10^{7}$ mi/min.

125. Strategy To find the time in seconds for the centrifuge to make one revolution, divide the time in seconds/min (6.0×10 s/m) by the number of revolutions per min (4×10^{8} rev/min).

Solution $\dfrac{6.0 \times 10}{4 \times 10^{8}} = 1.5 \times 10^{-7}$

The centrifuge makes one revolution in 1.5×10^{-7} s.

Applying the Concepts

127a.

$$1 + (1 + (1 + 2^{-1})^{-1})^{-1} = 1 + \left(1 + \left(1 + \frac{1}{2}\right)^{-1}\right)^{-1}$$
$$= 1 + \left(1 + \left(\frac{3}{2}\right)^{-1}\right)^{-1}$$
$$= 1 + \left(1 + \frac{2}{3}\right)^{-1}$$
$$= 1 + \left(\frac{5}{3}\right)^{-1}$$
$$= 1 + \frac{3}{5}$$
$$= \frac{8}{5}$$

b.

$$2 - (2 - (2 - 2^{-1})^{-1})^{-1} = 2 - \left(2 - \left(2 - \frac{1}{2}\right)^{-1}\right)^{-1}$$
$$= 2 - \left(2 - \left(\frac{3}{2}\right)^{-1}\right)^{-1}$$
$$= 2 - \left(2 - \frac{2}{3}\right)^{-1}$$
$$= 2 - \left(\frac{4}{3}\right)^{-1}$$
$$= 2 - \frac{3}{4}$$
$$= \frac{5}{4}$$

Section 5.2

Objective A Exercises

1. $P(3) = 3(3)^2 - 2(3) - 8$
$P(3) = 13$

3. $R(2) = 2(2)^3 - 3(2)^2 + 4(2) - 2$
$R(2) = 10$

5. $f(-1) = (-1)^4 - 2(-1)^2 - 10$
$f(-1) = -11$

7. Polynomial: (a) -1 (b) 8 (c) 2

9. Not a polynomial.

11. Not a polynomial.

13. Polynomial: (a) 3 (b) π (c) 5

15. Polynomial: (a) -5 (b) 2 (c) 3

17. Polynomial: (a) 14 (b) 14 (c) 0

19.

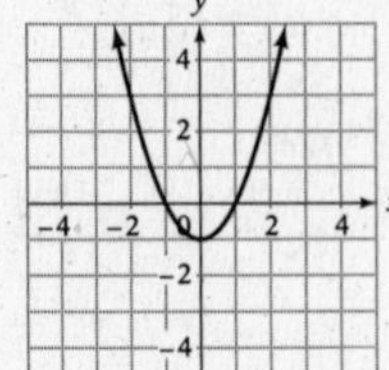

21.

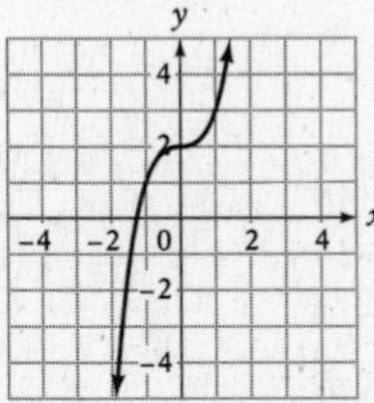

23.

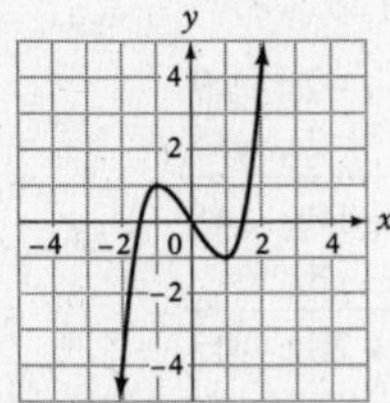

Objective B Exercises

25.
$$\begin{array}{r} 5x^2+2x-\ 7 \\ x^2-8x+12 \\ \hline 6x^2-6x+\ 5 \end{array}$$

27.
$$\begin{array}{r} x^2-3x+8 \\ -2x^2+3x-7 \\ \hline -x^2\qquad\ +1 \end{array}$$

29.
$$\begin{aligned} &(3y^2-7y)+(2y^2-8y+2) \\ &=(3y^2+2y^2)+(-7y-8y)+2 \\ &=5y^2-15y+2 \end{aligned}$$

31.
$$\begin{aligned} &(2a^2-3a-7)-(-5a^2-2a-9) \\ &=(2a^2+5a^2)+(-3a+2a)+(-7+9) \\ &=7a^2-a+2 \end{aligned}$$

33.
$$\begin{aligned} P(x)+R(x)&=(x^2-3xy+y^2)+(2x^2-3y^2) \\ &=(x^2+2x^2)-3xy+(y^2-3y^2) \\ &=3x^2-3xy-2y^2 \end{aligned}$$

35.
$$\begin{aligned} P(x)-R(x)&=(3x^2+2y^2)-(-5x^2+2xy-3y^2) \\ &=(3x^2+5x^2)-2xy+(2y^2+3y^2) \\ &=8x^2-2xy+5y^2 \end{aligned}$$

37.
$$\begin{aligned} S(x)&=(3x^4-3x^3-x^2)+(3x^3-7x^2+2x) \\ &=3x^4+(-3x^3+3x^3)+(-x^2-7x^2)+2x \\ S(x)&=3x^4-8x^2+2x \\ S(2)&=3(2)^4-8(2)^2+2(2) \\ S(2)&=20 \end{aligned}$$

Applying the Concepts

39a.
$$\begin{aligned} (2x^3+3x^2+kx+5)-(x^3+2x^2+3x+7)&=x^3+x^2+5x-2 \\ (2x^3-x^3)+(3x^2-2x^2)+(kx-3x)+(5-7)&=x^3+x^2+5x-2 \\ x^3+x^2+(k-3)x-2&=x^3+x^2+5x-2 \\ (k-3)x&=5x \\ k-3&=5 \\ k&=8 \end{aligned}$$

b.
$$\begin{aligned} (6x^3+kx^2-2x-1)-(4x^3-3x^2+1)&=2x^3-x^2-2x-2 \\ (6x^3-4x^3)+(kx^2+3x^2)+(-2x)+(-1-1)&=2x^3-x^2-2x-2 \\ 2x^3+(k+3)x^2-2x-2&=2x^3-x^2-2x-2 \\ (k+3)x^2&=-x^2 \\ k+3&=-1 \\ k&=-4 \end{aligned}$$

41. If $P(x)$ is a fifth-degree polynomial and $Q(x)$ is a fourth-degree polynomial, then $P(x)-Q(x)$ is a fifth-degree polynomial.
Example:
$$\begin{aligned} P(x)&=10x^5-x^4+3x^2-1 \\ Q(x)&=-x^4+x^3+x^2+2x+5 \\ P(x)-Q(x)&=10x^5-x^3+2x^2-2x-6 \end{aligned}$$
(a fifth-degree polynomial)

Section 5.3

Objective A Exercises

1. $2x(x-3) = 2x^2 - 6x$

3. $3x^2(2x^2 - x) = 6x^4 - 3x^3$

5. $3xy(2x - 3y) = 6x^2y - 9xy^2$

7. $x^n(x+1) = x^{n+1} + x^n$

9. $x^n(x^n + y^n) = x^{2n} + x^ny^n$

11. $2b + 4b(2-b) = 2b + 8b - 4b^2 = -4b^2 + 10b$

13. $-2a^2(3a^2 - 2a + 3) = -6a^4 + 4a^3 - 6a^2$

15. $(-3y^2 - 4y + 2)(y^2) = -3y^4 - 4y^3 + 2y^2$

17. $-5x^2(4 - 3x + 3x^2 + 4x^3)$
$= -20x^2 + 15x^3 - 15x^4 - 20x^5$

19. $-2x^2y(x^2 - 3xy + 2y^2)$
$= -2x^4y + 6x^3y^2 - 4x^2y^3$

21. $x^n(x^{2n} + x^n + x)$
$= x^{3n} + x^{2n} + x^{n+1}$

23. $a^{n+1}(a^n - 3a + 2)$
$= a^{2n+1} - 3a^{n+2} + 2a^{n+1}$

25. $2y^2 - y[3 - 2(y-4) - y]$
$= 2y^2 - y[3 - 2y + 8 - y]$
$= 2y^2 - y[11 - 3y]$
$= 2y^2 - 11y + 3y^2$
$= 5y^2 - 11y$

27. $2y - 3[y - 2y(y-3) + 4y]$
$= 2y - 3[y - 2y^2 + 6y + 4y]$
$= 2y - 3[11y - 2y^2]$
$= 2y - 33y + 6y^2$
$= 6y^2 - 31y$

29. $P(b) \cdot Q(b) = 3b(3b^4 - 3b^2 + 8)$
$= 9b^5 - 9b^3 + 24b$

Objective B Exercises

31. $(x-2)(x+7) = x^2 + 7x - 2x - 14$
$= x^2 + 5x - 14$

33. $(2y-3)(4y+7) = 8y^2 + 14y - 12y - 21$
$= 8y^2 + 2y - 21$

35. $2(2x - 3y)(2x + 5y)$
$= 2(4x^2 + 10xy - 6xy - 15y^2)$
$= 2(4x^2 + 4xy - 15y^2)$
$= 8x^2 + 8xy - 30y^2$

37. $(xy+4)(xy-3) = x^2y^2 - 3xy + 4xy - 12$
$= x^2y^2 + xy - 12$

39. $(2x^2 - 5)(x^2 - 5) = 2x^4 - 10x^2 - 5x^2 + 25$
$= 2x^4 - 15x^2 + 25$

41. $(5x^2 - 5y)(2x^2 - y) = 10x^4 - 5x^2y - 10x^2y + 5y^2$
$= 10x^4 - 15x^2y + 5y^2$

43. $(x^n + 2)(x^n - 3) = x^{2n} - 3x^n + 2x^n - 6$
$= x^{2n} - x^n - 6$

45. $(2a^n - 3)(3a^n + 5) = 6a^{2n} + 10a^n - 9a^n - 15$
$= 6a^{2n} + a^n - 15$

47. $(2a^n - b^n)(3a^n + 2b^n)$
$= 6a^{2n} + 4a^nb^n - 3a^nb^n - 2b^{2n}$
$= 6a^{2n} + a^nb^n - 2b^{2n}$

49.
$$\begin{array}{r} x^2 - 3x + 4 \\ x + 5 \\ \hline 5x^2 - 15x + 20 \\ x^3 - 3x^2 + 4x \\ \hline x^3 + 2x^2 - 11x + 20 \end{array}$$

51.
$$\begin{array}{r} 5a^2 - 6ab + 4b^2 \\ \times \quad 2a - 3b \\ \hline -15a^2b + 18ab^2 - 12b^3 \\ 10a^3 - 12a^2b + 8ab^2 \\ \hline 10a^3 - 27a^2b + 26ab^2 - 12b^3 \end{array}$$

53.
$$\begin{array}{r} 2x^4 - 3x^3 \qquad - 2x + 9 \\ \times \quad 2x - 5 \\ \hline -10x^4 + 15x^3 \qquad + 10x - 45 \\ 4x^5 - 6x^4 \qquad - 4x^2 + 18x \\ \hline 4x^5 - 16x^4 + 15x^3 - 4x^2 + 28x - 45 \end{array}$$

55.
$$\begin{array}{r} x^2 + 2x - 3 \\ \times \quad x^2 - 5x + 7 \\ \hline 7x^2 + 14x - 21 \\ -5x^3 - 10x^2 + 15x \\ x^4 + 2x^3 - 3x^2 \\ \hline x^4 - 3x^3 - 6x^2 + 29x - 21 \end{array}$$

57. $(a-2)(2a-3)(a+7)$
$= (2a^2 - 3a - 4a + 6)(a+7)$
$= (2a^2 - 7a + 6)(a+7)$
$$\begin{array}{r} 2a^2 - 7a + 6 \\ \times \quad a + 7 \\ \hline 14a^2 - 49a + 42 \\ 2a^3 - 7a^2 + 6a \\ \hline 2a^3 + 7a^2 - 43a + 42 \end{array}$$

59.
$$\begin{array}{r} x^{2n} + x^n + 1 \\ \times \quad x^n + 1 \\ \hline x^{2n} + x^n + 1 \\ x^{3n} + x^{2n} + x^n \\ \hline x^{3n} + 2x^{2n} + 2x^n + 1 \end{array}$$

61.
$$\begin{array}{r} x^n - 2x^ny^n + 3y^n \\ \times \quad x^n + y^n \\ \hline x^ny^n - 2x^ny^{2n} + 3y^{2n} \\ x^{2n} - 2x^{2n}y^n + 3x^ny^n \phantom{+3y^{2n}} \\ \hline x^{2n} - 2x^{2n}y^n + 4x^ny^n - 2x^ny^{2n} + 3y^{2n} \end{array}$$

63. $P(y) \cdot Q(y) = (2y^2 - 1)(y^3 - 5y^2 - 3)$

$$\begin{array}{r} y^3 - 5y^2 - 3 \\ 2y^2 - 1 \\ \hline - y^3 + 5y^2 + 3 \\ 2y^5 - 10y^4 \qquad - 6y^2 \qquad \\ \hline 2y^5 - 10y^4 - y^3 - y^2 + 3 \end{array}$$

Objective C Exercises

65. $(3x - 2)(3x + 2) = 9x^2 - 4$

67. $(6 - x)(6 + x) = 36 - x^2$

69. $(2a - 3b)(2a + 3b) = 4a^2 - 9b^2$

71. $(x^2 + 1)(x^2 - 1) = x^4 - 1$

73. $(x^n + 3)(x^n - 3) = x^{2n} - 9$

75. $(x - 5)^2 = x^2 - 10x + 25$

77. $(3a + 5b)^2 = 9a^2 + 30ab + 25b^2$

79. $(x^2 - 3)^2 = x^4 - 6x^2 + 9$

81. $(2x^2 - 3y^2)^2 = 4x^4 - 12x^2y^2 + 9y^4$

83. $(a^n - b^n)^2 = a^{2n} - 2a^nb^n + b^{2n}$

85. $y^2 - (x - y)^2 = y^2 - (x^2 - 2xy + y^2)$
$= y^2 - x^2 + 2xy - y^2$
$= -x^2 + 2xy$

87. $(x - y)^2 - (x + y)^2$
$= (x^2 - 2xy + y^2) - (x^2 + 2xy + y^2)$
$= x^2 - 2xy + y^2 - x^2 - 2xy - y^2$
$= -4xy$

Objective D Exercises

89. **Strategy** To find the area, replace the variables L and W in the equation $A = L \cdot W$ by the given values and solve for A.

Solution $A = L \cdot W$
$A = (3x - 2)(x + 4)$
$A = 3x^2 + 12x - 2x - 8$
$A = 3x^2 + 10x - 8$
The area is $(3x^2 + 10x - 8)$ ft^2.

91. **Strategy** To find the area, add the area of the small rectangle to the area of the large rectangle.
Large rectangle:
Length $= L_1 = x + 5$
Width $= W_1 = x - 2$
Small rectangle:
Length $= L_2 = 5$
Width $= W_2 = 2$

Solution $A =$ Area of the large rectangle + area of the small rectangle
$A = (L_1 \cdot W_1) + (L_2 \cdot W_2)$
$A = (x + 5)(x - 2) + (5)(2)$
$A = x^2 - 2x + 5x - 10 + 10$
$A = x^2 + 3x$
The area is $(x^2 + 3x)$ m^2.

93. **Strategy** To find the volume, replace the variable s in the equation $V = s^3$ with its given value and solve for V.

Solution $V = s^3$
$V = (x + 3)^3$
$V = (x + 3)(x + 3)(x + 3)$
$V = (x^2 + 6x + 9)(x + 3)$
$V = x^3 + 9x^2 + 27x + 27$
The volume is
$(x^3 + 9x^2 + 27x + 27)$ cm^3.

95. **Strategy** To find the volume, subtract the volume of the small rectangular solid from the volume of the large rectangular solid.
Large rectangular solid:
Length $= L_1 = x + 2$
Width $= W_1 = 2x$
Height $= h_1 = x$
Small rectangular solid:
Length $= L_2 = x$
Width $= W_2 = 2x$
Height $= h_2 = 2$

Solution $V = (L_1 \cdot W_1 \cdot h_1) - (L_2 \cdot W_2 \cdot h_2)$
$V = (x + 2)(2x)(x) - (x)(2x)(2)$
$V = (2x^2 + 4x)(x) - (2x^2)(2)$
$V = 2x^3 + 4x^2 - 4x^2$
$V = 2x^3$
The volume is $(2x^3)$ in^3.

97. **Strategy** To find the area, replace the variable r in the equation $A = \pi r^2$ by the given value and solve for A.

Solution $A = \pi r^2$
$A \approx 3.14(5x + 4)^2$
$A = 3.14(25x^2 + 40x + 16)$
$A = 78.5x^2 + 125.6x + 50.24$
The area is
$(78.5x^2 + 125.6x + 50.24)$ in^2.

Applying the Concepts

99a.

$$\begin{array}{r} a^2 + ab + b^2 \\ \times \qquad\qquad a - b \\ \hline -a^2b - ab^2 - b^3 \\ + a^3 + a^2b + ab^2 \qquad \\ \hline a^3 - b^3 \end{array}$$

b.
$$\begin{array}{r} x^2 - xy + y^2 \\ \times \qquad x + y \\ \hline x^2y - xy^2 + y^3 \\ x^3 - x^2y + xy^2 \qquad \\ \hline x^3 + y^3 \end{array}$$

101a.
$$\begin{aligned} (3x-k)(2x+k) &= 6x^2+5x-k^2 \\ 6x^2+xk-k^2 &= 6x^2+5x-k^2 \\ xk &= 5x \\ k &= 5 \end{aligned}$$

b.
$$\begin{aligned} (4x+k)^2 &= 16x^2+8x+k^2 \\ 16x^2+8xk+k^2 &= 16x^2+8x+k^2 \\ 8xk &= 8x \\ k &= 1 \end{aligned}$$

103. The product of $4a+b$ and $2a-b$ is
$(4a+b)(2a-b) = 8a^2-2ab-b^2$
Subtract $8a^2-2ab-b^2$ from $9a^2-2ab$.
$$\begin{array}{l} 9a^2-2ab \\ \underline{-(8a^2-2ab-b^2)} \\ \qquad a^2+b^2 \end{array}$$

Section 5.4

Objective A Exercises

1. $\dfrac{3x^2-6x}{3x} = \dfrac{3x^2}{3x} - \dfrac{6x}{3x}$
$= x-2$
check: $3x(x-2) = 3x^2-6x$

3. $\dfrac{5x^2-10x}{-5x} = \dfrac{5x^2}{-5x} - \dfrac{10x}{-5x}$
$= -x+2$
check: $-5x(-x+2) = 5x^2-10x$

5. $\dfrac{5x^2y^2+10xy}{5xy} = \dfrac{5x^2y^2}{5xy} + \dfrac{10xy}{5xy}$
$= xy+2$
check: $5xy(xy+2) = 5x^2y^2+10xy$

7. $\dfrac{x^3+3x^2-5x}{x} = \dfrac{x^3}{x} + \dfrac{3x^2}{x} - \dfrac{5x}{x}$
$= x^2+3x-5$
check: $x(x^2+3x-5) = x^3+3x^2-5x$

9. $\dfrac{9b^5+12b^4+6b^3}{3b^2} = \dfrac{9b^5}{3b^2} + \dfrac{12b^4}{3b^2} + \dfrac{6b^3}{3b^2}$
$= 3b^3+4b^2+2b$
check: $3b^2(3b^3+4b^2+2b) = 9b^5+12b^4+6b^3$

11. $\dfrac{a^5b-6a^3b+ab}{ab} = \dfrac{a^5b}{ab} - \dfrac{6a^3b}{ab} + \dfrac{ab}{ab}$
$= a^4-6a^2+1$
check: $ab(a^4-6a^2+1) = a^5b-6a^3b+ab$

Objective B Exercises

13.
$$\begin{array}{r} x+8 \\ x-5\overline{)x^2+3x-40} \\ \underline{x^2-5x} \qquad \\ 8x-40 \\ \underline{8x-40} \\ 0 \end{array}$$
$(x^2+3x-40) \div (x-3) = x+8$

15.
$$\begin{array}{r} x^2+3x+6 \\ x-3\overline{)x^3+0x^2-3x+2} \\ \underline{x^3-3x^2} \qquad\qquad \\ 3x^2-3x \qquad \\ \underline{3x^2-9x} \qquad \\ 6x+\ 2 \\ \underline{6x-18} \\ 20 \end{array}$$
$(x^3-3x+2) \div (x-3) = x^2+3x+6+\dfrac{20}{x-3}$

17.
$$\begin{array}{r} 3x+5 \\ 2x+1\overline{)6x^2+13x+8} \\ \underline{6x^2+\ 3x} \qquad \\ 10x+8 \\ \underline{10x+5} \\ 3 \end{array}$$
$(6x^2+13x+8) \div (2x+1) = 3x+5+\dfrac{3}{2x+1}$

19.
$$\begin{array}{r} 5x+7 \\ 2x-1\overline{)10x^2+9x-5} \\ \underline{10x^2-5x} \qquad \\ 14x-5 \\ \underline{14x-7} \\ 2 \end{array}$$
$(10x^2+9x-5) \div (2x-1) = 5x+7+\dfrac{2}{2x-1}$

21.
$$\begin{array}{r} 4x^2+6x+9 \\ 2x-3\overline{)8x^3+0x^2+0x-9} \\ \underline{8x^3-12x^2} \qquad\qquad \\ 12x^2+0 \qquad \\ \underline{12x^2-18x} \qquad \\ 18x-9 \\ \underline{18x-27} \\ 18 \end{array}$$
$(8x^3-9) \div (2x-3) = 4x^2+6x+9+\dfrac{18}{2x-3}$

23.
$$\begin{array}{r} 3x^2+1 \\ 2x^2-5\overline{)6x^4+0x^3-13x^2+0x-4} \\ \underline{6x^4 \qquad -15x^2} \qquad\qquad \\ 2x^2+0x-4 \\ \underline{2x^2 \qquad -5} \\ 1 \end{array}$$
$(6x^4-13x^2-4) \div (2x^2-5) = 3x^2+1+\dfrac{1}{2x^2-5}$

25.
$$\begin{array}{r}
x^2 - 3x - 10 \\
3x+1\overline{)3x^3 - 8x^2 - 33x - 10} \\
\underline{3x^3 + x^2} \\
-9x^2 - 33x \\
\underline{-9x^2 - 3x} \\
-30x - 10 \\
\underline{-30x - 10} \\
0
\end{array}$$
$$\frac{3x^3 - 8x^2 - 33x - 10}{3x+1} = x^2 - 3x - 10$$

27.
$$\begin{array}{r}
x^2 - 2x + 1 \\
x-3\overline{)x^3 - 5x^2 + 7x - 4} \\
\underline{x^3 - 3x^2} \\
-2x^2 + 7x \\
\underline{-2x^2 + 6x} \\
x - 4 \\
\underline{x - 3} \\
-1
\end{array}$$
$$\frac{x^3 - 5x^2 + 7x - 4}{x-3} = x^2 - 2x + 1 - \frac{1}{x-3}$$

29.
$$\begin{array}{r}
2x^3 - 3x^2 + x - 4 \\
x-5\overline{)2x^4 - 13x^3 + 16x^2 - 9x + 20} \\
\underline{2x^4 - 10x^3} \\
-3x^3 + 16x^2 \\
\underline{-3x^3 + 15x^2} \\
x^2 - 9x \\
\underline{x^2 - 5x} \\
-4x + 20 \\
\underline{-4x + 20} \\
0
\end{array}$$
$$\frac{2x^4 - 13x^3 + 16x^2 - 9x + 20}{x-5} = 2x^3 - 3x^2 + x - 4$$

31.
$$\begin{array}{r}
2x \\
x^2+2x-1\overline{)2x^3 + 4x^2 - x + 2} \\
\underline{2x^3 + 4x^2 - 2x} \\
x + 2
\end{array}$$
$$\frac{2x^3 + 4x^2 - x + 2}{x^2 + 2x - 1} = 2x + \frac{x+2}{x^2 + 2x - 1}$$

33.
$$\begin{array}{r}
x^2 + 4x + 6 \\
x^2-2x-1\overline{)x^4 + 2x^3 - 3x^2 - 6x + 2} \\
\underline{x^4 - 2x^3 - x^2} \\
4x^3 - 2x^2 - 6x \\
\underline{4x^3 - 8x^2 - 4x} \\
6x^2 - 2x + 2 \\
\underline{6x^2 - 12x - 6} \\
10x + 8
\end{array}$$
$$\frac{x^4 + 2x^3 - 3x^2 - 6x + 2}{x^2 - 2x - 1}$$
$$= x^2 + 4x + 6 + \frac{10x + 8}{x^2 - 2x - 1}$$

35. $\dfrac{P(x)}{Q(x)} = \dfrac{2x^3 + x^2 + 8x + 7}{2x+1}$
$$\begin{array}{r}
x^2 \quad + 4 \\
2x+1\overline{)2x^3 + x^2 + 8x + 7} \\
\underline{2x^3 + x^2} \\
8x + 7 \\
\underline{8x + 4} \\
3
\end{array}$$
$$\frac{2x^3 + x^2 + 8x + 7}{2x+1} = x^2 + 4 + \frac{3}{2x+1}$$

Objective C Exercises

37.
$$\begin{array}{r|rrr}
-1 & 2 & -6 & -8 \\
 & & -2 & 8 \\
\hline
 & 2 & -8 & 0
\end{array}$$
$(2x^2 - 6x - 8) \div (x+1) = 2x - 8$

39.
$$\begin{array}{r|rrr}
2 & 3 & -14 & 16 \\
 & & 6 & -16 \\
\hline
 & 3 & -8 & 0
\end{array}$$
$(3x^2 - 14x + 16) \div (x-2) = 3x - 8$

41.
$$\begin{array}{r|rrr}
1 & 3 & 0 & -4 \\
 & & 3 & 3 \\
\hline
 & 3 & 3 & -1
\end{array}$$
$(3x^2 - 4) \div (x-1) = 3x + 3 - \dfrac{1}{x-1}$

43.
$$\begin{array}{r|rrrr}
-1 & 2 & -1 & 6 & 9 \\
 & & -2 & 3 & -9 \\
\hline
 & 2 & -3 & 9 & 0
\end{array}$$
$(2x^3 - x^2 + 6x + 9) \div (x+1) = 2x^2 - 3x + 9$

45.
$$\begin{array}{r|rrrr}
2 & 4 & 0 & -1 & -18 \\
 & & 8 & 16 & 30 \\
\hline
 & 4 & 8 & 15 & 12
\end{array}$$
$(4x^3 - x - 18) \div (x-2) = 4x^2 + 8x + 15 + \frac{12}{x-2}$

47.
$$\begin{array}{r|rrrr}
-4 & 2 & 5 & -5 & 20 \\
 & & -8 & 12 & -28 \\
\hline
 & 2 & -3 & 7 & -8
\end{array}$$
$(2x^3 + 5x^2 - 5x + 20) \div (x+4)$
$= 2x^2 - 3x + 7 - \dfrac{8}{x+4}$

49.
$$\begin{array}{r|rrrrr}
2 & 3 & -4 & 8 & -5 & -5 \\
 & & 6 & 4 & 24 & 38 \\
\hline
 & 3 & 2 & 12 & 19 & 33
\end{array}$$
$$\frac{3x^4 - 4x^3 + 8x^2 - 5x - 5}{x-2}$$
$$= 3x^3 + 2x^2 + 12x + 19 + \frac{33}{x-2}$$

51.

$$\begin{array}{r|rrrrr} -1 & 3 & 3 & -1 & 3 & 2 \\ & & -3 & 0 & 1 & -4 \\ \hline & 3 & 0 & -1 & 4 & -2 \end{array}$$

$$\frac{3x^4+3x^3-x^2+3x+2}{x+1} = 3x^3-x+4-\frac{2}{x+1}$$

53. $\dfrac{P(x)}{Q(x)} = \dfrac{3x^2-5x+6}{x-2}$

$$\begin{array}{r|rrr} 2 & 3 & -5 & 6 \\ & & 6 & 2 \\ \hline & 3 & 1 & 8 \end{array}$$

$$\frac{P(x)}{Q(x)} = 3x+1+\frac{8}{x-2}$$

Objective D Exercises

55.

$$\begin{array}{r|rrr} 3 & 2 & -3 & -1 \\ & & 6 & 9 \\ \hline & 2 & 3 & 8 \end{array}$$

$P(3) = 8$

57.

$$\begin{array}{r|rrrr} 4 & 1 & -2 & 3 & -1 \\ & & 4 & 8 & 44 \\ \hline & 1 & 2 & 11 & 43 \end{array}$$

$R(4) = 43$

59.

$$\begin{array}{r|rrrr} -2 & 2 & -4 & 3 & -1 \\ & & -4 & 16 & -38 \\ \hline & 2 & -8 & 19 & -39 \end{array}$$

$P(-2) = -39$

61.

$$\begin{array}{r|rrrr} -3 & 2 & -1 & 0 & 3 \\ & & -6 & 21 & -63 \\ \hline & 2 & -7 & 21 & -60 \end{array}$$

$Z(-3) = -60$

63.

$$\begin{array}{r|rrrrr} 2 & 1 & 3 & -2 & 4 & -9 \\ & & 2 & 10 & 16 & 40 \\ \hline & 1 & 5 & 8 & 20 & 31 \end{array}$$

$Q(2) = 31$

65.

$$\begin{array}{r|rrrrr} -3 & 2 & -1 & 0 & 2 & -5 \\ & & -6 & 21 & -63 & 183 \\ \hline & 2 & -7 & 21 & -61 & 178 \end{array}$$

$F(-3) = 178$

67.

$$\begin{array}{r|rrrr} 5 & 1 & 0 & 0 & -3 \\ & & 5 & 25 & 125 \\ \hline & 1 & 5 & 25 & 122 \end{array}$$

$P(5) = 122$

69.

$$\begin{array}{r|rrrrr} -3 & 4 & 0 & -3 & 0 & 5 \\ & & -12 & 36 & -99 & 297 \\ \hline & 4 & -12 & 33 & -99 & 302 \end{array}$$

$R(-3) = 302$

71.

$$\begin{array}{r|rrrrrr} 2 & 1 & 0 & -4 & -2 & 5 & -2 \\ & & 2 & 4 & 0 & -4 & 2 \\ \hline & 1 & 2 & 0 & -2 & 1 & 0 \end{array}$$

$Q(2) = 0$

Applying the Concepts

73a.

$$\begin{array}{r} a^2-ab+b^2 \\ a+b\overline{)a^3 \qquad\qquad\quad +b^3} \\ \underline{a^3+a^2b} \qquad\qquad \\ -a^2b \qquad\qquad \\ \underline{-a^2b-ab^2} \qquad \\ ab^2+b^3 \\ \underline{ab^2+b^3} \\ 0 \end{array}$$

$$\frac{a^3+b^3}{a+b} = a^2-ab+b^2$$

b.

$$\begin{array}{r} x^4-x^3y+x^2y^2-xy^3+y^4 \\ x+y\overline{)x^5 \qquad\qquad\qquad\qquad\qquad +y^5} \\ \underline{x^5+x^4y} \qquad\qquad\qquad\qquad\quad \\ -x^4y \qquad\qquad\qquad\qquad\quad \\ \underline{-x^4y-x^3y^2} \qquad\qquad\qquad \\ x^3y^2 \qquad\qquad\qquad \\ \underline{x^3y^2+x^2y^3} \qquad\qquad \\ -x^2y^3 \qquad\qquad \\ \underline{-x^2y^3-xy^4} \qquad \\ xy^4+y^5 \\ \underline{xy^4+y^5} \\ 0 \end{array}$$

$$\frac{x^5+y^5}{x+y} = x^4-x^3y+x^2y^2-xy^3+y^4$$

c.

$$\begin{array}{r} x^5-x^4y+x^3y^2-x^2y^3+xy^4-y^5 \\ x+y\overline{)x^6 \qquad\qquad\qquad\qquad\qquad\qquad -y^6} \\ \underline{x^6+x^5y} \qquad\qquad\qquad\qquad\qquad \\ -x^5y \qquad\qquad\qquad\qquad\qquad \\ \underline{-x^5y-x^4y^2} \qquad\qquad\qquad\qquad \\ x^4y^2 \qquad\qquad\qquad\qquad \\ \underline{x^4y^2+x^3y^3} \qquad\qquad\qquad \\ -x^3y^3 \qquad\qquad\qquad \\ \underline{-x^3y^3-x^2y^4} \qquad\qquad \\ x^2y^4 \qquad\qquad \\ \underline{x^2y^4-xy^5} \qquad \\ -xy^5-y^6 \\ \underline{-xy^5-y^6} \\ 0 \end{array}$$

$$\frac{x^6-y^6}{x-y} = x^5-x^4y+x^3y^2-x^2y^3+xy^4-y^5$$

75. Synthetic division can be modified so that the divisor is in the form $ax+b$. Divide both the dividend and the divisor by a (or multiply both the dividend and the divisor by $\frac{1}{a}$). The divisor is now in the form $x+\frac{b}{a}$, and the expression $\frac{b}{a}$ can be used for a in the $x-a$ of synthetic division.

Section 5.5

Objective A Exercises

1. The GCF of $6a^2$ and $15a$ is $3a$.
$6a^2 - 15a = 3a(2a - 5)$

3. The GCF of $4x^3$ and $3x^2$ is x^2.
$4x^3 - 3x^2 = x^2(4x - 3)$

5. There is no common factor.
$3a^2 - 10b^3$ is nonfactorable over the integers.

7. The GCF of x^5, x^3, and x is x.
$x^5 - x^3 - x = x(x^4 - x^2 - 1)$

9. The GCF of $16x^2$, $12x$, and 24 is 4.
$16x^2 - 12 + 24 = 4(4x^2 - 3x + 6)$

11. The GCF of $5b^2$, $10b^3$, and $25b^4$ is $5b^2$.
$5b^2 - 10b^3 + 25b^4 = 5b^2(1 - 2b + 5b^2)$

13. The GCF of x^{2n} and x^n is x^n.
$x^{2n} - x^n = x^n(x^n - 1)$

15. The GCF of x^{3n} and x^{2n} is x^{2n}.
$x^{3n} - x^{2n} = x^{2n}(x^n - 1)$

17. The GCF of a^{2n+2} and a^2 is a^2.
$a^{2n+2} + a^2 = a^2(a^{2n} + 1)$

19. The GCF of $12x^2y^2$, $18x^3y$, and $24x^2y$ is $6x^2y$.
$12x^2y^2 - 18x^3y + 24x^2y = 6x^2y(2y - 3x + 4)$

21. The GCF of $24a^3b^2$, $4a^2b^2$, and $16a^2b^4$ is $4a^2b^2$.
$24a^3b^2 - 4a^2b^2 - 16a^2b^4$
$= 4a^2b^2(6a - 1 - 4b^2)$

23. The GCF of y^{2n+2}, y^{n+2}, and y^2 is y^2.
$y^{2n+2} + y^{n+2} - y^2 = y^2(y^{2n} + y^n - 1)$

Objective B Exercises

25. $x(a + 2) - 2(a + 2) = (a + 2)(x - 2)$

27. $a(x - 2) - b(2 - x) = a(x - 2) + b(x - 2)$
$= (x - 2)(a + b)$

29. $x(a - 2b) + y(2b - a) = x(a - 2b) - y(a - 2b)$
$= (a - 2b)(x - y)$

31. $xy + 4y - 2x - 8 = y(x + 4) - 2(x + 4)$
$= (x + 4)(y - 2)$

33. $ax + bx - ay - by = x(a + b) - y(a + b)$
$= (a + b)(x - y)$

35. $x^2y - 3x^2 - 2y + 6 = x^2(y - 3) - 2(y - 3)$
$= (y - 3)(x^2 - 2)$

37. $6 + 2y + 3x^2 + x^2y = 2(3 + y) + x^2(3 + y)$
$= (3 + y)(2 + x^2)$

39. $2ax^2 + bx^2 - 4ay - 2by = x^2(2a + b) - 2y(2a + b)$
$= (2a + b)(x^2 - 2y)$

41. $6xb + 3ax - 4by - 2ay = 3x(2b + a) - 2y(2b + a)$
$= (2b + a)(3x - 2y)$

43. $x^ny - 5x^n + y - 5 = x^n(y - 5) + (y - 5)$
$= (y - 5)(x^n + 1)$

45. $2x^3 - x^2 + 4x - 2 = x^2(2x - 1) + 2(2x - 1)$
$= (2x - 1)(x^2 + 2)$

Objective C Exercises

47. $x^2 + 12x + 20 = (x + 10)(x + 2)$

49. $a^2 + a - 72 = (a + 9)(a - 8)$

51. $a^2 + 7a + 6 = (a + 6)(a + 1)$

53. $y^2 - 18y + 72 = (y - 6)(y - 12)$

55. $x^2 + x - 132 = (x + 12)(x - 11)$

57. $x^2 + 15x + 50 = (x + 10)(x + 5)$

59. $b^2 - 6b - 16 = (b + 2)(b - 8)$

61. $a^2 - 3ab + 2b^2 = (a - b)(a - 2b)$

63. $a^2 + 8ab - 33b^2 = (a + 11b)(a - 3b)$

65. $x^2 + 5xy + 6y^2 = (x + 3y)(x + 2y)$

67. $2 + x - x^2 = (1 + x)(2 - x) = -(x - 2)(x + 1)$

69. $5 + 4x - x^2 = (1 + x)(5 - x) = -(x - 5)(x + 1)$

71. $x^2 - 5x + 6 = (x - 3)(x - 2)$

Objective D Exercises

73. $2x^2 + 7x + 3 = (2x + 1)(x + 3)$

75. $6y^2 + 5y - 6 = (2y + 3)(3y - 2)$

77. $6b^2 - b - 35 = (3b + 7)(2b - 5)$

79. $3y^2 - 22y + 39 = (3y - 13)(y - 3)$

81. There are no binomial factors whose product is $6a^2 - 26a + 15$. The trinomial is nonfactorable over the integers.

83. $4a^2 - a - 5 = (a + 1)(4a - 5)$

85. $10x^2 - 29x + 10 = (5x - 2)(2x - 5)$

87. There are no binomial factors whose product is $4x^2 - 6x + 1$. The trinomial is nonfactorable over the integers.

89. $6x^2 + 41xy - 7y^2 = (6x - y)(x + 7y)$

91. $7a^2 + 46ab - 21b^2 = (7a - 3b)(a + 7b)$

93. $18x^2 + 27xy + 10y^2 = (6x + 5y)(3x + 2y)$

95. $6 - 7x - 5x^2 = (2 + x)(3 - 5x) = -(5x - 3)(x + 2)$

97. There are no binomial factors whose product is $30 + 17a - 20a^2$. The trinomial is nonfactorable over the integers.

99. $35 - 6b - 8b^2 = (5 + 2b)(7 - 4b) = -(4b - 7)(2b + 5)$

101. The GCF of $5y^4$, $29y^3$, and $20y^2$ is y^2.
$5y^4 - 29y^3 + 20y^2 = y^2(5y^2 - 29y + 20)$
$= y^2(5y - 4)(y - 5)$

103. The GCF of $20x^2$, $38x^3$, and $30x^4$ is $2x^2$.
$20x^2 - 38x^3 - 30x^4 = 2x^2(10 - 19x - 15x^2)$
$= 2x^2(5 + 3x)(2 - 5x)$
$= -2x^2(5x - 2)(3x + 5)$

105. We are looking for two polynomials that when multiplied equal $2x^2 + 9x - 18$. We must factor $2x^2 + 9x - 18$.
$2x^2 + 9x - 18 = (2x - 3)(x + 6)$
$f(x) = 2x - 3;\ \ g(x) = x + 6$

107. We are looking for two polynomials that when multiplied equal $2x^2 - 9x - 5$. We must factor $2x^2 - 9x - 5$.
$2x^2 - 9x - 5 = (2x + 1)(x - 5)$
$f(x) = 2x + 1;\ \ g(x) = x - 5$

109. We are looking for two polynomials that when multiplied equal $4b^2 - 17b + 15$. We must factor $4b^2 - 17b + 15$.
$4b^2 - 17b + 15 = (4b - 5)(b - 3)$
$f(b) = 4b - 5;\ \ g(b) = b - 3$

111. We are looking for two polynomials that when multiplied equal $4x^2 - 12x - 7$. We must factor $4x^2 - 12x - 7$.
$4x^2 - 12x - 7 = (2x + 1)(2x - 7)$
$f(x) = 2x + 1;\ \ h(x) = 2x - 7$

113. We are looking for two polynomials that when multiplied equal $4x^2 + 23x + 15$. We must factor $4x^2 + 23x + 15$.
$4x^2 + 23x + 15 = (4x + 3)(x + 5)$
$f(x) = 4x + 3;\ \ g(x) = x + 5$

115. The GCF of $3y$, $16y^2$, and $16y^3$ is y.
$3y - 16y^2 + 16y^3 = y(16y^2 - 16y + 3)$
$= y(4y - 3)(4y - 1)$

Applying the Concepts

117a. $x^2 + kx + 8$
Find two positive or two negative factors of 8. Their sum is a value of k.

Factors	Sum
8, 1	9
2, 4	6
−1, −8	−9
−2, −4	−6

The values of k are −9, −6, 6, and 9.

b. $x^2 + kx - 6$
Find the factors of −6 with opposite signs. Their sum is a value of k.

Factors	Sum
−2, 3	1
−3, 2	−1
−6, 1	−5
−1, 6	5

The values of k are −5, −1, 1, and 5.

c. $2x^2 + kx + 3$
Find two positive or two negative factors of $6(2 \cdot 3)$. The sums of these factors are values of k.

Factors	Sum
2, 3	5
1, 6	7
−2, −3	−5
−1, −6	−7

The values of k are −7, −5, 5, and 7.

d. $2x^2 + kx - 5$
Find factors of −10 of opposite sign. The sums of these factors are values of k.

Factors	Sum
2, −5	−3
5, −2	3
−10, 1	−9
10, −1	9

The values of k are −9, −3, 3, and 9.

e. $3x^2 + kx + 5$
Find positive factors of 15. The sums of these factors are values of k.

Factors	Sum
3, 5	8
15, 1	16
−3, −5	−8
−15, −1	−16

The values of k are −16, −8, 8, and 16.

f. $2x^2 + kx - 3$
Find factors of −6 of opposite sign. The sums of these factors are values of k.

Factors	Sum
−3, 2	−1
3, −2	1
6, −1	5
−6, 1	−5

The values of k are −5, −1, 1, and 5.

Section 5.6

Objective A Exercises

1. perfect squares: 4; $25x^6$; $100x^4y^4$

3. $4z^4$

5. $9a^2b^3$

7. $x^2 - 16 = x^2 - 4^2$
$= (x + 4)(x - 4)$

9. $4x^2 - 1 = (2x)^2 - 1^2$
$= (2x + 1)(2x - 1)$

11. $16x^2 - 121 = (4x)^2 - 11^2$
$= (4x + 11)(4x - 11)$

13. $1 - 9a^2 = 1^2 - (3a)^2$
$= (1 + 3a)(1 - 3a)$

15. $x^2y^2 - 100 = (xy)^2 - 10^2$
$= (xy + 10)(xy - 10)$

17. $x^2 + 4$ is nonfactorable over the integers.

19. $25 - a^2b^2 = 5^2 - (ab)^2$
$= (5 + ab)(5 - ab)$

21. $a^{2n} - 1 = (a^n)^2 - 1^2$
$= (a^n + 1)(a^n - 1)$

23. $x^2 - 12x + 36 = (x - 6)^2$

25. $b^2 - 2b + 1 = (b - 1)^2$

27. $16x^2 - 40x + 25 = (4x - 5)^2$

29. $4a^2 + 4a - 1$ is nonfactorable over the integers.

31. $b^2 + 7b + 14$ is nonfactorable over the integers.

33. $x^2 + 6xy + 9y^2 = (x + 3y)^2$

35. $25a^2 - 40ab + 16b^2 = (5a - 4b)^2$

37. $x^{2n} + 6x^n + 9 = (x^n + 3)^2$

39. $(x - 4)^2 - 9 = [(x - 4) - 3][(x - 4) + 3]$
$= (x - 4 - 3)(x - 4 + 3)$
$= (x - 7)(x - 1)$

41. $(x - y)^2 - (a + b)^2$
$= [(x - y) - (a + b)][(x - y) + (a + b)]$
$= (x - y - a - b)(x - y + a + b)$

Objective B Exercises

43. perfect cubes: 8; x^9; $27c^{15}d^{18}$

45. $2x^3$

47. $4a^2b^6$

49. $x^3 - 27 = x^3 - 3^2$
$= (x - 3)(x^2 + 3x + 9)$

51. $8x^3 - 1 = (2x)^3 - 1^3$
$= (2x - 1)(4x^2 + 2x + 1)$

53. $x^3 - y^3 = (x - y)(x^2 + xy + y^2)$

55. $m^3 + n^3 = (m + n)(m^2 - mn + n^2)$

57. $64x^3 + 1 = (4x)^3 + 1^3$
$= (4x + 1)(16x^2 - 4x + 1)$

59. $27x^3 - 8y^3 = (3x)^3 - (2y)^3$
$= (3x - 2y)(9x^2 + 6xy + 4y^2)$

61. $x^3y^3 + 64 = (xy)^3 + 4^3$
$= (xy + 4)(x^2y^2 - 4xy + 16)$

63. $16x^3 - y^3$ is nonfactorable over the integers.

65. $8x^3 - 9y^3$ is nonfactorable over the integers.

67. $(a - b)^3 - b^3$
$= [(a - b) - b][(a - b)^2 + b(a - b) + b^2]$
$= (a - 2b)(a^2 - 2ab + b^2 + ab - b^2 + b^2)$
$= (a - 2b)(a^2 - ab + b^2)$

69. $x^{6n} + y^{3n} = (x^{2n})^3 + (y^n)^3$
$= (x^{2n} + y^n)(x^{4n} - x^{2n}y^n + y^{2n})$

71. $x^{3n} + 8 = (x^n)^3 + 2^3$
$= (x^n + 2)(x^{2n} - 2x^n + 4)$

Objective C Exercises

73. Let $u = xy$.
$x^2y^2 - 8xy + 15 = u^2 - 8u + 15$
$= (u - 5)(u - 3)$
$= (xy - 5)(xy - 3)$

75. Let $u = xy$.
$x^2y^2 - 17xy + 60 = u^2 - 17u + 60$
$= (u - 12)(u - 5)$
$= (xy - 12)(xy - 5)$

77. Let $u = x^2$.
$x^4 - 9x^2 + 18 = u^2 - 9u + 18$
$= (u - 3)(u - 6)$
$= (x^2 - 3)(x^2 - 6)$

79. Let $u = b^2$.
$b^4 - 13b^2 - 90 = u^2 - 13u - 90$
$= (u + 5)(u - 18)$
$= (b^2 + 5)(b^2 - 18)$

81. Let $u = x^2y^2$.
$x^4y^4 - 8x^2y^2 + 12 = u^2 - 8u + 12$
$= (u - 2)(u - 6)$
$= (x^2y^2 - 2)(x^2y^2 - 6)$

83. Let $u = x^n$.
$x^{2n} + 3x^n + 2 = u^2 + 3u + 2$
$= (u + 1)(u + 2)$
$= (x^n + 1)(x^n + 2)$

85. Let $u = xy$.
$3x^2y^2 - 14xy + 15 = 3u^2 - 14u + 15$
$= (3u - 5)(u - 3)$
$= (3xy - 5)(xy - 3)$

87. Let $u = ab$.
$6a^2b^2 - 23ab + 21 = 6u^2 - 23u + 21$
$= (2u - 3)(3u - 7)$
$= (2ab - 3)(3ab - 7)$

89. Let $u = x^2$.
$$\begin{aligned}2x^4 - 13x^2 - 15 &= 2u^2 - 13u - 15\\ &= (u+1)(2u-15)\\ &= (x^2+1)(2x^2-15)\end{aligned}$$

91. Let $u = x^n$.
$$\begin{aligned}2x^{2n} - 7x^n + 3 &= 2u^2 - 7u + 3\\ &= (2u-1)(u-3)\\ &= (2x^n-1)(x^n-3)\end{aligned}$$

93. Let $u = a^n$.
$$\begin{aligned}6a^{2n} + 19a^n + 10 &= 6u^2 + 19u + 10\\ &= (2u+5)(3u+2)\\ &= (2a^n+5)(3a^n+2)\end{aligned}$$

Objective D Exercises

95. $$\begin{aligned}12x^2 - 36x + 27 &= 3(4x^2 - 12x + 9)\\ &= 3(2x-3)^2\end{aligned}$$

97. $$\begin{aligned}27a^4 - a &= a(27a^3 - 1)\\ &= a(3a-1)(9a^2+3a+1)\end{aligned}$$

99. $$\begin{aligned}20x^2 - 5 &= 5(4x^2 - 1)\\ &= 5(2x+1)(2x-1)\end{aligned}$$

101. $$\begin{aligned}y^5 + 6y^4 - 55y^3 &= y^3(y^2 + 6y - 55)\\ &= y^3(y+11)(y-5)\end{aligned}$$

103. $$\begin{aligned}16x^4 - 81 &= (4x^2+9)(4x^2-9)\\ &= (4x^2+9)(2x+3)(2x-3)\end{aligned}$$

105. $$\begin{aligned}16a - 2a^4 &= 2a(8 - a^3)\\ &= 2a(2-a)(4+2a+a^2)\end{aligned}$$

107. $$\begin{aligned}a^3b^6 - b^3 &= b^3(a^3b^3 - 1)\\ &= b^3(ab-1)(a^2b^2+ab+1)\end{aligned}$$

109. $$\begin{aligned}8x^4 - 40x^3 + 50x^2 &= 2x^2(4x^2 - 20x + 25)\\ &= 2x^2(2x-5)^2\end{aligned}$$

111. $$\begin{aligned}x^4 - y^4 &= (x^2+y^2)(x^2-y^2)\\ &= (x^2+y^2)(x+y)(x-y)\end{aligned}$$

113. $$\begin{aligned}x^6 + y^6 &= (x^2)^3 + (y^2)^3\\ &= (x^2+y^2)(x^4 - x^2y^2 + y^4)\end{aligned}$$

115. $a^4 - 25a^2 - 144$ is nonfactorable over the integers.

117. $$\begin{aligned}16a^4 - 2a &= 2a(8a^3 - 1)\\ &= 2a(2a-1)(4a^2+2a+1)\end{aligned}$$

119. $$\begin{aligned}a^4b^2 - 8a^3b^3 - 48a^2b^4 &= a^2b^2(a^2 - 8ab - 48b^2)\\ &= a^2b^2(a+4b)(a-12b)\end{aligned}$$

121. $$\begin{aligned}&24a^2b^2 - 14ab^3 - 90b^4\\ &= 2b^2(12a^2 - 7ab - 45b^2)\\ &= 2b^2(3a+5b)(4a-9b)\end{aligned}$$

123. $$\begin{aligned}x^3 - 2x^2 - 4x + 8 &= x^2(x-2) - 4(x-2)\\ &= (x-2)(x^2-4)\\ &= (x-2)(x+2)(x-2)\\ &= (x-2)^2(x+2)\end{aligned}$$

125. $$\begin{aligned}&4x^4 - x^2 - 4x^2y^2 + y^2\\ &= x^2(4x^2-1) - y^2(4x^2-1)\\ &= (4x^2-1)(x^2-y^2)\\ &= (2x+1)(2x-1)(x+y)(x-y)\end{aligned}$$

127. $$\begin{aligned}x^{2n+1} + 2x^{n+1} + x &= x(x^{2n} + 2x^n + 1)\\ &= x(x^n+1)^2\end{aligned}$$

129. $$\begin{aligned}3b^{n+2} + 4b^{n+1} - 4b^n &= b^n(3b^2 + 4b - 4)\\ &= b^n(b+2)(3b-2)\end{aligned}$$

Applying the Concepts

131. If $x - 3$ and $x + 4$ are factors of $x^3 + 6x^2 - 7x - 60$, then $x^3 + 6x^2 - 7x - 60$ is divisible by $x - 3$ and $x + 4$. Divide $x^3 + 6x^2 - 7x - 60$ by $x - 3$. The quotient is $x^2 + 9x + 20$. Divide this quotient by $x + 4$. The quotient is $x + 5$. Therefore, $x + 5$ is a third first-degree factor of $x^3 + 6x^2 - 7x - 60$.

Section 5.7

Objective A Exercises

1. $(x-5)(x+3) = 0$

$x - 5 = 0 \quad x + 3 = 0$

$x = 5 \quad x = -3$

The solutions are -3 and 5.

3. $(x+7)(x-8) = 0$

$x + 7 = 0 \quad x - 8 = 0$

$x = -7 \quad x = 8$

The solutions are -7 and 8.

5. $2x(3x-2)(x+4) = 0$

$2x = 0 \quad 3x - 2 = 0 \quad x + 4 = 0$

$x = 0 \quad 3x = 2 \quad x = -4$

$x = \frac{2}{3}$

The solutions are -4, 0, and $\frac{2}{3}$.

7. $x^2 + 2x - 15 = 0$

$(x+5)(x-3) = 0$

$x + 5 = 0 \quad x - 3 = 0$

$x = -5 \quad x = 3$

The solutions are -5 and 3.

9. $z^2 - 4z + 3 = 0$

$(z-3)(z-1) = 0$

$z - 3 = 0 \quad z - 1 = 0$

$z = 3 \quad z = 1$

The solutions are 1 and 3.

11. $r^2 - 10 = 3r$

$r^2 - 3r - 10 = 0$

$(r-5)(r+2) = 0$

$r - 5 = 0 \quad r + 2 = 0$

$r = 5 \quad r = -2$

The solutions are -2 and 5.

13. $$\begin{aligned} 4t^2 &= 4t+3 \\ 4t^2-4t-3 &= 0 \\ (2t+1)(2t-3) &= 0 \end{aligned}$$
$2t+1=0 \quad 2t-3=0$
$2t=-1 \quad 2t=3$
$t=-\frac{1}{2} \quad t=\frac{3}{2}$
The solutions are $-\frac{1}{2}$ and $\frac{3}{2}$.

15. $4v^2-4v+1=0$
$(2v-1)(2v-1)=0$
$2v-1=0 \quad 2v-1=0$
$2v=1 \quad 2v=1$
$v=\frac{1}{2} \quad v=\frac{1}{2}$
The solution is $\frac{1}{2}$.

17. $x^2-9=0$
$(x+3)(x-3)=0$
$x+3=0 \quad x-3=0$
$x=-3 \quad x=3$
The solutions are −3 and 3.

19. $4y^2-1=0$
$(2y+1)(2y-1)=0$
$2y+1=0 \quad 2y-1=0$
$2y=-1 \quad 2y=1$
$y=-\frac{1}{2} \quad y=\frac{1}{2}$
The solutions are $-\frac{1}{2}$ and $\frac{1}{2}$.

21. $x+15=x(x-1)$
$x+15=x^2-x$
$0=x^2-2x-15$
$0=(x-5)(x+3)$
$x-5=0 \quad x+3=0$
$x=5 \quad x=-3$
The solutions are −3 and 5.

23. $v^2+v+5=(3v+2)(v-4)$
$v^2+v+5=3v^2-10v-8$
$0=2v^2-11v-13$
$0=(2v-13)(v+1)$
$2v-13=0 \quad v+1=0$
$2v=13 \quad v=-1$
$v=\frac{13}{2}$
The solutions are −1 and $\frac{13}{2}$.

25. $4x^2+x-10=(x-2)(x+1)$
$4x^2+x-10=x^2-x-2$
$3x^2+2x-8=0$
$(3x-4)(x+2)=0$
$3x-4=0 \quad x+2=0$
$3x=4 \quad x=-2$
$x=\frac{4}{3}$
The solutions are −2 and $\frac{4}{3}$.

27. $c^3+3c^2-10c=0$
$c(c^2+3c-10)=0$
$c(c+5)(c-2)=0$
$c=0 \quad c+5=0 \quad c-2=0$
$c=-5 \quad c=2$
The solutions are −5, 0, and 2.

29. $y^4-8y^2+16=0$
$(y^2-4)(y^2-4)=0$
$(y+2)(y-2)(y+2)(y-2)=0$
$y+2=0 \quad y-2=0 \quad y+2=0 \quad y-2=0$
$y=-2 \quad y=2 \quad y=-2 \quad y=2$
The solutions are −2 and 2.

31. $a^3+a^2-9a-9=0$
$a^2(a+1)-9(a+1)=0$
$(a^2-9)(a+1)=0$
$(a+3)(a-3)(a+1)=0$
$a+3=0 \quad a-3=0 \quad a+1=0$
$a=-3 \quad a=3 \quad a=-1$
The solutions are −3, −1, and 3.

33. $3x^3+2x^2-12x-8=0$
$x^2(3x+2)-4(3x+2)=0$
$(x^2-4)(3x+2)=0$
$(x+2)(x-2)(3x+2)=0$
$x+2=0 \quad x-2=0 \quad 3x+2=0$
$x=-2 \quad x=2 \quad 3x=-2$
$x=-\frac{2}{3}$
The solutions are −2, $-\frac{2}{3}$, and 2.

35. $5x^3+2x^2-20x-8=0$
$x^2(5x+2)-4(5x+2)=0$
$(x^2-4)(5x+2)=0$
$(x+2)(x-2)(5x+2)=0$
$x+2=0 \quad x-2=0 \quad 5x+2=0$
$x=-2 \quad x=2 \quad 5x=-2$
$x=-\frac{2}{5}$
The solutions are −2, $-\frac{2}{5}$, and 2.

Objective B Exercises

37. **Strategy**
- Number: x
  Square of the number: x^2
- Sum of the number and its square is 210.

Solution
$x+x^2=210$
$x^2+x-210=0$
$(x+15)(x-14)=0$
$x=-15 \quad x=14$
The number is 14 or −15.

39. Strategy
- Width of rectangle: w
  Length of the rectangle: $3w+8$
- Area of the rectangle is length · width $= w(3w+8)$.
  Area is 380 cm^2.

Solution
$$w(3w+8)=380$$
$$3w^2+8w-380=0$$
$$(3w+38)(w-10)=0$$
$$w=-\frac{38}{3} \qquad w=10$$
The width of the rectangle is 10 cm and the length is 38 cm.

41. Strategy
- Length of one leg: x
  Length of other leg: $2x+2$
- Square of the hypotenuse of triangle is the sum of the squares of the other two legs and is equal to $(2x+2)+1=2x+3$.

Solution
$$x^2+(2x+2)^2=(2x+3)^2$$
$$x^2+4x^2+8x+4=4x^2+12x+9$$
$$x^2-4x-5=0$$
$$(x-5)(x+1)=0$$
$$x=5 \qquad x=-1$$
The length of the hypotenuse is 13 ft.

43. Strategy
- Distance is d: 80 ft.
  Speed is v: 8 ft/s.

Solution
$$d=vt+16t^2$$
$$80=8t+16t^2$$
$$16t^2+8t-80=0$$
$$(16t+40)(t-2)=0$$
$$t=-2.5 \qquad t=2$$
The penny will hit the bottom of the wishing well in 2 s.

Applying the Concepts

45. Strategy
- Height of box: 2 in.
  Width of box: $w-4$
  Length of box: $(w+10)-4$
  $=w+6$
- The volume of the box is length · width · height and is equal to 112 in^3.

Solution
$$(w+6)(w-4)\cdot 2=112$$
$$(w+6)(w-4)=56$$
$$w^2+2w-24=56$$
$$w^2+2w-80=0$$
$$(w+10)(w-8)=0$$
$$w=-10 \qquad w=8$$
Width cannot be negative, so the width of the cardboard is 8 in. and the length is 18 in.

Chapter 5 Review Exercises

1. The GCF of $18a^5b^2-12a^3b^3+30a^2b$ is $6a^2b$.
$$18a^5b^2-12a^3b^3+30a^2b$$
$$=6a^2b(3a^3b-2ab^2+5)$$

2.
$$\begin{array}{r} 5x+4 \\ 3x-2\overline{)15x^2+2x-2} \\ \underline{15x^2-10x} \\ 12x-2 \\ \underline{12x-8} \\ 6 \end{array}$$
$$\frac{15x^2+2x-2}{3x-2}=5x+4+\frac{6}{3x-2}$$

3.
$$(2x^{-1}y^2z^5)^4(-3x^3yz^{-3})^2$$
$$=(16x^{-4}y^8z^{20})(9x^6y^2z^{-6})$$
$$=144x^2y^{10}z^{14}$$

4.
$$2ax+4bx-3ay-6by=2x(a+2b)-3y(a+2b)$$
$$=(a+2b)(2x-3y)$$

5. $12+x-x^2=(4-x)(3+x)=-(x-4)(x+3)$

6.
$$\begin{array}{r|rrrr} 2 & 1 & -2 & 3 & -5 \\ & & 2 & 0 & 6 \\ \hline & 1 & 0 & 3 & 1 \end{array}$$
$P(2)=1$

7.
$$(5x^2-8xy+2y^2)-(x^2-3y^2)$$
$$=(5x^2-x^2)-8xy+(2y^2+3y^2)$$
$$=4x^2-8xy+5y^2$$

8. $24x^2+38x+15=(6x+5)(4x+3)$

9. $4x^2+12xy+9y^2=(2x+3y)^2$

10. $(-2a^2b^4)(3ab^2)=-6a^3b^6$

11.
$$64a^3-27b^3=(4a)^3-(3b)^3$$
$$=(4a-3b)(16a^2+12ab+9b^2)$$

12.
$$\begin{array}{r|rrrr} -6 & 4 & 27 & 10 & 2 \\ & & -24 & -18 & 48 \\ \hline & 4 & 3 & -8 & 50 \end{array}$$
$$\frac{4x^3+27x^2+10x+2}{x+6}=4x^2+3x-8+\frac{50}{x+6}$$

13.
$$P(-2)=2(-2)^3-(-2)+7$$
$$P(-2)=-16+2+7$$
$$P(-2)=-7$$

14. $x^2-3x-40=(x-8)(x+5)$

15. Let $u=xy$.
$$x^2y^2-9=u^2-9$$
$$=(u+3)(u-3)$$
$$=(xy+3)(xy-3)$$

16.
$$4x^2y(3x^3y^2+2xy-7y^3)$$
$$=12x^5y^3+8x^3y^2-28x^2y^4$$

17. Let $u = x^n$.
$$x^{2n} - 12x^n + 36 = u^2 - 12u + 36 = (u-6)^2 = (x^n - 6)^2$$

18.
$$\begin{aligned} 6x^2 + 60 &= 39x \\ 6x^2 - 39x + 60 &= 0 \\ 3(2x^2 - 13x + 20) &= 0 \\ 3(2x-5)(x-4) &= 0 \end{aligned}$$
$$2x - 5 = 0 \quad x - 4 = 0$$
$$2x = 5 \quad x = 4$$
$$x = \frac{5}{2}$$
The solutions are $\frac{5}{2}$ and 4.

19.
$$\begin{aligned} &5x^2 - 4x[x - 3(3x+2) + x] \\ &= 5x^2 - 4x(x - 9x - 6 + x) \\ &= 5x^2 - 4x(-7x - 6) \\ &= 5x^2 + 28x^2 + 24x \\ &= 33x^2 + 24x \end{aligned}$$

20. $3a^6 - 15a^4 - 18a^2 = 3a^2(a^4 - 5a^2 - 6) = 3a^2(a^2 - 6)(a^2 + 1)$

21. $(4x - 3y)^2 = 16x^2 - 24xy + 9y^2$

22.
$$\begin{array}{r|rrrrr} 4 & 1 & 0 & 0 & 0 & -4 \\ & & 4 & 16 & 64 & 256 \\ \hline & 1 & 4 & 16 & 64 & 252 \end{array}$$
$$\frac{x^4 - 4}{x - 4} = x^3 + 4x^2 + 16x + 64 + \frac{252}{x - 4}$$

23.
$$\begin{array}{r} 3x^2 - 2x - 6 \\ -\ x^2 - 3x + 4 \\ \hline 2x^2 - 5x - 2 \end{array}$$

24. $(5x^2yz^4)(2xy^3z^{-1})(7x^{-2}y^{-2}z^3) = (10x^3y^4z^3)(7x^{-2}y^{-2}z^3) = 70xy^2z^6$

25.
$$\frac{3x^4yz^{-1}}{-12xy^3z^2} = -\frac{x^{4-1}y^{1-3}z^{-1-2}}{4} = -\frac{x^3y^{-2}z^{-3}}{4} = -\frac{x^3}{4y^2z^3}$$

26. $948{,}000{,}000 = 9.48 \times 10^8$

27. $\frac{3 \times 10^{-3}}{15 \times 10^2} = 0.2 \times 10^{-5} = 2 \times 10^{-6}$

28.
$$\begin{array}{r|rrrr} -3 & -2 & 2 & 0 & -4 \\ & & 6 & -24 & 72 \\ \hline & -2 & 8 & -24 & 68 \end{array}$$
$P(-3) = 68$

29.
$$\frac{16x^5 - 8x^3 + 20x}{4x} = \frac{16x^5}{4x} - \frac{8x^3}{4x} + \frac{20x}{4x} = 4x^4 - 2x^2 + 5$$

30.
$$\begin{array}{r} 2x - 3 \\ 6x + 1\overline{)12x^2 - 16x - 7} \\ \underline{12x^2 + 2x} \\ -18x - 7 \\ \underline{18x - 3} \\ -4 \end{array}$$
$$\frac{12x^2 - 16x - 7}{6x + 1} = 2x - 3 - \frac{4}{6x + 1}$$

31. $a^{2n+3}(a^n - 5a + 2) = a^{2n+3+n} - 5a^{2n+3+1} + 2a^{2n+3} = a^{3n+3} - 5a^{2n+4} + 2a^{2n+3}$

32.
$$\begin{array}{r} x^3 - 3x^2 - 5x + 1 \\ \times \qquad\qquad x + 6 \\ \hline 6x^3 - 18x^2 - 30x + 6 \\ x^4 - 3x^3 - 5x^2 + x \qquad \\ \hline x^4 + 3x^3 - 23x^2 - 29x + 6 \end{array}$$

33. $10a^3b^3 - 20a^2b^4 + 35ab^2 = 5ab^2(2a^2b - 4ab^2 + 7)$

34. $5x^{n+5} + x^{n+3} + 4x^2 = x^2(5x^{n+3} + x^{n+1} + 4)$

35. $x(y-3) + 4(3-y) = x(y-3) - 4(y-3) = (y-3)(x-4)$

36. $x^2 - 16x + 63 = (x-7)(x-9)$

37. $24x^2 + 61x - 8 = (8x - 1)(3x + 8)$

38. We are looking for two polynomials that when multiplied equal $5x^2 + 3x - 2$. We must factor $5x^2 + 3x - 2$.
$5x^2 + 3x - 2 = (5x - 2)(x + 1)$
$f(x) = 5x - 2$; $g(x) = x + 1$

39. $36 - a^{2n} = (6 + a^n)(6 - a^n)$

40. $8 - y^{3n} = 2^3 - (y^n)^3 = (2 - y^n)(4 + 2y^n + y^{2n})$

41. Let $u = x^4$.
$$36x^8 - 36x^4 + 5 = 36u^2 - 36u + 5 = (6u - 5)(6u - 1) = (6x^4 - 5)(6x^4 - 1)$$

42. $3a^4b - 3ab^4 = 3ab(a^3 - b^3) = 3ab(a - b)(a^2 + ab + b^2)$

43.
$$\begin{aligned} x^{4n} - 8x^{2n} + 16 &= (x^{2n} - 4)(x^{2n} - 4) \\ &= (x^n + 2)(x^n - 2)(x^n + 2)(x^n - 2) \\ &= (x^n + 2)^2(x^n - 2)^2 \end{aligned}$$

44.
$$\begin{aligned} x^3 - x^2 - 6x &= 0 \\ x(x^2 - x - 6) &= 0 \\ x(x-3)(x+2) &= 0 \end{aligned}$$
$$x = 0 \quad x - 3 = 0 \quad x + 2 = 0$$
$$x = 3 \quad x = -2$$
The solutions are -2, 0, and 3.

45.
$$\begin{aligned} x^3 - 16x &= 0 \\ x(x^2 - 16) &= 0 \\ x(x+4)(x-4) &= 0 \end{aligned}$$
$$x = 0 \qquad x + 4 = 0 \qquad x - 4 = 0$$
$$x = -4 \qquad x = 4$$
The solutions are −4, 0, and 4.

46.
$$\begin{aligned} y^3 - y^2 - 36y - 36 &= 0 \\ y^2(y+1) - 36(y+1) &= 0 \\ (y+1)(y^2 - 36) &= 0 \\ (y+1)(y+6)(y-6) &= 0 \end{aligned}$$
$$y + 1 = 0 \qquad y + 6 = 0 \qquad y - 6 = 0$$
$$y = -1 \qquad y = -6 \qquad y = 6$$
The solutions are −1, −6, and 6.

47. Let $u = x^2$.
$$\begin{aligned} 15x^4 + x^2 - 6 &= 15u^2 + u - 6 \\ &= (3u + 2)(5u - 3) \\ &= (3x^2 + 2)(5x^2 - 3) \end{aligned}$$

48.
$$\begin{aligned} \frac{(2a^4b^{-3}c^2)^3}{(2a^3b^2c^{-1})^4} &= \frac{8a^{12}b^{-9}c^6}{16a^{12}b^8c^{-4}} \\ &= \frac{1}{2}a^{12-12}b^{-9-8}c^{6-(-4)} \\ &= \frac{1}{2}b^{-17}c^{10} = \frac{c^{10}}{2b^{17}} \end{aligned}$$

49.
$$\begin{aligned} &(x-4)(3x+2)(2x-3) \\ &= (x-4)(6x^2 - 5x - 6) \\ &= 6x^3 - 5x^2 - 6x - 24x^2 + 20x + 24 \\ &= 6x^3 - 29x^2 + 14x + 24 \end{aligned}$$

50. Let $u = x^2y^2$.
$$\begin{aligned} 21x^4y^4 + 23x^2y^2 + 6 &= 21u^2 + 23u + 6 \\ &= (7u + 3)(3u + 2) \\ &= (7x^2y^2 + 3)(3x^2y^2 + 2) \end{aligned}$$

51.
$$\begin{aligned} x^3 + 16 &= x(x + 16) \\ x^3 + 16 &= x^2 + 16x \\ x^3 - x^2 - 16x + 16 &= 0 \\ x^2(x-1) - 16(x-1) &= 0 \\ (x-1)(x^2 - 16) &= 0 \\ (x-1)(x+4)(x-4) &= 0 \end{aligned}$$
$$x - 1 = 0 \qquad x + 4 = 0 \qquad x - 4 = 0$$
$$x = 1 \qquad x = -4 \qquad x = 4$$
The solutions are −4, 1, and 4.

52. $(5a + 2b)(5a - 2b) = 25a^2 - 4b^2$

53. $2.54 \times 10^{-3} = 0.00254$

54. $6x^2 - 31x + 18 = (3x - 2)(2x - 9)$

55. $y = x^2 + 1$

56a. 3

b. 8

c. 5

57. Strategy To find the mass of the moon, multiply the mass of the sun (2.19×10^{27} tons) by 3.7×10^{-8}.

Solution
$$\begin{aligned} &(2.19 \times 10^{27})(3.7 \times 10^{-8}) \\ &= 8.103 \times 10^{19} \end{aligned}$$
The mass of the moon is 8.103×10^{19} tons.

58. Strategy The unknown number: x
The square of the number: x^2

Solution
$$\begin{aligned} x + x^2 &= 56 \\ x^2 + x - 56 &= 0 \\ (x+8)(x-7) &= 0 \end{aligned}$$
$$x + 8 = 0 \qquad x - 7 = 0$$
$$x = -8 \qquad x = 7$$
The number is −8 or 7.

59. Strategy To find out how far Earth is from the Great Galaxy of Andromeda, use the equation $d = rt$ where $r = 6.7 \times 10^8$ mph and $t = 2.2 \times 10^6$ years.
$$\begin{aligned} &2.2 \times 10^6 \times 24 \times 365 \\ &= 1.9272 \times 10^{10} \text{ hours} \end{aligned}$$

Solution
$$\begin{aligned} d &= r \cdot t \\ &= (6.7 \times 10^8)(1.9272 \times 10^{10}) \\ &= 6.7 \times 1.9272 \times 10^{18} \\ &= 12.91224 \times 10^{18} \\ &= 1.291224 \times 10^{19} \end{aligned}$$
The distance from Earth to the Great Galaxy of Andromeda is 1.291224×10^{19} miles.

60. Strategy To find the area, replace the variables L and W in the equation $A = LW$ by the given values and solve for A.

Solution
$$\begin{aligned} A &= L \cdot W \\ A &= (5x + 3)(2x - 7) \\ &= 10x^2 - 29x - 21 \end{aligned}$$
The area is $(10x^2 - 29x - 21)$ cm^2.

Chapter 5 Test

1. $16t^2 + 24t + 9 = (4t + 3)^2$

2. $-6rs^2(3r - 2s - 3) = -18r^2s^2 + 12rs^3 + 18rs^2$

3.
$$\begin{aligned} P(2) &= 3(2)^2 - 8(2) + 1 \\ P(2) &= 12 - 16 + 1 \\ P(2) &= -3 \end{aligned}$$

4. $27x^3 - 8 = (3x)^3 - (2)^3 = (3x - 2)(9x^2 + 6x + 4)$

5. $16x^2 - 25 = (4x + 5)(4x - 5)$

6.
$$\begin{array}{r} 3t^3-4t^2+1 \\ \times \qquad 2t^2-5 \\ \hline -15t^3+20t^2-5 \\ 6t^5-8t^4 \qquad +2t^2 \qquad \\ \hline 6t^5-8t^4-15t^3+22t^2-5 \end{array}$$

7.
$$\begin{aligned} -5x[3-2(2x-4)-3x] &= -5x[3-4x+8-3x] \\ &= -5x[-7x+11] \\ &= 35x^2-55x \end{aligned}$$

8.
$$\begin{aligned} 12x^3+12x^2-45x &= 3x(4x^2+4x-15) \\ &= 3x(2x-3)(2x+5) \end{aligned}$$

9.
$$\begin{aligned} 6x^3+x^2-6x-1 &= 0 \\ x^2(6x+1)-1(6x+1) &= 0 \\ (x^2-1)(6x+1) &= 0 \\ (x+1)(x-1)(6x+1) &= 0 \end{aligned}$$
$$x+1=0 \quad x-1=0 \quad 6x+1=0$$
$$x=-1 \quad x=1 \quad 6x=-1$$
$$x=-\frac{1}{6}$$

The solutions are -1, $-\frac{1}{6}$, and 1.

10.
$$\begin{aligned} &(6x^3-7x^2+6x-7)-(4x^3-3x^2+7) \\ &= (6x^3-7x^2+6x-7)+(-4x^3+3x^2-7) \\ &= 2x^3-4x^2+6x-14 \end{aligned}$$

11. $0.000000501 = 5.01 \times 10^{-7}$

12.
$$\begin{array}{r} 2x+1 \\ 7x-3\overline{)14x^2+\ \ x+1} \\ \underline{14x^2-6x} \\ 7x+1 \\ \underline{7x-3} \\ 4 \end{array}$$
$$\frac{14x^2+x+1}{7x-3} = 2x+1+\frac{4}{7x-3}$$

13. $(7-5x)(7+5x) = 49-25x^2$

14. Let $u=a^2$.
$$\begin{aligned} 6a^4-13a^2-5 &= 6u^2-13u-5 \\ &= (2u-5)(3u+1) \\ &= (2a^2-5)(3a^2+1) \end{aligned}$$

15. $(3a+4b)(2a-7b) = 6a^2-13ab-28b^2$

16.
$$\begin{aligned} 3x^4-23x^2-36 &= (3x^2+4)(x^2-9) \\ &= (3x^2+4)(x-3)(x+3) \end{aligned}$$

17.
$$\begin{aligned} (-4a^2b)^3(-ab^4) &= -64a^6b^3(-ab^4) \\ &= 64a^7b^7 \end{aligned}$$

18.
$$\begin{aligned} 6x^2 &= x+1 \\ 6x^2-x-1 &= 0 \\ (2x-1)(3x+1) &= 0 \end{aligned}$$
$$2x-1=0 \quad 3x+1=0$$
$$2x=1 \quad 3x=-1$$
$$x=\frac{1}{2} \quad x=-\frac{1}{3}$$

The solutions are $-\frac{1}{3}$ and $\frac{1}{2}$.

19.
$$\begin{array}{r|rrrr} -2 & -1 & 0 & 4 & -8 \\ & & 2 & -4 & 0 \\ \hline & -1 & 2 & 0 & -8 \end{array}$$
$P(-2) = -8$

20.
$$\begin{aligned} \frac{(2a^{-4}b^2)^3}{4a^{-2}b^{-1}} &= \frac{8a^{-12}b^6}{4a^{-2}b^{-1}} \\ &= 2a^{-12-(-2)}b^{6-(-1)} \\ &= 2a^{-10}b^7 = \frac{2b^7}{a^{10}} \end{aligned}$$

21.
$$\begin{array}{r|rrrr} -3 & 1 & -2 & -5 & 7 \\ & & -3 & 15 & -30 \\ \hline & 1 & -5 & 10 & -23 \end{array}$$
$$\frac{x^3-2x^2-5x+7}{x+3} = x^2-5x+10-\frac{23}{x+3}$$

22.
$$\begin{aligned} 12-17x+6x^2 &= (3-2x)(4-3x) \\ &= (2x-3)(3x-4) \end{aligned}$$

23.
$$\begin{aligned} 6x^2-4x-3xa+2a &= 2x(3x-2)-a(3x-2) \\ &= (3x-2)(2x-a) \end{aligned}$$

24. **Strategy** To find the number of seconds in one week in scientific notation:
- Multiply the number of seconds in a minute (60) by the minutes in an hour (60) by the hours in a day (24) by the days in a week (7).
- Convert that product to scientific notation.

Solution
$$\begin{aligned} 60 \times 60 \times 24 \times 7 &= 604{,}800 \\ &= 6.048 \times 10^5 \end{aligned}$$

There are 6.048×10^5 s in a week.

25. **Strategy**
- The distance is h: 64 ft

Solution
$$\begin{aligned} h &= 32+48t-16t^2 \\ 64 &= 32+48t-16t^2 \\ 16t^2-48t+32 &= 0 \\ t^2-3t+2 &= 0 \\ (t-1)(t-2) &= 0 \end{aligned}$$
$$t-1=0 \quad t-2=0$$
$$t=1 \quad t=2$$

The arrow will be 64 ft above the ground at 1 s and at 2 s after the arrow has been released.

26. **Strategy** To find the area, replace the variables L and W in the equation $A = L \cdot W$ by the given values and solve for A.

Solution
$$\begin{aligned} A &= L \cdot W \\ A &= (5x+1)(2x-1) \\ &= 10x^2-3x-1 \end{aligned}$$

The area is $(10x^2-3x-1)\text{ft}^2$.

Cumulative Review Exercises

1. $8-2[-3-(-1)]^2+4 = 8-2[-3+1]^2+4$
$= 8-2[-2]^2+4$
$= 8-2(4)+4$
$= 8-8+4$
$= 0+4 = 4$

2. $\dfrac{2a-b}{b-c} = \dfrac{2(4)-(-2)}{-2-6}$
$= \dfrac{8+2}{-8}$
$= \dfrac{10}{-8} = -\dfrac{5}{4}$

3. The Inverse Property of Addition

4. $2x-4[x-2(3-2x)+4] = 2x-4[x-6+4x+4]$
$= 2x-4(5x-2)$
$= 2x-20x+8$
$= -18x+8$

5. $\dfrac{2}{3}-y = \dfrac{5}{6}$
$\dfrac{2}{3}-y-\dfrac{2}{3} = \dfrac{5}{6}-\dfrac{2}{3}$
$-y = \dfrac{1}{6}$
$(-1)(-y) = (-1)\dfrac{1}{6}$
$y = -\dfrac{1}{6}$

6. $8x-3-x = -6+3x-8$
$7x-3 = 3x-14$
$7x-3-3x = 3x-14-3x$
$4x-3 = -14$
$4x-3+3 = -14+3$
$4x = -11$
$\dfrac{4x}{4} = -\dfrac{11}{4}$
$x = -\dfrac{11}{4}$

7.
```
3 | 1  0  0  -3
  |    3  9  27
  ---------------
    1  3  9  24
```
$\dfrac{x^3-3}{x-3} = x^2+3x+9+\dfrac{24}{x-3}$

8. $3-|2-3x| = -2$
$-|2-3x| = -5$
$|2-3x| = 5$
$2-3x = 5 \qquad 2-3x = -5$
$-3x = 3 \qquad -3x = -7$
$x = -1 \qquad x = \dfrac{7}{3}$

The solutions are -1 and $\dfrac{7}{3}$.

9. $P(-2) = 3(-2)^2-2(-2)+2$
$P(-2) = 12+4+2$
$P(-2) = 18$

10. $x = -2$

11. Range $= \{-4, -1, 8\}$

12. $M = \dfrac{y_2-y_1}{x_2-x_1} = \dfrac{2-3}{4-(-2)} = -\dfrac{1}{6}$

The slope is $-\dfrac{1}{6}$.

13. Use the point-slope form.
$y-y_1 = m(x-x_1)$
$y-2 = -\dfrac{3}{2}[x-(-1)]$
$y-2 = -\dfrac{3}{2}x-\dfrac{3}{2}$
$y = -\dfrac{3}{2}x+\dfrac{1}{2}$

14. First find the x- and y-intercepts. Then use them to find the slope of the line.
x-intercept $= \left(\dfrac{4}{3}, 0\right)$
y-intercept $= (0, 2)$
$m = \dfrac{y_2-y_1}{x_2-x_1} = \dfrac{2-0}{0-\dfrac{4}{3}} = -\dfrac{3}{2}\left(-\dfrac{3}{2}\right)$

The perpendicular line will have a slope that is the negative reciprocal of $-\dfrac{3}{2}$.
$m = \dfrac{2}{3}$
Now use the point-slope form of the equation to find the equation of the line.
$y-y_1 = m(x-x_1)$
$y-4 = \dfrac{2}{3}[x-(-2)]$
$y-4 = \dfrac{2}{3}x+\dfrac{4}{3}$
$y = \dfrac{2}{3}x+\dfrac{16}{3}$
The equation of the perpendicular line is $y = \dfrac{2}{3}x+\dfrac{16}{3}$.

15. $2x - 3y = 2$
$x + y = -3$

$D = \begin{vmatrix} 2 & -3 \\ 1 & 1 \end{vmatrix} = 5$

$D_x = \begin{vmatrix} 2 & -3 \\ -3 & 1 \end{vmatrix} = -7$

$D_y = \begin{vmatrix} 2 & 2 \\ 1 & -3 \end{vmatrix} = -8$

$x = \frac{D_x}{D} = -\frac{7}{5}$

$y = \frac{D_y}{D} = -\frac{8}{5}$

The solution is $\left(-\frac{7}{5}, -\frac{8}{5}\right)$.

16. (1) $x - y + z = 0$
(2) $2x + y - 3z = -7$
(3) $-x + 2y + 2z = 5$

Add Equations (1) and (3) to eliminate x.

$$\begin{array}{r} x - y + z = 0 \\ \underline{-x + 2y + 2z = 5} \\ y + 3z = 5 \end{array}$$

Add −2 times Equation (1) and Equation (2) to eliminate x.

$$\begin{array}{r} -2x + 2y - 2z = 0 \\ \underline{2x + y - 3z = -7} \\ 3y - 5z = -7 \end{array}$$

Now solve the system in two variables.

$y + 3z = 5$
$3y - 5z = -7$

Add −3 times the first of these equations to the second.

$$\begin{array}{r} -3y - 9z = -15 \\ \underline{3y - 5z = -7} \\ -14z = -22 \end{array}$$

$z = \frac{11}{7}$

Next find y.

$y + 3z = 5$

$y + 3\left(\frac{11}{7}\right) = 5$

$y = \frac{2}{7}$

Replace y and z in Equation (1) and solve for x.

$x - \frac{2}{7} + \frac{11}{7} = 0$

$x = -\frac{9}{7}$

The solution is $\left(-\frac{9}{7}, \frac{2}{7}, \frac{11}{7}\right)$.

17.

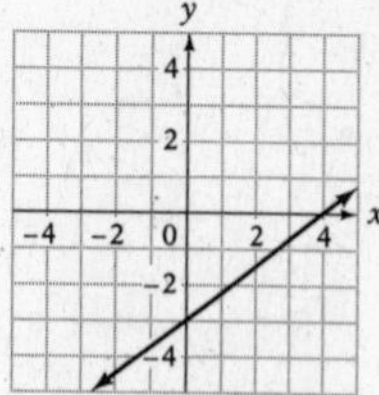

18.

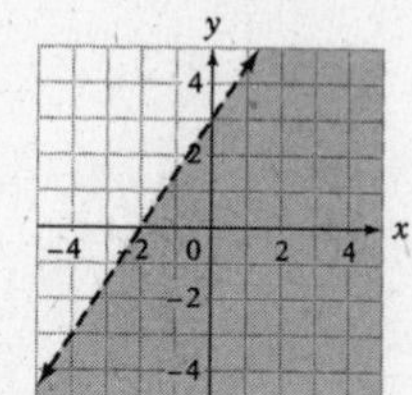

19.

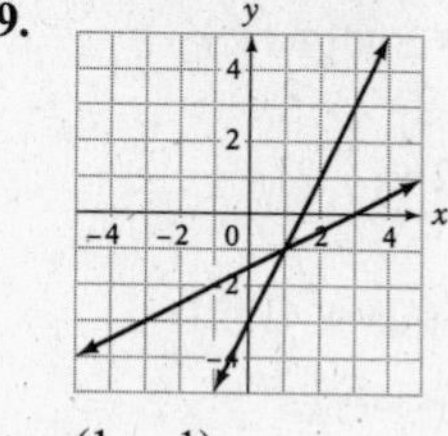

(1, −1)

20. Solve each inequality.

$2x + y < 3$
$y < 3 - 2x$

$2x - y \geq 1$
$-y \geq -2x + 1$
$y \leq 2x - 1$

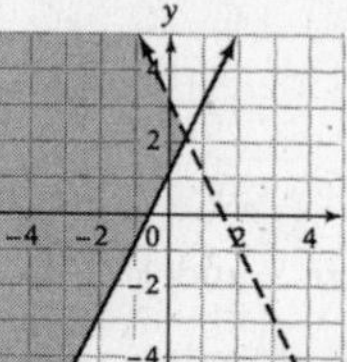

21. $(4a^{-2}b^3)(2a^3b^{-1})^{-2} = 4a^{-2}b^3(2^{-2}a^{-6}b^2)$
$= 4 \cdot 2^{-2}a^{-2-6}b^{3+2}$
$= 4 \cdot \frac{1}{4}a^{-8}b^5$
$= \frac{b^5}{a^8}$

22. $\frac{(5x^3y^{-3}z)^{-2}}{y^4z^{-2}} = \frac{5^{-2}x^{-6}y^6z^{-2}}{y^4z^{-2}}$
$= 5^{-2}x^{-6}y^{6-4}z^{-2-(-2)}$
$= \frac{1}{25}x^{-6}y^2$
$= \frac{y^2}{25x^6}$

23.
$$\begin{aligned} 3-(3-3^{-1})^{-1} &= 3-\left(3-\frac{1}{3}\right)^{-1} \\ &= 3-\left(\frac{8}{3}\right)^{-1} \\ &= 3-\frac{3}{8}=\frac{21}{8} \end{aligned}$$

24.
$$\begin{array}{r} 2x^2-3x+1 \\ \times \qquad 2x+3 \\ \hline 6x^2-9x+3 \\ 4x^3-6x^2+2x \quad \\ \hline 4x^3 \qquad -7x+3 \end{array}$$

25. $-4x^3+14x^2-12x = -2x(2x^2-7x+6)$
$= -2x(2x-3)(x-2)$

26. $a(x-y)-b(y-x) = a(x-y)+b(x-y)$
$= (x-y)(a+b)$

27. $x^4-16 = (x^2+4)(x^2-4)$
$= (x^2+4)(x+2)(x-2)$

28. $2x^2-16 = 2(x^3-8)$
$= 2(x-2)(x^2+2x+4)$

29. Strategy • Smaller integer: x
Larger integer: $24-x$
• The difference between four times the smaller and nine is 3 less than twice the larger.
$4x-9=2(24-x)-3$

Solution
$$\begin{aligned} 4x-9&=2(24-x)-3 \\ 4x-9&=48-2x-3 \\ 4x-9&=45-2x \\ 6x-9&=45 \\ 6x&=54 \\ x&=9 \\ 24-x&=15 \end{aligned}$$
The integers are 9 and 15.

30. Strategy • The number of ounces of pure gold: x

	Amount	Cost	Value
Pure gold	x	360	$360x$
Alloy	80	120	80(120)
Mixture	$x+80$	200	$200(x+80)$

• The sum of the values before mixing equals the value after mixing.

Solution
$$\begin{aligned} 360x+80(120)&=200(x+80) \\ 360x+9600&=200x+16{,}000 \\ 160x+9600&=16{,}000 \\ 160x&=6400 \\ x&=40 \end{aligned}$$
40 oz of pure gold must be mixed with the alloy.

31. Strategy • Faster cyclist: x
Slower cyclist: $\frac{2}{3}x$

	Rate	Time	Distance
Faster cyclist	x	2	$2x$
Slower cyclist	$\frac{2}{3}x$	2	$2(\frac{2}{3}x)$

• The sum of the distances is 25 miles.

Solution
$$\begin{aligned} 2x+2\left(\frac{2}{3}x\right)&=25 \\ 2x+\frac{4}{3}x&=25 \\ \frac{10}{3}x&=25 \\ x&=7.5 \\ \frac{2}{3}x&=5 \end{aligned}$$
The slower cyclist travels at 5 mph, the faster cyclist at 7.5 mph.

32. Strategy To find the time:
• Use the equation $d=rt$, where r is the speed of the space vehicle and d is the distance from Earth to the moon.

Solution
$$\begin{aligned} d&=rt \\ 2.4\times10^5&=(2\times10^4)t \\ \frac{2.4\times10^5}{2\times10^4}&=t \\ 1.2\times10^1&=t \end{aligned}$$
The vehicle will reach the moon in 12 h.

33. $m=\dfrac{y_2-y_1}{x_2-x_1}=\dfrac{300-100}{6-2}$
$m=50$
The average speed is 50 mph.

Chapter 6: Rational Expressions

Prep Test

1. Multiples of 10 are 10, 20, 30, 40, <u>50</u>, 60, ...
Multiples of 25 are 25, <u>50</u>, 75, ...
LCM is 50.

2. $-\frac{3}{8}\cdot\frac{4}{9}=-\frac{\overset{1}{\cancel{3}}\cdot\overset{1}{\cancel{2}}\cdot\overset{1}{\cancel{2}}}{\underset{1}{\cancel{2}}\cdot\underset{1}{\cancel{2}}\cdot 2\cdot\underset{1}{\cancel{3}}\cdot 3}=-\frac{1}{6}$

3. $-\frac{4}{5}\div\frac{8}{15}=-\frac{4}{5}\cdot\frac{15}{8}=\frac{\overset{1}{\cancel{2}}\cdot\overset{1}{\cancel{2}}\cdot 3\cdot\overset{1}{\cancel{5}}}{\underset{1}{\cancel{5}}\cdot\underset{1}{\cancel{2}}\cdot\underset{1}{\cancel{2}}\cdot 2}=-\frac{3}{2}$

4. $-\frac{5}{6}+\frac{7}{8}=-\frac{20}{24}+\frac{21}{24}=\frac{1}{24}$

5. $-\frac{3}{8}-\left(-\frac{7}{12}\right)=-\frac{3}{8}+\frac{7}{12}=-\frac{9}{24}+\frac{14}{24}=\frac{5}{24}$

6. $\frac{\frac{2}{3}-\frac{1}{4}}{\frac{1}{8}-\frac{2}{1}}=\frac{\frac{8}{12}-\frac{3}{12}}{\frac{1}{8}-\frac{16}{8}}=\frac{\frac{5}{12}}{-\frac{15}{8}}=\frac{5}{12}\div-\frac{15}{8}=\frac{5}{12}\cdot-\frac{8}{15}$

$=-\frac{\overset{1}{\cancel{5}}\cdot\overset{1}{\cancel{2}}\cdot\overset{1}{\cancel{2}}\cdot 2}{\underset{1}{\cancel{2}}\cdot\underset{1}{\cancel{2}}\cdot 3\cdot\underset{1}{\cancel{5}}\cdot 3}=-\frac{2}{9}$

7. $\frac{2(2)-3}{(2)^2-(2)+1}=\frac{4-3}{4-2+1}=\frac{1}{3}$

8. $4(2x+1)=3(x-2)$
$8x+4=3x-6$
$\left(\frac{1}{5}\right)5x=-10\left(\frac{1}{5}\right)$
$x=-2$

9. $10\left(\frac{t}{2}+\frac{t}{5}\right)=10(1)$
$\frac{10t}{2}+\frac{10t}{5}=10$
$5t+2t=10$
$7t=10$
$t=\frac{10}{7}$

10.

	Rate (r)	Time (t)	Distance (d)
Plane 1	$r-20$	2	$2(r-20)$
Plane 2	r	2	$2r$

Since the planes are 480 mi apart, the total distance the two planes have traveled is

$2(r-20)+2r=480$
$2r-40+2r=480$
$4r-40=480$
$4r=520$
$r=130$ mph, rate of Plane 2
$r-20=130-20=110$ mph, rate of Plane 1

Go Figure

12 boxes of cereal filled in 7 minutes by 6 machines is 0.286 boxes filled per machine per minute.

$\frac{12}{7\cdot 6}\approx 0.286$

14 machines can fill 48 boxes in 12 minutes.

$(0.286)\cdot 14\cdot 12=48$

Answer: 48 boxes

Section 6.1

Objective A Exercises

1. A rational function is a function that is written as an expression in which the numerator and denominator are polynomials. An example is $f(x)=\frac{x^2-2x+3}{7x-4}$.

3. $f(x)=\frac{2}{x-3}$
$f(4)=\frac{2}{4-3}=\frac{2}{1}=2$

5. $f(x)=\frac{x-2}{x+4}$
$f(-2)=\frac{-2-2}{-2+4}=\frac{-4}{2}=-2$

7. $f(x)=\frac{1}{x^2-2x+1}$
$f(-2)=\frac{1}{(-2)^2-2(-2)+1}=\frac{1}{9}$

9. $f(x)=\frac{x-2}{2x^2+3x+8}$
$f(-3)=\frac{3-2}{2(3)^2+3(3)+8}=\frac{1}{35}$

11. $f(x)=\frac{x^2-2x}{x^3-x+4}$
$f(-1)=\frac{(-1)^2-2(-1)}{(-1)^3-(-1)+4}=\frac{3}{4}$

13. $f(x)=\frac{4}{x-3}$
$x-3=0$
$x=3$
The domain is $\{x|x\neq 3\}$.

15. $H(x)=\frac{x}{x+4}$
$x+4=0$
$x=-4$
The domain is $\{x|x\neq -4\}$.

17. $h(x) = \frac{5x}{3x+9}$

$3x + 9 = 0$
$3x = -9$
$x = -3$

The domain is $\{x|x \neq -3\}$.

19. $q(x) = \frac{4-x}{(x-4)(3x-2)}$

$(x-4)(3x-2) = 0$
$x - 4 = 0 \quad 3x - 2 = 0$
$x = 4 \quad 3x = 2$
$x = \frac{2}{3}$

The domain is $\left\{x|x \neq \frac{2}{3}, 4\right\}$.

21. $f(x) = \frac{2x-1}{x^2+x-6}$

$x^2 + x - 6 = 0$
$(x+3)(x-2) = 0$
$x + 3 = 0 \quad x - 2 = 0$
$x = -3 \quad x = 2$

The domain is $\{x|x \neq -3, 2\}$.

23. $f(x) = \frac{x+1}{x^2+1}$

The domain is $\{x|x \in \text{real numbers}\}$.

Objective B Exercises

25. A rational expression is in simplest form when the numerator and denominator have no common factors other than 1.

27. $\frac{4-8x}{4} = \frac{4(1-2x)}{4} = 1 - 2x$

29. $\frac{6x^2-2x}{2x} = \frac{2x(3x-1)}{2x} = 3x - 1$

31. $\frac{8x^2(x-3)}{4x(x-3)} = \frac{8x^2}{4x} = 2x$

33. $\frac{-36a^2-48a}{18a^3+24a^2} = \frac{-12a(3a+4)}{6a^2(3a+4)}$
$= \frac{-12a}{6a^2} = -\frac{2}{a}$

35. $\frac{3x-6}{x^2+2x} = \frac{3(x-2)}{x(x+2)}$

The expression is in simplest form.

37. $\frac{3x^3y^3 - 12x^2y^2 + 15xy}{3xy} = \frac{3xy(x^2y^2 - 4xy + 5)}{3xy}$
$= x^2y^2 - 4xy + 5$

39. $\frac{x^{2n} + x^ny^n}{x^{2n} - y^{2n}} = \frac{x^n(x^n + y^n)}{(x^n + y^n)(x^n - y^n)}$
$= \frac{x^n}{x^n - y^n}$

41. $\frac{x^2 - 7x + 12}{x^2 - 9x + 20} = \frac{(x-3)(x-4)}{(x-4)(x-5)}$
$= \frac{x-3}{x-5}$

43. $\frac{3x^2 + 10x - 8}{8 - 14x + 3x^2} = \frac{(3x-2)(x+4)}{(2-3x)(4-x)}$
$= -\frac{x+4}{4-x} = \frac{x+4}{x-4}$

45. $\frac{a^2 - b^2}{a^3 + b^3} = \frac{(a+b)(a-b)}{(a+b)(a^2 - ab + b^2)}$
$= \frac{a-b}{a^2 - ab + b^2}$

47. $\frac{8x^3 - y^3}{4x^2 - y^2} = \frac{(2x-y)(4x^2 + 2xy + y^2)}{(2x+y)(2x-y)}$
$= \frac{4x^2 + 2xy + y^2}{2x + y}$

49. $\frac{x^2(a-2) - a + 2}{ax^2 - ax} = \frac{x^2(a-2) - (a-2)}{ax(x-1)}$
$= \frac{(a-2)(x^2-1)}{ax(x-1)}$
$= \frac{(a-2)(x+1)(x-1)}{ax(x-1)}$
$= \frac{(a-2)(x+1)}{ax}$

51. $\frac{x^4 - 2x^2 - 3}{x^4 + 2x^2 + 1} = \frac{(x^2+1)(x^2-3)}{(x^2+1)(x^2+1)}$
$= \frac{x^2-3}{x^2+1}$

53. $\frac{6x^2y^2 + 11xy + 4}{9x^2y^2 + 9xy - 4} = \frac{(2xy+1)(3xy+4)}{(3xy+4)(3xy-1)}$
$= \frac{2xy+1}{3xy-1}$

55. $\frac{a^{2n} + a^n - 12}{a^{2n} - 2a^n - 3} = \frac{(a^n+4)(a^n-3)}{(a^n+1)(a^n-3)}$
$= \frac{a^n+4}{a^n+1}$

57. $\frac{a^{2n} + 2a^nb^n + b^{2n}}{a^{2n} - b^{2n}} = \frac{(a^n+b^n)(a^n+b^n)}{(a^n+b^n)(a^n-b^n)}$
$= \frac{a^n+b^n}{a^n-b^n}$

59. $\frac{x^2(a+b) + a + b}{x^4 - 1} = \frac{x^2(a+b) + 1(a+b)}{(x^2+1)(x^2-1)}$
$= \frac{(a+b)(x^2+1)}{(x^2+1)(x+1)(x-1)}$
$= \frac{a+b}{(x+1)(x-1)}$

Objective C Exercises

61. $\frac{15x^2y^4}{24ab^3} \cdot \frac{28a^2b^4}{35xy^4} = \frac{15 \cdot 28a^2b^4x^2y^4}{24 \cdot 35ab^3xy^4}$
$= \frac{2^2 \cdot 3 \cdot 5 \cdot 7a^2b^4x^2y^4}{2^3 \cdot 3 \cdot 5 \cdot 7ab^3xy^4}$
$= \frac{abx}{2}$

63. $\dfrac{2x^2+4x}{8x^2-40x}\cdot\dfrac{6x^3-30x^2}{3x^2+6x}=\dfrac{2x(x+2)}{8x(x-5)}\cdot\dfrac{6x^2(x-5)}{3x(x+2)}$

$=\dfrac{2x(x+2)\cdot 6x^2(x-5)}{8x(x-5)\cdot 3x(x+2)}$

$=\dfrac{2x\cdot 6x^2}{8x\cdot 3x}$

$=\dfrac{x}{2}$

65. $\dfrac{2x^2-5x+3}{x^6y^3}\cdot\dfrac{x^4y^4}{2x^2-x-3}$

$=\dfrac{(2x-3)(x-1)}{x^6y^3}\cdot\dfrac{x^4y^4}{(2x-3)(x+1)}$

$=\dfrac{(2x-3)(x-1)\cdot x^4y^4}{x^6y^3(2x-3)(x+1)}$

$=\dfrac{(x-1)\cdot x^4y^4}{x^6y^3(x+1)}$

$=\dfrac{y(x-1)}{x^2(x+1)}$

67. $\dfrac{x^2+x-6}{12+x-x^2}\cdot\dfrac{x^2+x-20}{x^2-4x+4}$

$=\dfrac{(x+3)(x-2)}{(3+x)(4-x)}\cdot\dfrac{(x+5)(x-4)}{(x-2)(x-2)}$

$=\dfrac{(x+3)(x-2)(x+5)(x-4)}{(3+x)(4-x)(x-2)(x-2)}$

$=-\dfrac{x+5}{x-2}$

69. $\dfrac{x^{2n}+2x^n}{x^{n+1}+2x}\cdot\dfrac{x^2-3x}{x^{n+1}-3x^n}$

$=\dfrac{x^n(x^n+2)}{x(x^n+2)}\cdot\dfrac{x(x-3)}{x^n(x-3)}$

$=\dfrac{x^n(x^n+2)\cdot x(x-3)}{x(x^n+2)\cdot x^n(x-3)}$

$=\dfrac{x^n\cdot x}{x\cdot x^n}$

$=1$

71. $\dfrac{x^{2n}+3x^n+2}{x^{2n}-x^n-6}\cdot\dfrac{x^{2n}+x^n-12}{x^{2n}-1}$

$=\dfrac{(x^n+1)(x^n+2)}{(x^n+2)(x^n-3)}\cdot\dfrac{(x^n+4)(x^n-3)}{(x^n+1)(x^n-1)}$

$=\dfrac{(x^n+1)(x^n+2)(x^n+4)(x^n-3)}{(x^n+2)(x^n-3)(x^n+1)(x^n-1)}$

$=\dfrac{x^n+4}{x^n-1}$

73. $\dfrac{x^4-5x^2+4}{3x^2-4x-4}\cdot\dfrac{3x^2-10x-8}{x^2-4}$

$=\dfrac{(x^2-1)(x^2-4)}{(3x+2)(x-2)}\cdot\dfrac{(3x+2)(x-4)}{(x^2-4)}$

$=\dfrac{(x+1)(x-1)(x^2-4)(3x+2)(x-4)}{(3x+2)(x-2)(x^2-4)}$

$=\dfrac{(x+1)(x-1)(x-4)}{x-2}$

75. $\dfrac{x^2-y^2}{x^2+xy+y^2}\cdot\dfrac{x^2-xy}{3x^2-3xy}\cdot\dfrac{x^3-y^3}{x^2-2xy+y^2}$

$=\dfrac{(x+y)(x-y)}{(x^2+xy+y^2)}\cdot\dfrac{x(x-y)}{3x(x-y)}\cdot\dfrac{(x-y)(x^2+xy+y^2)}{(x-y)(x-y)}$

$=\dfrac{(x+y)(x-y)\cdot x(x-y)(x-y)(x^2+xy+y^2)}{(x^2+xy+y^2)\cdot 3x(x-y)(x-y)(x-y)}$

$=\dfrac{x+y}{3}$

Objective D Exercises

77. $\dfrac{12a^4b^7}{13x^2y^2}\div\dfrac{18a^5b^6}{26xy^3}$

$=\dfrac{12a^4b^7}{13x^2y^2}\cdot\dfrac{26xy^3}{18a^5b^6}$

$=\dfrac{12\cdot 26a^4b^7xy^3}{13\cdot 18a^5b^6x^2y^2}$

$=\dfrac{2^3\cdot 3\cdot 13a^4b^7xy^3}{2\cdot 3^2\cdot 13a^5b^6x^2y^2}$

$=\dfrac{2^2by}{3ax}=\dfrac{4by}{3ax}$

79. $\dfrac{4x^2-4y^2}{6x^2y^2}\div\dfrac{3x^2+3xy}{2x^2y-2xy^2}$

$=\dfrac{4x^2-4y^2}{6x^2y^2}\cdot\dfrac{2x^2y-2xy^2}{3x^3+3xy}$

$=\dfrac{4(x+y)(x-y)}{6x^2y^2}\cdot\dfrac{2xy(x-y)}{3x(x+y)}$

$=\dfrac{4(x+y)(x-y)\cdot 2xy(x-y)}{6x^2y^2\cdot 3x(x+y)}$

$=\dfrac{4(x-y)\cdot 2xy(x-y)}{6x^2y^2\cdot 3x}$

$=\dfrac{4(x-y)^2}{9x^2y}$

81. $\dfrac{8x^3+12x^2y}{4x^2-9y^2} \div \dfrac{16x^2y^2}{4x^2-12xy+9y^2}$

$= \dfrac{8x^3+12x^2y}{4x^2-9y^2} \cdot \dfrac{4x^2-12xy+9y^2}{16x^2y^2}$

$= \dfrac{4x^2(2x+3y)}{(2x+3y)(2x-3y)} \cdot \dfrac{(2x-3y)(2x-3y)}{16x^2y^2}$

$= \dfrac{4x^2(2x+3y)(2x-3y)(2x-3y)}{(2x+3y)(2x-3y)16x^2y^2}$

$= \dfrac{4x^2(2x-3y)}{16x^2y^2}$

$= \dfrac{2x-3y}{4y^2}$

83. $\dfrac{2x^2+13x+20}{8-10x-3x^2} \div \dfrac{6x^2-13x-5}{9x^2-3x-2}$

$= \dfrac{2x^2+13x+20}{8-10x-3x^2} \cdot \dfrac{9x^2-3x-2}{6x^2-13x-5}$

$= \dfrac{(2x+5)(x+4)}{(4+x)(2-3x)} \cdot \dfrac{(3x+1)(3x-2)}{(3x+1)(2x-5)}$

$= \dfrac{(2x+5)(x+4)(3x+1)(3x-2)}{(4+x)(2-3x)(3x+1)(2x-5)}$

$= -\dfrac{2x+5}{2x-5}$

85. $\dfrac{x^{2n}-4}{4x^n+8} \div \dfrac{x^{n+1}-2x}{4x^3-12x^2}$

$= \dfrac{x^{2n}-4}{4x^n+8} \cdot \dfrac{4x^3-12x^2}{x^{n+1}-2x}$

$= \dfrac{(x^n+2)(x^n-2)}{4(x^n+2)} \cdot \dfrac{4x^2(x-3)}{x(x^n-2)}$

$= \dfrac{(x^n+2)(x^n-2)\cdot 4x^2(x-3)}{4(x^n+2)\cdot x(x^n-2)}$

$= x(x-3)$

87. $\dfrac{16x^2-9}{6-5x-4x^2} \div \dfrac{16x^2+24x+9}{4x^2+11x+6}$

$= \dfrac{16x^2-9}{6-5x-4x^2} \cdot \dfrac{4x^2+11x+6}{16x^2+24x+9}$

$= \dfrac{(4x+3)(4x-3)}{(2+x)(3-4x)} \cdot \dfrac{(x+2)(4x+3)}{(4x+3)(4x+3)}$

$= \dfrac{(4x+3)(4x-3)(x+2)(4x+3)}{(2+x)(3-4x)(4x+3)(4x+3)}$

$= -1$

89. $\dfrac{x^{4n}-1}{x^{2n}+x^n-2} \div \dfrac{x^{2n}+1}{x^{2n}+3x^n+2}$

$= \dfrac{x^{4n}-1}{x^{2n}+x^n-2} \cdot \dfrac{x^{2n}+3x^n+2}{x^{2n}+1}$

$= \dfrac{(x^{2n}+1)(x^{2n}-1)}{(x^n+2)(x^n-1)} \cdot \dfrac{(x^n+1)(x^n+2)}{(x^{2n}+1)}$

$= \dfrac{(x^{2n}+1)(x^n+1)(x^n-1)}{(x^n+2)(x^n-1)} \cdot \dfrac{(x^n+1)(x^n+2)}{(x^{2n}+1)}$

$= (x^n+1)^2$

91. $\dfrac{x^3+y^3}{2x^3+2x^2y} \div \dfrac{3x^3-3x^2y+3xy^2}{6x^2-6y^2}$

$= \dfrac{x^3+y^3}{2x^3+2x^2y} \cdot \dfrac{6x^2-6y^2}{3x^3-3x^2y+3xy^2}$

$= \dfrac{(x+y)(x^2-xy+y^2)}{2x^2(x+y)} \cdot \dfrac{6(x^2-y^2)}{3x(x^2-xy+y^2)}$

$= \dfrac{(x+y)(x^2-xy+y^2)}{2x^2(x+y)} \cdot \dfrac{6(x+y)(x-y)}{3x(x^2-xy+y^2)}$

$= \dfrac{6(x+y)(x-y)}{2x^2\cdot 3x}$

$= \dfrac{(x+y)(x-y)}{x^3}$

93. $\dfrac{3x^2+10x-8}{x^2+4x+3} \cdot \dfrac{x^2+6x+9}{3x^2-5x+2} \div \dfrac{x^2+x-12}{x^2-1}$

$= \dfrac{3x^2+10x-8}{x^2+4x+3} \cdot \dfrac{x^2+6x+9}{3x^2-5x+2} \cdot \dfrac{x^2-1}{x^2+x-12}$

$= \dfrac{(3x-2)(x+4)}{(x+3)(x+1)} \cdot \dfrac{(x+3)(x+3)}{(3x-2)(x-1)} \cdot \dfrac{(x+1)(x-1)}{(x+4)(x-3)}$

$= \dfrac{(3x-2)(x+4)(x+3)(x+3)(x+1)(x-1)}{(x+3)(x+1)(3x-2)(x-1)(x+4)(x-3)}$

$= \dfrac{x+3}{x-3}$

Applying the Concepts

95. $\dfrac{5y^2-20}{3y^2-12y} \cdot \dfrac{9y^3+6y}{2y^2-4y} \div \dfrac{y^3+2y^2}{2y^2-8y}$

$= \dfrac{5y^2-20}{3y^2-12y} \cdot \dfrac{9y^3+6y}{2y^2-4y} \cdot \dfrac{2y^2-8y}{y^3+2y^2}$

$= \dfrac{5(y-2)(y+2)}{3y(y-4)} \cdot \dfrac{3y(3y^2+2)}{2y(y-2)} \cdot \dfrac{2y(y-4)}{y^2(y+2)}$

$= \dfrac{5(y-2)(y+2)\cdot 3y(3y^2+2)\cdot 2y(y-4)}{3y(y-4)\cdot 2y(y-2)\cdot y^2(y+2)}$

$= \dfrac{5(3y^2+2)}{y^2}$

Section 6.2

Objective A Exercises

1. The LCM is $12x^2y^4$.

$\dfrac{3}{4x^2y} = \dfrac{3}{4x^2y} \cdot \dfrac{3y^3}{3y^3} = \dfrac{9y^3}{12x^2y^4}$

$\dfrac{17}{12xy^4} = \dfrac{17}{12xy^4} \cdot \dfrac{x}{x} = \dfrac{17x}{12x^2y^4}$

3. The LCM is $6x^2(x-2)$.

$\dfrac{x-2}{3x(x-2)} = \dfrac{x-2}{3x(x-2)} \cdot \dfrac{2x}{2x} = \dfrac{2x^2-4x}{6x^2(x-2)}$

$\dfrac{3}{6x^2} = \dfrac{3}{6x^2} \cdot \dfrac{x-2}{x-2} = \dfrac{3x-6}{6x^2(x-2)}$

5. The LCM is $2x(x-5)$.
$$\frac{3x-1}{2x^2-10x}=\frac{3x-1}{2x(x-5)}$$
$$-3x=\frac{-3x}{1}\cdot\frac{2x(x-5)}{2x(x-5)}=-\frac{6x^3-30x^2}{2x(x-5)}$$

7. The LCM is $(2x+3)(2x-3)$.
$$\frac{3x}{2x-3}=\frac{3x}{2x-3}\cdot\frac{2x+3}{2x+3}=\frac{6x^2+9x}{(2x+3)(2x-3)}$$
$$\frac{5x}{2x+3}=\frac{5x}{2x+3}\cdot\frac{2x-3}{2x-3}=\frac{10x^2-15x}{(2x+3)(2x-3)}$$

9. $x^2-9=(x+3)(x-3)$
The LCM is $(x+3)(x-3)$.
$$\frac{2x}{x^2-9}=\frac{2x}{(x+3)(x-3)}$$
$$\frac{x+1}{x-3}=\frac{x+1}{x-3}\cdot\frac{x+3}{x+3}=\frac{x^2+4x+3}{(x+3)(x-3)}$$

11. $3x^2-12y^2=3(x+2y)(x-2y)$;
$6x-12y=6(x-2y)$
The LCM is $6(x+2y)(x-2y)$.
$$\frac{3}{3x^2-12y^2}=\frac{3}{3(x+2y)(x-2y)}\cdot\frac{2}{2}=\frac{6}{6(x+2y)(x-2y)}$$
$$\frac{5}{6x-12y}=\frac{5}{6(x-2y)}\cdot\frac{x+2y}{x+2y}=\frac{5x+10y}{6(x+2y)(x-2y)}$$

13. $x^2-1=(x+1)(x-1)$;
$x^2-2x+1=(x-1)^2$
The LCM is $(x+1)(x-1)^2$.
$$\frac{3x}{x^2-1}=\frac{3x}{(x+1)(x-1)}\cdot\frac{x-1}{x-1}=\frac{3x^2-3x}{(x+1)(x-1)^2}$$
$$\frac{5x}{x^2-2x+1}=\frac{5x}{(x-1)(x-1)}\cdot\frac{x+1}{x+1}=\frac{5x^2+5x}{(x+1)(x-1)^2}$$

15. $8-x^3=-(x^3-8)=-(x-2)(x^2+2x+4)$
The LCM is $(x-2)(x^2+2x+4)$.
$$\frac{x-3}{8-x^3}=-\frac{x-3}{(x-2)(x^2+2x+4)}$$
$$\frac{2}{4+2x+x^2}=\frac{2}{4+2x+x^2}\cdot\frac{x-2}{x-2}=\frac{2x-4}{(x-2)(x^2+2x+4)}$$

17. $x^2+2x-3=(x+3)(x-1)$;
$x^2+6x+9=(x+3)^2$
The LCM is $(x-1)(x+3)^2$.
$$\frac{2x}{x^2+2x-3}=\frac{2x}{(x+3)(x-1)}\cdot\frac{x+3}{x+3}=\frac{2x^2+6x}{(x-1)(x+3)^2}$$
$$\frac{-x}{x^2+6x+9}=\frac{-x}{(x+3)(x+3)}\cdot\frac{x-1}{x-1}=-\frac{x^2-x}{(x-1)(x+3)^2}$$

19. $4x^2-16x+15=(2x-3)(2x-5)$;
$6x^2-19x+10=(2x-5)(3x-2)$
The LCM is $(2x-3)(2x-5)(3x-2)$.
$$\frac{-4x}{4x^2-16x+15}=-\frac{4x}{(2x-3)(2x-5)}\cdot\frac{3x-2}{3x-2}=-\frac{12x^2-8x}{(2x-3)(2x-5)(3x-2)}$$
$$\frac{3x}{6x^2-19x+10}=\frac{3x}{(2x-5)(3x-2)}\cdot\frac{2x-3}{2x-3}=\frac{6x^2-9x}{(2x-3)(2x-5)(3x-2)}$$

21. $6x^2-17x+12=(3x-4)(2x-3)$;
$4-3x=-(3x-4)$
The LCM is $(3x-4)(2x-3)$.
$$\frac{5}{6x^2-17x+12}=\frac{5}{(3x-4)(2x-3)}$$
$$\frac{2x}{4-3x}=-\frac{2x}{3x-4}\cdot\frac{2x-3}{2x-3}=-\frac{4x^2-6x}{(3x-4)(2x-3)}$$
$$\frac{x+1}{2x-3}=\frac{x+1}{2x-3}\cdot\frac{3x-4}{3x-4}=\frac{3x^2-x-4}{(3x-4)(2x-3)}$$

23. $15-2x-x^2=-(x^2+2x-15)$
$=-(x+5)(x-3)$
The LCM is $(x+5)(x-3)$.
$$\frac{2x}{x-3}=\frac{2x}{x-3}\cdot\frac{x+5}{x+5}=\frac{2x^2+10x}{(x+5)(x-3)}$$
$$\frac{-2}{x+5}=-\frac{2}{x+5}\cdot\frac{x-3}{x-5}=-\frac{2x-6}{(x+5)(x-3)}$$
$$\frac{x-1}{15-2x-x^2}=-\frac{x-1}{(x+5)(x-3)}$$

25. $x^{2n}+3x^n+2=(x^n+2)(x^n+1)$
The LCM is $(x^n+1)(x^n+2)$.
$$\frac{x-5}{x^{2n}+3x^n+2}=\frac{x-5}{(x^n+1)(x^n+2)}$$
$$\frac{2x}{x^n+2}=\frac{2x}{x^n+2}\cdot\frac{x^n+1}{x^n+1}=\frac{2x^{n+1}+2x}{(x^n+1)(x^n+2)}$$

Objective B Exercises

27. The LCM is $4x^2$.

$$-\frac{3}{4x^2}+\frac{8}{4x^2}-\frac{3}{4x^2}=\frac{-3+8-3}{4x^2}$$
$$=\frac{2}{4x^2}=\frac{1}{2x^2}$$

29. The LCM is $3x^2+x-10$.

$$\frac{3x}{3x^2+x-10}-\frac{5}{3x^2+x-10}=\frac{3x-5}{3x^2+x-10}$$
$$=\frac{1}{x+2}$$

31. The LCM is $30a^2b^2$.

$$\frac{2}{5ab}-\frac{3}{10a^2b}+\frac{4}{15ab^2}$$
$$=\frac{2}{5ab}\cdot\frac{6ab}{6ab}-\frac{3}{10a^2b}\cdot\frac{3b}{3b}+\frac{4}{15ab^2}\cdot\frac{2a}{2a}$$
$$=\frac{12ab-9b+8a}{30a^2b^2}$$

33. The LCM is $40ab$.

$$\frac{3}{4ab}-\frac{2}{5a}+\frac{3}{10b}-\frac{5}{8ab}$$
$$=\frac{3}{4ab}\cdot\frac{10}{10}-\frac{2}{5a}\cdot\frac{8b}{8b}+\frac{3}{10b}\cdot\frac{4a}{4a}-\frac{5}{8ab}\cdot\frac{5}{5}$$
$$=\frac{30-16b+12a-25}{40ab}$$
$$=\frac{5-16b+12a}{40ab}$$

35. The LCM is $12x$.

$$\frac{3x-4}{6x}-\frac{2x-5}{4x}=\frac{3x-4}{6x}\cdot\frac{2}{2}-\frac{2x-5}{4x}\cdot\frac{3}{3}$$
$$=\frac{(6x-8)-(6x-15)}{12x}$$
$$=\frac{6x-8-6x+15}{12x}=\frac{7}{12x}$$

37. The LCM is $10x^2y^2$.

$$\frac{2y-4}{5xy^2}+\frac{3-2x}{10x^2y}=\frac{2y-4}{5xy^2}\cdot\frac{2x}{2x}+\frac{3-2x}{10x^2y}\cdot\frac{y}{y}$$
$$=\frac{4xy-8x+(3y-2xy)}{10x^2y^2}$$
$$=\frac{4xy-8x+3y-2xy}{10x^2y^2}$$
$$=\frac{2xy-8x+3y}{10x^2y^2}$$

39. The LCM is $(a-2)(a+1)$.

$$\frac{3a}{a-2}-\frac{5a}{a+1}=\frac{3a}{a-2}\cdot\frac{a+1}{a+1}-\frac{5a}{a+1}\cdot\frac{a-2}{a-2}$$
$$=\frac{(3a^2+3a)-(5a^2-10a)}{(a-2)(a+1)}$$
$$=\frac{3a^2+3a-5a^2+10a}{(a+1)(a-2)}$$
$$=\frac{-2a^2+13a}{(a-2)(a+1)}$$
$$=\frac{-(2a^2-13a)}{(a+1)(a-2)}$$
$$=-\frac{a(2a-13)}{(a+1)(a-2)}$$

41. The LCM is $(2x-5)(5x-2)$.

$$\frac{x}{2x-5}-\frac{2}{5x-2}=\frac{x}{2x-5}\cdot\frac{5x-2}{5x-2}-\frac{2}{5x-2}\cdot\frac{2x-5}{2x-5}$$
$$=\frac{(5x^2-2x)-(4x-10)}{(2x-5)(5x-2)}$$
$$=\frac{5x^2-2x-4x+10}{(5x-2)(2x-5)}$$
$$=\frac{5x^2-6x+10}{(5x-2)(2x-5)}$$

43. The LCM is $b(a-b)$.

$$\frac{1}{a-b}+\frac{1}{b}=\frac{1}{a-b}\cdot\frac{b}{b}+\frac{1}{b}\cdot\frac{a-b}{a-b}$$
$$=\frac{b+(a-b)}{b(a-b)}$$
$$=\frac{b+a-b}{b(a-b)}$$
$$=\frac{a}{b(a-b)}$$

45. The LCM is $a(a-3)$.

$$\frac{6a}{a-3}-5+\frac{3}{a}=\frac{6a}{a-3}\cdot\frac{a}{a}-5\cdot\frac{a(a-3)}{a(a-3)}+\frac{3}{a}\cdot\frac{a-3}{a-3}$$
$$=\frac{6a^2-5a(a-3)+3a-9}{a(a-3)}$$
$$=\frac{6a^2-5a^2+15a+3a-9}{a(a-3)}$$
$$=\frac{a^2+18a-9}{a(a-3)}$$

47. $5 - 6x = -(6x - 5)$
The LCM is $x(6x - 5)$.

$$\frac{5}{x} - \frac{5x}{5 - 6x} + 2$$
$$= \frac{5}{x} \cdot \frac{6x - 5}{6x - 5} - \frac{(-5x)}{6x - 5} \cdot \frac{x}{x} + 2 \cdot \frac{x(6x - 5)}{x(6x - 5)}$$
$$= \frac{30x - 25 + 5x^2 + 2x(6x - 5)}{x(6x - 5)}$$
$$= \frac{30x - 25 + 5x^2 + 12x^2 - 10x}{x(6x - 5)}$$
$$= \frac{17x^2 + 20x - 25}{x(6x - 5)}$$

49. $x^2 - 6x + 9 = (x - 3)^2$
$x^2 - 9 = (x + 3)(x - 3)$
The LCM is $(x + 3)(x - 3)^2$.

$$\frac{1}{x^2 - 6x + 9} - \frac{1}{x^2 - 9}$$
$$= \frac{1}{(x - 3)^2} \cdot \frac{x + 3}{x + 3} - \frac{1}{(x + 3)(x - 3)} \cdot \frac{x - 3}{x - 3}$$
$$= \frac{x + 3 - (x - 3)}{(x + 3)(x - 3)^2}$$
$$= \frac{x + 3 - x + 3}{(x - 3)^2(x + 3)}$$
$$= \frac{6}{(x - 3)^2(x + 3)}$$

51. $x^2 + 4x + 4 = (x + 2)^2$
The LCM is $(x + 2)^2$.

$$\frac{1}{x + 2} - \frac{3x}{x^2 + 4x + 4} = \frac{1}{x + 2} \cdot \frac{x + 2}{x + 2} - \frac{3x}{(x + 2)^2}$$
$$= \frac{x + 2 - 3x}{(x + 2)^2}$$
$$= \frac{-2x + 2}{(x + 2)^2}$$
$$= \frac{-(2x - 2)}{(x + 2)^2} = -\frac{2(x - 1)}{(x + 2)^2}$$

53. $x^2 + 2x - 8 = (x + 4)(x - 2)$
The LCM is $(x + 4)(x - 2)$.

$$\frac{-3x^2 + 8x + 2}{x^2 + 2x - 8} - \frac{2x - 5}{x + 4}$$
$$= \frac{-3x^2 + 8x + 2}{(x + 4)(x - 2)} - \frac{2x - 5}{x + 4} \cdot \frac{x - 2}{x - 2}$$
$$= \frac{-3x^2 + 8x + 2 - (2x - 5)(x - 2)}{(x + 4)(x - 2)}$$
$$= \frac{-3x^2 + 8x + 2 - (2x^2 - 9x + 10)}{(x + 4)(x - 2)}$$
$$= \frac{-3x^2 + 8x + 2 - 2x^2 + 9x - 10}{(x + 4)(x - 2)}$$
$$= \frac{-5x^2 + 17x - 8}{(x + 4)(x - 2)}$$
$$= \frac{-(5x^2 - 17x + 8)}{(x + 4)(x - 2)}$$
$$= -\frac{5x^2 - 17x + 8}{(x + 4)(x - 2)}$$

55. $x^{2n} - 1 = (x^n + 1)(x^n - 1)$
The LCM is $(x^n + 1)(x^n - 1)$.

$$\frac{2}{x^n - 1} + \frac{x^n}{x^{2n} - 1}$$
$$= \frac{2}{x^n - 1} \cdot \frac{x^n + 1}{x^n + 1} + \frac{x^n}{(x^n + 1)(x^n - 1)}$$
$$= \frac{2x^n + 2 + x^n}{(x^n + 1)(x^n - 1)}$$
$$= \frac{3x^n + 2}{(x^n + 1)(x^n - 1)}$$

57. $x^{2n} - x^n - 6 = (x^n - 3)(x^n + 2)$
The LCM is $(x^n - 3)(x^n + 2)$.

$$\frac{2x^n - 6}{x^{2n} - x^n - 6} + \frac{x^n}{x^n + 2}$$
$$= \frac{2x^n - 6}{(x^n - 3)(x^n + 2)} + \frac{x^n}{x^n + 2} \cdot \frac{x^n - 3}{x^n - 3}$$
$$= \frac{2x^n - 6 + x^{2n} - 3x^n}{(x^n - 3)(x^n + 2)}$$
$$= \frac{x^{2n} - x^n - 6}{(x^n - 3)(x^n + 2)} = 1$$

59. $4x^2-36=4(x^2-9)=4(x+3)(x-3)$
The LCM is $4(x+3)(x-3)$.

$$\frac{x^2+4}{4x^2-36}-\frac{13}{x+3}$$
$$=\frac{x^2+4}{4(x+3)(x-3)}-\frac{13}{x+3}\cdot\frac{4(x-3)}{4(x-3)}$$
$$=\frac{x^2+4-52(x-3)}{4(x+3)(x-3)}$$
$$=\frac{x^2+4-52x+156}{4(x+3)(x-3)}$$
$$=\frac{x^2-52x+160}{4(x+3)(x-3)}$$

61. $4x^2+9x+2=(4x+1)(x+2)$
The LCM is $(4x+1)(x+2)$.

$$\frac{3x-4}{4x+1}+\frac{3x+6}{4x^2+9x+2}$$
$$=\frac{3x-4}{4x+1}\cdot\frac{x+2}{x+2}+\frac{3x+6}{(4x+1)(x+2)}$$
$$=\frac{3x^2+2x-8+3x+6}{(4x+1)(x+2)}$$
$$=\frac{3x^2+5x-2}{(4x+1)(x+2)}$$
$$=\frac{(3x-1)(x+2)}{(4x+1)(x+2)}$$
$$=\frac{3x-1}{4x+1}$$

63. $x^2+x-12=(x+4)(x-3)$
$x^2+7x+12=(x+4)(x+3)$
The LCM is $(x+4)(x+3)(x-3)$.

$$\frac{x+1}{x^2+x-12}-\frac{x-3}{x^2+7x+12}$$
$$=\frac{x+1}{(x+4)(x-3)}\cdot\frac{x+3}{x+3}-\frac{x-3}{(x+4)(x+3)}\cdot\frac{x-3}{x-3}$$
$$=\frac{(x+1)(x+3)-(x-3)(x-3)}{(x+4)(x-3)(x+3)}$$
$$=\frac{(x^2+4x+3)-(x^2-6x+9)}{(x+4)(x-3)(x+3)}$$
$$=\frac{x^2+4x+3-x^2+6x-9}{(x+4)(x-3)(x+3)}$$
$$=\frac{2(5x-3)}{(x+4)(x-3)(x+3)}$$

65. $x^2-2x-15=(x-5)(x+3)$
$5-x=-(x-5)$
The LCM is $(x-5)(x+3)$.

$$\frac{2x^2-2x}{x^2-2x-15}-\frac{2}{x+3}+\frac{x}{5-x}$$
$$=\frac{2x^2-2x}{(x-5)(x+3)}-\frac{2}{x+3}\cdot\frac{x-5}{x-5}+\frac{-x}{x-5}\cdot\frac{x+3}{x+3}$$
$$=\frac{2x^2-2x-2x+10-x^2-3x}{(x-5)(x+3)}$$
$$=\frac{x^2-7x+10}{(x-5)(x+3)}$$
$$=\frac{(x-2)(x-5)}{(x-5)(x+3)}$$
$$=\frac{x-2}{x+3}$$

67. $3x^2-11x-20=(3x+4)(x-5)$
The LCM is $(3x+4)(x-5)$.

$$\frac{x}{3x+4}+\frac{3x+2}{x-5}-\frac{7x^2+24x+28}{3x^2-11x-20}$$
$$=\frac{x}{3x+4}\cdot\frac{x-5}{x-5}+\frac{3x+2}{x-5}\cdot\frac{3x+4}{3x+4}-\frac{7x^2+24x+28}{(3x+4)(x-5)}$$
$$=\frac{x^2-5x+9x^2+18x+8-(7x^2+24x+28)}{(3x+4)(x-5)}$$
$$=\frac{x^2-5x+9x^2+18x+8-7x^2-24x-28}{(3x+4)(x-5)}$$
$$=\frac{3x^2-11x-20}{(3x+4)(x-5)}$$
$$=1$$

69. $1-2x=-(2x-1)$
$8x^2-10x+3=(2x-1)(4x-3)$
The LCM is $(2x-1)(4x-3)$.

$$\frac{x+1}{1-2x}-\frac{x+3}{4x-3}+\frac{10x^2+7x-9}{8x^2-10x+3}$$
$$=\frac{-(x+1)}{2x-1}\cdot\frac{4x-3}{4x-3}-\frac{x+3}{4x-3}\cdot\frac{2x-1}{2x-1}+\frac{10x^2+7x-9}{8x^2-10x+3}$$
$$=\frac{-(x+1)(4x-3)-(x+3)(2x-1)+10x^2+7x-9}{(2x-1)(4x-3)}$$
$$=\frac{-4x^2-x+3-2x^2-5x+3+10x^2+7x-9}{(2x-1)(4x-3)}$$
$$=\frac{4x^2+x-3}{(2x-1)(4x-3)}$$
$$=\frac{(4x-3)(x+1)}{(2x-1)(4x-3)}$$
$$=\frac{x+1}{2x-1}$$

71. $8x^3 - 1 = (2x-1)(4x^2+2x+1)$
The LCM is $(2x-1)(4x^2+2x+1)$.

$$\frac{2x}{4x^2+2x+1} + \frac{4x+1}{8x^3-1}$$
$$= \frac{2x}{4x^2+2x+1} \cdot \frac{2x-1}{2x-1} + \frac{4x+1}{(2x-1)(4x^2+2x+1)}$$
$$= \frac{4x^2-2x+4x+1}{(2x-1)(4x^2+2x+1)}$$
$$= \frac{1}{2x-1}$$

73. $x^4 - 16 = (x^2+4)(x+2)(x-2)$
$x^2 - 4 = (x+2)(x-2)$
The LCM is $(x^2+4)(x+2)(x-2)$.

$$\frac{x^2-12}{x^4-16} + \frac{1}{x^2-4} - \frac{1}{x^2+4}$$
$$= \frac{x^2-12}{(x^2+4)(x+2)(x-2)} + \frac{1}{(x+2)(x-2)} \cdot \frac{x^2+4}{x^2+4} - \frac{1}{x^2+4} \cdot \frac{(x+2)(x-2)}{(x+2)(x-2)}$$
$$= \frac{x^2-12+x^2+4-(x^2-4)}{(x^2+4)(x+2)(x-2)}$$
$$= \frac{x^2-12+x^2+4-x^2+4}{(x^2+4)(x+2)(x-2)}$$
$$= \frac{x^2-4}{(x^2+4)(x+2)(x-2)}$$
$$= \frac{1}{x^2+4}$$

75. The LCM of $\frac{a-3}{a^2}$ and $\frac{a-3}{9}$ is $9a^2$.

$$\left[\frac{a-3}{a^2} - \frac{a-3}{9}\right] \div \frac{a^2-9}{3a}$$
$$= \left[\frac{a-3}{a^2} \cdot \frac{9}{9} - \frac{a-3}{9} \cdot \frac{a^2}{a^2}\right] \div \frac{a^2-9}{3a}$$
$$= \left[\frac{9(a-3)}{9a^2} - \frac{a^2(a-3)}{9a^2}\right] \div \frac{a^2-9}{3a}$$
$$= \frac{9(a-3) - a^2(a-3)}{9a^2} \div \frac{a^2-9}{3a}$$
$$= \frac{(9-a^2)(a-3)}{9a^2} \cdot \frac{3a}{a^2-9}$$
$$= \frac{(3-a)(3+a)(a-3)}{9a^2} \cdot \frac{3a}{(a-3)(a+3)}$$
$$= \frac{3-a}{3a}$$

77.
$$\frac{x^2-4x+4}{2x+1} \cdot \frac{2x^2+x}{x^3-4x} - \frac{3x-2}{x+1}$$
$$= \frac{(x-2)(x-2)}{2x+1} \cdot \frac{x(2x+1)}{x(x-2)(x+2)} - \frac{3x-2}{x+1}$$
$$= \frac{(x-2)(x-2)(x)(2x+1)}{(2x+1)(x)(x-2)(x+2)} - \frac{3x-2}{x+1}$$
$$= \frac{x-2}{x+2} - \frac{3x-2}{x+1}$$

The LCM is $(x+2)(x+1)$.

$$\frac{x-2}{x+2} - \frac{3x-2}{x+1} = \frac{x-2}{x+2} \cdot \frac{x+1}{x+1} - \frac{3x-2}{x+1} \cdot \frac{x+2}{x+2}$$
$$= \frac{x^2-x-2-(3x^2+4x-4)}{(x+2)(x+1)}$$
$$= \frac{x^2-x-2-3x^2-4x+4}{(x+2)(x+1)}$$
$$= \frac{-2x^2-5x+2}{(x+2)(x+1)}$$
$$= -\frac{2x^2+5x-2}{(x+2)(x+1)}$$

79. The LCM is ab.

$$\left[\frac{a-2b}{b} + \frac{b}{a}\right] \div \left[\frac{b+a}{a} - \frac{2a}{b}\right]$$
$$= \left[\frac{a-2b}{b} \cdot \frac{a}{a} + \frac{b}{a} \cdot \frac{b}{b}\right] \div \left[\frac{b+a}{a} \cdot \frac{b}{b} - \frac{2a}{b} \cdot \frac{a}{a}\right]$$
$$= \frac{a^2-2ab+b^2}{ab} \div \frac{b^2+ab-2a^2}{ab}$$
$$= \frac{a^2-2ab+b^2}{ab} \cdot \frac{ab}{b^2+ab-2a^2}$$
$$= \frac{(a-b)(a-b)ab}{ab(b+2a)(b-a)}$$
$$= \frac{a-b}{-(b+2a)}$$
$$= \frac{-(a-b)}{b+2a} = \frac{b-a}{b+2a}$$

81.
$$\frac{2x}{x^2-x-6} - \frac{6x-6}{2x^2-9x+9} \div \frac{x^2+x-2}{2x-3}$$
$$= \frac{2x}{(x-3)(x+2)} - \frac{6(x-1)}{(2x-3)(x-3)} \div \frac{(x+2)(x-1)}{2x-3}$$
$$= \frac{2x}{(x-3)(x+2)} - \frac{6(x-1)}{(2x-3)(x-3)} \cdot \frac{2x-3}{(x+2)(x-1)}$$
$$= \frac{2x}{(x-3)(x+2)} - \frac{6(x-1)(2x-3)}{(2x-3)(x-3)(x+2)(x-1)}$$
$$= \frac{2x}{(x-3)(x+2)} - \frac{6}{(x-3)(x+2)}$$
$$= \frac{2x-6}{(x-3)(x+2)}$$
$$= \frac{2(x-3)}{(x-3)(x+2)}$$
$$= \frac{2}{x+2}$$

Applying the Concepts

83a. $\left(\frac{b}{6}-\frac{6}{b}\right) \div \left(\frac{6}{b}-4+\frac{b}{2}\right)$

$$\frac{b}{6}-\frac{6}{b}=\frac{b}{6}\cdot\frac{b}{b}-\frac{6}{b}\cdot\frac{6}{6}$$
$$=\frac{b^2-36}{6b}$$
$$=\frac{(b-6)(b+6)}{6b}$$
$$\frac{6}{b}-4+\frac{b}{2}=\frac{6}{b}\cdot\frac{2}{2}-\frac{4}{1}\cdot\frac{2b}{2b}+\frac{b}{2}\cdot\frac{b}{b}$$
$$=\frac{12-8b+b^2}{2b}$$
$$=\frac{(b-6)(b-2)}{2b}$$
$$\left(\frac{(b-6)(b+6)}{6b}\right) \div \left(\frac{(b-6)(b-2)}{2b}\right)$$
$$=\frac{(b-6)(b+6)}{6b}\cdot\frac{2b}{(b-6)(b-2)}$$
$$=\frac{b+6}{3(b-2)}$$

b. $\left(\frac{x+1}{2x-1}-\frac{x-1}{2x+1}\right)\cdot\left(\frac{2x-1}{x}-\frac{2x-1}{x^2}\right)$

$$\frac{x+1}{2x-1}-\frac{x-1}{2x+1}=\frac{x+1}{2x-1}\cdot\frac{2x+1}{2x+1}-\frac{x-1}{2x+1}\cdot\frac{2x-1}{2x-1}$$
$$=\frac{(x+1)(2x+1)-(x-1)(2x-1)}{(2x-1)(2x+1)}$$
$$=\frac{2x^2+3x+1-(2x^2-3x+1)}{(2x+1)(2x-1)}$$
$$=\frac{2x^2+3x+1-2x^2+3x-1}{(2x+1)(2x-1)}$$
$$=\frac{6x}{(2x+1)(2x-1)}$$
$$\frac{2x-1}{x}-\frac{2x-1}{x^2}=\frac{2x-1}{x}\cdot\frac{x}{x}-\frac{2x-1}{x^2}$$
$$=\frac{x(2x-1)-(2x-1)}{x^2}$$
$$=\frac{(2x-1)(x-1)}{x^2}$$
$$\left(\frac{x+1}{2x-1}-\frac{x-1}{2x+1}\right)\cdot\left(\frac{2x-1}{x}-\frac{2x-1}{x^2}\right)$$
$$=\frac{6x}{(2x+1)(2x-1)}\cdot\frac{(2x-1)(x-1)}{x^2}$$
$$=\frac{6(x-1)}{x(2x+1)}$$

85a.
$$f(x)=\frac{x}{x+2}$$
$$f(4)=\frac{2}{3}$$
$$g(x)=\frac{4}{x-3}$$
$$g(4)=4$$
$$S(x)=\frac{x^2+x+8}{x^2-x-6}$$
$$S(4)=4\frac{2}{3}$$
$$f(4)+g(4)=\frac{2}{3}+4=\frac{14}{3}=S(4)$$

Yes, $f(4)+g(4)=S(4)$.

b. $S(a)=f(a)+g(a)$

Section 6.3

Objective A Exercises

1. A complex fraction is a fraction whose numerator or denominator contains one or more fractions.

3. The LCM is 3.

$$\frac{2-\frac{1}{3}}{4+\frac{11}{3}}=\frac{2-\frac{1}{3}}{4+\frac{11}{3}}\cdot\frac{3}{3}$$
$$=\frac{2\cdot 3-\frac{1}{3}\cdot 3}{4\cdot 3+\frac{11}{3}\cdot 3}$$
$$=\frac{6-1}{12+11}$$
$$=\frac{5}{23}$$

5. The LCM is 6.

$$\frac{3-\frac{2}{3}}{5+\frac{5}{6}}=\frac{3-\frac{2}{3}}{5+\frac{5}{6}}\cdot\frac{6}{6}$$
$$=\frac{3\cdot 6-\frac{2}{3}\cdot 6}{5\cdot 6+\frac{5}{6}\cdot 6}$$
$$=\frac{18-4}{30+5}$$
$$=\frac{14}{35}=\frac{2}{5}$$

7. The LCM of x and x^2 is x^2.

$$\frac{1+\frac{1}{x}}{1-\frac{1}{x^2}} = \frac{1+\frac{1}{x}}{1-\frac{1}{x^2}} \cdot \frac{x^2}{x^2}$$
$$= \frac{1 \cdot x^2 + \frac{1}{x} \cdot x^2}{1 \cdot x^2 - \frac{1}{x^2} \cdot x^2}$$
$$= \frac{x^2+x}{x^2-1}$$
$$= \frac{x(x+1)}{(x+1)(x-1)}$$
$$= \frac{x}{x-1}$$

9. The LCM is a.

$$\frac{a-2}{\frac{4}{a}-a} = \frac{a-2}{\frac{4}{a}-a} \cdot \frac{a}{a}$$
$$= \frac{a \cdot a - 2 \cdot a}{\frac{4}{a} \cdot a - a \cdot a}$$
$$= \frac{a^2-2a}{4-a^2}$$
$$= \frac{a(a-2)}{(2+a)(2-a)}$$
$$= -\frac{a}{a+2}$$

11. The LCM of a^2 and a is a^2.

$$\frac{\frac{1}{a^2}-\frac{1}{a}}{\frac{1}{a^2}+\frac{1}{a}} = \frac{\frac{1}{a^2}-\frac{1}{a}}{\frac{1}{a^2}+\frac{1}{a}} \cdot \frac{a^2}{a^2}$$
$$= \frac{\frac{1}{a^2} \cdot a^2 - \frac{1}{a} \cdot a^2}{\frac{1}{a^2} \cdot a^2 + \frac{1}{a} \cdot a^2}$$
$$= \frac{1-a}{1+a}$$
$$= \frac{-(a-1)}{a+1}$$
$$= -\frac{a-1}{a+1}$$

13. The LCM is $x+2$.

$$\frac{2-\frac{4}{x+2}}{5-\frac{10}{x+2}} = \frac{2-\frac{4}{x+2}}{5-\frac{10}{x+2}} \cdot \frac{x+2}{x+2}$$
$$= \frac{2(x+2) - \frac{4}{x+2} \cdot (x+2)}{5(x+2) - \frac{10}{x+2} \cdot (x+2)}$$
$$= \frac{2x+4-4}{5x+10-10}$$
$$= \frac{2x}{5x} = \frac{2}{5}$$

15. The LCM is $2a-3$.

$$\frac{\frac{3}{2a-3}+2}{\frac{-6}{2a-3}-4} = \frac{\frac{3}{2a-3}+2}{\frac{-6}{2a-3}-4} \cdot \frac{2a-3}{2a-3}$$
$$= \frac{\frac{3}{2a-3} \cdot (2a-3) + 2(2a-3)}{\frac{-6}{2a-3} \cdot (2a-3) - 4(2a-3)}$$
$$= \frac{3+4a-6}{-6-8a+12}$$
$$= \frac{4a-3}{-8a+6}$$
$$= -\frac{1}{2}$$

17. The LCM of x and $x+1$ is $x(x+1)$.

$$\frac{\frac{x}{x+1}-\frac{1}{x}}{\frac{x}{x+1}+\frac{1}{x}} = \frac{\frac{x}{x+1}-\frac{1}{x}}{\frac{x}{x+1}+\frac{1}{x}} \cdot \frac{x(x+1)}{x(x+1)}$$
$$= \frac{\frac{x}{x+1} \cdot x(x+1) - \frac{1}{x} \cdot x(x+1)}{\frac{x}{x+1} \cdot x(x+1) + \frac{1}{x} \cdot x(x+1)}$$
$$= \frac{x^2-(x+1)}{x^2+(x+1)}$$
$$= \frac{x^2-x-1}{x^2+x+1}$$

19. The LCM of x and x^2 is x^2.

$$\frac{1-\frac{1}{x}-\frac{6}{x^2}}{1-\frac{4}{x}+\frac{3}{x^2}} = \frac{1-\frac{1}{x}-\frac{6}{x^2}}{1-\frac{4}{x}+\frac{3}{x^2}} \cdot \frac{x^2}{x^2}$$
$$= \frac{1 \cdot x^2 - \frac{1}{x} \cdot x^2 - \frac{6}{x^2} \cdot x^2}{1 \cdot x^2 - \frac{4}{x} \cdot x^2 + \frac{3}{x^2} \cdot x^2}$$
$$= \frac{x^2-x-6}{x^2-4x+3}$$
$$= \frac{(x+2)(x-3)}{(x-1)(x-3)}$$
$$= \frac{x+2}{x-1}$$

21. The LCM of x and x^2 is x^2.

$$\frac{1+\frac{1}{x}-\frac{12}{x^2}}{\frac{9}{x^2}+\frac{3}{x}-2} = \frac{1+\frac{1}{x}-\frac{12}{x^2}}{\frac{9}{x^2}+\frac{3}{x}-2} \cdot \frac{x^2}{x^2}$$
$$= \frac{1 \cdot x^2 + \frac{1}{x} \cdot x^2 - \frac{12}{x^2} \cdot x^2}{\frac{9}{x^2} \cdot x^2 + \frac{3}{x} \cdot x^2 - 2 \cdot x^2}$$
$$= \frac{x^2+x-12}{9+3x-2x^2}$$
$$= \frac{(x+4)(x-3)}{(3+2x)(3-x)}$$
$$= -\frac{x+4}{2x+3}$$

23. $a+\dfrac{a}{a+\frac{1}{a}}=a+\dfrac{a}{a+\frac{1}{a}}\cdot\dfrac{a}{a}$

$=a+\dfrac{a\cdot a}{a\cdot a+\frac{1}{a}\cdot a}$

$=a+\dfrac{a^2}{a^2+1}$

The LCM is a^2+1.

$a+\dfrac{a^2}{a^2+1}=a\cdot\dfrac{a^2+1}{a^2+1}+\dfrac{a^2}{a^2+1}$

$=\dfrac{a(a^2+1)+a^2}{a^2+1}$

$=\dfrac{a^3+a+a^2}{a^2+1}$

$=\dfrac{a(a^2+a+1)}{a^2+1}$

25. $\dfrac{1-\frac{1}{x-4}}{1-\frac{6}{x+1}}$

The LCM is $(x-4)(x+1)$.

$\dfrac{1-\frac{1}{x-4}}{1-\frac{6}{x+1}}\cdot\dfrac{(x-4)(x+1)}{(x-4)(x+1)}=\dfrac{(x-4)(x+1)-(x+1)}{(x-4)(x+1)-6(x-4)}$

$=\dfrac{x^2-3x-4-x-1}{x^2-3x-4-6x+24}$

$=\dfrac{x^2-4x-5}{x^2-9x+20}$

$=\dfrac{(x-5)(x+1)}{(x-5)(x-4)}$

$=\dfrac{x+1}{x-4}$

27. $\dfrac{x-\frac{1}{x}}{x+\frac{1}{x}}$

The LCM is x.

$\dfrac{x-\frac{1}{x}}{x+\frac{1}{x}}\cdot\dfrac{x}{x}=\dfrac{(x-1)(x+1)}{x^2+1}$

29. $\dfrac{\frac{1}{x+h}-\frac{1}{x}}{h}$

The LCM is $x(x+h)$.

$\dfrac{\frac{1}{x+h}-\frac{1}{x}}{h}\cdot\dfrac{x(x+h)}{x(x+h)}=\dfrac{x-(x+h)}{hx(x+h)}$

$=\dfrac{h}{hx(x+h)}$

$=-\dfrac{1}{x(x+h)}$

31. $\dfrac{1-\frac{2}{x-3}}{1+\frac{3}{2-x}}$

The LCM is $(x-3)(2-x)$.

$\dfrac{1-\frac{2}{x-3}}{1+\frac{3}{2-x}}\cdot\dfrac{(x-3)(2-x)}{(x-3)(2-x)}=\dfrac{(x-3)(2-x)-2(2-x)}{(x-3)(2-x)+3(x-3)}$

$=\dfrac{-x^2+5x-6-4+2x}{-x^2+5x-6+3x-9}$

$=\dfrac{-x^2+7x-10}{-x^2+8x-15}$

$=\dfrac{-(x^2-7x+10)}{-(x^2-8x+15)}$

$=\dfrac{(x-5)(x-2)}{(x-5)(x-3)}$

$=\dfrac{x-2}{x-3}$

33. The LCM is $2x+3$.

$\dfrac{x-4+\frac{9}{2x+3}}{x+3-\frac{5}{2x+3}}=\dfrac{x-4+\frac{9}{2x+3}}{x+3-\frac{5}{2x+3}}\cdot\dfrac{2x+3}{2x+3}$

$=\dfrac{(x-4)(2x+3)+\frac{9}{2x+3}\cdot(2x+3)}{(x+3)(2x+3)-\frac{5}{2x+3}\cdot(2x+3)}$

$=\dfrac{2x^2-5x-12+9}{2x^2+9x+9-5}$

$=\dfrac{2x^2-5x-3}{2x^2+9x+4}$

$=\dfrac{(2x+1)(x-3)}{(2x+1)(x+4)}$

$=\dfrac{x-3}{x+4}$

35. The LCM is $2x-1$.

$\dfrac{3x-2-\frac{5}{2x-1}}{x-6+\frac{9}{2x-1}}=\dfrac{3x-2-\frac{5}{2x-1}}{x-6+\frac{9}{2x-1}}\cdot\dfrac{2x-1}{2x-1}$

$=\dfrac{(3x-2)(2x-1)-\frac{5}{2x-1}\cdot(2x-1)}{(x-6)(2x-1)+\frac{9}{2x-1}\cdot(2x-1)}$

$=\dfrac{6x^2-7x+2-5}{2x^2-13x+6+9}$

$=\dfrac{6x^2-7x-3}{2x^2-13x+15}$

$=\dfrac{(3x+1)(2x-3)}{(2x-3)(x-5)}$

$=\dfrac{3x+1}{x-5}$

37. The LCM is $a(a-2)$.

$$\frac{\frac{1}{a}-\frac{3}{a-2}}{\frac{2}{a}+\frac{5}{a-2}}=\frac{\frac{1}{a}-\frac{3}{a-2}}{\frac{2}{a}+\frac{5}{a-2}}\cdot\frac{a(a-2)}{a(a-2)}$$
$$=\frac{\frac{1}{a}\cdot a(a-2)-\frac{3}{a-2}\cdot a(a-2)}{\frac{2}{a}\cdot a(a-2)+\frac{5}{a-2}\cdot a(a-2)}$$
$$=\frac{a-2-3a}{2a-4+5a}$$
$$=\frac{-2a-2}{7a-4}$$
$$=\frac{-(2a+2)}{7a-4}$$
$$=-\frac{2a+2}{7a-4}$$

39. The LCM of y^2, xy, and x^2 is x^2y^2.

$$\frac{\frac{1}{y^2}-\frac{1}{xy}-\frac{2}{x^2}}{\frac{1}{y^2}-\frac{3}{xy}+\frac{2}{x^2}}=\frac{\frac{1}{y^2}-\frac{1}{xy}-\frac{2}{x^2}}{\frac{1}{y^2}-\frac{3}{xy}+\frac{2}{x^2}}\cdot\frac{x^2y^2}{x^2y^2}$$
$$=\frac{\frac{1}{y^2}\cdot x^2y^2-\frac{1}{xy}\cdot x^2y^2-\frac{2}{x^2}\cdot x^2y^2}{\frac{1}{y^2}\cdot x^2y^2-\frac{3}{xy}\cdot x^2y^2+\frac{2}{x^2}\cdot x^2y^2}$$
$$=\frac{x^2-xy-2y^2}{x^2-3xy+2y^2}$$
$$=\frac{(x+y)(x-2y)}{(x-y)(x-2y)}$$
$$=\frac{x+y}{x-y}$$

41. The LCM is $(x+1)(x-1)$.

$$\frac{\frac{x-1}{x+1}-\frac{x+1}{x-1}}{\frac{x-1}{x+1}+\frac{x+1}{x-1}}=\frac{\frac{x-1}{x+1}-\frac{x+1}{x-1}}{\frac{x-1}{x+1}+\frac{x+1}{x-1}}\cdot\frac{(x+1)(x-1)}{(x+1)(x-1)}$$
$$=\frac{\frac{x-1}{x+1}\cdot(x+1)(x-1)-\frac{x+1}{x-1}\cdot(x+1)(x-1)}{\frac{x-1}{x+1}\cdot(x+1)(x-1)+\frac{x+1}{x-1}\cdot(x+1)(x-1)}$$
$$=\frac{(x-1)(x-1)-(x+1)(x+1)}{(x-1)(x-1)+(x+1)(x+1)}$$
$$=\frac{x^2-2x+1-(x^2+2x+1)}{x^2-2x+1+x^2+2x+1}$$
$$=\frac{x^2-2x+1-x^2-2x-1}{2x^2+2}$$
$$=\frac{-4x}{2(x^2+1)}$$
$$=-\frac{2x}{x^2+1}$$

43. The LCM is $1-a$.

$$a-\frac{a}{1-\frac{a}{1-a}}=a-\frac{a}{1-\frac{a}{1-a}}\cdot\frac{1-a}{1-a}$$
$$=a-\frac{a\cdot(1-a)}{1\cdot(1-a)-\frac{a}{1-a}\cdot(1-a)}$$
$$=a-\frac{a-a^2}{1-a-a}$$
$$=a-\frac{a-a^2}{1-2a}$$

The LCM is $1-2a$.

$$a-\frac{a-a^2}{1-2a}=a\cdot\frac{1-2a}{1-2a}-\frac{a-a^2}{1-2a}$$
$$=\frac{a(1-2a)-(a-a^2)}{1-2a}$$
$$=\frac{a-2a^2-a+a^2}{1-2a}$$
$$=\frac{-a^2}{1-2a}$$
$$=-\frac{a^2}{1-2a}$$

45.
$$3-\frac{2}{1-\frac{2}{3-\frac{2}{x}}}=3-\frac{2}{1-\frac{2}{3-\frac{2}{x}}\cdot\frac{x}{x}}$$
$$=3-\frac{2}{1-\frac{2\cdot x}{3\cdot x-\frac{2}{x}\cdot x}}$$
$$=3-\frac{2}{1-\frac{2x}{3x-2}}$$

The LCM is $3x-2$.

$$3-\frac{2}{\frac{3x-2}{3x-2}-\frac{2x}{3x-2}}=3-\frac{2}{\frac{3x-2-2x}{3x-2}}$$
$$=3-\frac{2}{\frac{x-2}{3x-2}}$$
$$=3-\frac{2}{\frac{x-2}{3x-2}}\cdot\frac{3x-2}{3x-2}$$
$$=3-\frac{2(3x-2)}{x-2}$$

The LCM is $x-2$.

$$3\cdot\frac{x-2}{x-2}-\frac{2(3x-2)}{x-2}=\frac{3(x-2)-2(3x-2)}{x-2}$$
$$=\frac{3x-6-6x+4}{x-2}$$
$$=\frac{-3x-2}{x-2}$$
$$=-\frac{3x+2}{x-2}$$

Applying the Concepts

47. Strategy
- The first even integer: n
  The second consecutive even integer: $n+2$
  The third consecutive even integer: $n+4$
- Add the reciprocals of the three integers.

Solution
$$\frac{1}{n}+\frac{1}{n+2}+\frac{1}{n+4}=\frac{1}{n}\cdot\frac{(n+2)(n+4)}{(n+2)(n+4)}+\frac{1}{n+2}\cdot\frac{n(n+4)}{n(n+4)}+\frac{1}{n+4}\cdot\frac{n(n+2)}{n(n+2)}$$
$$=\frac{(n+2)(n+4)+n(n+4)+n(n+2)}{n(n+2)(n+4)}$$
$$=\frac{n^2+6n+8+n^2+4n+n^2+2n}{n(n+2)(n+4)}$$
$$=\frac{3n^2+12n+8}{n(n+2)(n+4)}$$

49a.
$$\frac{x^{-1}}{x^{-1}+2^{-1}}=\frac{\frac{1}{x}}{\frac{1}{x}+\frac{1}{2}}$$
The LCM is $2x$.
$$\frac{\frac{1}{x}}{\frac{1}{x}+\frac{1}{2}}=\frac{\frac{1}{x}}{\frac{1}{x}+\frac{1}{2}}\cdot\frac{2x}{2x}$$
$$=\frac{\frac{2x}{x}}{\frac{1}{x}(2x)+\frac{1}{2}(2x)}$$
$$=\frac{2}{2+x}$$

b.
$$\frac{-x^{-1}+x}{x^{-1}-x}=\frac{-\frac{1}{x}+x}{\frac{1}{x}-x}$$
The LCM is x.
$$\frac{-\frac{1}{x}+x}{\frac{1}{x}-x}=\frac{-\frac{1}{x}+x}{\frac{1}{x}-x}\cdot\frac{x}{x}$$
$$=\frac{-\frac{1}{x}(x)+x\cdot x}{\frac{1}{x}(x)-x\cdot x}$$
$$=\frac{-1+x^2}{1-x^2}$$
$$=\frac{x^2-1}{-1\cdot(x^2-1)}$$
$$=\frac{1}{-1}=-1$$

c.
$$\frac{x^{-1}-x^{-2}-6x^{-3}}{x^{-1}-4x^{-3}}=\frac{\frac{1}{x}-\frac{1}{x^2}-\frac{6}{x^3}}{\frac{1}{x}-\frac{4}{x^3}}$$
The LCM is x^3.
$$\frac{\frac{1}{x}-\frac{1}{x^2}-\frac{6}{x^3}}{\frac{1}{x}-\frac{4}{x^3}}=\frac{\frac{1}{x}-\frac{1}{x^2}-\frac{6}{x^3}}{\frac{1}{x}-\frac{4}{x^3}}\cdot\frac{x^3}{x^3}$$
$$=\frac{x^2-x-6}{x^2-4}$$
$$=\frac{(x-3)(x+2)}{(x-2)(x+2)}$$
$$=\frac{x-3}{x-2}$$

Section 6.4

Objective A Exercises

1. A ratio is the quotient of two quantities that have the same unit. A rate is the quotient of two quantities that have different units.

3.
$$\frac{x}{30}=\frac{3}{10}$$
$$\frac{x}{30}\cdot 30=\frac{3}{10}\cdot 30$$
$$x=3\cdot 3$$
$$x=9$$

5.
$$\frac{2}{x}=\frac{8}{30}$$
$$\frac{2}{x}\cdot 30x=\frac{8}{30}\cdot 30x$$
$$2\cdot 30=8x$$
$$60=8x$$
$$x=\frac{15}{2}$$

7. $$\frac{x+1}{10}=\frac{2}{5}$$
$$\frac{x+1}{10}\cdot 10=\frac{2}{5}\cdot 10$$
$$x+1=2\cdot 2$$
$$x+1=4$$
$$x=3$$

9. $$\frac{4}{x+2}=\frac{3}{4}$$
$$\frac{4}{x+2}\cdot 4(x+2)=\frac{3}{4}\cdot 4(x+2)$$
$$4\cdot 4=3(x+2)$$
$$16=3x+6$$
$$10=3x$$
$$x=\frac{10}{3}$$

11. $$\frac{x}{4}=\frac{x-2}{8}$$
$$\frac{x}{4}\cdot 8=\frac{x-2}{8}\cdot 8$$
$$x\cdot 2=x-2$$
$$2x=x-2$$
$$x=-2$$

13. $$\frac{16}{2-x}=\frac{4}{x}$$
$$\frac{16}{2-x}\cdot x(2-x)=\frac{4}{x}\cdot x(2-x)$$
$$16x=4(2-x)$$
$$16x=8-4x$$
$$20x=8$$
$$x=\frac{8}{20}$$
$$x=\frac{2}{5}$$

15. $$\frac{8}{x-2}=\frac{4}{x+1}$$
$$\frac{8}{x-2}\cdot(x-2)(x+1)=\frac{4}{x+1}\cdot(x-2)(x+1)$$
$$8(x+1)=4(x-2)$$
$$8x+8=4x-8$$
$$4x+8=-8$$
$$4x=-16$$
$$x=-4$$

17. $$\frac{x}{3}=\frac{x+1}{7}$$
$$\frac{x}{3}\cdot 21=\frac{x+1}{7}\cdot 21$$
$$7x=(x+1)3$$
$$7x=3x+3$$
$$4x=3$$
$$x=\frac{3}{4}$$

19. $$\frac{8}{3x-2}=\frac{2}{2x+1}$$
$$\frac{8}{3x-2}\cdot(3x-2)(2x+1)=\frac{2}{2x+1}\cdot(3x-2)(2x+1)$$
$$8(2x+1)=2(3x-2)$$
$$16x+8=6x-4$$
$$10x=-12$$
$$x=\frac{-12}{10}$$
$$x=-\frac{6}{5}$$

21. $$\frac{3x+1}{3x-4}=\frac{x}{x-2}$$
$$\frac{3x+1}{3x-4}\cdot(3x-4)(x-2)=\frac{x}{x-2}\cdot(3x-4)(x-2)$$
$$(3x+1)(x-2)=x(3x-4)$$
$$3x^2-5x-2=3x^2-4x$$
$$-5x-2=-4x$$
$$x=-2$$

Objective B Exercises

23. **Strategy** To find how many grams of protein are in a 454-gram box of pasta, write and solve a proportion using x to represent the number grams of protein in a 454-gram box of pasta.

Solution
$$\frac{56}{7}=\frac{454}{x}$$
$$\frac{56}{7}\cdot 7x=\frac{454}{x}\cdot 7x$$
$$56x=454\cdot 7$$
$$56x=3178$$
$$x=56.75$$

There are 56.75 g of protein in a 454 g box of the pasta.

25. **Strategy** To find the number of squawfish, write and solve a proportion using x to represent the number of squawfish in the area.

Solution
$$\frac{2400}{x}=\frac{9}{225}$$
$$\frac{2400}{x}=\frac{1}{25}$$
$$\frac{2400}{x}\cdot 25x=\frac{1}{25}\cdot 25x$$
$$2400\cdot 25=x$$
$$60{,}000=x$$

There would be 60,000 squawfish in the area.

27. Strategy To find the number of computers with defective CD-ROM drives, write and solve a proportion using x to represent the number of computers with defective CD-ROM drives.

Solution

$$\frac{250}{45}=\frac{2800}{x}$$
$$\frac{50}{9}=\frac{2800}{x}$$
$$\frac{50}{9}\cdot 9x=\frac{2800}{x}\cdot 9x$$
$$50x=2800\cdot 9$$
$$50x=25{,}200$$
$$x=504$$

There would be 504 computers with defective CD-ROM drives.

29. Strategy To find the dimensions of the room, write and solve two proportions using L to represent the length of the room and W to represent the width of the room.

Solution

$$\frac{\frac{1}{4}}{1}=\frac{4\frac{1}{2}}{W} \qquad \frac{\frac{1}{4}}{1}=\frac{6}{L}$$
$$\frac{\frac{1}{4}}{1}\cdot W=\frac{4\frac{1}{2}}{W}\cdot W \qquad \frac{\frac{1}{4}}{1}\cdot L=\frac{6}{L}\cdot L$$
$$\frac{1}{4}W=4\frac{1}{2} \qquad \frac{1}{4}L=6$$
$$W=18 \qquad L=24$$

The dimensions of the room are 18 ft by 24 ft.

31. Strategy To find the additional amount of medicine, write and solve a proportion using x to represent the additional amount of medicine. Then, $x+1.5$ is the total amount of medicine.

Solution

$$\frac{1.5}{140}=\frac{x+1.5}{210}$$
$$\frac{1.5}{140}\cdot 420=\frac{x+1.5}{210}\cdot 420$$
$$1.5\cdot 3=(x+1.5)2$$
$$4.5=2x+3$$
$$1.5=2x$$
$$0.75=x$$

0.75 additional ounces of medicine are required.

33. Strategy To find how out how many miles a person would need to walk to lose 1 pound, write and solve a proportion using x to represent the number of miles a person needs to walk to lose 1 pound.

Solution

$$\frac{\frac{4}{2}}{650}=\frac{\frac{x}{2}}{3500}$$
$$\frac{1}{325}\cdot 45500=\frac{\frac{x}{2}}{3500}\cdot 45500$$
$$140=13\left(\frac{x}{2}\right)$$
$$\left(\frac{2}{1}\right)\cdot 140=13\left(\frac{x}{2}\right)\cdot\left(\frac{2}{1}\right)$$
$$280=13x$$
$$21.54\approx x$$

A person must walk 21.54 mi to lose 1 pound.

Applying the Concepts

35. $\frac{a}{b}=\frac{c}{d}$

Add 1 to each side.

$\frac{a}{b}+1=\frac{c}{d}+1$

Rewrite each side of the equation as a single fraction.

$$\frac{a}{b}+1\cdot\frac{b}{b}=\frac{c}{d}+1\cdot\frac{d}{d}$$
$$\frac{a+b}{b}=\frac{c+d}{d}$$

Section 6.5

Objective A Exercises

1.

$$\frac{x}{2}+\frac{5}{6}=\frac{x}{3}$$
$$12\left(\frac{x}{2}+\frac{5}{6}\right)=12\left(\frac{x}{3}\right)$$
$$12\cdot\frac{x}{2}+12\cdot\frac{5}{6}=4x$$
$$6x+10=4x$$
$$10=-2x$$
$$x=-5$$

3.

$$\frac{8}{2x-1}=2$$
$$(2x-1)\cdot\frac{8}{2x-1}=(2x-1)2$$
$$8=4x-2$$
$$10=4x$$
$$x=\frac{5}{2}$$

5.
$$1-\frac{3}{y}=4$$
$$y\left(1-\frac{3}{y}\right)=y\cdot 4$$
$$y\cdot 1-y\cdot\frac{3}{y}=4y$$
$$y-3=4y$$
$$-3=3y$$
$$y=-1$$

7.
$$\frac{3}{x-2}=\frac{4}{x}$$
$$x(x-2)\cdot\frac{3}{x-2}=x(x-2)\cdot\frac{4}{x}$$
$$3x=(x-2)4$$
$$3x=4x-8$$
$$-x=-8$$
$$x=8$$

9.
$$\frac{6}{2y+3}=\frac{6}{y}$$
$$y(2y+3)\cdot\frac{6}{2y+3}=y(2y+3)\cdot\frac{6}{y}$$
$$6y=(2y+3)6$$
$$6y=12y+18$$
$$-6y=18$$
$$y=-3$$

11.
$$\frac{5}{y+3}-2=\frac{7}{y+3}$$
$$(y+3)\left(\frac{5}{y+3}-2\right)=(y+3)\cdot\frac{7}{y+3}$$
$$(y+3)\cdot\frac{5}{y+3}-(y+3)2=7$$
$$5-2y-6=7$$
$$-2y-1=7$$
$$-2y=8$$
$$y=-4$$

13.
$$\frac{-4}{a-4}=3-\frac{a}{a-4}$$
$$(a-4)\frac{-4}{a-4}=(a-4)\left(3-\frac{a}{a-4}\right)$$
$$-4=(a-4)3-(a-4)\frac{a}{a-4}$$
$$-4=3a-12-a$$
$$-4=2a-12$$
$$8=2a$$
$$4=a$$

4 does not check as a solution.
The equation has no solution.

15.
$$\frac{2x}{x+2}+3x=\frac{-5}{x+2}$$
$$(x+2)\left(\frac{2x}{x+2}+3x\right)=\frac{-5}{x+2}(x+2)$$
$$2x+3x(x+2)=-5$$
$$2x+3x^2+6x=-5$$
$$3x^2+8x+5=0$$
$$(3x+5)(x+1)=0$$

$3x+5=0$ $\quad x+1=0$
$3x=-5$ $\quad x=-1$
$x=-\frac{5}{3}$

$-\frac{5}{3}$ and -1

17.
$$\frac{x}{2x-9}-3x=\frac{10}{9-2x}$$
$$(2x-9)\left(\frac{x}{2x-9}-3x\right)=\frac{10}{(9-2x)}(2x-9)$$
$$x-3x(2x-9)=-10$$
$$x-6x^2+27x=-10$$
$$-6x^2+28x+10=0$$
$$-2(3x^2-14x-5)=0$$
$$(3x+1)(x-5)=0$$

$3x+1=0$ $\quad x-5=0$
$3x=-1$ $\quad x=5$
$x=-\frac{1}{3}$

$-\frac{1}{3}$ and 5

19.
$$\frac{5}{x-2}-\frac{2}{x+2}=\frac{3}{x^2-4}$$
$$\frac{5}{x-2}-\frac{2}{x+2}=\frac{3}{(x+2)(x-2)}$$
$$(x+2)(x-2)\left(\frac{5}{x-2}-\frac{2}{x+2}\right)=(x+2)(x-2)\frac{3}{(x+2)(x-2)}$$
$$(x+2)(x-2)\frac{5}{x-2}-(x+2)(x-2)\frac{2}{x+2}=3$$
$$(x+2)5-(x-2)2=3$$
$$5x+10-2x+4=3$$
$$3x+14=3$$
$$3x=-11$$
$$x=-\frac{11}{3}$$

21.
$$\frac{9}{x^2+7x+10}=\frac{5}{x+2}-\frac{3}{x+5}$$
$$\frac{9}{(x+2)(x+5)}=\frac{5}{x+2}-\frac{3}{x+5}$$
$$(x+2)(x+5)\frac{9}{(x+2)(x+5)}=(x+2)(x+5)\left(\frac{5}{x+2}-\frac{3}{x+5}\right)$$
$$9=(x+2)(x+5)\frac{5}{x+2}-(x+2)(x+5)\frac{3}{x+5}$$
$$9=(x+5)5-(x+2)3$$
$$9=5x+25-3x-6$$
$$9=2x+19$$
$$-10=2x$$
$$-5=x$$

-5 does not check as a solution. The equation has no solution.

23.
$$\frac{P_1V_1}{T_1}=\frac{P_2V_2}{T_2}$$
$$\frac{P_1V_1T_2}{T_1}=P_2V_2$$
$$P_2=\frac{P_1V_1T_2}{V_2T_1}$$

25.
$$\frac{1}{f}=\frac{1}{a}+\frac{1}{b}$$
$$abf\left(\frac{1}{f}\right)=abf\left(\frac{1}{a}+\frac{1}{b}\right)$$
$$ab=abf\frac{1}{a}+abf\frac{1}{b}$$
$$ab=bf+af$$
$$ab-bf=af$$
$$b(a-f)=af$$
$$b=\frac{af}{a-f}$$

Objective B Exercises

27. **Strategy**
- Time for the experienced bricklayer to do the job: t
  Time for the inexperienced bricklayer to do the job: $2t$

	Rate	Time	Part
Experienced bricklayer	$\frac{1}{t}$	6	$\frac{6}{t}$
Inexperienced bricklayer	$\frac{1}{2t}$	16	$\frac{16}{2t}$

- The sum of the parts of the task completed by each bricklayer must equal 1.

Solution
$$\frac{6}{t}+\frac{16}{2t}=1$$
$$2t\left(\frac{6}{t}+\frac{16}{2t}\right)=2t(1)$$
$$12+16=2t$$
$$28=2t$$
$$14=t$$

Working alone, the experienced bricklayer can do the job in 14 h.

29. Strategy

- Time for the slower machine to transmit the fax.

	Rate	Time	Part
Faster fax	$\frac{1}{40}$	35	$\frac{35}{40}$
Slower fax	$\frac{1}{t}$	20	$\frac{20}{t}$

- The sum of the part sent by the slower fax and the part sent by the faster fax is 1.

Solution

$$\frac{35}{40}+\frac{20}{t}=1$$
$$40t\left(\frac{35}{40}+\frac{20}{t}\right)=40t(1)$$
$$35t+40(20)=40t$$
$$35t+800=40t$$
$$800=5t$$
$$160=t$$

The slower machine working alone would have taken 160 min to send the fax.

31. Strategy

- Unknown time to fill the bottles working together: t

	Rate	Time	Part
First machine	$\frac{1}{10}$	t	$\frac{t}{10}$
Second machine	$\frac{1}{12}$	t	$\frac{t}{12}$
Third machine	$\frac{1}{15}$	t	$\frac{t}{50}$

- The sum of the parts of the task completed by each machine must equal 1.

Solution

$$\frac{t}{10}+\frac{t}{12}+\frac{t}{15}=1$$
$$60\left(\frac{t}{10}+\frac{t}{12}+\frac{t}{15}\right)=60(1)$$
$$6t+5t+4t=60$$
$$15t=60$$
$$t=4$$

When all three machines are working, it would take 4 h to fill the bottles.

33. Strategy

- Unknown time to empty the tank working together: t

	Rate	Time	Part
Inlet pipe	$\frac{1}{45}$	t	$\frac{t}{45}$
Outlet pipe	$\frac{1}{30}$	t	$\frac{1}{30}$

- The part of the task completed by the outlet pipe minus the part of the task completed by the inlet pipe is 1.

Solution

$$\frac{t}{30}-\frac{t}{45}=1$$
$$90\left(\frac{t}{30}-\frac{t}{45}\right)=90(1)$$
$$3t-2t=90$$
$$t=90$$

It would take 90 min to empty the tank.

35. Strategy

- Time for the clowns to blow up 76 balloons: t

	Rate	Time	Part
First clown	$\frac{1}{2}$	t	$\frac{t}{2}$
Second clown	$\frac{1}{3}$	t	$\frac{t}{3}$
Balloon popping	$\frac{1}{5}$	t	$\frac{t}{5}$

- The sum of the tasks completed by the two clowns minus the part completed by the balloons popping is equal to 76. This is the number of balloons filled after t minutes.

Solution

$$\frac{t}{2}+\frac{t}{3}-\frac{t}{5}=76$$
$$30\left(\frac{t}{2}+\frac{t}{3}-\frac{t}{5}\right)=30(76)$$
$$15t+10t-6t=2280$$
$$19t=2280$$
$$t=120$$

It will take 120 min to have 76 balloons.

37. Strategy • Unknown time to address 140 envelopes: t

	Time to address envelope	Time to address 140 envelopes	Rate	Time	Part
First clerk	$30 \text{ sec} = \frac{1}{2}$ min	70 min	$\frac{1}{70}$	t	$\frac{t}{70}$
Second clerk	$40 \text{ sec} = \frac{2}{3}$ min	$\frac{280}{3}$ min	$\frac{3}{280}$	t	$\frac{3t}{280}$

• The sum of the parts of the task completed by the two clerks is 1.

$$\frac{t}{70}+\frac{3t}{280}=1$$

Solution

$$\frac{t}{70}+\frac{3t}{280}=1$$
$$280\left(\frac{t}{70}+\frac{3t}{280}\right)=(1)280$$
$$4t+3t=280$$
$$7t=280$$
$$t=40 \text{ min} = 2400 \text{ seconds}$$

The two clerks, working together, can complete the task in 40 min (or 2400 s).

Objective C Exercises

39. Strategy • Rate of the runner: r
Rate of the bicyclist: $r+7$

	Distance	Rate	Time
Runner	16	r	$\frac{16}{r}$
Bicyclist	30	$r+7$	$\frac{30}{r+7}$

• The time the runner travels equals the time the bicyclist travels.

Solution

$$\frac{16}{r}=\frac{30}{r+7}$$
$$r(r+7)\frac{16}{r}=r(r+7)\frac{30}{r+7}$$
$$(r+7)(16)=30r$$
$$16r+112=30r$$
$$112=14r$$
$$8=r$$

The rate of the runner is 8 mph.

41. Strategy • Rate of the tortoise: r
Rate of the hare: $180r$

	Distance	Rate	Time
Tortoise	360	r	$\frac{360}{r}$
Hare	360	$180r$	$\frac{360}{180r}$

• The time for the tortoise is 14 min 55 s (895 s) more than the hare.

Solution

$$\frac{360}{180r}+895=\frac{360}{r}$$
$$\frac{2}{r}+895=\frac{360}{r}$$
$$r\left(\frac{2}{r}+895\right)=r\left(\frac{360}{r}\right)$$
$$2+895r=360$$
$$895r=358$$
$$r=0.4$$

$180r=180(0.4)=72$
The rate of the tortoise is 0.4 ft/s.
The rate of the hare is 72 ft/s.

43. Strategy • Rate of the jogger: r
Rate of the cyclist: $2r$

	Distance	Rate	Time
Jogger	30	r	$\frac{30}{r}$
Cyclist	30	$2r$	$\frac{30}{2r}$

• The time for the jogger is 3 h more than the time for the cyclist.

Solution

$$\frac{30}{2r}+3=\frac{30}{r}$$
$$2r\left(\frac{30}{2r}+3\right)=2r\left(\frac{30}{r}\right)$$
$$30+6r=60$$
$$6r=30$$
$$r=5$$

$2r=2(5)=10$
The rate of the cyclist is 10 mph.

45. Strategy • Rate of helicopter: r
• Rate of commercial jet: $4r$

	Distance	Rate	Time
Helicopter	105	r	$\frac{105}{r}$
Commercial jet	735	$4r$	$\frac{735}{4r}$

• The total time was 2.2 h

Solution

$$\frac{105}{r}+\frac{735}{4r}=2.2$$
$$4r\left(\frac{105}{r}+\frac{735}{4r}\right)=4r(2.2)$$
$$420+735=8.8r$$
$$1155=8.8r$$
$$131.25=r$$

$4r=4(131.25)=525$

The rate of the jet was 525 mph.

47. Strategy • Rate of the current: r

	Distance	Rate	Time
With current	20	$7+r$	$\frac{20}{7+r}$
Against current	8	$7-r$	$\frac{8}{7-r}$

• The time traveling with the current equals the time traveling against the current.

Solution

$$\frac{20}{7+r}=\frac{8}{7-r}$$
$$(7+r)(7-r)\frac{20}{7+r}=(7+r)(7-r)\frac{8}{7-r}$$
$$(7-r)20=(7+r)8$$
$$140-20r=56+8r$$
$$140=56+28r$$
$$84=28r$$
$$3=r$$

The rate of the current is 3 mph.

49. Strategy • Rate of the wind: r

	Distance	Rate	Time
With wind	3059	$550+r$	$\frac{3059}{550+r}$
Against wind	2450	$550-r$	$\frac{2450}{550-r}$

• The time flying with the wind equals the time flying against the wind.

Solution

$$\frac{3059}{550+r}=\frac{2450}{550-r}$$
$$(550+r)(550-r)\frac{3059}{550+r}=(550+r)(550-r)\frac{2450}{550-r}$$
$$(550+r)(3059)=(550+r)(2450)$$
$$1{,}682{,}450-3059r=1{,}347{,}500+2450r$$
$$1{,}682{,}450=1{,}347{,}500+5509r$$
$$334{,}950=5509r$$
$$60.80\approx r$$

The rate of the wind is 60.80 mph.

51. Strategy • Rate of the current: c

	Distance	Rate	Time
With wind	16	$6+c$	$\frac{16}{6+c}$
Against wind	16	$6-c$	$\frac{16}{6-c}$

• The sum of the times is 6 hours.

$$\frac{16}{6+c}+\frac{16}{6-c}=6$$

Solution

$$\frac{16}{6+c}+\frac{16}{6-c}=6$$
$$(6+c)(6-c)\left(\frac{16}{6+c}-\frac{16}{6-c}\right)=(6)(6+c)(6-c)$$
$$16(6-c)+16(6+c)=6(36-c^2)$$
$$96-16c+96+16c=216-6c^2$$
$$192=216-6c^2$$
$$0=24-6c^2$$
$$0=6(4-c^2)$$
$$0=6(2-c)(2+c)$$

$2-c=0 \quad 2+c=0$

$2=c \quad c=-2$

The rate of the current is 2 mph.

Applying the Concepts

53a.

$$\frac{x-2}{y}=\frac{x+2}{5y}$$
$$\frac{x-2}{y}\cdot 5y=\frac{x+2}{5y}\cdot 5y$$
$$(x-2)\cdot 5=x+2$$
$$5x-10=x+2$$
$$4x=12$$
$$x=3$$

b.

$$\frac{x}{x+y}=\frac{2x}{4y}$$
$$\frac{x}{x+y}\cdot 4y(x+y)=\frac{2x}{4y}\cdot 4y(x+y)$$
$$4xy=2x(x+y)$$
$$4xy=2x^2+2xy$$
$$2xy=2x^2$$
$$2x^2-2xy=0$$
$$x^2-xy=0$$
$$x(x-y)=0$$

$x=0 \quad x-y=0$
$x=y$

c.

$$\frac{x-y}{x}=\frac{2x}{9y}$$
$$\frac{x-y}{x}\cdot 9xy=\frac{2x}{9y}\cdot 9xy$$
$$(x-y)\cdot 9y=2x^2$$
$$9xy-9y^2=2x^2$$
$$2x^2-9xy+9y^2=0$$
$$(2x-3y)(x-3y)=0$$

$2x-3y=0 \quad x-3y=0$
$2x=3y \quad x=3y$
$x=\frac{3}{2}y$

55. Strategy • Rate of the bus: r

	Distance	Rate	Time
Usual conditions	165	r	$\frac{165}{r}$
Bad weather	165	$r-5$	$\frac{165}{r-5}$

• The difference between the times is 15 minutes, or $\frac{1}{4}$ hr.

Solution

$$\frac{165}{r-5}-\frac{165}{r}=\frac{1}{4}$$
$$4r\cdot(r-5)\left(\frac{165}{r-5}-\frac{165}{r}\right)=\frac{1}{4}\cdot 4r(r-5)$$
$$660r-660(r-5)=r^2-5r$$
$$660r-660r+3300=r^2-5r$$
$$r^2-5r-3300=0$$
$$(r-60)(r+55)=0$$

$r-60=0 \quad r+55=0$
$r=60 \quad r=-55$

The rate of the bus cannot be negative, so the usual rate of the bus is 60 mph.

Section 6.6

Objective A Exercises

1. Strategy To find the profit:
• Write the basic direct variation equation, replace the variables by the given values, and solve for k.
• Write the direct variation equation, replacing k by its value. Substitute 5000 for s and solve for P.

Solution
$P=ks$
$4000=k(250)$
$16=k$
$P=16s=16(5000)=80{,}000$
When the company sells 5000 products, the profit is $80,000.

3. Strategy To find the pressure:
• Write the basic direct variation equation, replace the variables by the given values, and solve for k.
• Write the direct variation equation, replacing k by its value. Substitute 15 for d and solve for p.

Solution
$p=kd$
$4.5=k(10)$
$0.45=k$
$p=0.45d=0.45(15)=6.75$
The pressure is 6.75 lb/in^2.

5. Strategy To find how far the object will fall:
• Write the basic direct variation equation, replace the variables by the given values, and solve for k.
• Write the direct variation equation, replace k by its value. Substitute 10 for t and solve for d.

Solution
$d=kt^2$
$144=k(3)^2$
$144=9k$
$16=k$
$d=16t^2=16(10)^2=16(100)=1600$
In 10 s, the object will fall 1600 ft.

7. Strategy To find the distance:
• Write the basic direct variation equation, replace the variables by the given values, and solve for k.
• Write the direct variation equation, replacing k by its value. Substitute 3 for t and solve for s.

Solution
$s=kt^2$
$6=k(1)^2$
$6=k$
$s=6t^2=6(3)^2=6(9)=54$
In 3 s, the ball will roll 54 ft.

9. Strategy To find the length of the rectangle:
- Write the basic inverse variation equation, replace the variables by the given values, and solve for k.
- Write the inverse variation equation, replacing k by its value. Substitute 4 for w and solve for L.

Solution
$$L = \frac{k}{w}$$
$$8 = \frac{k}{5}$$
$$40 = k$$
$$L = \frac{40}{w} = \frac{40}{4} = 10$$
When the width is 4 ft, the length is 10 ft.

11. Strategy To find the speed:
- Write the basic inverse variation equation, replace the variables by the given values, and solve for k.
- Write the inverse variation equation, replacing k by its value. Substitute 36 for t and solve for v.

Solution
$$v = \frac{k}{t}$$
$$24 = \frac{k}{45}$$
$$1080 = k$$
$$v = \frac{1080}{t} = \frac{1080}{36} = 30$$
The gear that has 36 teeth will make 30 revolutions/min.

13. Strategy To find the current:
- Write the basic combined variation equation, replace the variables by the given values, and solve for k.
- Write the combined variation equation, replacing k by its value. Substitute 180 for v and 24 for r.

Solution
$$I = \frac{kv}{r}$$
$$10 = \frac{k(110)}{11}$$
$$110 = 110k$$
$$1 = k$$
$$I = \frac{v}{r} = \frac{180}{24} = 7.5$$
The current is 7.5 amps.

15. Strategy To find the intensity:
- Write the basic inverse variation equation, replace the variables by the given values, and solve for k.
- Write the inverse variation equation, replacing k by its value. Substitute 5 for d and solve for I.

Solution
$$I = \frac{k}{d^2}$$
$$12 = \frac{k}{10^2}$$
$$12 = \frac{k}{100}$$
$$1200 = k$$
$$I = \frac{1200}{d^2} = \frac{1200}{5^2} = \frac{1200}{25} = 48$$
The intensity is 48 foot-candles when the distance is 5 ft.

Applying the Concepts

17. Strategy To find the effect on y, replace x with $2x$.

Solution $y = kx \quad y = k(2x)$
$$y = 2kx$$
If x is doubled, then y is doubled.

19. inversely

21. inversely

Chapter 6 Review Exercises

1. $\dfrac{a^6b^4 + a^4b^6}{a^5b^4 - a^4b^4} \cdot \dfrac{a^2 - b^2}{a^4 - b^4}$
$$= \frac{a^4b^4(a^2+b^2)}{a^4b^4(a-1)} \cdot \frac{(a+b)(a-b)}{(a^2+b^2)(a+b)(a-b)}$$
$$= \frac{a^4b^4(a^2+b^2)(a+b)(a-b)}{a^4b^4(a-1)(a^2+b^2)(a+b)(a-b)}$$
$$= \frac{1}{a-1}$$

2. The LCM is $(x-3)(x+2)$.
$$\frac{x}{x-3} - 4 - \frac{2x-5}{x+2}$$
$$= \frac{x}{x-3} \cdot \frac{x+2}{x+2} - 4 \cdot \frac{(x-3)(x+2)}{(x-3)(x+2)} - \frac{2x-5}{x+2} \cdot \frac{x-3}{x-3}$$
$$= \frac{x^2 + 2x - 4(x^2 - x - 6) - (2x^2 - 11x + 15)}{(x-3)(x+2)}$$
$$= \frac{x^2 + 2x - 4x^2 + 4x + 24 - 2x^2 + 11x - 15}{(x-3)(x+2)}$$
$$= \frac{-5x^2 + 17x + 9}{(x-3)(x+2)}$$
$$= \frac{-(5x^2 - 17x - 9)}{(x-3)(x+2)}$$
$$= -\frac{5x^2 - 17x - 9}{(x-3)(x+2)}$$

3. $P(x) = \dfrac{x}{x-3}$
$$P(4) = \frac{4}{4-3} = \frac{4}{1}$$
$$P(4) = 4$$

4.
$$\frac{3x-2}{x+6}=\frac{3x+1}{x+9}$$
$$\frac{3x-2}{x+6}\cdot(x+6)(x+9)=\frac{3x+1}{x+9}\cdot(x+6)(x+9)$$
$$(3x-2)(x+9)=(3x+1)(x+6)$$
$$3x^2+25x-18=3x^2+19x+6$$
$$25x-18=19x+6$$
$$6x-18=6$$
$$6x=24$$
$$x=4$$

The solution is 4.

5. $\dfrac{\frac{3x+4}{3x-4}+\frac{3x-4}{3x+4}}{\frac{3x-4}{3x+4}-\frac{3x+4}{3x-4}}$

The LCM is $(3x-4)(3x+4)$.

$$\frac{\frac{3x+4}{3x-4}+\frac{3x-4}{3x+4}}{\frac{3x-4}{3x+4}-\frac{3x+4}{3x-4}}\cdot\frac{(3x-4)(3x+4)}{(3x-4)(3x+4)}$$
$$=\frac{(3x+4)^2+(3x-4)^2}{(3x-4)^2-(3x+4)^2}$$
$$=\frac{9x^2+24x+16+9x^2-24x+16}{9x^2-24x+16-(9x^2+24x+16)}$$
$$=\frac{18x^2+32}{9x^2-24x+16-9x^2-24x-16}$$
$$=\frac{18x^2+32}{-48x}$$
$$=\frac{2(9x^2+16)}{2(-24x)}$$
$$=-\frac{9x^2+16}{24x}$$

6. The LCM is $(4x-1)(4x+1)$.

$$\frac{4x}{4x-1}\cdot\frac{4x+1}{4x+1}=\frac{16x^2+4x}{(4x+1)(4x-1)}$$
$$\frac{3x-1}{4x+1}\cdot\frac{4x-1}{4x-1}=\frac{12x^2-7x+1}{(4x+1)(4x-1)}$$

7.
$$S=\frac{a}{1-r}$$
$$S(1-r)=\frac{a}{1-r}(1-r)$$
$$S-Sr=a$$
$$-Sr=a-S$$
$$r=\frac{a-S}{-S}$$
$$r=\frac{S-a}{S}$$

8.
$$P(x)=\frac{x^2-2}{3x^2-2x+5}$$
$$P(-2)=\frac{(-2)^2-2}{3(-2)^2-2(-2)+5}=\frac{4-2}{12+4+5}=\frac{2}{21}$$
$$P(-2)=\frac{2}{21}$$

9.
$$\frac{10}{5x+3}=\frac{2}{10x-3}$$
$$\frac{10}{5x+3}\cdot(5x+3)(10x-3)=\frac{2}{10x-3}\cdot(5x+3)(10x-3)$$
$$10(10x-3)=2(5x+3)$$
$$100x-30=10x+6$$
$$90x-30=6$$
$$90x=36$$
$$x=\frac{36}{90}=\frac{2}{5}$$

The solution is $\frac{2}{5}$.

10. $f(x)=\dfrac{2x-7}{3x^2+3x-18}$

$$3x^2+3x-18=0$$
$$3(x^2+x-6)=0$$
$$3(x+3)(x-2)=0$$
$$x+3=0 \qquad x-2=0$$
$$x=-3 \qquad x=2$$

The domain is $\{x|x\neq -3, 2\}$.

11.
$$\frac{3x^4+11x^2-4}{3x^4+13x^2+4}=\frac{(3x^2-1)(x^2+4)}{(3x^2+1)(x^2+4)}$$
$$=\frac{3x^2-1}{3x^2+1}$$

12. $q(x)=\dfrac{2x}{x-3}$

$$x-3=0$$
$$x=3$$

The domain is $\{x|x\neq 3\}$.

13.
$$\frac{x^3-8}{x^3+2x^2+4x}\cdot\frac{x^3+2x^2}{x^2-4}$$
$$=\frac{(x-2)(x^2+2x+4)}{x(x^2+2x+4)}\cdot\frac{x^2(x+2)}{(x+2)(x-2)}$$
$$=\frac{(x-2)(x^2+2x+4)x^2(x+2)}{x(x^2+2x+4)(x+2)(x-2)}=x$$

14. The LCM is x^2-4.

$$\frac{3x^2+2}{x^2-4}-\frac{9x-x^2}{x^2-4}=\frac{3x^2+2-(9x-x^2)}{x^2-4}$$
$$=\frac{3x^2+2-9x+x^2}{x^2-4}$$
$$=\frac{4x^2-9x+2}{x^2-4}$$
$$=\frac{(4x-1)(x-2)}{(x+2)(x-2)}$$
$$=\frac{4x-1}{x+2}$$

15.

$$\begin{aligned}
Q &= \frac{N-S}{N} \\
Q \cdot N &= \frac{N-S}{N} \cdot N \\
QN &= N - S \\
QN - N &= -S \\
N(Q-1) &= -S \\
N &= \frac{-S}{Q-1} \\
N &= \frac{S}{1-Q}
\end{aligned}$$

16.

$$\begin{aligned}
\frac{30}{x^2+5x+4} + \frac{10}{x+4} &= \frac{4}{x+1} \\
\frac{30}{(x+4)(x+1)} + \frac{10}{x+4} &= \frac{4}{x+1} \\
(x+4)(x+1)\left(\frac{30}{(x+4)(x+1)} + \frac{10}{x+4}\right) &= \frac{4}{x+1} \cdot (x+4)(x+1) \\
30 + 10(x+1) &= 4(x+4) \\
30 + 10x + 10 &= 4x + 16 \\
10x + 40 &= 4x + 16 \\
6x + 40 &= 16 \\
6x &= -24 \\
x &= -4
\end{aligned}$$

-4 does not check as a solution.
The equation has no solution.

17.

$$\begin{aligned}
x + \frac{\frac{4}{x} - 1}{\frac{1}{x} - \frac{3}{x^2}} &= x + \frac{\frac{4}{x} - 1}{\frac{1}{x} - \frac{3}{x^2}} \cdot \frac{x^2}{x^2} \\
&= x + \frac{4x - x^2}{x-3}
\end{aligned}$$

The LCM is $x-3$.

$$\begin{aligned}
x \cdot \frac{x-3}{x-3} + \frac{4x-x^2}{x-3} &= \frac{x^2 - 3x + 4x - x^2}{x-3} \\
&= \frac{x}{x-3}
\end{aligned}$$

18. $x^2 - 9x + 20 = (x-5)(x-4)$;
$4 - x = -(x-4)$
The LCM is $(x-5)(x-4)$.

$$\frac{x-3}{x-5} \cdot \frac{x-4}{x-4} = \frac{x^2 - 7x + 12}{(x-5)(x-4)}$$

$$\frac{x}{x^2 - 9x + 20} = \frac{x}{(x-5)(x-4)}$$

$$\begin{aligned}
\frac{1}{4-x} &= \frac{-1}{x-4} \cdot \frac{x-5}{x-5} = \frac{-x+5}{(x-5)(x-4)} \\
&= \frac{-(x-5)}{(x-5)(x-4)} \\
&= -\frac{x-5}{(x-5)(x-4)}
\end{aligned}$$

19.
$$\frac{6}{2x-3}=\frac{5}{x+5}+\frac{5}{2x^2+7x-15}$$
$$\frac{6}{2x-3}=\frac{5}{x+5}+\frac{5}{(2x-3)(x+5)}$$
$$(2x-3)(x+5)\left(\frac{6}{2x-3}\right)=\left(\frac{5}{x+5}+\frac{5}{(2x-3)(x+5)}\right)(2x-3)(x+5)$$
$$6(x+5)=5(2x-3)+5$$
$$6x+30=10x-15+5$$
$$6x+30=10x-10$$
$$-4x=-40$$
$$x=10$$

20.
$$\frac{x^{n+1}+x}{x^{2n}-1}\div\frac{x^{n+2}-x^2}{x^{2n}-2x^n+1}=\frac{x^{n+1}+x}{x^{2n}-1}\cdot\frac{x^{2n}-2x^n+1}{x^{n+2}-x^2}$$
$$=\frac{x(x^n+1)}{(x^n+1)(x^n-1)}\cdot\frac{(x^n-1)(x^n-1)}{x^2(x^n-1)}$$
$$=\frac{x(x^n+1)(x^n-1)(x^n-1)}{x^2(x^n+1)(x^n-1)(x^n-1)}=\frac{1}{x}$$

21.
$$\frac{27x^3-8}{9x^3+6x^2+4x}\div\frac{9x^2-12x+4}{9x^2-4}=\frac{27x^3-8}{9x^3+6x^2+4x}\cdot\frac{9x^2-4}{9x^2-12x+4}$$
$$=\frac{(3x-2)(9x^2+6x+4)}{x(9x^2+6x+4)}\cdot\frac{(3x+2)(3x-2)}{(3x-2)(3x-2)}$$
$$=\frac{(3x-2)(9x^2+6x+4)(3x+2)(3x-2)}{x(9x^2+6x+4)(3x-2)(3x-2)}$$
$$=\frac{3x+2}{x}$$

22. $3x^2-7x+2=(3x-1)(x-2)$
The LCM is $(3x-1)(x-2)$.
$$\frac{6x}{3x^2-7x+2}-\frac{2}{3x-1}+\frac{3x}{x-2}$$
$$=\frac{6x}{(3x-1)(x-2)}-\frac{2}{3x-1}\cdot\frac{x-2}{x-2}+\frac{3x}{x-2}\cdot\frac{3x-1}{3x-1}$$
$$=\frac{6x-2(x-2)+3x(3x-1)}{(3x-1)(x-2)}$$
$$=\frac{6x-2x+4+9x^2-3x}{(3x-1)(x-2)}$$
$$=\frac{9x^2+x+4}{(3x-1)(x-2)}$$

23.
$$\frac{x^3-27}{x^2-9}=\frac{(x-3)(x^2+3x+9)}{(x-3)(x+3)}$$
$$=\frac{x^2+3x+9}{x+3}$$

24. The LCM is $24a^2b^4$.
$$\frac{5}{3a^2b^3}+\frac{7}{8ab^4}=\frac{5}{3a^2b^3}\cdot\frac{8b}{8b}+\frac{7}{8ab^4}\cdot\frac{3a}{3a}$$
$$=\frac{21a+40b}{24a^2b^4}$$

25. Since $3x^2+4>0$ for all x, the domain is $\{x|x\in\text{ real numbers}\}$.

26. $$\frac{12a^{4n}+9a^{3n}-6a^{2n}}{3a^{2n}}=\frac{12a^{4n}}{3a^{2n}}+\frac{9a^{3n}}{3a^{2n}}-\frac{6a^{2n}}{3a^{2n}}$$
$$=4a^{2n}+3a^{n}-2$$

27. $$\frac{16-x^2}{x^3-2x^2-8x}=-\frac{x^2-16}{x(x^2-2x-8)}$$
$$=-\frac{(x+4)\cancel{(x-4)}}{x\cancel{(x-4)}(x+2)}$$
$$=-\frac{x+4}{x(x+2)}$$

28. $$\frac{8x^3-27}{4x^2-9}=\frac{\cancel{(2x-3)}(4x^2+6x+9)}{\cancel{(2x-3)}(2x+3)}$$
$$=\frac{4x^2+6x+9}{2x+3}$$

29. $$\frac{16-x^2}{6x+12}\cdot\frac{x^2+5x+6}{x^2-8x+16}=-\frac{(x+4)\cancel{(x-4)}}{6\cancel{(x+2)}}\cdot\frac{\cancel{(x+2)}(x+3)}{\cancel{(x-4)}(x-4)}$$
$$=-\frac{(x+4)(x+3)}{6(x-4)}$$

30. $$\frac{x^{2n}-5x^n+4}{x^{2n}-2x^n-8}\div\frac{x^{2n}-4x^n+3}{x^{2n}+8x^n+12}$$
$$=\frac{x^{2n}-5x^n+4}{x^{2n}-2x^n-8}\cdot\frac{x^{2n}+8x^n+12}{x^{2n}-4x^n+3}$$
$$=\frac{\cancel{(x^n-4)}\cancel{(x^n-1)}}{\cancel{(x^n-4)}\cancel{(x^n+2)}}\cdot\frac{\cancel{(x^n+2)}(x^n+6)}{(x^n-3)\cancel{(x^n-1)}}$$
$$=\frac{x^n+6}{x^n-3}$$

31. $$\frac{8x^3-64}{4x^3+4x^2+x}\div\frac{x^2+2x+4}{4x^2-1}$$
$$=\frac{8(x^3-8)}{x(4x^2+4x+1)}\cdot\frac{4x^2-1}{x^2+2x+4}$$
$$=\frac{8(x-2)\cancel{(x^2+2x+4)}}{x(2x+1)\cancel{(2x+1)}}\cdot\frac{\cancel{(2x+1)}(2x-1)}{\cancel{x^2+2x+4}}$$
$$=\frac{8(x-2)(2x-1)}{x(2x+1)}$$

32. $$\frac{3-x}{x^2+3x+9}\div\frac{x^2-9}{x^3-27}=-\frac{x-3}{x^2+3x+9}\cdot\frac{x^3-27}{x^2-9}$$
$$=-\frac{x-3}{\cancel{x^2+3x+9}}=\frac{\cancel{(x-3)}\cancel{(x^2+3x+9)}}{(x+3)\cancel{(x-3)}}$$
$$=-\frac{x-3}{x+3}$$

33. $$\frac{3}{2a^3b^2}+\frac{5}{6a^2b}=\frac{9}{6a^3b^2}+\frac{5ab}{6a^3b^2}=\frac{9+5ab}{6a^3b^2}$$

34. $\frac{8}{9x^2-4}+\frac{5}{3x-2}-\frac{4}{3x+2}$

$=\frac{8}{(3x-2)(3x+2)}+\frac{5(3x+2)}{(3x-2)(3x+2)}-\frac{4(3x-2)}{(3x+2)(3x-2)}$

$=\frac{8+5(3x+2)-4(3x-2)}{(3x-2)(3x+2)}$

$=\frac{8+15x+10-12x+8}{(3x-2)(3x+2)}$

$=\frac{3x+26}{(3x-2)(3x+2)}$

35. $\frac{x-6+\frac{6}{x-1}}{x+3-\frac{12}{x-1}}=\frac{x-6+\frac{6}{x-1}}{x+3-\frac{12}{x-1}}\cdot\frac{x-1}{x-1}$

$=\frac{(x-6)(x-1)+6}{(x+3)(x-1)-12}$

$=\frac{x^2-7x+12}{x^2+2x-15}$

$=\frac{(x-4)(x-3)}{(x+5)(x-3)}$

$=\frac{x-4}{x+5}$

36. $\frac{x+\frac{3}{x-4}}{3+\frac{x}{x-4}}=\frac{x+\frac{3}{x-4}}{3+\frac{x}{x-4}}\cdot\frac{x-4}{x-4}$

$=\frac{x(x-4)+3}{3(x-4)+x}$

$=\frac{x^2-4x+3}{4x-12}$

$=\frac{(x-3)(x-1)}{4(x-3)}$

$=\frac{x-1}{4}$

37. $\frac{x+2}{x-3}=\frac{2x-5}{x+1}$

$(x+1)(x-3)\cdot\frac{x+2}{x-3}=(x+1)(x-3)\cdot\frac{2x-5}{x+1}$

$(x+1)(x+2)=(x-3)(2x-5)$

$x^2+3x+2=2x^2-11x+15$

$0=x^2-14x+13$

$0=(x-1)(x-13)$

$x-1=0 \qquad x-13=0$

$x=1 \qquad x=13$

The solutions are 1 and 13.

38. $\frac{5x}{2x-3}+4=\frac{3}{2x-3}$

$(2x-3)\left[\frac{5x}{2x-3}+4\right]=(2x-3)\cdot\frac{3}{2x-3}$

$5x+4(2x-3)=3$

$5x+8x-12=3$

$13x=15$

$x=\frac{15}{13}$

39. $I=\frac{1}{R}V$

$RI=R\left(\frac{1}{R}V\right)$

$RI=V$

$R=\frac{V}{I}$

40. $\frac{6}{x-3}-\frac{1}{x+3}=\frac{51}{x^2-9}$

$(x-3)(x+3)\left[\frac{6}{x-3}-\frac{1}{x+3}\right]=(x^2-9)\cdot\frac{51}{x^2-9}$

$6(x+3)-(x-3)=51$

$6x+18-x+3=51$

$5x=30$

$x=6$

41. Strategy • Unknown time to empty tub: t

	Rate	Time	Part
Inlet pipe	$\frac{1}{24}$	t	$\frac{t}{24}$
Drain pipe	$\frac{1}{15}$	t	$\frac{t}{15}$

• The tub is empty when the difference between the drain pipe part and the inlet pipe part equals 1.

Solution $\frac{t}{15}-\frac{t}{24}=1$

$120\left(\frac{t}{15}-\frac{t}{24}\right)=120\cdot 1$

$8t-5t=120$

$3t=120$

$t=40$

It would take 40 min to empty the tub with both pipes open.

42. Strategy

- Rate of cyclist: r
  Rate of bus: $3r$

	Distance	Rate	Time
Cyclist	90	r	$\frac{90}{r}$
Bus	90	$3r$	$\frac{90}{3r}$

- The difference in the timing is 4 h.

Solution

$$\frac{90}{r} - \frac{90}{3r} = 4$$
$$\frac{90}{r} - \frac{30}{r} = 4$$
$$\frac{60}{r} = 4$$
$$60 = 4r$$
$$15 = r$$
$$3r = 45$$

The rate of the bus is 45 mph.

43. Strategy

- Rate of the helicopter: r
  Rate of the airplane: $r + 20$

	Distance	Rate	Time
Helicopter	9	r	$\frac{9}{r}$
Airplane	10	$r+20$	$\frac{10}{r+20}$

- The time the helocopter travels equals the time the airplane travels.

Solution

$$\frac{9}{r} = \frac{10}{r+20}$$
$$\frac{9}{r}[r(r+20)] = \frac{10}{r+20}[r(r+20)]$$
$$9(r+20) = 10r$$
$$9r + 180 = 10r$$
$$180 = r$$

The rate of the helicopter is 180 mph.

44. Strategy To find how long it will take to do the reading, write and solve a proportion using x to represent the time.

Solution

$$\frac{2}{5} = \frac{150}{x}$$
$$\frac{2}{5} \cdot 5x = \frac{150}{x} \cdot 5x$$
$$2x = 750$$
$$x = 375$$

To read 150 pages, it will take 375 min.

45. Strategy To find the current:

- Write the basic inverse variation equation, replace the variables by the given values, and solve for k.
- Write the inverse variation equation, replacing k by its value. Substitute 100 for R and solve for I.

Solution

$$I = \frac{k}{R} \qquad I = \frac{k}{R}$$
$$4 = \frac{k}{50} \qquad = \frac{200}{R}$$
$$200 = k \qquad = \frac{200}{100}$$
$$= 2$$

The current is 2 amps.

46. Strategy To find the number of miles represented, write and solve a proportion using x to represent the number of miles.

Solution

$$\frac{2.5}{10} = \frac{12}{x}$$
$$\frac{2.5}{10} \cdot 10x = \frac{12}{x} \cdot 10x$$
$$2.5x = 120$$
$$x = 48$$

48 mi would be represented.

47. Strategy To find the stopping distance:

- Write the basic direct variation equation, replace the variables by the given values, and solve for k.
- Write the direct variation equation, replacing k by its value. Substitute 65 for v and solve for s.

Solution

$$s = kv^2 \qquad s = kv^2$$
$$170 = k(50)^2 \qquad = 0.068v^2$$
$$170 = k(2500) \qquad = 0.068(65)^2$$
$$0.068 = k \qquad = 0.068(4225)$$
$$= 287.3$$

The stopping distance for a car traveling at 65 mph is 287.3 ft.

48. Strategy • Unknown time for apprentice, working alone, to install fan: t

	Rate	Time	Part
Electrician	$\frac{1}{65}$	40	$\frac{40}{65}$
Apprentice	$\frac{1}{t}$	40	$\frac{40}{t}$

• The sum of the part of the task completed by the electrician and the part of the task completed by the apprentice is 1.

Solution

$$\frac{40}{65}+\frac{40}{t}=1$$
$$65t\left(\frac{40}{65}+\frac{40}{t}\right)=1\cdot 65t$$
$$40t+2600=65t$$
$$2600=25t$$
$$104=t$$

It would take the apprentice 104 min to complete the job alone.

Chapter 6 Test

1.
$$\frac{3}{x+1}=\frac{2}{x}$$
$$\frac{3}{x+1}\cdot x(x+1)=\frac{2}{x}\cdot x(x+1)$$
$$3x=2(x+1)$$
$$3x=2x+2$$
$$x=2$$

2.
$$\frac{x^2+x-6}{x^2+7x+12}\div\frac{x^2-3x+2}{x^2+6x+8}$$
$$=\frac{x^2+x-6}{x^2+7x+12}\cdot\frac{x^2+6x+8}{x^2-3x+2}$$
$$=\frac{(x+3)(x-2)}{(x+3)(x+4)}\cdot\frac{(x+4)(x+2)}{(x-2)(x-1)}$$
$$=\frac{(x+3)(x-2)(x+4)(x+2)}{(x+3)(x+4)(x-2)(x-1)}$$
$$=\frac{x+2}{x-1}$$

3. The LCM is $(x+2)(x-3)$.
$$\frac{2x-1}{x+2}-\frac{x}{x-3}=\frac{2x-1}{x+2}\cdot\frac{x-3}{x-3}-\frac{x}{x-3}\cdot\frac{x+2}{x+2}$$
$$=\frac{2x^2-7x+3-(x^2+2x)}{(x-3)(x+2)}$$
$$=\frac{x^2-9x+3}{(x-3)(x+2)}$$

4.
$$x^2+x-6=(x+3)(x-2)$$
$$x^2-9=(x+3)(x-3)$$
The LCM is $(x+3)(x-3)(x-2)$.
$$\frac{x+1}{x^2+x-6}=\frac{x+1}{(x+3)(x-2)}\cdot\frac{x-3}{x-3}$$
$$=\frac{x^2-2x-3}{(x+3)(x-3)(x-2)}$$
$$\frac{2x}{x^2-9}=\frac{2x}{(x+3)(x-3)}\cdot\frac{x-2}{x-2}$$
$$=\frac{2x^2-4x}{(x+3)(x-3)(x-2)}$$

5.
$$\frac{4x}{2x-1}=2-\frac{1}{2x-1}$$
$$(2x-1)\frac{4x}{2x-1}=\left(2-\frac{1}{2x-1}\right)2x-1$$
$$4x=2(2x-1)-1$$
$$4x=4x-2-1$$
$$4x=4x-3$$
$$0=-3$$
There is no solution.

6.
$$\frac{v^3-4v}{2v^2-5v+2}=\frac{v(v^2-4)}{(2v-1)(v-2)}$$
$$=\frac{v(v+2)(v-2)}{(2v-1)(v-2)}$$
$$=\frac{v(v+2)}{2v-1}$$

7.
$$\frac{3x^2-12}{5x-15}\cdot\frac{2x^2-18}{x^2+5x+6}$$
$$=\frac{3(x+2)(x-2)}{5(x-3)}\cdot\frac{2(x+3)(x-3)}{(x+3)(x+2)}$$
$$=\frac{3(x+2)(x-2)2(x+3)(x-3)}{5(x-3)(x+3)(x+2)}$$
$$=\frac{6(x-2)}{5}$$

8.
$$f(x)=\frac{3x^2-x+1}{x^2-9}$$
$$x^2-9=0$$
$$(x+3)(x-3)=0$$
$$x+3=0 \qquad x-3=0$$
$$x=-3 \qquad x=3$$
The domain is $\{x|x\neq 3,\ -3\}$.

9.
$$\frac{1-\frac{1}{x}-\frac{12}{x^2}}{1+\frac{6}{x}+\frac{9}{x^2}}=\frac{1-\frac{1}{x}-\frac{12}{x^2}}{1+\frac{6}{x}+\frac{9}{x^2}}\cdot\frac{x^2}{x^2}$$
$$=\frac{x^2-x-12}{x^2+6x+9}$$
$$=\frac{(x-4)(x+3)}{(x+3)(x+3)}=\frac{x-4}{x+3}$$

10. $\dfrac{1-\frac{1}{x+2}}{1-\frac{3}{x+4}}=\dfrac{1-\frac{1}{x+2}}{1-\frac{3}{x+4}}\cdot\dfrac{(x+2)(x+4)}{(x+2)(x+4)}$

$=\dfrac{(x+2)(x+4)-(x+4)}{(x+2)(x+4)-3(x+2)}$

$=\dfrac{x^2+6x+8-x-4}{x^2+6x+8-3x-6}$

$=\dfrac{x^2+5x+4}{x^2+3x+2}$

$=\dfrac{(x+4)(x+1)}{(x+2)(x+1)}$

$=\dfrac{x+4}{x+2}$

11. $\dfrac{2x^2-x-3}{2x^2-5x+3}\div\dfrac{3x^2-x-4}{x^2-1}$

$=\dfrac{2x^2-x-3}{2x^2-5x+3}\cdot\dfrac{x^2-1}{3x^2-x-4}$

$=\dfrac{(2x-3)(x+1)}{(2x-3)(x-1)}\cdot\dfrac{(x+1)(x-1)}{(3x-4)(x+1)}$

$=\dfrac{(2x-3)(x+1)(x+1)(x-1)}{(2x-3)(x-1)(3x-4)(x+1)}$

$=\dfrac{x+1}{3x-4}$

12. $\dfrac{4x}{x+1}-x=\dfrac{2}{x+1}$

$(x+1)\left[\dfrac{4x}{x+1}-x\right]=\dfrac{2}{x+1}(x+1)$

$4x-x(x+1)=2$

$4x-x^2-x=2$

$3x-x^2=2$

$0=x^2-3x+2$

$0=(x-2)(x-1)$

$0=x-2 \quad x-1=0$

$2=x \quad x=1$

The solutions are 1 and 2.

13. $\dfrac{2a^2-8a+8}{4+4a-3a^2}=\dfrac{2(a^2-4a+4)}{(2-a)(2+3a)}$

$=\dfrac{2(a-2)(a-2)}{(2-a)(2+3a)}$

$=-\dfrac{2(a-2)}{3a+2}$

14. $\dfrac{1}{r}=\dfrac{1}{2}-\dfrac{2}{t}$

$2rt\left(\dfrac{1}{r}\right)=\left(\dfrac{1}{2}-\dfrac{2}{t}\right)2rt$

$2t=rt-4r$

$2t-rt=-4r$

$t(2-r)=-4r$

$t=\dfrac{-4r}{2-r}$

$t=\dfrac{4r}{r-2}$

15. $f(x)=\dfrac{3-x^2}{x^3-2x^2+4}$

$f(-1)=\dfrac{3-(-1)^2}{(-1)^3-2(-1)^2+4}$

$f(-1)=2$

16. $x^2+3x-4=(x+4)(x-1)$

$x^2-1=(x+1)(x-1)$

The LCM is $(x+4)(x-1)(x+1)$.

$\dfrac{x+2}{x^2+3x-4}-\dfrac{2x}{x^2-1}$

$=\dfrac{x+2}{(x+4)(x-1)}\cdot\dfrac{x+1}{x+1}-\dfrac{2x}{(x+1)(x-1)}\cdot\dfrac{x+4}{x+4}$

$=\dfrac{x^2+3x+2-2x(x+4)}{(x+4)(x-1)(x+1)}$

$=\dfrac{x^2+3x+2-2x^2-8x}{(x+4)(x-1)(x+1)}$

$=\dfrac{-x^2-5x+2}{(x+1)(x+4)(x-1)}$

17. $x(2x+5)\cdot\dfrac{x+1}{2x+5}=\dfrac{x-3}{x}\cdot x(2x+5)$

$x(x+1)=(x-3)(2x+5)$

$x^2+x=2x^2-x-15$

$0=x^2-2x-15$

$0=(x-5)(x+3)$

$x-5=0 \quad x+3=0$

$x=5 \quad x=-3$

The solutions are -3 and 5.

18. **Strategy**

- Rate of hiker: r
  Rate of cyclist: $r+7$

	Distance	Rate	Time
Hiker	6	r	$\frac{6}{r}$
Cyclist	20	$r+7$	$\frac{20}{r+7}$

- The time the hiker hikes equals the time the cyclist cycles.

Solution

$\dfrac{6}{r}=\dfrac{20}{r+7}$

$\dfrac{6}{r}[r(r+7)]=\dfrac{20}{r+7}\cdot[r(r+7)]$

$6(r+7)=20r$

$6r+42=20r$

$42=14r$

$3=r$

$10=r+7$

The rate of the cyclist is 10 mph.

19. Strategy To find the resistance:

- Write the basic combined variation equation, replace the variables with the given values, and solve for k.
- Write the combined variation equation, replacing k with its value. Substitute 8000 for l and d for $\frac{1}{2}$, and solve for r.

Solution

$$r = \frac{kl}{d^2}$$
$$3.2 = \frac{k \cdot 16{,}000}{\left(\frac{1}{4}\right)^2}$$
$$k = 0.0000125$$
$$r = \frac{0.0000125l}{d^2}$$
$$= \frac{0.00000125(8000)}{\left(\frac{1}{2}\right)^2}$$
$$r = 0.4$$

The resistance is 0.4 ohm.

20. Strategy To find the number of rolls of wallpaper, write and solve a proportion using x to represent the number of rolls.

Solution

$$\frac{2}{45} = \frac{x}{315}$$
$$\left(\frac{2}{45}\right) \cdot 315 = \frac{x}{315} \cdot 315$$
$$14 = x$$

The office requires 14 rolls of wallpaper.

21. Strategy

- Unknown time for both landscapers working together: t

	Rate	Time	Part
First landscaper	$\frac{1}{30}$	t	$\frac{t}{30}$
Second landscaper	$\frac{1}{15}$	t	$\frac{t}{15}$

- The sum of the part of the task completed by the first landscaper and the part of the task completed by the second landscaper is 1.

Solution

$$\frac{t}{30} + \frac{t}{15} = 1$$
$$30\left(\frac{t}{30} + \frac{t}{15}\right) = 1(30)$$
$$t + 2t = 30$$
$$3t = 30$$
$$t = 10$$

Working together, the landscapers can complete the task in 10 min.

22. Strategy To find the current:

- Write the basic inverse variation equation, replace the variables by the given values, and solve for k.
- Write the inverse variation equation, replacing k by its value. Substitute 1.25 for I and solve for R.

Solution

$$I = \frac{k}{R} \qquad I = \frac{k}{R}$$
$$0.25 = \frac{k}{8} \qquad 1.25 = \frac{2}{R}$$
$$2 = k \qquad 2 = 1.25R$$
$$1.6 = R$$

The resistance is 1.6 ohms.

Cumulative Review Exercises

1. $8 - 4[-3 - (-2)]^2 \div 5 = 8 - 4[-3 + 2]^2 \div 5$
$= 8 - 4[-1]^2 \div 5$
$= 8 - 4(1) \div 5$
$= 8 - 4 \div 5$
$= 8 - \frac{4}{5} = \frac{36}{5}$

2.

$$\frac{2x-3}{6} - \frac{x}{9} = \frac{x-4}{3}$$
$$18\left(\frac{2x-3}{6} - \frac{x}{9}\right) = \left(\frac{x-4}{3}\right)18$$
$$3(2x-3) - 2x = 6(x-4)$$
$$6x - 9 - 2x = 6x - 24$$
$$4x - 9 = 6x - 24$$
$$-2x = -15$$
$$x = \frac{15}{2}$$

3.

$$5 - |x-4| = 2$$
$$-|x-4| = -3$$
$$|x-4| = 3$$
$$x - 4 = 3 \qquad x - 4 = -3$$
$$x = 7 \qquad x = 1$$

The solutions are 1 and 7.

4.

$$\frac{x}{x-3}$$
$$x - 3 = 0$$
$$x = 3$$

The domain is $\{x | x \neq 3\}$.

5.

$$P(x) = \frac{x-1}{2x-3}$$
$$P(-2) = \frac{-2-1}{2(-2)-3} = \frac{-3}{-4-3} = \frac{-3}{-7}$$
$$P(-2) = \frac{3}{7}$$

6. $0.000000035 = 3.5 \times 10^{-8}$

7. $\frac{(2a^{-2}b^3)^{-2}}{(4a)^{-1}} = \frac{2^{-2}a^4b^{-6}}{4^{-1}a^{-1}}$
$= 2^{-2}4a^{4-(-1)}b^{-6}$
$= \frac{1}{4}4a^5b^{-6} = \frac{a^5}{b^6}$

8. $x - 3(1 - 2x) \geq 1 - 4(2 - 2x)$
$x - 3 + 6x \geq 1 - 8 + 8x$
$7x - 3 \geq 8x - 7$
$-x - 3 \geq -7$
$-x \geq -4$
$(-1)(-x) \leq (-1)(-4)$
$x \leq 4$
$(-\infty, 4]$

9. $(2a^2 - 3a + 1)(-2a^2) = -4a^4 + 6a^3 - 2a^2$

10. Let $x^n = u$
$2x^{2n} + 3x^n - 2 = 2u^2 + 3u - 2$
$= (2u - 1)(u + 2)$
$= (2x^n - 1)(x^n + 2)$

11. $x^3y^3 - 27 = (xy)^3 - (3)^3$
$= (xy - 3)(x^2y^2 + 3xy + 9)$

12. $\frac{x^4 + x^3y - 6x^2y^2}{x^3 - 2x^2y} = \frac{x^2(x^2 + xy - 6y^2)}{x^2(x - 2y)}$
$= \frac{x^2(x + 3y)(x - 2y)}{x^2(x - 2y)}$
$= x + 3y$

13. $3x - 2y = 6$
$-2y = -3x + 6$
$y = \frac{3}{2}x - 3$
$m = \frac{3}{2}$
$y - y_1 = m(x - x_1)$
$y - (-1) = \frac{3}{2}[x - (-2)]$
$y + 1 = \frac{3}{2}(x + 2)$
$y + 1 = \frac{3}{2}x + 3$
$y = \frac{3}{2}x + 2$
The equation of the line is $y = \frac{3}{2}x + 2$.

14. $8x^2 - 6x - 9 = 0$
$(4x + 3)(2x - 3) = 0$
$4x + 3 = 0 \quad 2x - 3 = 0$
$4x = -3 \quad 2x = 3$
$x = -\frac{3}{4} \quad x = \frac{3}{2}$
The solutions are $-\frac{3}{4}$ and $\frac{3}{2}$.

15. $\frac{4x^3 + 2x^2 - 10x + 1}{x - 2}$

2	4	2	−10	1
		8	20	20
	4	10	10	21

The simplified form is $4x^2 + 10x + 10 + \frac{21}{x-2}$.

16. $\frac{16x^2 - 9y^2}{16x^2y - 12xy^2} \div \frac{4x^2 - xy - 3y^2}{12x^2y^2}$
$= \frac{16x^2 - 9y^2}{16x^2y - 12xy^2} \cdot \frac{12x^2y^2}{4x^2 - xy - 3y^2}$
$= \frac{(4x - 3y)(4x + 3y)}{4xy(4x - 3y)} \cdot \frac{12x^2y^2}{(4x + 3y)(x - y)}$
$= \frac{(4x - 3y)(4x + 3y)12x^2y^2}{4xy(4x - 3y)(4x + 3y)(x - y)}$
$= \frac{3xy}{x - y}$

17. $2x^2 + 2x = 2x(x + 1)$
$2x^4 - 2x^3 - 4x^2 = 2x^2(x^2 - x - 2)$
$= 2x^2(x - 2)(x + 1)$
The LCM $= 2x^2(x - 2)(x + 1)$.
$\frac{xy}{2x^2 + 2x} = \frac{xy}{2x(x + 1)} \cdot \frac{x(x - 2)}{x(x - 2)}$
$= \frac{x^2y(x - 2)}{2x^2(x + 1)(x - 2)}$
$= \frac{x^3y - 2x^2y}{2x^2(x + 1)(x - 2)}$
$\frac{2}{2x^4 - 2x^3 - 4x^2} = \frac{2}{2x^2(x - 2)(x + 1)}$

18. $3x^2 - x - 2 = (3x + 2)(x - 1)$
$x^2 - 1 = (x + 1)(x - 1)$
The LCM is $(3x + 2)(x + 1)(x - 1)$.
$\frac{5x}{3x^2 - x - 2} - \frac{2x}{x^2 - 1}$
$= \frac{5x}{(3x + 2)(x - 1)} \cdot \frac{x + 1}{x + 1} - \frac{2x}{(x + 1)(x - 1)} \cdot \frac{3x + 2}{3x + 2}$
$= \frac{5x^2 + 5x - (6x^2 + 4x)}{(3x + 2)(x - 1)(x + 1)}$
$= \frac{-x^2 + x}{(3x + 2)(x - 1)(x + 1)}$
$= \frac{-x(x - 1)}{(3x + 2)(x - 1)(x + 1)}$
$= -\frac{x}{(3x + 2)(x + 1)}$

19. $-3x + 5y = -15$
x-intercept: (5, 0)
y-intercept: (0, −3)

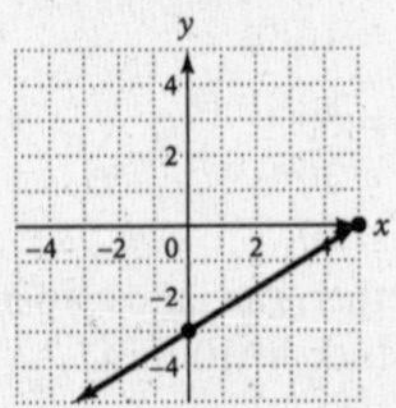

20. $x + y \le 3 \qquad -2x + y > 4$

$y \le 3 - x \qquad y > 4 + 2x$

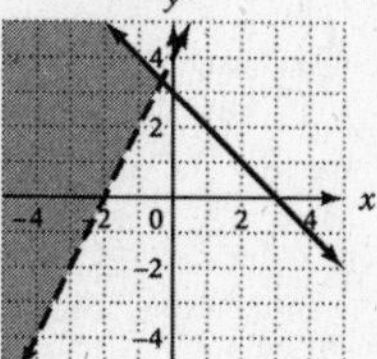

21. $\begin{vmatrix} 6 & 5 \\ 2 & -3 \end{vmatrix} = 6(-3) - 5 \cdot 2 = -18 - 10 = -28$

22.
$$\frac{x - 4 + \frac{5}{x+2}}{x + 2 - \frac{1}{x+2}} = \frac{x - 4 + \frac{5}{x+2}}{x + 2 - \frac{1}{x+2}} \cdot \frac{x+2}{x+2}$$
$$= \frac{(x-4)(x+2) + 5}{(x+2)^2 - 1}$$
$$= \frac{x^2 - 2x - 8 + 5}{x^2 + 4x + 4 - 1}$$
$$= \frac{x^2 - 2x - 3}{x^2 + 4x + 3}$$
$$= \frac{(x-3)(x+1)}{(x+3)(x+1)}$$
$$= \frac{x-3}{x+3}$$

23.
$$x + y + z = 3$$
$$-2x + y + 3z = 2$$
$$2x - 4y + z = -1$$
$$D = \begin{vmatrix} 1 & 1 & 1 \\ -2 & 1 & 3 \\ 2 & -4 & 1 \end{vmatrix} = 27$$
$$D_x = \begin{vmatrix} 3 & 1 & 1 \\ 2 & 1 & 3 \\ -1 & -4 & 1 \end{vmatrix} = 27$$
$$D_y = \begin{vmatrix} 1 & 3 & 1 \\ -2 & 2 & 3 \\ 2 & -1 & 1 \end{vmatrix} = 27$$
$$D_z = \begin{vmatrix} 1 & 1 & 3 \\ -2 & 1 & 2 \\ 2 & -4 & -1 \end{vmatrix} = 27$$
$$x = \frac{D_x}{D} = \frac{27}{27} = 1$$
$$y = \frac{D_y}{D} = \frac{27}{27} = 1$$
$$z = \frac{D_z}{D} = \frac{27}{27} = 1$$

The solution is (1, 1, 1).

24. $|3x - 2| > 4$

$3x - 2 < -4 \qquad 3x - 2 > 4$

$3x < -2 \qquad 3x > 6$

$x < -\frac{2}{3} \qquad x > 2$

$$\left\{x \middle| x < \frac{2}{3}\right\} \cup \{x | x > 2\} = \left\{x \middle| x < \frac{2}{3} \text{ or } x > 2\right\}$$

25.
$$\frac{2}{x-3} = \frac{5}{2x-3}$$
$$\left(\frac{2}{x-3}\right)(x-3)(2x-3) = \left(\frac{5}{2x-3}\right)(x-3)(2x-3)$$
$$2(2x-3) = 5(x-3)$$
$$4x - 6 = 5x - 15$$
$$-x - 6 = -15$$
$$-x = -9$$
$$x = 9$$

26.
$$\frac{3}{x^2 - 36} = \frac{2}{x-6} - \frac{5}{x+6}$$
$$\frac{3}{(x+6)(x-6)} = \frac{2}{x-6} - \frac{5}{x+6}$$
$$(x+6)(x-6)\left(\frac{3}{(x+6)(x-6)}\right) = \left(\frac{2}{x-6} - \frac{5}{x+6}\right)(x+6)(x-6)$$
$$3 = 2(x+6) - 5(x-6)$$
$$3 = 2x + 12 - 5x + 30$$
$$3 = -3x + 42$$
$$-39 = -3x$$
$$13 = x$$

27.
$$I = \frac{E}{R+r}$$
$$I \cdot (R+r) = \frac{E}{R+r}(R+r)$$
$$IR + Ir = E$$
$$Ir = E - IR$$
$$r = \frac{E - IR}{I}$$

28.
$$(1 - x^{-1})^{-1} = \left(1 - \frac{1}{x}\right)^{-1}$$
$$= \left(\frac{x-1}{x}\right)^{-1}$$
$$= \frac{x}{x-1}$$

29. **Strategy**
- Smaller integer: x

  Larger integer: $15 - x$

  $5x = 5 + 2(15 - x)$

Solution

$5x = 5 + 2(15 - x)$

$5x = 5 + 30 - 2x$

$7x = 35$

$x = 5$

$15 - x = 10$

The smaller integer is 5 and the larger integer is 10.

30. Strategy • The unknown number of pounds of almonds: x

	Amount	Cost	Value
Almonds	x	5.40	$5.40x$
Peanuts	50	2.60	50(2.60)
Mixture	$x+50$	4.00	$4(x+50)$

• The sum of the values before mixing equals the value after mixing.
$5.40x + 50(2.60) = 4(x+50)$

Solution

$$\begin{aligned} 5.40x + 50(2.60) &= 4(x+50) \\ 5.4x + 130 &= 4x + 200 \\ 1.4x + 130 &= 200 \\ 1.4x &= 70 \\ x &= 50 \end{aligned}$$

50 lb of almonds must be mixed.

31. Strategy • To find the number of people expected to vote, write and solve a proportion using x to represent the number of people expected to vote.

Solution

$$\begin{aligned} \frac{3}{5} &= \frac{x}{125{,}000} \\ \frac{3}{5} \cdot 125{,}000 &= \frac{x}{125{,}000} \cdot 125{,}000 \\ 75{,}000 &= x \end{aligned}$$

75,000 people are expected to vote.

32. Strategy • Time it takes older computer: $6r$
Time it takes new computer: r

	Rate	Time	Part
Older computer	$\frac{1}{6r}$	12	$\frac{12}{6r}$
New Computer	$\frac{1}{r}$	12	$\frac{12}{r}$

• The sum of the part of the task completed by the older computer and the part of the task completed by the new computer is 1.

Solution

$$\begin{aligned} \frac{12}{6r} + \frac{12}{r} &= 1 \\ 6r\left(\frac{12}{6r} + \frac{12}{r}\right) &= (1)6r \\ 12 + 72 &= 6r \\ 84 &= 6r \\ 14 &= r \end{aligned}$$

It would take the new computer 14 min to do the job working alone.

33. Strategy • Unknown rate of the wind: r

	Distance	Rate	Time
With the wind	900	$300+r$	$\frac{900}{300+r}$
Against the wind	600	$300-r$	$\frac{600}{300-r}$

• The time traveled with the wind equals the time traveled against the wind.

$$\frac{900}{300+r}=\frac{600}{300-r}$$

Solution

$$\begin{aligned}\frac{900}{300+r}&=\frac{600}{300-r}\\(300+r)(300-r)\left(\frac{900}{300+r}\right)&=\left(\frac{600}{300-r}\right)(300+r)(300-r)\\(300-r)(900)&=600(300+r)\\270{,}000-900r&=180{,}000+600r\\-1500r&=-90{,}000\\r&=60\end{aligned}$$

The rate of the wind is 60 mph.

34. Strategy To find the distance apart, calculate the number of times around the track each person has gone and find the difference in distance.

Solution The walker travels 3 miles in 1 hour, or 3 miles ÷ 0.25 miles/cycle = 12 complete cycles.

The jogger travels 5 miles in 1 hour, or 5 miles ÷ 0.25 miles/cycle = 20 complete cycles.

Because the walker and jogger have completed integer multiples of cycles, they are both at the starting point after 1 hour.

They have no distance between them after 1 hour.

Chapter 7: Exponents and Radicals

Prep Test

1. $48 = ? \cdot 3$
$\left(\frac{1}{3}\right)48 = ? \cdot 3\left(\frac{1}{3}\right)$
$16 = ?$

2. $2^5 = 2 \cdot 2 \cdot 2 \cdot 2 \cdot 2 = 32$

3. $\frac{6}{1}\left(\frac{3}{2}\right) = \frac{3 \cdot \cancel{2}}{1}\left(\frac{3}{\cancel{2}}\right) = \frac{3 \cdot 3}{1 \cdot 1} = 9$

4. $\frac{1}{2} - \frac{2}{3} + \frac{1}{4} = \frac{6}{12} - \frac{8}{12} + \frac{3}{12} = \frac{6 - 8 + 3}{12} = \frac{1}{12}$

5. $(3 - 7x) - (4 - 2x)$
$= 3 - 7x - 4 + 2x$
$= -5x - 1$

6. $\frac{3x^5y^6}{12x^4y} = \frac{\cancel{3}x^{5-4}y^{6-1}}{\cancel{3} \cdot 2 \cdot 2} = \frac{xy^5}{4}$

7. $(3x - 2)^2$
$= (3x)^2 + 2(3x)(-2) + (-2)^2$
$= 9x^2 - 12x + 4$

8. $(2 + 4x)(5 - 3x)$
$= 10 - 6x + 20x - 12x^2$
$= -12x^2 + 14x + 10$

9. $(6x - 1)(6x + 1)$
$= 36x^2 + 6x - 6x - 1$
$= 36x^2 - 1$

10. $x^2 - 14x - 5 = 10$
$x^2 - 14x - 5 - 10 = 0$
$x^2 - 14x - 15 = 0$
$(x - 15)(x + 1) = 0$
$x - 15 = 0 \quad x + 1 = 0$
$x = 15 \quad x = -1$
The solutions are -1 and 15.

Go Figure

Let x represent the number of tables. If there are 5 people at each table with only 2 people at the last table, this means 3 people are missing from the last table. This can be expressed as $5x - 3$. If there are 3 people at each table and 9 people are left over, this can be expressed as $3x + 9$. Solving: $5x - 3 = 3x + 9$; $2x = 12$; $x = 6$. There are 6 tables; therefore, by substituting 6 back into either equation, one can determine that there will be 27 guests at the party.

Answer: 27 guests

Section 7.1

Objective A Exercises

1. $8^{1/3} = (2^3)^{1/3} = 2$

3. $9^{3/2} = (3^2)^{3/2} = 3^3 = 27$

5. $27^{-2/3} = (3^3)^{-2/3} = 3^{-2} = \frac{1}{3^2} = \frac{1}{9}$

7. $32^{2/5} = (2^5)^{2/5} = 2^2 = 4$

9. $(-25)^{5/2}$
The base of the exponential expression is a negative number, while the denominator of the exponent is a positive even number. Therefore, $(-25)^{5/2}$ is not a real number.

11. $\left(\frac{25}{49}\right)^{-3/2} = \left(\frac{5^2}{7^2}\right)^{-3/2}$
$= \left[\left(\frac{5}{7}\right)^2\right]^{-3/2}$
$= \left(\frac{5}{7}\right)^{-3}$
$= \frac{5^{-3}}{7^{-3}} = \frac{7^3}{5^3} = \frac{343}{125}$

13. $x^{1/2}x^{1/2} = x$

15. $y^{-1/4}y^{3/4} = y^{1/2}$

17. $x^{-2/3} \cdot x^{3/4} = x^{1/12}$

19. $a^{1/3} \cdot a^{3/4} \cdot a^{-1/2} = a^{7/12}$

21. $\frac{a^{1/2}}{a^{3/2}} = a^{-1} = \frac{1}{a}$

23. $\frac{y^{-3/4}}{y^{1/4}} = y^{-1} = \frac{1}{y}$

25. $\frac{y^{2/3}}{y^{-5/6}} = y^{9/6} = y^{3/2}$

27. $(x^2)^{-1/2} = x^{-1} = \frac{1}{x}$

29. $(x^{-2/3})^6 = x^{-4} = \frac{1}{x^4}$

31. $(a^{-1/2})^{-2} = a$

33. $(x^{-3/8})^{-4/5} = x^{3/10}$

35. $(a^{1/2} \cdot a)^2 = (a^{3/2})^2 = a^3$

37. $(x^{-1/2}x^{3/4})^{-2} = (x^{1/4})^{-2} = x^{-1/2} = \frac{1}{x^{1/2}}$

39. $(y^{-1/2}y^{2/3})^{2/3} = (y^{1/6})^{2/3} = y^{1/9}$

41. $(x^8y^2)^{1/2} = x^4y$

43. $(x^4y^2z^6)^{3/2} = x^6y^3z^9$

45. $(x^{-3}y^6)^{-1/3} = xy^{-2} = \dfrac{x}{y^2}$

47. $(x^{-2}y^{1/3})^{-3/4} = x^{3/2}y^{-1/4} = \dfrac{x^{3/2}}{y^{1/4}}$

49. $\left(\dfrac{x^{1/2}}{y^2}\right)^4 = \dfrac{x^2}{y^8}$

51. $\dfrac{x^{1/4} \cdot x^{-1/2}}{x^{2/3}} = \dfrac{x^{-1/4}}{x^{2/3}} = x^{-11/12} = \dfrac{1}{x^{11/12}}$

53. $\left(\dfrac{y^{2/3} \cdot y^{-5/6}}{y^{1/9}}\right)^9 = \left(\dfrac{y^{-1/6}}{y^{1/9}}\right)^9$
$= (y^{-5/18})^9$
$= y^{-5/2} = \dfrac{1}{y^{5/2}}$

55. $\left(\dfrac{b^2 \cdot b^{-3/4}}{b^{-1/2}}\right)^{-1/2} = \left(\dfrac{b^{5/4}}{b^{-1/2}}\right)^{-1/2}$
$= (b^{7/4})^{-1/2}$
$= b^{-7/8} = \dfrac{1}{b^{7/8}}$

57. $(a^{2/3}b^2)^6(a^3b^3)^{1/3} = (a^4b^{12})(ab) = a^5b^{13}$

59. $(16m^{-2}n^4)^{-1/2}(mn^{1/2}) = (2^4)^{-1/2}mn^{-2} \cdot mn^{1/2}$
$= 2^{-2}m^2n^{-3/2}$
$= \dfrac{m^2}{2^2n^{3/2}}$
$= \dfrac{m^2}{4n^{3/2}}$

61. $\left(\dfrac{x^{1/2}y^{-3/4}}{y^{2/3}}\right)^{-6} = (x^{1/2}y^{-17/12})^{-6}$
$= x^{-3}y^{17/2}$
$= \dfrac{y^{17/2}}{x^3}$

63. $\left(\dfrac{2^{-6}b^{-3}}{a^{-1/2}}\right)^{-2/3} = \dfrac{2^4b^2}{a^{1/3}} = \dfrac{16b^2}{a^{1/3}}$

65. $y^{3/2}(y^{1/2} - y^{-1/2}) = y^{4/2} - y^{2/2} = y^2 - y$

67. $a^{-1/4}(a^{5/4} - a^{9/4}) = a^1 - a^2 = a - a^2$

69. $x^n \cdot x^{3n} = x^{4n}$

71. $x^n \cdot x^{n/2} = x^{3n/2}$

73. $\dfrac{y^{n/2}}{y^{-n}} = y^{3n/2}$

75. $(x^{2n})^n = x^{2n^2}$

77. $(x^{n/4}y^{n/8})^8 = x^{2n}y^n$

79. $(x^{n/5}y^{n/10})^{20} = x^{4n}y^{2n}$

Objective B Exercises

81. $3^{1/4} = \sqrt[4]{3}$

83. $a^{3/2} = (a^3)^{1/2} = \sqrt{a^3}$

85. $(2t)^{5/2} = \sqrt{(2t)^5} = \sqrt{32t^5}$

87. $-2x^{2/3} = -2(x^2)^{1/3} = -2\sqrt[3]{x^2}$

89. $(a^2b)^{2/3} = \sqrt[3]{(a^2b)^2} = \sqrt[3]{a^4b^2}$

91. $(a^2b^4)^{3/5} = \sqrt[5]{(a^2b^4)^3} = \sqrt[5]{a^6b^{12}}$

93. $(4x-3)^{3/4} = \sqrt[4]{(4x-3)^3}$

95. $x^{-2/3} = \dfrac{1}{x^{2/3}} = \dfrac{1}{\sqrt[3]{x^2}}$

97. $\sqrt{14} = 14^{1/2}$

99. $\sqrt[3]{x} = x^{1/3}$

101. $\sqrt[3]{x^4} = x^{4/3}$

103. $\sqrt[5]{b^3} = b^{3/5}$

105. $\sqrt[3]{2x^2} = (2x^2)^{1/3}$

107. $-\sqrt{3x^5} = -(3x^5)^{1/2}$

109. $3x\sqrt[3]{y^2} = 3xy^{2/3}$

111. $\sqrt{a^2 - 2} = (a^2 - 2)^{1/2}$

Objective C Exercises

113. $\sqrt{x^{16}} = x^8$

115. $-\sqrt{x^8} = -x^4$

117. $\sqrt[3]{x^3y^9} = xy^3$

119. $-\sqrt[3]{x^{15}y^3} = -x^5y$

121. $\sqrt{16a^4b^{12}} = \sqrt{2^4a^4b^{12}} = 2^2a^2b^6 = 4a^2b^6$

123. $\sqrt{-16x^4y^2}$
The square root of a negative number is not a real number, since the square of a real number must be positive.

125. $\sqrt[3]{27x^9} = \sqrt[3]{3^3x^9} = 3x^3$

127. $\sqrt[3]{-64x^9y^{12}} = \sqrt[3]{(-4)^3x^9y^{12}} = -4x^3y^4$

129. $-\sqrt[4]{x^8y^{12}} = -x^2y^3$

131. $\sqrt[5]{x^{20}y^{10}} = x^4y^2$

133. $\sqrt[4]{81x^4y^{20}} = \sqrt[4]{3^4x^4y^{20}} = 3xy^5$

135. $\sqrt[5]{32a^5b^{10}} = \sqrt[5]{2^5a^5b^{10}} = 2ab^2$

Applying the Concepts

137a. $\sqrt{(-2)^2} = -2$, false
$\sqrt{(-2)^2} = \sqrt{4} = 2$

b. $\sqrt[3]{(-3)^3} = -3$, true

c. $\sqrt[n]{a} = a^{1/n}$ true

d. $\sqrt[n]{a^n + b^n} = a + b$, false
$\sqrt[n]{a^n + b^n} = (a^n + b^n)^{1/n}$

e. $(a^{1/2} + b^{1/2})^2 = a + b$, false
$(a^{1/2} + b^{1/2})^2 = a + 2a^{1/2}b^{1/2} + b$

f. $\sqrt[m]{a^n} = a^{mn}$, false
$\sqrt[m]{a^n} = a^{n/m}$

139. No. If $x \geq 0$, the statement is true. However, if $x < 0$, then $\sqrt{x^2} = |x|$. For example, if $x = -2$, then $\sqrt{x^2} = \sqrt{(-2)^2} = \sqrt{4} = 2$, not -2.

Section 7.2

Objective A Exercises

1. $\sqrt{x^4y^3z^5} = \sqrt{x^4y^2z^4(yz)}$
$= \sqrt{x^4y^2z^4}\sqrt{yz}$
$= x^2yz^2\sqrt{yz}$

3. $\sqrt{8a^3b^8} = \sqrt{2^3a^3b^8}$
$= \sqrt{2^2a^2b^8(2a)}$
$= \sqrt{2^2a^2b^8}\sqrt{2a}$
$= 2ab^4\sqrt{2a}$

5. $\sqrt{45x^2y^3z^5} = \sqrt{3^2 \cdot 5x^2y^3z^5}$
$= \sqrt{3^2x^2y^2z^4(5yz)}$
$= \sqrt{3^2x^2y^2z^4}\sqrt{5yz}$
$= 3xyz^2\sqrt{5yz}$

7. $\sqrt{-9x^3}$
The square root of a negative number is not a real number, since the square of a real number must be positive. Therefore, $\sqrt{-9x^3}$ is not a real number.

9. $\sqrt[3]{a^{16}b^8} = \sqrt[3]{a^{15}b^6(ab^2)}$
$= \sqrt[3]{a^{15}b^6}\sqrt[3]{ab^2}$
$= a^5b^2\sqrt[3]{ab^2}$

11. $\sqrt[3]{-125x^2y^4} = \sqrt[3]{(-5)^3x^2y^4}$
$= \sqrt[3]{(-5)^3y^3(x^2y)}$
$= \sqrt[3]{(-5)^3y^3}\sqrt[3]{x^2y}$
$= -5y\sqrt[3]{x^2y}$

13. $\sqrt[3]{a^4b^5c^6} = \sqrt[3]{a^3b^3c^6(ab^2)}$
$= \sqrt[3]{a^3b^3c^6}\sqrt[3]{ab^2}$
$= abc^2\sqrt[3]{ab^2}$

15. $\sqrt[4]{16x^9y^5} = \sqrt[4]{2^4x^9y^5}$
$= \sqrt[4]{2^4x^8y^4(xy)}$
$= \sqrt[4]{2^4x^8y^4}\sqrt[4]{xy}$
$= 2x^2y\sqrt[4]{xy}$

Objective B Exercises

17. $2\sqrt{x} - 8\sqrt{x} = -6\sqrt{x}$

19. $\sqrt{8} - \sqrt{32} = \sqrt{2^3} - \sqrt{2^5}$
$= \sqrt{2^2}\sqrt{2} - \sqrt{2^4}\sqrt{2}$
$= 2\sqrt{2} - 2^2\sqrt{2}$
$= 2\sqrt{2} - 4\sqrt{2}$
$= -2\sqrt{2}$

21. $\sqrt{18b} + \sqrt{75b} = \sqrt{2 \cdot 3^2b} + \sqrt{3 \cdot 5^2b}$
$= \sqrt{3^2}\sqrt{2b} + \sqrt{5^2}\sqrt{3b}$
$= 3\sqrt{2b} + 5\sqrt{3b}$

23. $3\sqrt{8x^2y^3} - 2x\sqrt{32y^3}$
$= 3\sqrt{2^3x^2y^3} - 2x\sqrt{2^5y^3}$
$= 3\sqrt{2^2x^2y^2}\sqrt{2y} - 2x\sqrt{2^4y^2}\sqrt{2y}$
$= 3 \cdot 2xy\sqrt{2y} - 2x \cdot 2^2y\sqrt{2y}$
$= 6xy\sqrt{2y} - 8xy\sqrt{2y}$
$= -2xy\sqrt{2y}$

25. $2a\sqrt{27ab^5} + 3b\sqrt{3a^3b}$
$= 2a\sqrt{3^3ab^5} + 3b\sqrt{3a^3b}$
$= 2a\sqrt{3^2b^4}\sqrt{3ab} + 3b\sqrt{a^2}\sqrt{3ab}$
$= 2a \cdot 3b^2\sqrt{3ab} + 3ab\sqrt{3ab}$
$= 6ab^2\sqrt{3ab} + 3ab\sqrt{3ab}$

27. $\sqrt[3]{16} - \sqrt[3]{54} = \sqrt[3]{2^4} - \sqrt[3]{2 \cdot 3^3}$
$= \sqrt[3]{2^3}\sqrt[3]{2} - \sqrt[3]{3^3}\sqrt[3]{2}$
$= 2\sqrt[3]{2} - 3\sqrt[3]{2}$
$= -\sqrt[3]{2}$

29. $2b\sqrt[3]{16b^2} + \sqrt[3]{128b^5}$
$= 2b\sqrt[3]{2^4b^2} + \sqrt[3]{2^7b^5}$
$= 2b\sqrt[3]{2^3}\sqrt[3]{2b^2} + \sqrt[3]{2^6b^3}\sqrt[3]{2b^2}$
$= 2b \cdot 2\sqrt[3]{2b^2} + 2^2b\sqrt[3]{2b^2}$
$= 4b\sqrt[3]{2b^2} + 4b\sqrt[3]{2b^2}$
$= 8b\sqrt[3]{2b^2}$

31. $3\sqrt[4]{32a^5} - a\sqrt[4]{162a}$
$= 3\sqrt[4]{2^5a^5} - a\sqrt[4]{2 \cdot 3^4a}$
$= 3\sqrt[4]{2^4a^4}\sqrt[4]{2a} - a\sqrt[4]{3^4}\sqrt[4]{2a}$
$= 3 \cdot 2a\sqrt[4]{2a} - 3a\sqrt[4]{2a}$
$= 6a\sqrt[4]{2a} - 3a\sqrt[4]{2a}$
$= 3a\sqrt[4]{2a}$

33. $2\sqrt{50} - 3\sqrt{125} + \sqrt{98}$
$= 2\sqrt{2 \cdot 5^2} - 3\sqrt{5^3} + \sqrt{2 \cdot 7^2}$
$= 2\sqrt{5^2}\sqrt{2} - 3\sqrt{5^2}\sqrt{5} + \sqrt{7^2}\sqrt{2}$
$= 2 \cdot 5\sqrt{2} - 3 \cdot 5\sqrt{5} + 7\sqrt{2}$
$= 10\sqrt{2} - 15\sqrt{5} + 7\sqrt{2}$
$= 17\sqrt{2} - 15\sqrt{5}$

35. $\sqrt{9b^3} - \sqrt{25b^3} + \sqrt{49b^3}$
$= \sqrt{3^2b^3} - \sqrt{5^2b^3} + \sqrt{7^2b^3}$
$= \sqrt{3^2b^2}\sqrt{b} - \sqrt{5^2b^2}\sqrt{b} + \sqrt{7^2b^2}\sqrt{b}$
$= 3b\sqrt{b} - 5b\sqrt{b} + 7b\sqrt{b}$
$= 5b\sqrt{b}$

37. $2x\sqrt{8xy^2} - 3y\sqrt{32x^3} + \sqrt{4x^3y^3}$
$= 2x\sqrt{2^3xy^2} - 3y\sqrt{2^5x^3} + \sqrt{2^2x^3y^3}$
$= 2x\sqrt{2^2y^2}\sqrt{2x} - 3y\sqrt{2^4x^2}\sqrt{2x} + \sqrt{2^2x^2y^2}\sqrt{xy}$
$= 2x \cdot 2y\sqrt{2x} - 3y \cdot 2^2x\sqrt{2x} + 2xy\sqrt{xy}$
$= 4xy\sqrt{2x} - 12xy\sqrt{2x} + 2xy\sqrt{xy}$
$= -8xy\sqrt{2x} + 2xy\sqrt{xy}$

39. $\sqrt[3]{54xy^3} - 5\sqrt[3]{2xy^3} + y\sqrt[3]{128x}$
$= \sqrt[3]{2 \cdot 3^3xy^3} - 5\sqrt[3]{2xy^3} + y\sqrt[3]{2^7x}$
$= \sqrt[3]{3^3y^3}\sqrt[3]{2x} - 5\sqrt[3]{y^3}\sqrt[3]{2x} + y\sqrt[3]{2^6}\sqrt[3]{2x}$
$= 3y\sqrt[3]{2x} - 5y\sqrt[3]{2x} + 2^2y\sqrt[3]{2x}$
$= 3y\sqrt[3]{2x} - 5y\sqrt[3]{2x} + 4y\sqrt[3]{2x}$
$= 2y\sqrt[3]{2x}$

41. $2a\sqrt[4]{32b^5} - 3b\sqrt[4]{162a^4b} + \sqrt[4]{2a^4b^5}$
$= 2a\sqrt[4]{2^5b^5} - 3b\sqrt[4]{2 \cdot 3^4a^4b} + \sqrt[4]{2a^4b^5}$
$= 2a\sqrt[4]{2^4b^4}\sqrt[4]{2b} - 3b\sqrt[4]{3^4a^4}\sqrt[4]{2b} + \sqrt[4]{a^4b^4}\sqrt[4]{2b}$
$= 2a \cdot 2b\sqrt[4]{2b} - 3b \cdot 3a\sqrt[4]{2b} + ab\sqrt[4]{2b}$
$= 4ab\sqrt[4]{2b} - 9ab\sqrt[4]{2b} + ab\sqrt[4]{2b}$
$= -4ab\sqrt[4]{2b}$

Objective C Exercises

43. $\sqrt{8}\sqrt{32} = \sqrt{256} = \sqrt{2^8} = 2^4 = 16$

45. $\sqrt[3]{4}\sqrt[3]{8} = \sqrt[3]{32} = \sqrt[3]{2^5} = \sqrt[3]{2^3}\sqrt[3]{2^2} = 2\sqrt[3]{4}$

47. $\sqrt{x^2y^5}\sqrt{xy} = \sqrt{x^3y^6} = \sqrt{x^2y^6}\sqrt{x} = xy^3\sqrt{x}$

49. $\sqrt{2x^2y}\sqrt{32xy} = \sqrt{64x^3y^2}$
$= \sqrt{2^6x^3y^2}$
$= \sqrt{2^6x^2y^2}\sqrt{x}$
$= 2^3xy\sqrt{x} = 8xy\sqrt{x}$

51. $\sqrt[3]{x^2y}\sqrt[3]{16x^4y^2} = \sqrt[3]{16x^6y^3}$
$= \sqrt[3]{2^4x^6y^3}$
$= \sqrt[3]{2^3x^6y^3}\sqrt[3]{2}$
$= 2x^2y\sqrt[3]{2}$

53. $\sqrt[4]{12ab^3}\sqrt[4]{4a^5b^2} = \sqrt[4]{48a^6b^5}$
$= \sqrt[4]{2^4 \cdot 3a^6b^5}$
$= \sqrt[4]{2^4a^4b^4}\sqrt[4]{3a^2b}$
$= 2ab\sqrt[4]{3a^2b}$

55. $\sqrt{3}(\sqrt{27} - \sqrt{3}) = \sqrt{81} - \sqrt{9}$
$= \sqrt{3^4} - \sqrt{3^2}$
$= 3^2 - 3$
$= 9 - 3 = 6$

57. $\sqrt{x}(\sqrt{x} - \sqrt{2}) = \sqrt{x^2} - \sqrt{2x} = x - \sqrt{2x}$

59. $\sqrt{2x}(\sqrt{8x} - \sqrt{32}) = \sqrt{16x^2} - \sqrt{64x}$
$= \sqrt{2^4x^2} - \sqrt{2^6x}$
$= 2^2x - 2^3\sqrt{x}$
$= 4x - 8\sqrt{x}$

61. $(\sqrt{x} - 3)^2 = (\sqrt{x})^2 - 3\sqrt{x} - 3\sqrt{x} + 9$
$= x - 6\sqrt{x} + 9$

63. $(4\sqrt{5} + 2)^2 = (4\sqrt{5})^2 + 8\sqrt{5} + 8\sqrt{5} + 4$
$= 16 \cdot 5 + 16\sqrt{5} + 4$
$= 80 + 16\sqrt{5} + 4$
$= 84 + 16\sqrt{5}$

65. $2\sqrt{14xy} \cdot 4\sqrt{7x^2y} \cdot 3\sqrt{8xy^2} = 24\sqrt{784x^4y^4}$
$= 24\sqrt{2^4 \cdot 7^2x^4y^4}$
$= 24 \cdot 2^2 \cdot 7x^2y^2$
$= 672x^2y^2$

67. $\sqrt[3]{2a^2b}\sqrt[3]{4a^3b^2}\sqrt[3]{8a^5b^6} = \sqrt[3]{64a^{10}b^9}$
$= \sqrt[3]{2^6a^{10}b^9}$
$= \sqrt[3]{2^6a^9b^9}\sqrt[3]{a}$
$= 2^2a^3b^3\sqrt[3]{a}$
$= 4a^3b^3\sqrt[3]{a}$

69. $(\sqrt{5} - 5)(2\sqrt{5} + 2) = 2\sqrt{5^2} + 2\sqrt{5} - 10\sqrt{5} - 10$
$= 2 \cdot 5 - 8\sqrt{5} - 10$
$= 10 - 8\sqrt{5} - 10$
$= -8\sqrt{5}$

71. $(\sqrt{x} - y)(\sqrt{x} + y) = \sqrt{x^2} - y^2 = x - y^2$

73. $(2\sqrt{3x} - \sqrt{y})(2\sqrt{3x} + \sqrt{y}) = 4\sqrt{3^2x^2} - \sqrt{y^2}$
$= 4 \cdot 3x - y$
$= 12x - y$

Objective D Exercises

75. To rationalize the denominator of a radical expression means to rewrite the expression with no radicals in the denominator. It is accomplished by multiplying both the numerator and the denominator by the same expression, one that removes the radical(s) from the denominator of the original expression.

77. $\dfrac{\sqrt{60y^4}}{\sqrt{12y}} = \sqrt{\dfrac{60y^4}{12y}}$
$= \sqrt{5y^3}$
$= \sqrt{y^2}\sqrt{5y}$
$= y\sqrt{5y}$

79. $\dfrac{\sqrt{65ab^4}}{\sqrt{5ab}} = \sqrt{\dfrac{65ab^4}{5ab}}$
$= \sqrt{13b^3}$
$= \sqrt{b^2}\sqrt{13b} = b\sqrt{13b}$

81. $\dfrac{1}{\sqrt{2}} = \dfrac{1}{\sqrt{2}} \cdot \dfrac{\sqrt{2}}{\sqrt{2}} = \dfrac{\sqrt{2}}{\sqrt{2^2}} = \dfrac{\sqrt{2}}{2}$

83. $\dfrac{2}{\sqrt{3y}} = \dfrac{2}{\sqrt{3y}} \cdot \dfrac{\sqrt{3y}}{\sqrt{3y}} = \dfrac{2\sqrt{3y}}{\sqrt{3^2y^2}} = \dfrac{2\sqrt{3y}}{3y}$

85. $\dfrac{9}{\sqrt{3a}} = \dfrac{9}{\sqrt{3a}} \cdot \dfrac{\sqrt{3a}}{\sqrt{3a}} = \dfrac{9\sqrt{3a}}{\sqrt{3^2a^2}} = \dfrac{9\sqrt{3a}}{3a} = \dfrac{3\sqrt{3a}}{a}$

87. $\sqrt{\dfrac{y}{2}} = \dfrac{\sqrt{y}}{\sqrt{2}} = \dfrac{\sqrt{y}}{\sqrt{2}} \cdot \dfrac{\sqrt{2}}{\sqrt{2}} = \dfrac{\sqrt{2y}}{\sqrt{2^2}} = \dfrac{\sqrt{2y}}{2}$

89. $\dfrac{5}{\sqrt[3]{9}} = \dfrac{5}{\sqrt[3]{3^2}} \cdot \dfrac{\sqrt[3]{3}}{\sqrt[3]{3}} = \dfrac{5\sqrt[3]{3}}{\sqrt[3]{3^3}} = \dfrac{5\sqrt[3]{3}}{3}$

91. $\dfrac{5}{\sqrt[3]{3y}} = \dfrac{5}{\sqrt[3]{3y}} \cdot \dfrac{\sqrt[3]{3^2y^2}}{\sqrt[3]{3^2y^2}} = \dfrac{5\sqrt[3]{3^2y^2}}{\sqrt[3]{3^3y^3}} = \dfrac{5\sqrt[3]{9y^2}}{3y}$

93. $\dfrac{\sqrt{15a^2b^5}}{\sqrt{30a^5b^3}} = \sqrt{\dfrac{15a^2b^5}{30a^5b^3}}$
$= \sqrt{\dfrac{b^2}{2a^3}}$
$= \dfrac{b}{\sqrt{a^2}\sqrt{2a}}$
$= \dfrac{b}{a\sqrt{2a}} \cdot \dfrac{\sqrt{2a}}{\sqrt{2a}}$
$= \dfrac{b\sqrt{2a}}{a\sqrt{2^2a^2}}$
$= \dfrac{b\sqrt{2a}}{a \cdot 2a} = \dfrac{b\sqrt{2a}}{2a^2}$

95. $\dfrac{\sqrt{12x^3y}}{\sqrt{20x^4y}} = \sqrt{\dfrac{12x^3y}{20x^4y}}$
$= \sqrt{\dfrac{3}{5x}}$
$= \dfrac{\sqrt{3}}{\sqrt{5x}} \cdot \dfrac{\sqrt{5x}}{\sqrt{5x}}$
$= \dfrac{\sqrt{15x}}{\sqrt{5^2x^2}} = \dfrac{\sqrt{15x}}{5x}$

97. $\dfrac{-2}{1-\sqrt{2}} = \dfrac{-2}{1-\sqrt{2}} \cdot \dfrac{1+\sqrt{2}}{1+\sqrt{2}}$
$= \dfrac{-2-2\sqrt{2}}{1^2-(\sqrt{2})^2}$
$= \dfrac{-2-2\sqrt{2}}{1-2}$
$= \dfrac{-2-2\sqrt{2}}{-1} = 2+2\sqrt{2}$

99. $\dfrac{-4}{3-\sqrt{2}} = \dfrac{-4}{3-\sqrt{2}} \cdot \dfrac{3+\sqrt{2}}{3+\sqrt{2}}$
$= \dfrac{-12-4\sqrt{2}}{3^2-(\sqrt{2})^2}$
$= \dfrac{-12-4\sqrt{2}}{9-2}$
$= \dfrac{-12-4\sqrt{2}}{7} = -\dfrac{12+4\sqrt{2}}{7}$

101. $\dfrac{5}{2-\sqrt{7}} = \dfrac{5}{2-\sqrt{7}} \cdot \dfrac{2+\sqrt{7}}{2+\sqrt{7}}$
$= \dfrac{10+5\sqrt{7}}{2^2-(\sqrt{7})^2}$
$= \dfrac{10+5\sqrt{7}}{4-7}$
$= \dfrac{10+5\sqrt{7}}{-3} = -\dfrac{10+5\sqrt{7}}{3}$

103. $\dfrac{-7}{\sqrt{x}-3} = -\dfrac{7}{\sqrt{x}-3} \cdot \dfrac{\sqrt{x}+3}{\sqrt{x}+3}$
$= -\dfrac{7\sqrt{x}+21}{(\sqrt{x})^2-3^2}$
$= -\dfrac{7\sqrt{x}+21}{x-9}$

105. $\dfrac{\sqrt{3}+\sqrt{4}}{\sqrt{2}+\sqrt{3}} = \dfrac{\sqrt{3}+\sqrt{2^2}}{\sqrt{2}+\sqrt{3}} = \dfrac{\sqrt{3}+2}{\sqrt{2}+\sqrt{3}} \cdot \dfrac{\sqrt{2}-\sqrt{3}}{\sqrt{2}-\sqrt{3}}$
$= \dfrac{\sqrt{6}-\sqrt{3^2}+2\sqrt{2}-2\sqrt{3}}{(\sqrt{2})^2-(\sqrt{3})^2}$
$= \dfrac{\sqrt{6}-3+2\sqrt{2}-2\sqrt{3}}{2-3}$
$= \dfrac{\sqrt{6}-3+2\sqrt{2}-2\sqrt{3}}{-1}$
$= -\sqrt{6}+3-2\sqrt{2}+2\sqrt{3}$

107. $\dfrac{2+3\sqrt{5}}{1-\sqrt{5}} = \dfrac{2+3\sqrt{5}}{1-\sqrt{5}} \cdot \dfrac{1+\sqrt{5}}{1+\sqrt{5}}$

$$= \frac{2+2\sqrt{5}+3\sqrt{5}+3(\sqrt{5})^2}{1-(\sqrt{5})^2}$$
$$= \frac{2+5\sqrt{5}+3(5)}{1-5}$$
$$= \frac{2+5\sqrt{5}+15}{-4}$$
$$= \frac{17+5\sqrt{5}}{-4}$$
$$= -\frac{17+5\sqrt{5}}{4}$$

109. $\dfrac{2\sqrt{a}-\sqrt{b}}{4\sqrt{a}+3\sqrt{b}}$

$$= \frac{2\sqrt{a}-\sqrt{b}}{4\sqrt{a}+3\sqrt{b}} \cdot \frac{4\sqrt{a}-3\sqrt{b}}{4\sqrt{a}-3\sqrt{b}}$$
$$= \frac{8(\sqrt{a})^2 - 6\sqrt{ab} - 4\sqrt{ab} + 3(\sqrt{b})^2}{16(\sqrt{a})^2 - 9(\sqrt{b})^2}$$
$$= \frac{8a - 10\sqrt{ab} + 3b}{16a - 9b}$$

111. $\dfrac{3\sqrt{y}-y}{\sqrt{y}+2y} = \dfrac{3\sqrt{y}-y}{\sqrt{y}+2y} \cdot \dfrac{\sqrt{y}-2y}{\sqrt{y}-2y}$

$$= \frac{3(\sqrt{y})^2 - 6y\sqrt{y} - y\sqrt{y} + 2y^2}{(\sqrt{y})^2 - 4y^2}$$
$$= \frac{3y - 7y\sqrt{y} + 2y^2}{y - 4y^2}$$
$$= \frac{3 - 7\sqrt{y} + 2y}{1 - 4y}$$

Applying the Concepts

113a. $\sqrt[2]{3} \cdot \sqrt[3]{4} = \sqrt[5]{12}$, false
$\sqrt[2]{3} \cdot \sqrt[3]{4} = \sqrt[6]{432}$

b. $\sqrt{3} \cdot \sqrt{3} = 3$, true

c. $\sqrt[3]{x} \cdot \sqrt[3]{x} = x$, false
$\sqrt[3]{x} \cdot \sqrt[3]{x} = x^{1/3}x^{1/3} = x^{2/3}$

d. $\sqrt{x} + \sqrt{y} = \sqrt{x+y}$, false
$\sqrt{x} + \sqrt{y}$ is in an irreducible form.

e. $\sqrt[2]{2} + \sqrt[3]{3} = \sqrt[5]{2+3}$, false
$\sqrt[2]{2} + \sqrt[3]{3}$ is an irreducible form.

f. $8\sqrt[5]{a} - 2\sqrt[5]{a} = 6\sqrt[5]{a}$, true

115. $\dfrac{\sqrt[4]{(a+b)^3}}{\sqrt{a+b}} = \dfrac{(a+b)^{3/4}}{(a+b)^{1/2}} = (a+b)^{1/4} = \sqrt[4]{a+b}$

Section 7.3

Objective A Exercises

1.
$$\begin{aligned} \sqrt[3]{4x} &= -2 \\ (\sqrt[3]{4x})^3 &= (-2)^3 \\ 4x &= -8 \\ x &= -2 \end{aligned}$$

Check:
$$\begin{array}{c|c} \sqrt[3]{4x} = -2 & \\ \hline \sqrt[3]{4(-2)} & -2 \\ \sqrt[3]{-8} & -2 \\ -2 = -2 & \end{array}$$

The solution is -2.

3.
$$\begin{aligned} \sqrt{3x-2} &= 5 \\ (\sqrt{3x-2})^2 &= 5^2 \\ 3x-2 &= 25 \\ 3x &= 27 \\ x &= 9 \end{aligned}$$

Check:
$$\begin{array}{c|c} \sqrt{3x-2} = 5 & \\ \hline \sqrt{3(9)-2} & 5 \\ \sqrt{27-2} & 5 \\ 5 = 5 & \end{array}$$

The solution is 9.

5.
$$\begin{aligned} \sqrt{4x-3} - 5 &= 0 \\ \sqrt{4x-3} &= 5 \\ (\sqrt{4x-3})^2 &= 5^2 \\ 4x-3 &= 25 \\ 4x &= 28 \\ x &= 7 \end{aligned}$$

Check:
$$\begin{array}{c|c} \sqrt{4x-3} - 5 = 0 & \\ \hline \sqrt{4(7)-3} - 5 & 0 \\ \sqrt{28-3} - 5 & 0 \\ \sqrt{25} - 5 & 0 \\ 5 - 5 & 0 \\ 0 = 0 & \end{array}$$

The solution is 7.

7. $\sqrt{2x+4} = \sqrt{5x-9}$
$(\sqrt{2x+4})^2 = (\sqrt{5x-9})^2$
$2x+4 = 5x-9$
$-3x+4 = -9$
$-3x = -13$
$x = \frac{13}{3}$

Check:

$\sqrt{2x+4} = \sqrt{5x-9}$	
$\sqrt{2\left(\frac{13}{3}\right)+4}$	$\sqrt{5\left(\frac{13}{3}\right)-9}$
$\sqrt{\frac{26}{3}+4}$	$\sqrt{\frac{65}{3}-9}$

$\sqrt{\frac{38}{3}} = \sqrt{\frac{38}{3}}$

The solution is $\frac{13}{3}$.

9. $\sqrt[3]{2x-6} = 4$
$(\sqrt[3]{2x-6})^3 = 4^3$
$2x-6 = 64$
$2x = 70$
$x = 35$

Check:

$\sqrt[3]{2x-6} = 4$	
$\sqrt[3]{2(35)-6}$	4
$\sqrt[3]{70-6}$	4
$\sqrt[3]{64}$	4

$4 = 4$

The solution is 35.

11. $\sqrt[3]{x-12} = \sqrt[3]{5x+16}$
$(\sqrt[3]{x-12})^3 = (\sqrt[3]{5x+16})^3$
$x-12 = 5x+16$
$-4x-12 = 16$
$-4x = 28$
$x = -7$

Check:

$\sqrt[3]{x-12} = \sqrt[3]{5x+16}$	
$\sqrt[3]{-7-12}$	$\sqrt[3]{5(-7)+16}$
$\sqrt[3]{-19}$	$\sqrt[3]{-35+16}$

$\sqrt[3]{-19} = \sqrt[3]{-19}$

The solution is -7.

13. $\sqrt[3]{2x-3}+5 = 2$
$\sqrt[3]{2x-3} = -3$
$(\sqrt[3]{2x-3})^3 = (-3)^3$
$2x-3 = -27$
$2x = -24$
$x = -12$

Check:

$\sqrt[3]{2x-3}+5 = 2$	
$\sqrt[3]{2(-12)-3}+5$	2
$\sqrt[3]{-24-3}+5$	2
$\sqrt[3]{-27}+5$	2
$-3+5$	2

$2 = 2$

The solution is -12.

15. $\sqrt{x}+\sqrt{x-5} = 5$
$\sqrt{x} = 5-\sqrt{x-5}$
$(\sqrt{x})^2 = (5-\sqrt{x-5})^2$
$x = 25-10\sqrt{x-5}+x-5$
$0 = 20-10\sqrt{x-5}$
$-20 = -10\sqrt{x-5}$
$2 = \sqrt{x-5}$
$(2)^2 = (\sqrt{x-5})^2$
$4 = x-5$
$9 = x$

Check:

$\sqrt{x}+\sqrt{x-5} = 5$	
$\sqrt{9}+\sqrt{9-5}$	5
$3+\sqrt{4}$	5
$3+2$	5

$5 = 5$

The solution is 9.

17. $\sqrt{2x+5}-\sqrt{2x} = 1$
$\sqrt{2x+5} = 1+\sqrt{2x}$
$(\sqrt{2x+5})^2 = (1+\sqrt{2x})^2$
$2x+5 = 1+2\sqrt{2x}+2x$
$5 = 1+2\sqrt{2x}$
$4 = 2\sqrt{2x}$
$2 = \sqrt{2x}$
$2^2 = (\sqrt{2x})^2$
$4 = 2x$
$2 = x$

Check:

$\sqrt{2x+5}-\sqrt{2x} = 1$	
$\sqrt{2\cdot 2+5}-\sqrt{2\cdot 2}$	1
$\sqrt{9}-\sqrt{4}$	1
$3-2$	1

$1 = 1$

The solution is 2.

19. $\sqrt{2x} - \sqrt{x-1} = 1$

$$\begin{aligned}
\sqrt{2x} &= 1 + \sqrt{x-1} \\
\left(\sqrt{2x}\right)^2 &= \left(1 + \sqrt{x-1}\right)^2 \\
2x &= 1 + 2\sqrt{x-1} + x - 1 \\
x &= 2\sqrt{x-1} \\
(x)^2 &= \left(2\sqrt{x-1}\right)^2 \\
x^2 &= 4(x-1) \\
x^2 &= 4x - 4 \\
x^2 - 4x + 4 &= 0 \\
(x-2)(x-2) &= 0
\end{aligned}$$

$x - 2 = 0 \quad x - 2 = 0$

$x = 2 \quad\quad x = 2$

Check:

$\sqrt{2x} - \sqrt{x-1} = 1$	
$\sqrt{2 \cdot 2} - \sqrt{2-1}$	1
$\sqrt{4} - \sqrt{1}$	1
$2 - 1$	1
$1 = 1$	

The solution is 2.

21. $\sqrt{2x+2} + \sqrt{x} = 3$

$$\begin{aligned}
\sqrt{2x+2} &= 3 - \sqrt{x} \\
\left(\sqrt{2x+2}\right)^2 &= \left(3 - \sqrt{x}\right)^2 \\
2x + 2 &= 9 - 6\sqrt{x} + x \\
x - 7 &= -6\sqrt{x} \\
(x-7)^2 &= \left(-6\sqrt{x}\right)^2 \\
x^2 - 14x + 49 &= 36x \\
x^2 - 50x + 49 &= 0 \\
(x-49)(x-1) &= 0
\end{aligned}$$

$x - 49 = 0 \quad x - 1 = 0$

$x = 49 \quad\quad x = 1$

Check:

$\sqrt{2x+2} + \sqrt{x} = 3$	
$\sqrt{2(49)+2} + \sqrt{49}$	3
$\sqrt{100} + \sqrt{49}$	3
$10 + 7$	3
$17 \neq 3$	

Check:

$\sqrt{2x+2} + \sqrt{x} = 3$	
$\sqrt{2(1)+2} + \sqrt{1}$	3
$\sqrt{4} + \sqrt{1}$	3
$2 + 1$	3
$3 \neq 3$	

The solution is 1.

Objective B Exercises

23. Strategy To find the distance the object will fall, replace t and g in the equation with the given values and solve for d.

Solution

$$\begin{aligned}
t &= \sqrt{\frac{2d}{g}} \\
3 &= \sqrt{\frac{2d}{5.5}} \\
3^2 &= \left(\sqrt{\frac{2d}{5.5}}\right)^2 \\
9 &= \frac{2d}{5.5} \\
49.5 &= 2d \\
24.75 &= d
\end{aligned}$$

On the moon, an object will fall 24.75 ft in 3 s.

25. Strategy To find the difference in width:

- Use the Pythagorean Theorem to find the width of the screen on the regular television. The hypotenuse is the diagonal. The height is one leg.
- Use the Pythagorean Theorem to find the width of the HDTV. The hypotenuse is the diagonal and the height is one leg.
- Subtract the width of the regular TV from the width of the HDTV.

Solution

$$\begin{aligned}
c^2 &= a^2 + b^2 \\
27^2 &= 16.2^2 + b^2 \\
729 &= 262.44 + b^2 \\
466.56 &= b^2 \\
(466.56)^{1/2} &= (b^2)^{1/2} \\
21.6 &= b
\end{aligned}$$

$$\begin{aligned}
33^2 &= 16.2^2 + b^2 \\
1089 &= 262.44 + b^2 \\
826.56 &= b^2 \\
(826.56)^{1/2} &= (b^2)^{1/2} \\
28.75 &\approx b
\end{aligned}$$

$28.75 - 21.6 = 7.15$

The HDTV is approximately 7.15 in. wider.

27. Strategy To find the length of the pendulum, replace T in the equation with the given value and solve for L.

Solution

$$T = 2\pi\sqrt{\frac{L}{32}}$$
$$3 = 2\pi\sqrt{\frac{L}{32}}$$
$$\frac{3}{2\pi} = \sqrt{\frac{L}{32}}$$
$$\left(\frac{3}{2\pi}\right)^2 = \left(\sqrt{\frac{L}{32}}\right)^2$$
$$\left(\frac{3}{2\pi}\right)^2 = \frac{L}{32}$$
$$32\left(\frac{3}{2\pi}\right)^2 = L$$
$$7.30 \approx L$$

The length of the pendulum is 7.30 ft.

Applying the Concepts

29.

$$\sqrt{3x-2} = \sqrt{2x-3} + \sqrt{x-1}$$
$$\left(\sqrt{3x-2}\right)^2 = \left(\sqrt{2x-3} + \sqrt{x-1}\right)^2$$
$$3x-2 = 2x-3+2\sqrt{2x-3}\cdot\sqrt{x-1}+x-1$$
$$3x-2 = 3x-4+2\sqrt{2x-3}\cdot\sqrt{x-1}$$
$$2 = 2\sqrt{2x-3}\cdot\sqrt{x-1}$$
$$(1)^2 = \left(\sqrt{2x-3}\cdot\sqrt{x-1}\right)^2$$
$$1 = (2x-3)(x-1)$$
$$1 = 2x^2-5x+3$$
$$2x^2-5x+2 = 0$$
$$(2x-1)(x-2) = 0$$

$x = \frac{1}{2}$ $\quad x = 2$

Check:

$$\sqrt{3x-2} = \sqrt{2x-3} + \sqrt{x-1}$$
$$\sqrt{3\left(\frac{1}{2}\right)-2} \;\Big|\; \sqrt{2\left(\frac{1}{2}\right)-3} + \sqrt{\frac{1}{2}-1}$$
$$\sqrt{\frac{3}{2}-2} \;\Big|\; \sqrt{1-3} + \sqrt{-\frac{1}{2}}$$
$$\sqrt{-\frac{1}{2}} = \sqrt{-1} + \sqrt{-\frac{1}{2}}$$

Not real numbers

Check:

$$\sqrt{3x-2} = \sqrt{2x-3} + \sqrt{x-1}$$
$$\sqrt{3(2)-2} \;\Big|\; \sqrt{2(2)-3} + \sqrt{2-1}$$
$$\sqrt{4} \;\Big|\; \sqrt{1} + \sqrt{1}$$
$$2 = 1+1$$

The solution is 2.

31.

$$V = \frac{4}{3}\pi r^3$$
$$V\cdot\frac{3}{4\pi} = \frac{4}{3}\pi r^3\cdot\frac{3}{4\pi}$$
$$\frac{3V}{4\pi} = r^3$$
$$r = \sqrt[3]{\frac{3V}{4\pi}}$$

Section 7.4

Objective A Exercises

1. An imaginary number is a number whose square is a negative number. Imaginary numbers are defined in terms of i, the number whose square is -1. A complex number is a number of the form $a + bi$, where a and b are real numbers and $i = \sqrt{-1}$. The number a is the real part and the number b is the imaginary part of the complex number.

3. $\sqrt{-4} = i\sqrt{4} = i\sqrt{2^2} = 2i$

5. $\sqrt{-98} = i\sqrt{98} = i\sqrt{2\cdot 7^2} = 7i\sqrt{2}$

7. $\sqrt{-27} = i\sqrt{27} = i\sqrt{3^2\cdot 3} = 3i\sqrt{3}$

9.
$$\sqrt{16}+\sqrt{-4} = \sqrt{16}+i\sqrt{4}$$
$$= \sqrt{2^4}+i\sqrt{2^2} = 4+2i$$

11.
$$\sqrt{12}-\sqrt{-18} = \sqrt{12}-i\sqrt{18}$$
$$= \sqrt{2^2\cdot 3}-i\sqrt{3^2\cdot 2}$$
$$= 2\sqrt{3}-3i\sqrt{2}$$

13.
$$\sqrt{160}-\sqrt{-147} = \sqrt{160}-i\sqrt{147}$$
$$= \sqrt{2^4\cdot 2\cdot 5}-i\sqrt{7^2\cdot 3}$$
$$= 4\sqrt{10}-7i\sqrt{3}$$

Objective B Exercises

15. $(2+4i)+(6-5i) = 8-i$

17. $(-2-4i)-(6-8i) = -8+4i$

19.
$$(8-\sqrt{-4})-(2+\sqrt{-16}) = (8-i\sqrt{4})-(2+i\sqrt{16})$$
$$= \left(8-i\sqrt{2^2}\right)-\left(2+i\sqrt{2^4}\right)$$
$$= (8-2i)-(2+4i)$$
$$= 6-6i$$

21.
$$(12-\sqrt{-50})+(7-\sqrt{-8})$$
$$= (12-i\sqrt{50})+(7-i\sqrt{8})$$
$$= \left(12-i\sqrt{5^2\cdot 2}\right)+\left(7-i\sqrt{2^2\cdot 2}\right)$$
$$= (12-5i\sqrt{2})+(7-2i\sqrt{2})$$
$$= 19-7i\sqrt{2}$$

23. $(\sqrt{8}+\sqrt{-18})+(\sqrt{32}-\sqrt{-72})$
$= (\sqrt{8}+i\sqrt{18})+(\sqrt{32}-i\sqrt{72})$
$= \left(\sqrt{2^2\cdot 2}+i\sqrt{3^2\cdot 2}\right)+\left(\sqrt{2^4\cdot 2}-i\sqrt{2^2\cdot 3^2\cdot 2}\right)$
$= (2\sqrt{2}+3i\sqrt{2})+(4\sqrt{2}-6i\sqrt{2})$
$= 6\sqrt{2}-3i\sqrt{2}$

Objective C Exercises

25. $(7i)(-9i) = -63i^2 = -63(-1) = 63$

27. $\sqrt{-2}\sqrt{-8} = i\sqrt{2}\cdot i\sqrt{8} = i^2\sqrt{16} = -\sqrt{2^4} = -4$

29. $\sqrt{-3}\sqrt{-6} = i\sqrt{3}\cdot i\sqrt{6}$
$= i^2\sqrt{18}$
$= -\sqrt{3^2\cdot 2} = -3\sqrt{2}$

31. $2i(6+2i) = 12i+4i^2$
$= 12i+4(-1)$
$= -4+12i$

33. $\sqrt{-2}(\sqrt{8}+\sqrt{-2}) = i\sqrt{2}(\sqrt{8}+i\sqrt{2})$
$= i\sqrt{16}+i^2\sqrt{2^2}$
$= i\sqrt{2^4}-\sqrt{2^2}$
$= 4i-2 = -2+4i$

35. $(5-2i)(3+i) = 15+5i-6i-2i^2$
$= 15-i-2i^2$
$= 15-i-2(-1)$
$= 17-i$

37. $(6+5i)(3+2i) = 18+12i+15i+10i^2$
$= 18+27i+10i^2$
$= 18+27i+10(-1)$
$= 8+27i$

39. $(1-i)\left(\frac{1}{2}+\frac{1}{2}i\right) = \frac{1}{2}+\frac{1}{2}i-\frac{1}{2}i-\frac{1}{2}i^2$
$= \frac{1}{2}-\frac{1}{2}i^2$
$= \frac{1}{2}-\frac{1}{2}(-1)$
$= \frac{1}{2}+\frac{1}{2} = 1$

41. $\left(\frac{6}{5}+\frac{3}{5}i\right)\left(\frac{2}{3}-\frac{1}{3}i\right) = \frac{4}{5}-\frac{2}{5}i+\frac{2}{5}i-\frac{1}{5}i^2$
$= \frac{4}{5}-\frac{1}{5}i^2$
$= \frac{4}{5}-\frac{1}{5}(-1)$
$= \frac{4}{5}+\frac{1}{5} = 1$

Objective D Exercises

43. $\frac{3}{i} = \frac{3}{i}\cdot\frac{i}{i} = \frac{3i}{i^2} = \frac{3i}{-1} = -3i$

45. $\frac{2-3i}{-4i} = \frac{2-3i}{-4i}\cdot\frac{i}{i}$
$= \frac{2i-3i^2}{-4i^2}$
$= \frac{2i-3(-1)}{-4(-1)}$
$= \frac{3+2i}{4} = \frac{3}{4}+\frac{1}{2}i$

47. $\frac{4}{5+i} = \frac{4}{5+i}\cdot\frac{5-i}{5-i}$
$= \frac{20-4i}{25+1}$
$= \frac{20-4i}{26}$
$= \frac{10-2i}{13} = \frac{10}{13}-\frac{2}{13}i$

49. $\frac{2}{2-i} = \frac{2}{2-i}\cdot\frac{2+i}{2+i}$
$= \frac{4+2i}{4+1}$
$= \frac{4+2i}{5}$
$= \frac{4}{5}+\frac{2}{5}i$

51. $\frac{1-3i}{3+i} = \frac{1-3i}{3+i}\cdot\frac{3-i}{3-i}$
$= \frac{3-i-9i+3i^2}{9+1}$
$= \frac{3-10i+3i^2}{10}$
$= \frac{3-10i+3(-1)}{10}$
$= \frac{-10i}{10} = -i$

53. $\dfrac{\sqrt{-10}}{\sqrt{8}-\sqrt{-2}}=\dfrac{i\sqrt{10}}{\sqrt{8}-i\sqrt{2}}$
$=\dfrac{i\sqrt{10}}{\sqrt{2^2\cdot 2}-i\sqrt{2}}$
$=\dfrac{i\sqrt{10}}{2\sqrt{2}-i\sqrt{2}}\cdot\dfrac{2\sqrt{2}+i\sqrt{2}}{2\sqrt{2}+i\sqrt{2}}$
$=\dfrac{2i\sqrt{20}+i^2\sqrt{20}}{(2\sqrt{2})^2+(\sqrt{2})^2}$
$=\dfrac{2i\sqrt{2^2\cdot 5}-\sqrt{2^2\cdot 5}}{8+2}$
$=\dfrac{2i\cdot 2\sqrt{5}-2\sqrt{5}}{10}$
$=\dfrac{-2\sqrt{5}+4i\sqrt{5}}{10}$
$=\dfrac{-\sqrt{5}+2i\sqrt{5}}{5}$
$=-\dfrac{\sqrt{5}}{5}+\dfrac{2\sqrt{5}}{5}i$

55. $\dfrac{2-3i}{3+i}=\dfrac{2-3i}{3+i}\cdot\dfrac{3-i}{3-i}$
$=\dfrac{6-2i-9i+3i^2}{9+1}$
$=\dfrac{6-11i+3i^2}{10}$
$=\dfrac{6-11i+3(-1)}{10}$
$=\dfrac{3-11i}{10}=\dfrac{3}{10}-\dfrac{11}{10}i$

57. $\dfrac{5+3i}{3-i}=\dfrac{5+3i}{3-i}\cdot\dfrac{3+i}{3+i}$
$=\dfrac{15+5i+9i+3i^2}{9+1}$
$=\dfrac{15+14i+3i^2}{10}$
$=\dfrac{15+14i+3(-1)}{10}$
$=\dfrac{12+14i}{10}$
$=\dfrac{6+7i}{5}=\dfrac{6}{5}+\dfrac{7}{5}i$

Applying the Concepts

59a. $2x^2+18=0$; plug in $3i$.
$2(3i)^2+18=2(-9)+18=0$
$0=0$
Yes, $3i$ is a solution.

b. $x^2-6x+10=0$; plug in $3+i$.
$(3+i)^2-6(3+i)+10=9+6i+i^2-18-6i+10$
$=-9+(-1)+10=0$
$0=0$
Yes, $3+i$ is a solution.

Chapter 7 Review Exercises

1. $(16x^{-4}y^{12})^{1/4}(100x^6y^{-2})^{1/2}$
$=(2^4)^{1/4}x^{-1}y^3\cdot(10^2)^{1/2}x^3y^{-1}$
$=20x^2y^2$

2. $\sqrt[4]{3x-5}=2$
$\left(\sqrt[4]{3x-5}\right)^4=2^4$
$3x-5=16$
$3x=21$
$x=7$

Check:

$\sqrt[4]{3x-5}=2$	
$\sqrt[4]{3\cdot 7-5}$	2
$\sqrt[4]{21-5}$	2
$\sqrt[4]{16}$	2
$2=2$	

The solution is 7.

3. $(6-5i)(4+3i)=24+18i-20i-15i^2$
$=24-2i-15(-1)$
$=24+15-2i$
$=39-2i$

4. $7y\sqrt[3]{x^2}=7x^{2/3}y$

5. $(\sqrt{3}+8)(\sqrt{3}-2)=\sqrt{3^2}+6\sqrt{3}-16$
$=3+6\sqrt{3}-16$
$=6\sqrt{3}-13$

6. $\sqrt{4x+9}+10=11$
$\sqrt{4x+9}=1$
$\left(\sqrt{4x+9}\right)^2=1^2$
$4x+9=1$
$4x=-8$
$x=-2$

Check:

$\sqrt{4x+9}+10=11$	
$\sqrt{4(-2)+9}+10$	11
$\sqrt{1}+10$	11
$1+10$	11
$11=11$	

The solution is -2.

7. $\dfrac{x^{-3/2}}{x^{7/2}}=x^{-10/2}=x^{-5}=\dfrac{1}{x^5}$

8. $\dfrac{8}{\sqrt{3y}}=\dfrac{8}{\sqrt{3y}}\cdot\dfrac{\sqrt{3y}}{\sqrt{3y}}=\dfrac{8\sqrt{3y}}{\sqrt{3^2y^2}}=\dfrac{8\sqrt{3y}}{3y}$

9. $\sqrt[3]{-8a^6b^{12}}=\sqrt[3]{(-2)^3a^6b^{12}}=-2a^2b^4$

10. $\sqrt{50a^4b^3}-ab\sqrt{18a^2b}$
$=\sqrt{5^2a^4b^2(2b)}-ab\sqrt{3^2a^2(2b)}$
$=5a^2b\sqrt{2b}-3a^2b\sqrt{2b}$
$=2a^2b\sqrt{2b}$

11. $\dfrac{x+2}{\sqrt{x}+\sqrt{2}}=\dfrac{x+2}{\sqrt{x}+\sqrt{2}}\cdot\dfrac{\sqrt{x}-\sqrt{2}}{\sqrt{x}-\sqrt{2}}$
$=\dfrac{x\sqrt{x}-x\sqrt{2}+2\sqrt{x}-2\sqrt{2}}{\sqrt{x^2}-\sqrt{2^2}}$
$=\dfrac{x\sqrt{x}-x\sqrt{2}+2\sqrt{x}-2\sqrt{2}}{x-2}$

12. $\dfrac{5+2i}{3i}=\dfrac{5+2i}{3i}\cdot\dfrac{-3i}{-3i}$
$=\dfrac{-15i-6i^2}{-9i^2}$
$=\dfrac{-15i-6(-1)}{-9(-1)}$
$=\dfrac{-15i+6}{9}$
$=\dfrac{6}{9}-\dfrac{15}{9}i$
$=\dfrac{2}{3}-\dfrac{5}{3}i$

13. $\sqrt{18a^3b^6}=\sqrt{3^2a^2b^6(2a)}=3ab^3\sqrt{2a}$

14. $(\sqrt{50}+\sqrt{-72})-(\sqrt{162}-\sqrt{-8})$
$=\left(\sqrt{5^2\cdot 2}+i\sqrt{6^2\cdot 2}\right)-\left(\sqrt{9^2\cdot 2}-i\sqrt{2^2\cdot 2}\right)$
$=(5\sqrt{2}+6i\sqrt{2})-(9\sqrt{2}-2i\sqrt{2})$
$=-4\sqrt{2}+8i\sqrt{2}$

15. $3x\sqrt[3]{54x^8y^{10}}-2x^2y\sqrt[3]{16x^5y^7}$
$=3x\sqrt[3]{3^3x^6y^9(2x^2y)}-2x^2y\sqrt[3]{2^3x^3y^6(2x^2y)}$
$=9x^3y^3\sqrt[3]{2x^2y}-4x^3y^3\sqrt[3]{2x^2y}$
$=5x^3y^3\sqrt[3]{2x^2y}$

16. $\sqrt[3]{16x^4y}\sqrt[3]{4xy^5}=\sqrt[3]{64x^5y^6}$
$=\sqrt[3]{4^3x^3y^6(x^2)}$
$=4xy^2\sqrt[3]{x^2}$

17. $i(3-7i)=3i-7i^2$
$=3i-7(-1)$
$=7+3i$

18. $3x^{3/4}=3\sqrt[4]{x^3}$

19. $\sqrt[5]{-64a^8b^{12}}=\sqrt[5]{(-2)^5a^5b^{10}(2a^3b^2)}$
$=-2ab^2\sqrt[5]{2a^3b^2}$

20. $\dfrac{5+9i}{1-i}=\dfrac{5+9i}{1-i}\cdot\dfrac{1+i}{1+i}$
$=\dfrac{5+14i+9i^2}{1+1}$
$=\dfrac{5+14i-9}{2}$
$=\dfrac{-4+14i}{2}$
$=-2+7i$

21. $\sqrt{-12}\sqrt{-6}=i\sqrt{12}\cdot i\sqrt{6}$
$=i^2\sqrt{72}$
$=(-1)\sqrt{6^2\cdot 2}=-6\sqrt{2}$

22. $\sqrt{x-5}+\sqrt{x+6}=11$
$\sqrt{x-5}=11-\sqrt{x+6}$
$\left(\sqrt{x-5}\right)^2=\left(11-\sqrt{x+6}\right)^2$
$x-5=121-22\sqrt{x+6}+x+6$
$-11=121-22\sqrt{x+6}$
$-132=-22\sqrt{x+6}$
$6=\sqrt{x+6}$
$6^2=\left(\sqrt{x+6}\right)^2$
$36=x+6$
$30=x$

Check:

$\sqrt{x-5}+\sqrt{x+6}=11$	
$\sqrt{30-5}+\sqrt{30+6}$	11
$\sqrt{25}+\sqrt{36}$	11
$5+6$	11
$11=11$	

The solution is 30.

23. $\sqrt[4]{81a^8b^{12}}=\sqrt[4]{3^4a^8b^{12}}=3a^2b^3$

24. $\sqrt{-50}=i\sqrt{50}=i\sqrt{5^2\cdot 2}=5i\sqrt{2}$

25. $(-8+3i)-(4-7i)=-12+10i$

26. $\left(5-\sqrt{6}\right)^2=25-10\sqrt{6}+\sqrt{6^2}$
$=25-10\sqrt{6}+6$
$=31-10\sqrt{6}$

27. $4x\sqrt{12x^2y}+\sqrt{3x^4y}-x^2\sqrt{27y}$
$=4x\sqrt{2^2x^2(3y)}+\sqrt{x^4(3y)}-x^2\sqrt{3^2(3y)}$
$=8x^2\sqrt{3y}+x^2\sqrt{3y}-3x^2\sqrt{3y}$
$=6x^2\sqrt{3y}$

28. $81^{-1/4}=\left(3^4\right)^{-1/4}=3^{-1}=\dfrac{1}{3}$

29. $\left(a^{16}\right)^{-5/8}=a^{-10}=\dfrac{1}{a^{10}}$

30. $-\sqrt{49x^6y^{16}}=-\sqrt{7^2(x^3)^2(y^8)^2}=-7x^3y^8$

31. $4a^{2/3}=4\sqrt[3]{a^2}$

32. $\sqrt[8]{b^5}=b^{5/8}$

33. $\sqrt[4]{x^6y^8z^{10}}=\sqrt[4]{x^4y^8z^8\cdot x^2z^2}=xy^2z^2\sqrt[4]{x^2z^2}$

34. $\sqrt{54}+\sqrt{24}=\sqrt{9\cdot 6}+\sqrt{4\cdot 6}=3\sqrt{6}+2\sqrt{6}=5\sqrt{6}$

35. $\sqrt{48x^5y}-x\sqrt{80x^2y}$
$=\sqrt{4^2x^4\cdot 3xy}-x\sqrt{4^2x^2\cdot 5y}$
$=4x^2\sqrt{3xy}-4x^2\sqrt{5y}$

36. $\sqrt{32}\sqrt{50}=\sqrt{1600}=\sqrt{40^2}=40$

37. $\sqrt{3x}\left(3+\sqrt{3x}\right) = 3\sqrt{3x} + 3x$ or $3x + 3\sqrt{3x}$

38. $\dfrac{\sqrt{125x^6}}{\sqrt{5x^3}} = \sqrt{\dfrac{125x^6}{5x^3}} = \sqrt{25x^3} = 5x\sqrt{x}$

39. $$\begin{aligned}\frac{\sqrt{x}+\sqrt{y}}{\sqrt{x}-\sqrt{y}} &= \frac{\sqrt{x}+\sqrt{y}}{\sqrt{x}-\sqrt{y}}\cdot\frac{\sqrt{x}+\sqrt{y}}{\sqrt{x}+\sqrt{y}} \\ &= \frac{x+2\sqrt{xy}+y}{x-y}\end{aligned}$$

40. $\sqrt{-36} = i\sqrt{36} = i\sqrt{6^2} = 6i$

41. $$\begin{aligned}\sqrt{49}-\sqrt{-16} &= \sqrt{49}-i\sqrt{16} \\ &= 7-4i\end{aligned}$$

42. $$\begin{aligned}\sqrt{200}+\sqrt{-12} &= \sqrt{100\cdot 2}-i\sqrt{12} \\ &= 10\sqrt{2}+2i\sqrt{3}\end{aligned}$$

43. $(5+2i)+(4-3i) = 9-i$

44. $$\begin{aligned}&\left(9-\sqrt{-16}\right)+\left(5+\sqrt{-36}\right) \\ &= \left(9-i\sqrt{16}\right)+\left(5+i\sqrt{36}\right) \\ &= (9-4i)+(5+6i) \\ &= 14+2i\end{aligned}$$

45. $(3-9i)-7 = -4-9i$

46. $(8i)(2i) = 16i^2 = 16(-1) = -16$

47. $\dfrac{-6}{i} = \dfrac{-6}{i}\cdot\dfrac{-i}{-i} = \dfrac{6i}{-i^2} = \dfrac{6i}{-(-1)} = 6i$

48. $\dfrac{7}{2-i} = \dfrac{7}{2-i}\cdot\dfrac{2+i}{2+i} = \dfrac{14+7i}{4+1} = \dfrac{14}{5}+\dfrac{7}{5}i$

49. $$\begin{aligned}\frac{\sqrt{16}}{\sqrt{4}-\sqrt{-4}} &= \frac{\sqrt{16}}{\sqrt{4}-i\sqrt{4}} = \frac{4}{2-2i} \\ &= \frac{2}{1-i}\cdot\frac{1+i}{1+i} \\ &= \frac{2(1+i)}{2} \\ &= 1+i\end{aligned}$$

50. $$\begin{aligned}\sqrt[3]{9x} &= -6 \\ 9x &= -216 \\ x &= -24\end{aligned}$$

Check:

$$\begin{array}{c|c} \sqrt[3]{9x} & -6 \\ \hline \sqrt[3]{9(-22)} & -6 \\ \sqrt[3]{-216} & -6 \\ -6 & -6 \end{array}$$

The solution is -24.

51. **Strategy** To find the width, use the Pythagorean Theorem. The hypotenuse is the diagonal (13) and the length is one leg (12).

Solution $$\begin{aligned}c^2 &= a^2+b^2 \\ 13^2 &= 12^2+b^2 \\ 25 &= b^2 \\ 25^{1/2} &= (b^2)^{1/2} \\ 5 &= b\end{aligned}$$

The width is 5 in.

52. **Strategy** To find the amount of power, replace v in the equation with the given value and solve for P.

Solution $$\begin{aligned}v &= 4.05\sqrt[3]{P} \\ 20 &= 4.05\sqrt[3]{P} \\ 4.94 &\approx \sqrt[3]{P} \\ (4.94)^3 &= (\sqrt[3]{P})^3 \\ 120 &\approx P\end{aligned}$$

The amount of power is 120 watts.

53. **Strategy** To find the distance required, replace v and a in the equation with the given values and solve for s.

Solution $$\begin{aligned}v &= \sqrt{2as} \\ 88 &= \sqrt{2\cdot 16s} \\ 88^2 &= (\sqrt{32s})^2 \\ 7744 &= 32s \\ 242 &= s\end{aligned}$$

The distance required is 242 ft.

54. **Strategy** To find the distance, use the Pythagorean Theorem. The hypotenuse is the length of the ladder (12 ft). One leg is the height on the building that the ladder reaches (10 ft). The distance from the bottom of the ladder to the building is the other leg.

Solution $$\begin{aligned}c^2 &= a^2+b^2 \\ 12^2 &= 10^2+b^2 \\ 144 &= 100+b^2 \\ 44 &= b^2 \\ 44^{1/2} &= (b^2)^{1/2} \\ \sqrt{44} &= b \\ 6.63 &\approx b\end{aligned}$$

The distance is 6.63 ft.

Chapter 7 Test

1. $\dfrac{1}{2}\sqrt[4]{x^3} = \dfrac{1}{2}x^{3/4}$

2. $$\begin{aligned}&\sqrt[3]{54x^7y^3} - x\sqrt[3]{128x^4y^3} - x^2\sqrt[3]{2xy^3} \\ &= \sqrt[3]{3^3x^6y^3(2x)} - x\sqrt[3]{4^3x^3y^3(2x)} - x^2\sqrt[3]{y^3(2x)} \\ &= 3x^2y\sqrt[3]{2x} - 4x^2y\sqrt[3]{2x} - x^2y\sqrt[3]{2x} \\ &= -2x^2y\sqrt[3]{2x}\end{aligned}$$

3. $3y^{2/5} = 3\sqrt[5]{y^2}$

4. $$\begin{aligned}(2+5i)(4-2i) &= 8-4i+20i-10i^2\\ &= 8+16i-10(-1)\\ &= 8+16i+10\\ &= 18+16i\end{aligned}$$

5. $$\begin{aligned}\left(2\sqrt{x}+\sqrt{y}\right)^2 &= 4\sqrt{x^2}+4\sqrt{xy}+\sqrt{y^2}\\ &= 4x+4\sqrt{xy}+y\end{aligned}$$

6. $\dfrac{r^{2/3}r^{-1}}{r^{-1/2}} = \dfrac{r^{-1/3}}{r^{-1/2}} = r^{1/6}$

7. $$\begin{aligned}\sqrt{x+12}-\sqrt{x} &= 2\\ \sqrt{x+12} &= 2+\sqrt{x}\\ \left(\sqrt{x+12}\right)^2 &= (2+\sqrt{x})^2\\ x+12 &= 4+4\sqrt{x}+x\\ 12 &= 4+4\sqrt{x}\\ 8 &= 4\sqrt{x}\\ 2 &= \sqrt{x}\\ 2^2 &= (\sqrt{x})^2\\ 4 &= x\end{aligned}$$

Check:

$$\begin{array}{r|l} \sqrt{x+12}-\sqrt{x} = & 2\\ \hline \sqrt{4+12}-\sqrt{4} & 2\\ \sqrt{16}-\sqrt{4} & 2\\ 4-2 & 2\\ 2 = & 2\end{array}$$

The solution is 4.

8. $\sqrt[3]{8x^3y^6} = \sqrt[3]{2^3x^3y^6} = 2xy^2$

9. $$\begin{aligned}\sqrt{3x}\left(\sqrt{x}-\sqrt{25x}\right) &= \sqrt{3x^2}-\sqrt{75x^2}\\ &= \sqrt{x^2(3)}-\sqrt{5^2x^2(3)}\\ &= x\sqrt{3}-5x\sqrt{3} = -4x\sqrt{3}\end{aligned}$$

10. $(5-2i)-(8-4i) = -3+2i$

11. $\sqrt{32x^4y^7} = \sqrt{2^4x^4y^6(2y)} = 4x^2y^3\sqrt{2y}$

12. $$\begin{aligned}\left(2\sqrt{3}+4\right)\left(3\sqrt{3}-1\right) &= 6\sqrt{3^2}-2\sqrt{3}+12\sqrt{3}-4\\ &= 18+10\sqrt{3}-4\\ &= 14+10\sqrt{3}\end{aligned}$$

13. $$\begin{aligned}(2+i)+(2-i)(3+2i) &= 2+i+6+4i-3i-2i^2\\ &= 8+2i-2(-1)\\ &= 8+2+2i\\ &= 10+2i\end{aligned}$$

14. $$\begin{aligned}\frac{4-2\sqrt{5}}{2-\sqrt{5}} &= \frac{4-2\sqrt{5}}{2-\sqrt{5}}\cdot\frac{2+\sqrt{5}}{2+\sqrt{5}}\\ &= \frac{8+4\sqrt{5}-4\sqrt{5}-2\sqrt{5^2}}{2^2-\sqrt{5^2}}\\ &= \frac{8-2\cdot 5}{4-5}\\ &= \frac{8-10}{-1} = \frac{-2}{-1} = 2\end{aligned}$$

15. $$\begin{aligned}\sqrt{18a^3}+a\sqrt{50a} &= \sqrt{3^2a^2(2a)}+a\sqrt{5^2(2a)}\\ &= 3a\sqrt{2a}+5a\sqrt{2a}\\ &= 8a\sqrt{2a}\end{aligned}$$

16. $$\begin{aligned}&\left(\sqrt{a}-3\sqrt{b}\right)\left(2\sqrt{a}+5\sqrt{b}\right)\\ &= 2\sqrt{a^2}+5\sqrt{ab}-6\sqrt{ab}-15\sqrt{b^2}\\ &= 2a-\sqrt{ab}-15b\end{aligned}$$

17. $\dfrac{(2x^{1/3}y^{-2/3})^6}{(x^{-4}y^8)^{1/4}} = \dfrac{2^6x^2y^{-4}}{x^{-1}y^2} = 2^6x^3y^{-6} = \dfrac{64x^3}{y^6}$

18. $$\begin{aligned}\frac{\sqrt{x}}{\sqrt{x}-\sqrt{y}} &= \frac{\sqrt{x}}{\sqrt{x}-\sqrt{y}}\cdot\frac{\sqrt{x}+\sqrt{y}}{\sqrt{x}+\sqrt{y}}\\ &= \frac{\sqrt{x^2}+\sqrt{xy}}{\sqrt{x^2}-\sqrt{y^2}}\\ &= \frac{x+\sqrt{xy}}{x-y}\end{aligned}$$

19. $$\begin{aligned}\frac{2+3i}{1-2i} &= \frac{2+3i}{1-2i}\cdot\frac{1+2i}{1+2i}\\ &= \frac{2+4i+3i+6i^2}{1+4}\\ &= \frac{2+7i+6(-1)}{5}\\ &= \frac{2-6+7i}{5}\\ &= \frac{-4+7i}{5} = -\frac{4}{5}+\frac{7}{5}i\end{aligned}$$

20. $$\begin{aligned}\sqrt[3]{2x-2}+4 &= 2\\ \sqrt[3]{2x-2} &= -2\\ (\sqrt[3]{2x-2})^3 &= (-2)^3\\ 2x-2 &= -8\\ 2x &= -6\\ x &= -3\end{aligned}$$

Check:

$$\begin{array}{r|l} \sqrt[3]{2x-2}+4 = & 2\\ \hline \sqrt[3]{2(-3)-2}+4 & 2\\ \sqrt[3]{-8}+4 & 2\\ -2+4 & 2\\ 2 = & 2\end{array}$$

The solution is −3.

21. $$\begin{aligned}\left(\frac{4a^4}{b^2}\right)^{-3/2} &= \frac{4^{-3/2}a^{-6}}{b^{-3}}\\ &= (2^2)^{-3/2}a^{-6}b^3\\ &= 2^{-3}a^{-6}b^3\\ &= \frac{b^3}{8a^6}\end{aligned}$$

22. $\sqrt[3]{27a^4b^3c^7} = \sqrt[3]{3^3a^3b^3c^6(ac)} = 3abc^2\sqrt[3]{ac}$

23. $\dfrac{\sqrt{32x^5y}}{\sqrt{2xy^3}} = \sqrt{\dfrac{32x^5y}{2xy^3}} = \sqrt{\dfrac{16x^4}{y^2}} = \sqrt{\dfrac{4^2x^4}{y^2}} = \dfrac{4x^2}{y}$

24. $$\begin{aligned}\left(\sqrt{-8}\right)\left(\sqrt{-2}\right) &= i\sqrt{8}\cdot i\sqrt{2} = i^2\sqrt{16}\\ &= -1\cdot 4 = -4\end{aligned}$$

25. Strategy To find the distance, replace v in the formula and solve for d.

Solution
$$v = \sqrt{64d}$$
$$192 = \sqrt{64d}$$
$$(192)^2 = \left(\sqrt{64d}\right)^2$$
$$36864 = 64d$$
$$576 = d$$
The distance is 576 ft.

Cumulative Review Exercises

1. The Distributive Property

2. $f(-3) = 3(-3)^2 - 2(-3) + 1$
$f(-3) = 3 \cdot 9 + 6 + 1$
$f(-3) = 34$

3. $$5 - \frac{2}{3}x = 4$$
$$5 - \frac{2}{3}x - 5 = 4 - 5$$
$$-\frac{2}{3}x = -1$$
$$\left(-\frac{3}{2}\right)\left(-\frac{2}{3}\right)x = -1\left(-\frac{3}{2}\right)$$
$$x = \frac{3}{2}$$
The solution is $\frac{3}{2}$.

4. $$2[4 - 2(3 - 2x)] = 4(1 - x)$$
$$2[4 - 6 + 4x] = 4 - 4x$$
$$2[-2 + 4x] = 4 - 4x$$
$$-4 + 8x = 4 - 4x$$
$$-4 + 8x + 4x = 4 - 4x + 4x$$
$$12x - 4 = 4$$
$$12x - 4 + 4 = 4 + 4$$
$$12x = 8$$
$$\left(\frac{1}{12}\right)12x = \frac{1}{12}(8)$$
$$x = \frac{2}{3}$$
The solution is $\frac{2}{3}$.

5. $$2 + |4 - 3x| = 5$$
$$|4 - 3x| = 3$$
$4 - 3x = 3$ $\quad$ $4 - 3x = -3$
$-3x = -1$ $\quad$ $-3x = -7$
$x = \frac{1}{3}$ $\quad$ $x = \frac{7}{3}$
The solutions are $\frac{1}{3}$ and $\frac{7}{3}$.

6. $$6x - 3(2x + 2) > 3 - 3(x + 2)$$
$$6x - 6x - 6 > 3 - 3x - 6$$
$$-6 > -3 - 3x$$
$$-6 + 3x > -3 - 3x + 3x$$
$$3x - 6 > -3$$
$$3x > 3$$
$$\left(\frac{1}{3}\right)3x > \frac{1}{3}(3)$$
$$x > 1$$
$$\{x|x > 1\}$$

7. $$|2x + 3| \le 9$$
$$-9 \le 2x + 3 \le 9$$
$$-9 - 3 \le 2x + 3 - 3 \le 9 - 3$$
$$-12 \le 2x \le 6$$
$$\frac{1}{2}(-12) \le \frac{1}{2}(2x) \le \frac{1}{2}(6)$$
$$-6 \le x \le 3$$
$$\{x| -6 \le x \le 3\}$$

8. $81x^2 - y^2 = (9x + y)(9x - y)$

9. $$x^5 + 2x^3 - 3x = x(x^4 + 2x^2 - 3)$$
$$= x(x^2 + 3)(x^2 - 1)$$
$$= x(x^2 + 3)(x + 1)(x - 1)$$

10. First find the slope of the line.
$$m = \frac{y_2 - y_1}{x_2 - x_1} = \frac{2 - 3}{-1 - 2} = \frac{-1}{-3} = \frac{1}{3}$$
Use the point-slope form to find the equation of the line.
$$y - y_1 = m(x - x_1)$$
$$y - 3 = \frac{1}{3}(x - 2)$$
$$y - 3 = \frac{1}{3}x - \frac{2}{3}$$
$$y = \frac{1}{3}x + \frac{7}{3}$$
The equation of the line is $y = \frac{1}{3}x + \frac{7}{3}$.

11. $$\begin{vmatrix} 1 & 2 & -3 \\ 0 & -1 & 2 \\ 3 & 1 & -2 \end{vmatrix} = 1 \cdot \begin{vmatrix} -1 & 2 \\ 1 & -2 \end{vmatrix} - 2\begin{vmatrix} 0 & 2 \\ 3 & -2 \end{vmatrix} - 3\begin{vmatrix} 0 & -1 \\ 3 & 1 \end{vmatrix}$$
$$= 1 \cdot 0 - 2(-6) - 3 \cdot 3$$
$$= 3$$

12. $$P = \frac{R - C}{n}$$
$$P \cdot n = \frac{R - C}{n} \cdot n$$
$$nP = R - C$$
$$nP + C = R$$
$$C = R - nP$$

13. $$(2^{-1}x^2y^{-6})(2^{-1}y^{-4})^{-2} = (2^{-1}x^2y^{-6})(2^2y^8)$$
$$= 2^{-1+2}x^2y^{-6+8}$$
$$= 2^1x^2y^2$$
$$= 2x^2y^2$$

14. $\dfrac{x^2y^3}{x^2+2x-8}\cdot\dfrac{2x^2-7x+6}{xy^4}$

$=\dfrac{x^2y^3}{(x+4)(x-2)}\cdot\dfrac{(2x-3)(x-2)}{xy^4}$

$=\dfrac{x^2y^3(2x-3)(x-2)}{(x+4)(x-2)xy^4}$

$=\dfrac{x(2x-3)}{y(x+4)}$

15. $\sqrt{40x^3}-x\sqrt{90x}=\sqrt{2^2x^2(10x)}-x\sqrt{3^2(10x)}$

$=2x\sqrt{10x}-3x\sqrt{10x}$

$=-x\sqrt{10x}$

16. $\dfrac{x}{x-2}-2x=\dfrac{-3}{x-2}$

$(x-2)\left(\dfrac{x}{x-2}-2x\right)=(x-2)\left(\dfrac{-3}{x-2}\right)$

$x-(x-2)(2x)=-3$

$x-2x^2+4x=-3$

$-2x^2+5x+3=0$

$2x^2-5x-3=0$

$(2x+1)(x-3)=0$

$2x+1=0 \qquad x-3=0$

$2x=-1 \qquad x=3$

$x=-\dfrac{1}{2}$

The solutions are $-\dfrac{1}{2}$ and 3.

17. Find the y-intercept at $x=0$.

$3(0)-2y=-6$

$y=3$

The y-intercept is (0, 3).

To find the slope, find the x-intercept and use it to get the slope.

$3x-2(0)=-6$

$x=-2$

The x-intercept is $(-2, 0)$.

$m=\dfrac{y_2-y_1}{x_2-x_1}$

$m=\dfrac{3-0}{0-(-2)}=\dfrac{3}{2}$

The slope is $\dfrac{3}{2}$, and the y-intercept is (0, 3).

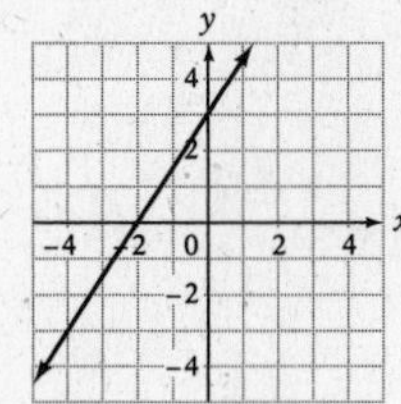

18. First graph the line $3x+2y=4$:

Find the y-intercept at $x=0$.

$3(0)+2y=4$

$y=2$

The y-intercept is (0, 2).

Find the x-intercept at $y=0$.

$3x+2(0)=4$

$x=\dfrac{4}{3}$

The x-intercept is $\left(\dfrac{4}{3}, 0\right)$.

Determine the shading: (0, 0) is a solution, so shade below the line.

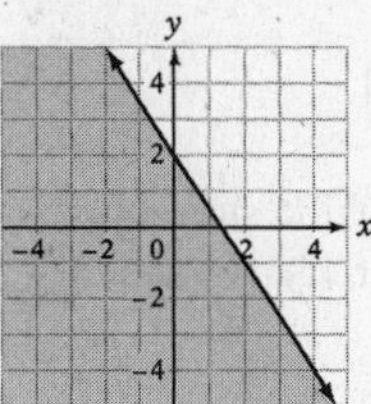

19. $\dfrac{2i}{3-i}=\dfrac{2i}{3-i}\cdot\dfrac{3+i}{3+i}$

$=\dfrac{6i+2i^2}{9+1}$

$=\dfrac{6i+2(-1)}{10}$

$=\dfrac{-2+6i}{10}=-\dfrac{1}{5}+\dfrac{3}{5}i$

20. $\sqrt[3]{3x-4}+5=1$

$\sqrt[3]{3x-4}=-4$

$\left(\sqrt[3]{3x-4}\right)^3=(-4)^3$

$3x-4=-64$

$3x=-60$

$x=-20$

Check:

$\sqrt[3]{3x-4}+5=1$	
$\sqrt[3]{3(-20)-4}+5$	1
$\sqrt[3]{-64}+5$	1
$-4+5$	1
$1=1$	

The solution is -20.

21. The LCM is $(2x-3)(x+4)$.

$\dfrac{x}{2x-3}+\dfrac{4}{x+4}=\dfrac{x}{2x-3}\cdot\dfrac{x+4}{x+4}+\dfrac{4}{x+4}\cdot\dfrac{2x-3}{2x-3}$

$=\dfrac{x(x+4)+4(2x-3)}{(2x-3)(x+4)}$

$=\dfrac{x^2+4x+8x-12}{(2x-3)(x+4)}$

$=\dfrac{x^2+12x-12}{(2x-3)(x+4)}$

22. $2x - y = 4$
$-2x + 3y = 5$

$$D = \begin{vmatrix} 2 & -1 \\ -2 & 3 \end{vmatrix} = 4$$

$$D_x = \begin{vmatrix} 4 & -1 \\ 5 & 3 \end{vmatrix} = 17$$

$$D_y = \begin{vmatrix} 2 & 4 \\ -2 & 5 \end{vmatrix} = 18$$

$$x = \frac{D_x}{D} = \frac{17}{4}$$

$$y = \frac{D_y}{D} = \frac{18}{4} = \frac{9}{2}$$

The solution is $\left(\frac{17}{4}, \frac{9}{2}\right)$.

23. Strategy • Number of 18¢ stamps: x
Number of 13¢ stamps: $30 - x$

Stamps	Number	Value	Total Value
18¢	x	18	$18x$
13¢	$30 - x$	13	$13(30 - x)$

• The sum of the total values of each type of stamp equals the total value of the stamps (485¢).
$18x + 13(30 - x) = 485$

Solution

$$18x + 13(30 - x) = 485$$
$$18x + 390 - 13x = 485$$
$$5x + 390 = 485$$
$$5x = 95$$
$$x = 19$$

There are nineteen 18¢ stamps.

24. Strategy • Unknown rate of the car: x
Unknown rate of the plane: $5x$

	Distance	Rate	Time
Car	25	x	$\frac{25}{x}$
Plane	625	$5x$	$\frac{625}{5x}$

• The total time of the trip was 3 h.

$$\frac{25}{x} + \frac{625}{5x} = 3$$

Solution

$$5x\left(\frac{25}{x} + \frac{625}{5x}\right) = 3(5x)$$
$$125 + 625 = 15x$$
$$750 = 15x$$
$$50 = x$$
$$250 = 5x$$

The rate of the plane is 250 mph.

25. Strategy • To find the time it takes light to travel from the Earth to the moon, use the formula $RT = D$, substituting for R and D and solving for T.

Solution

$$RT = D$$
$$1.86 \times 10^5 \cdot T = 232{,}500$$
$$1.86 \times 10^5 \cdot T = 2.325 \times 10^5$$
$$T = 1.25 \times 10^0$$
$$T = 1.25$$

The time is 1.25 s.

26. Strategy • To find the height of the periscope, replace d in the given equation and solve for h.

Solution

$$d = \sqrt{1.5h}$$
$$7 = \sqrt{1.5h}$$
$$7^2 = \left(\sqrt{1.5h}\right)^2$$
$$49 = 1.5h$$
$$32.7 \approx h$$

The height of the periscope is 32.7 ft.

27. Slope $m = \frac{y_2 - y_1}{x_2 - x_1} = \frac{400 - 0}{5000 - 0} = \frac{400}{5000} = 0.08$
The annual income is 8% of the investment.

Chapter 8: Quadratic Equations

Prep Test

1. $\sqrt{18} = \sqrt{9 \cdot 2} = 3\sqrt{2}$

2. $\sqrt{-9}$; Not a real number

3. $\frac{3x-2}{x-1} - \frac{1}{1} = \frac{3x-2}{x-1} - \frac{x-1}{x-1} = \frac{3x-2-x+1}{x-1}$
$= \frac{2x-1}{x-1}$

4. $(-4)^2 - 4(2)(1)$
$= 16 - 8$
$= 8$

5. Because $4x^2$ is a perfect square and 49 is a perfect square, we will try factoring the trinomial as a square of a binomial:
$4x^2 + 28x + 49 \stackrel{?}{=} (2x+7)^2$
Check: $(2x+7)^2 = (2x+7)(2x+7)$
$= 4x^2 + 14x + 14x + 49$
$= 4x^2 + 28x + 49$
The check verifies that
$4x^2 + 28x + 49 = (2x+7)^2$. Yes

6. $4x^2 - 4x + 1 = (2x-1)(2x-1) = (2x-1)^2$

7. $9x^2 - 4 = (3x+2)(3x-2)$

8. –5 –4 –3 –2 –1 0 1 2 3 4 5

9. $x(x-1) = x + 15$
$x^2 - x = x + 15$
$x^2 - 2x - 15 = 0$
$(x-5)(x+3) = 0$

$x - 5 = 0 \quad x + 3 = 0$
$x = 5 \quad x = -3$
The solutions are −3 and 5.

10. $x(x-3) \cdot \frac{4}{x-3} = \frac{16}{x} \cdot x(x-3)$

$4x = 16(x-3)$
$4x = 16x - 48$

$\left(-\frac{1}{12}\right) - 12x = -48\left(-\frac{1}{12}\right)$
$x = 4$

The solution is 4.

Go Figure

Answer: 9

Section 8.1

Objective A Exercises

1. If $a = 0$ in the equation $ax^2 + bx + c = 0$, then there is no second-degree term in the equation, and it is therefore not a quadratic equation.

3. $2x^2 - 4x - 5 = 0$; $a = 2$, $b = -4$, $c = -5$

5. $4x^2 - 5x + 6 = 0$; $a = 4$, $b = -5$, $c = 6$

7. $x^2 - 4x = 0$
$x(x-4) = 0$
$x = 0 \quad x - 4 = 0$
$x = 4$

The solutions are 0 and 4.

9. $t^2 - 25 = 0$
$(t-5)(t+5) = 0$
$t - 5 = 0 \quad t + 5 = 0$
$t = 5 \quad t = -5$

The solutions are −5 and 5.

11. $s^2 - s - 6 = 0$
$(s-3)(s+2) = 0$
$s - 3 = 0 \quad s + 2 = 0$
$s = 3 \quad s = -2$

The solutions are −2 and 3.

13. $y^2 - 6y + 9 = 0$
$(y-3)(y-3) = 0$
$y - 3 = 0 \quad y - 3 = 0$
$y = 3 \quad y = 3$

The solution is 3.

15. $9z^2 - 18z = 0$
$9z(z-2) = 0$
$9z = 0 \quad z - 2 = 0$
$z = 0 \quad z = 2$
The solutions are 0 and 2.

17. $r^2 - 3r = 10$
$r^2 - 3r - 10 = 0$
$(r-5)(r+2) = 0$
$r - 5 = 0 \quad r + 2 = 0$
$r = 5 \quad r = -2$
The solutions are −2 and 5.

19. $v^2 + 10 = 7v$
$v^2 - 7v + 10 = 0$
$(v-2)(v-5) = 0$
$v - 2 = 0 \quad v - 5 = 0$
$v = 2 \quad v = 5$
The solutions are 2 and 5.

21. $2x^2 - 9x - 18 = 0$
$(x-6)(2x+3) = 0$
$x - 6 = 0 \quad 2x + 3 = 0$
$x = 6 \quad 2x = -3$
$x = -\frac{3}{2}$

The solutions are $-\frac{3}{2}$ and 6.

23. $4z^2 - 9z + 2 = 0$
$(z-2)(4z-1) = 0$
$z - 2 = 0 \quad 4z - 1 = 0$
$z = 2 \quad 4z = 1$
$z = \frac{1}{4}$

The solutions are $\frac{1}{4}$ and 2.

25. $3w^2 + 11w = 4$
$3w^2 + 11w - 4 = 0$
$(3w-1)(w+4) = 0$
$3w - 1 = 0 \quad w + 4 = 0$
$3w = 1 \quad w = -4$
$w = \frac{1}{3}$

The solutions are -4 and $\frac{1}{3}$.

27. $6x^2 = 23x + 18$
$6x^2 - 23x - 18 = 0$
$(2x-9)(3x+2) = 0$
$2x - 9 = 0 \quad 3x + 2 = 0$
$2x = 9 \quad 3x = -2$
$x = \frac{9}{2} \quad x = -\frac{2}{3}$

The solutions are $-\frac{2}{3}$ and $\frac{9}{2}$.

29. $4 - 15u - 4u^2 = 0$
$(1-4u)(4+u) = 0$
$1 - 4u = 0 \quad 4 + u = 0$
$-4u = -1 \quad u = -4$
$u = \frac{1}{4}$

The solutions are -4 and $\frac{1}{4}$.

31. $x + 18 = x(x-6)$
$x + 18 = x^2 - 6x$
$0 = x^2 - 7x - 18$
$0 = (x-9)(x+2)$
$x - 9 = 0 \quad x + 2 = 0$
$x = 9 \quad x = -2$

The solutions are -2 and 9.

33. $4s(s+3) = s - 6$
$4s^2 + 12s = s - 6$
$4s^2 + 11s + 6 = 0$
$(s+2)(4s+3) = 0$
$s + 2 = 0 \quad 4s + 3 = 0$
$s = -2 \quad 4s = -3$
$s = -\frac{3}{4}$

The solutions are -2 and $-\frac{3}{4}$.

35. $u^2 - 2u + 4 = (2u-3)(u+2)$
$u^2 - 2u + 4 = 2u^2 + u - 6$
$0 = u^2 + 3u - 10$
$0 = (u-2)(u+5)$
$u - 2 = 0 \quad u + 5 = 0$
$u = 2 \quad u = -5$

The solutions are -5 and 2.

37. $(3x-4)(x+4) = x^2 - 3x - 28$
$3x^2 + 8x - 16 = x^2 - 3x - 28$
$2x^2 + 11x + 12 = 0$
$(x+4)(2x+3) = 0$
$x + 4 = 0 \quad 2x + 3 = 0$
$x = -4 \quad 2x = -3$
$x = -\frac{3}{2}$

The solutions are -4 and $-\frac{3}{2}$.

39. $x^2 - 9bx + 14b^2 = 0$
$(x-2b)(x-7b) = 0$
$x - 2b = 0 \quad x - 7b = 0$
$x = 2b \quad x = 7b$

The solutions are $2b$ and $7b$.

41. $x^2 - 6cx - 7c^2 = 0$
$(x-7c)(x+c) = 0$
$x - 7c = 0 \quad x + c = 0$
$x = 7c \quad x = -c$

The solutions are $-c$ and $7c$.

43. $2x^2 + 3bx + b^2 = 0$
$(2x+b)(x+b) = 0$
$2x + b = 0 \quad x + b = 0$
$2x = -b \quad x = -b$
$x = -\frac{b}{2}$

The solutions are $-b$ and $-\frac{b}{2}$.

45. $3x^2 - 14ax + 8a^2 = 0$
$(x - 4a)(3x - 2a) = 0$
$x - 4a = 0 \quad 3x - 2a = 0$
$x = 4a \quad 3x = 2a$
$x = \frac{2a}{3}$
The solutions are $\frac{2a}{3}$ and $4a$.

47. $3x^2 - 8ax - 3a^2 = 0$
$(3x + a)(x - 3a) = 0$
$3x + a = 0 \quad x - 3a = 0$
$3x = -a \quad x = 3a$
$x = -\frac{a}{3}$
The solutions are $-\frac{a}{3}$ and $3a$.

49. $4x^2 + 8xy + 3y^2 = 0$
$(2x + 3y)(2x + y) = 0$
$2x + 3y = 0 \quad 2x + y = 0$
$2x = -3y \quad 2x = -y$
$x = -\frac{3y}{2} \quad x = -\frac{y}{2}$
The solutions are $-\frac{3y}{2}$ and $-\frac{y}{2}$.

51. $6x^2 + 11ax + 4a^2 = 0$
$(2x + a)(3x + 4a) = 0$
$2x + a = 0 \quad 3x + 4a = 0$
$2x = -a \quad 3x = -4a$
$x = -\frac{a}{2} \quad x = -\frac{4a}{3}$
The solutions are $-\frac{4a}{3}$ and $-\frac{a}{2}$.

Objective B Exercises

53. $(x - r_1)(x - r_2) = 0$
$(x - 2)(x - 5) = 0$
$x^2 - 7x + 10 = 0$

55. $(x - r_1)(x - r_2) = 0$
$[x - (-2)][x - (-4)] = 0$
$(x + 2)(x + 4) = 0$
$x^2 + 6x + 8 = 0$

57. $(x - r_1)(x - r_2) = 0$
$(x - 6)[x - (-1)] = 0$
$(x - 6)(x + 1) = 0$
$x^2 - 5x - 6 = 0$

59. $(x - r_1)(x - r_2) = 0$
$(x - 3)[x - (-3)] = 0$
$(x - 3)(x + 3) = 0$
$x^2 - 9 = 0$

61. $(x - r_1)(x - r_2) = 0$
$(x - 4)(x - 4) = 0$
$x^2 - 8x + 16 = 0$

63. $(x - r_1)(x - r_2) = 0$
$(x - 0)(x - 5) = 0$
$x(x - 5) = 0$
$x^2 - 5x = 0$

65. $(x - r_1)(x - r_2) = 0$
$(x - 0)(x - 3) = 0$
$x(x - 3) = 0$
$x^2 - 3x = 0$

67. $(x - r_1)(x - r_2) = 0$
$(x - 3)\left(x - \frac{1}{2}\right) = 0$
$x^2 - \frac{7}{2}x + \frac{3}{2} = 0$
$2\left(x^2 - \frac{7}{2}x + \frac{3}{2}\right) = 2 \cdot 0$
$2x^2 - 7x + 3 = 0$

69. $(x - r_1)(x - r_2) = 0$
$\left[x - \left(-\frac{3}{4}\right)\right](x - 2) = 0$
$\left(x + \frac{3}{4}\right)(x - 2) = 0$
$x^2 - \frac{5}{4}x - \frac{3}{2} = 0$
$4\left(x^2 - \frac{5}{4}x - \frac{3}{2}\right) = 4 \cdot 0$
$4x^2 - 5x - 6 = 0$

71. $(x - r_1)(x - r_2) = 0$
$\left[x - \left(-\frac{5}{3}\right)\right][x - (-2)] = 0$
$\left(x + \frac{5}{3}\right)(x + 2) = 0$
$x^2 + \frac{11}{3}x + \frac{10}{3} = 0$
$3\left(x^2 + \frac{11}{3}x + \frac{10}{3}\right) = 3 \cdot 0$
$3x^2 + 11x + 10 = 0$

73. $(x - r_1)(x - r_2) = 0$
$\left[x - \left(-\frac{2}{3}\right)\right]\left(x - \frac{2}{3}\right) = 0$
$\left(x + \frac{2}{3}\right)\left(x - \frac{2}{3}\right) = 0$
$x^2 - \frac{4}{9} = 0$
$9\left(x^2 - \frac{4}{9}\right) = 9 \cdot 0$
$9x^2 - 4 = 0$

75.
$$(x-r_1)(x-r_2)=0$$
$$\left(x-\frac{1}{2}\right)\left(x-\frac{1}{3}\right)=0$$
$$x^2-\frac{5}{6}x+\frac{1}{6}=0$$
$$6\left(x^2-\frac{5}{6}x+\frac{1}{6}\right)=6\cdot 0$$
$$6x^2-5x+1=0$$

77.
$$(x-r_1)(x-r_2)=0$$
$$\left(x-\frac{6}{5}\right)\left[x-\left(-\frac{1}{2}\right)\right]=0$$
$$\left(x-\frac{6}{5}\right)\left(x+\frac{1}{2}\right)=0$$
$$x^2-\frac{7}{10}x-\frac{3}{5}=0$$
$$10\left(x^2-\frac{7}{10}x-\frac{3}{5}\right)=10\cdot 0$$
$$10x^2-7x-6=0$$

79.
$$(x-r_1)(x-r_2)=0$$
$$\left[x-\left(-\frac{1}{4}\right)\right]\left[x-\left(-\frac{1}{2}\right)\right]=0$$
$$\left(x+\frac{1}{4}\right)\left(x+\frac{1}{2}\right)=0$$
$$x^2+\frac{3}{4}x+\frac{1}{8}=0$$
$$8\left(x^2+\frac{3}{4}x+\frac{1}{8}\right)=8\cdot 0$$
$$8x^2+6x+1=0$$

81.
$$(x-r_1)(x-r_2)=0$$
$$\left(x-\frac{3}{5}\right)\left[x-\left(-\frac{1}{10}\right)\right]=0$$
$$\left(x-\frac{3}{5}\right)\left(x+\frac{1}{10}\right)=0$$
$$x^2-\frac{1}{2}x-\frac{3}{50}=0$$
$$50\left(x^2-\frac{1}{2}x-\frac{3}{50}\right)=50\cdot 0$$
$$50x^2-25x-3=0$$

Objective C Exercises

83.
$$y^2=49$$
$$\sqrt{y^2}=\sqrt{49}$$
$$y=\pm\sqrt{49}=\pm 7$$
The solutions are -7 and 7.

85.
$$z^2=-4$$
$$\sqrt{z^2}=\sqrt{-4}$$
$$z=\pm\sqrt{-4}=\pm 2i$$
The solutions are $-2i$ and $2i$.

87.
$$s^2-4=0$$
$$s^2=4$$
$$\sqrt{s^2}=\sqrt{4}$$
$$s=\pm\sqrt{4}=\pm 2$$
The solutions are -2 and 2.

89.
$$4x^2-81=0$$
$$4x^2=81$$
$$x^2=\frac{81}{4}$$
$$\sqrt{x^2}=\sqrt{\frac{81}{4}}$$
$$x=\pm\sqrt{\frac{81}{4}}=\pm\frac{9}{2}$$
The solutions are $-\frac{9}{2}$ and $\frac{9}{2}$.

91.
$$y^2+49=0$$
$$y^2=-49$$
$$\sqrt{y^2}=\sqrt{-49}$$
$$y=\pm\sqrt{-49}=\pm 7i$$
The solutions are $-7i$ and $7i$.

93.
$$v^2-48=0$$
$$v^2=48$$
$$\sqrt{v^2}=\sqrt{48}$$
$$v=\pm\sqrt{48}=\pm 4\sqrt{3}$$
The solutions are $-4\sqrt{3}$ and $4\sqrt{3}$.

95.
$$r^2-75=0$$
$$r^2=75$$
$$\sqrt{r^2}=\sqrt{75}$$
$$r=\pm\sqrt{75}=\pm 5\sqrt{3}$$
The solutions are $-5\sqrt{3}$ and $5\sqrt{3}$.

97.
$$z^2+18=0$$
$$z^2=-18$$
$$\sqrt{z^2}=\sqrt{-18}$$
$$z=\pm\sqrt{-18}=\pm 3i\sqrt{2}$$
The solutions are $-3i\sqrt{2}$ and $3i\sqrt{2}$.

99.
$$(x-1)^2=36$$
$$\sqrt{(x-1)^2}=\sqrt{36}$$
$$x-1=\pm\sqrt{36}=\pm 6$$
$$x-1=6 \quad x-1=-6$$
$$x=7 \quad\quad x=-5$$
The solutions are -5 and 7.

101.
$$3(y+3)^2=27$$
$$(y+3)^2=9$$
$$\sqrt{(y+3)^2}=\sqrt{9}$$
$$y+3=\pm\sqrt{9}=\pm 3$$
$$y+3=3 \quad y+3=-3$$
$$y=0 \quad\quad y=-6$$
The solutions are -6 and 0.

103. $5(z+2)^2 = 125$
$(z+2)^2 = 25$
$\sqrt{(z+2)^2} = \sqrt{25}$
$z+2 = \pm\sqrt{25} = \pm 5$
$z+2 = 5 \quad z+2 = -5$
$z = 3 \quad z = -7$
The solutions are -7 and 3.

105. $\left(v - \frac{1}{2}\right)^2 = \frac{1}{4}$
$\sqrt{\left(v - \frac{1}{2}\right)^2} = \sqrt{\frac{1}{4}}$
$v - \frac{1}{2} = \pm\sqrt{\frac{1}{4}} = \pm\frac{1}{2}$
$v - \frac{1}{2} = \frac{1}{2} \quad v - \frac{1}{2} = -\frac{1}{2}$
$v = 1 \quad v = 0$
The solutions are 0 and 1.

107. $(x+5)^2 - 6 = 0$
$(x+5)^2 = 6$
$\sqrt{(x+5)^2} = \sqrt{6}$
$x+5 = \pm\sqrt{6}$
$x+5 = \sqrt{6} \quad x+5 = -\sqrt{6}$
$x = -5+\sqrt{6} \quad x = -5-\sqrt{6}$
The solutions are $-5-\sqrt{6}$ and $-5+\sqrt{6}$.

109. $(v-3)^2 + 45 = 0$
$(v-3)^2 = -45$
$\sqrt{(v-3)^2} = \sqrt{-45}$
$v-3 = \pm\sqrt{-45} = \pm 3i\sqrt{5}$
$v-3 = 3i\sqrt{5} \quad v-3 = -3i\sqrt{5}$
$v = 3+3i\sqrt{5} \quad v = 3-3i\sqrt{5}$
The solutions are $3-3i\sqrt{5}$ and $3+3i\sqrt{5}$.

111. $\left(u+\frac{2}{3}\right)^2 - 18 = 0$
$\left(u+\frac{2}{3}\right)^2 = 18$
$\sqrt{\left(u+\frac{2}{3}\right)^2} = \sqrt{18}$
$u+\frac{2}{3} = \pm\sqrt{18} = \pm 3\sqrt{2}$
$u+\frac{2}{3} = 3\sqrt{2} \quad u+\frac{2}{3} = -3\sqrt{2}$
$u = -\frac{2}{3}+3\sqrt{2} \quad u = -\frac{2}{3}-3\sqrt{2}$
$u = -\frac{2-9\sqrt{2}}{3} \quad u = -\frac{2+9\sqrt{2}}{3}$
The solutions are $-\frac{2-9\sqrt{2}}{3}$ and $-\frac{2+9\sqrt{2}}{3}$.

Applying the Concepts

113. $(x-r_1)(x-r_2) = 0$
$(x-\sqrt{2})[x-(-\sqrt{2})] = 0$
$(x-\sqrt{2})(x+\sqrt{2}) = 0$
$x^2 - 2 = 0$

115. $(x-r_1)(x-r_2) = 0$
$(x-i)[x-(-i)] = 0$
$(x-i)(x+i) = 0$
$x^2 + 1 = 0$

117. $(x-r_1)(x-r_2) = 0$
$(x-2\sqrt{2})[x-(-2\sqrt{2})] = 0$
$(x-2\sqrt{2})(x+2\sqrt{2}) = 0$
$x^2 - 8 = 0$

119. $(x-r_1)(x-r_2) = 0$
$(x-i\sqrt{2})[x-(-i\sqrt{2})] = 0$
$(x-i\sqrt{2})(x+i\sqrt{2}) = 0$
$x^2 + 2 = 0$

121. $4a^2x^2 = 36b^2$
$x^2 = \frac{36b^2}{4a^2}$
$x^2 = \frac{9b^2}{a^2}$
$\sqrt{x^2} = \sqrt{\frac{9b^2}{a^2}}$
$x = \pm\sqrt{\frac{9b^2}{a^2}} = \pm\frac{3b}{a}$
Since $a > 0$ and $b > 0$, the solutions are $-\frac{3b}{a}$ and $\frac{3b}{a}$.

123. $(x+a)^2 - 4 = 0$
$(x+a)^2 = 4$
$\sqrt{(x+a)^2} = \sqrt{4}$
$x+a = \pm\sqrt{4} = \pm 2$
$x+a = 2 \quad x+a = -2$
$x = -a+2 \quad x = -a-2$
The solutions are $-a-2$ and $-a+2$.

125. $(2x-1)^2 = (2x+3)^2$
$\sqrt{(2x-1)^2} = \sqrt{(2x+3)^2}$
$2x-1 = \pm\sqrt{(2x+3)^2} = \pm(2x+3)$
$2x-1 = 2x+3 \quad 2x-1 = -(2x+3)$
$-1 = 3 \quad 4x = -2$
$x = -\frac{1}{2}$
The solution is $-\frac{1}{2}$.

127. **a.** $ax^2 + bx = 0,\ a > 0,\ b > 0$
$$x(ax+b) = 0$$
$$x = 0 \quad ax + b = 0$$
$$ax = -b$$
$$x = -\frac{b}{a}$$
The solutions are $-\frac{b}{a}$ and 0.

b. $ax^2 + c = 0,\ a > 0,\ c > 0$
$$ax^2 = -c$$
$$x^2 = -\frac{c}{a}$$
$$\sqrt{x^2} = \sqrt{-\frac{c}{a}}$$
$$x = \pm i\sqrt{\frac{c}{a}}$$
$$x = \pm i\sqrt{\frac{c}{a}\cdot\frac{a}{a}}$$
$$x = \pm i\sqrt{\frac{ca}{a^2}}$$
$$x = \pm i\frac{\sqrt{ca}}{a}$$
$$x = \pm\frac{\sqrt{ca}}{a}i$$
The solutions are $-\frac{\sqrt{ca}}{a}i$ and $\frac{\sqrt{ca}}{a}i$.

Section 8.2

Objective A Exercises

1. $x^2 - 4x - 5 = 0$
$$x^2 - 4x = 5$$
Complete the square.
$$x^2 - 4x + 4 = 5 + 4$$
$$(x-2)^2 = 9$$
$$\sqrt{(x-2)^2} = \sqrt{9}$$
$$x - 2 = \pm\sqrt{9} = \pm 3$$
$$x - 2 = 3 \quad x - 2 = -3$$
$$x = 5 \quad x = -1$$
The solutions are -1 and 5.

3. $v^2 + 8v - 9 = 0$
$$v^2 + 8v = 9$$
Complete the square.
$$v^2 + 8v + 16 = 9 + 16$$
$$(v+4)^2 = 25$$
$$\sqrt{(v+4)^2} = \sqrt{25}$$
$$v + 4 = \pm\sqrt{25} = \pm 5$$
$$v + 4 = 5 \quad v + 4 = -5$$
$$v = 1 \quad v = -9$$
The solutions are -9 and 1.

5. $z^2 - 6z + 9 = 0$
$$z^2 - 6z = -9$$
Complete the square.
$$z^2 - 6z + 9 = -9 + 9$$
$$(z-3)^2 = 0$$
$$\sqrt{(z-3)^2} = \sqrt{0}$$
$$z - 3 = 0$$
$$z = 3$$
The solution is 3.

7. $r^2 + 4r - 7 = 0$
$$r^2 + 4r = 7$$
Complete the square.
$$r^2 + 4r + 4 = 7 + 4$$
$$(r+2)^2 = 11$$
$$\sqrt{(r+2)^2} = \sqrt{11}$$
$$r + 2 = \pm\sqrt{11}$$
$$r + 2 = \sqrt{11} \quad r + 2 = -\sqrt{11}$$
$$r = -2 + \sqrt{11} \quad r = -2 - \sqrt{11}$$
The solutions are $-2 - \sqrt{11}$ and $-2 + \sqrt{11}$.

9. $x^2 - 6x + 7 = 0$
$$x^2 - 6x = -7$$
Complete the square.
$$x^2 - 6x + 9 = -7 + 9$$
$$(x-3)^2 = 2$$
$$\sqrt{(x-3)^2} = \sqrt{2}$$
$$x - 3 = \pm\sqrt{2}$$
$$x - 3 = \sqrt{2} \quad x - 3 = -\sqrt{2}$$
$$x = 3 + \sqrt{2} \quad x = 3 - \sqrt{2}$$
The solutions are $3 - \sqrt{2}$ and $3 + \sqrt{2}$.

11. $z^2 - 2z + 2 = 0$
$$z^2 - 2z = -2$$
Complete the square.
$$z^2 - 2z + 1 = -2 + 1$$
$$(z-1)^2 = -1$$
$$\sqrt{(z-1)^2} = \sqrt{-1}$$
$$z - 1 = \pm i$$
$$z - 1 = i \quad z - 1 = -i$$
$$z = 1 + i \quad z = 1 - i$$
The solutions are $1 - i$ and $1 + i$.

13. $s^2 - 5s - 24 = 0$
$s^2 - 5s = 24$
Complete the square.
$s^2 - 5s + \frac{25}{4} = 24 + \frac{25}{4}$
$\left(s - \frac{5}{2}\right)^2 = \frac{121}{4}$
$\sqrt{\left(s - \frac{5}{2}\right)^2} = \sqrt{\frac{121}{4}}$
$s - \frac{5}{2} = \pm\frac{11}{2}$
$s - \frac{5}{2} = \frac{11}{2}$ $\quad s - \frac{5}{2} = -\frac{11}{2}$
$s = \frac{5}{2} + \frac{11}{2}$ $\quad s = \frac{5}{2} - \frac{11}{2}$
$s = \frac{16}{2} = 8$ $\quad s = -\frac{6}{2} = -3$
The solutions are −3 and 8.

15. $x^2 + 5x - 36 = 0$
$x^2 + 5x = 36$
Complete the square.
$x^2 + 5x + \frac{25}{4} = 36 + \frac{25}{4}$
$\left(x + \frac{5}{2}\right)^2 = \frac{169}{4}$
$\sqrt{\left(x + \frac{5}{2}\right)^2} = \sqrt{\frac{169}{4}}$
$x + \frac{5}{2} = \pm\frac{13}{2}$
$x + \frac{5}{2} = \frac{13}{2}$ $\quad x + \frac{5}{2} = -\frac{13}{2}$
$x = -\frac{5}{2} + \frac{13}{2}$ $\quad x = -\frac{5}{2} - \frac{13}{2}$
$x = \frac{8}{2} = 4$ $\quad x = -\frac{18}{2} = -9$
The solutions are −9 and 4.

17. $p^2 - 3p + 1 = 0$
$p^2 - 3p = -1$
Complete the square.
$p^2 - 3p + \frac{9}{4} = -1 + \frac{9}{4}$
$\left(p - \frac{3}{2}\right)^2 = \frac{5}{4}$
$\sqrt{\left(p - \frac{3}{2}\right)^2} = \sqrt{\frac{5}{4}}$
$p - \frac{3}{2} = \pm\frac{\sqrt{5}}{2}$
$p - \frac{3}{2} = \frac{\sqrt{5}}{2}$ $\quad p - \frac{3}{2} = -\frac{\sqrt{5}}{2}$
$p = \frac{3}{2} + \frac{\sqrt{5}}{2}$ $\quad p = \frac{3}{2} - \frac{\sqrt{5}}{2}$
The solutions are $\frac{3 - \sqrt{5}}{2}$ and $\frac{3 + \sqrt{5}}{2}$.

19. $t^2 - t - 1 = 0$
$t^2 - t = 1$
Complete the square.
$t^2 - t + \frac{1}{4} = 1 + \frac{1}{4}$
$\left(t - \frac{1}{2}\right)^2 = \frac{5}{4}$
$\sqrt{\left(t - \frac{1}{2}\right)^2} = \sqrt{\frac{5}{4}}$
$t - \frac{1}{2} = \pm\frac{\sqrt{5}}{2}$
$t - \frac{1}{2} = \frac{\sqrt{5}}{2}$ $\quad t - \frac{1}{2} = -\frac{\sqrt{5}}{2}$
$t = \frac{1}{2} + \frac{\sqrt{5}}{2}$ $\quad t = \frac{1}{2} - \frac{\sqrt{5}}{2}$
The solutions are $\frac{1 - \sqrt{5}}{2}$ and $\frac{1 + \sqrt{5}}{2}$.

21. $y^2 - 6y = 4$
Complete the square.
$y^2 - 6y + 9 = 4 + 9$
$(y - 3)^2 = 13$
$\sqrt{(y - 3)^2} = \sqrt{13}$
$y - 3 = \pm\sqrt{13}$
$y - 3 = \sqrt{13}$ $\quad y - 3 = -\sqrt{13}$
$y = 3 + \sqrt{13}$ $\quad y = 3 - \sqrt{13}$
The solutions are $3 - \sqrt{13}$ and $3 + \sqrt{13}$.

23. $x^2 = 8x - 15$
$x^2 - 8x = -15$
Complete the square.
$x^2 - 8x + 16 = -15 + 16$
$(x - 4)^2 = 1$
$\sqrt{(x - 4)^2} = \sqrt{1}$
$x - 4 = \pm 1$
$x - 4 = 1$ $\quad x - 4 = -1$
$x = 5$ $\quad x = 3$
The solutions are 3 and 5.

25. $v^2 = 4v - 13$
$v^2 - 4v = -13$
Complete the square.
$v^2 - 4v + 4 = -13 + 4$
$(v - 2)^2 = -9$
$\sqrt{(v - 2)^2} = \sqrt{-9}$
$v - 2 = \pm 3i$
$v - 2 = 3i$ $\quad v - 2 = -3i$
$v = 2 + 3i$ $\quad v = 2 - 3i$
The solutions are $2 - 3i$ and $2 + 3i$.

27. $p^2 + 6p = -13$
Complete the square.
$p^2 + 6p + 9 = -13 + 9$
$(p+3)^2 = -4$
$\sqrt{(p+3)^2} = \sqrt{-4}$
$p + 3 = \pm 2i$
$p + 3 = 2i \qquad p + 3 = -2i$
$p = -3 + 2i \qquad p = -3 - 2i$
The solutions are $-3 - 2i$ and $-3 + 2i$.

29. $y^2 - 2y = 17$
Complete the square.
$y^2 - 2y + 1 = 17 + 1$
$(y-1)^2 = 18$
$\sqrt{(y-1)^2} = \sqrt{18}$
$y - 1 = \pm 3\sqrt{2}$
$y - 1 = 3\sqrt{2} \qquad y - 1 = -3\sqrt{2}$
$y = 1 + 3\sqrt{2} \qquad y = 1 - 3\sqrt{2}$
The solutions are $1 - 3\sqrt{2}$ and $1 + 3\sqrt{2}$.

31. $z^2 = z + 4$
$z^2 - z = 4$
Complete the square.
$z^2 - z + \frac{1}{4} = 4 + \frac{1}{4}$
$\left(z - \frac{1}{2}\right)^2 = \frac{17}{4}$
$\sqrt{\left(z - \frac{1}{2}\right)^2} = \sqrt{\frac{17}{4}}$
$z - \frac{1}{2} = \pm\frac{\sqrt{17}}{2}$
$z - \frac{1}{2} = \frac{\sqrt{17}}{2} \qquad z - \frac{1}{2} = -\frac{\sqrt{17}}{2}$
$z = \frac{1}{2} + \frac{\sqrt{17}}{2} \qquad z = \frac{1}{2} - \frac{\sqrt{17}}{2}$
The solutions are $\frac{1 - \sqrt{17}}{2}$ and $\frac{1 + \sqrt{17}}{2}$.

33. $x^2 + 13 = 2x$
$x^2 - 2x = -13$
Complete the square.
$x^2 - 2x + 1 = -13 + 1$
$(x-1)^2 = -12$
$\sqrt{(x-1)^2} = \sqrt{-12}$
$x - 1 = \pm 2i\sqrt{3}$
$x - 1 = 2i\sqrt{3} \qquad x - 1 = -2i\sqrt{3}$
$x = 1 + 2i\sqrt{3} \qquad x = 1 - 2i\sqrt{3}$
The solutions are $1 - 2i\sqrt{3}$ and $1 + 2i\sqrt{3}$.

35. $4x^2 - 4x + 5 = 0$
$4x^2 - 4x = -5$
$\frac{1}{4}(4x^2 - 4x) = \frac{1}{4}(-5)$
$x^2 - x = -\frac{5}{4}$
Complete the square.
$x^2 - x + \frac{1}{4} = -\frac{5}{4} + \frac{1}{4}$
$\left(x - \frac{1}{2}\right)^2 = -1$
$\sqrt{\left(x - \frac{1}{2}\right)^2} = \sqrt{-1}$
$x - \frac{1}{2} = \pm i$
$x - \frac{1}{2} = i \qquad x - \frac{1}{2} = -i$
$x = \frac{1}{2} + i \qquad x = \frac{1}{2} - i$
The solutions are $\frac{1}{2} - i$ and $\frac{1}{2} + i$.

37. $9x^2 - 6x + 2 = 0$
$9x^2 - 6x = -2$
$\frac{1}{9}(9x^2 - 6x) = \frac{1}{9}(-2)$
$x^2 - \frac{2}{3}x = -\frac{2}{9}$
Complete the square.
$x^2 - \frac{2}{3}x + \frac{1}{9} = -\frac{2}{9} + \frac{1}{9}$
$\left(x - \frac{1}{3}\right)^2 = -\frac{1}{9}$
$\sqrt{\left(x - \frac{1}{3}\right)^2} = \sqrt{-\frac{1}{9}}$
$x - \frac{1}{3} = \pm\frac{1}{3}i$
$x - \frac{1}{3} = \frac{1}{3}i \qquad x - \frac{1}{3} = -\frac{1}{3}i$
$x = \frac{1}{3} + \frac{1}{3}i \qquad x = \frac{1}{3} - \frac{1}{3}i$
The solutions are $\frac{1}{3} - \frac{1}{3}i$ and $\frac{1}{3} + \frac{1}{3}i$.

39.
$$2s^2 = 4s + 5$$
$$2s^2 - 4s = 5$$
$$\frac{1}{2}(2s^2 - 4s) = \frac{1}{2}(5)$$
$$s^2 - 2s = \frac{5}{2}$$
Complete the square.
$$s^2 - 2s + 1 = \frac{5}{2} + 1$$
$$(s-1)^2 = \frac{7}{2}$$
$$\sqrt{(s-1)^2} = \sqrt{\frac{7}{2}}$$
$$s - 1 = \pm\sqrt{\frac{7}{2}} = \pm\frac{\sqrt{14}}{2}$$
$$s - 1 = \frac{\sqrt{14}}{2} \qquad s - 1 = -\frac{\sqrt{14}}{2}$$
$$s = \frac{2}{2} + \frac{\sqrt{14}}{2} \qquad s = \frac{2}{2} - \frac{\sqrt{14}}{2}$$
The solutions are $\frac{2-\sqrt{14}}{2}$ and $\frac{2+\sqrt{14}}{2}$.

41.
$$2r^2 = 3 - r$$
$$2r^2 + r = 3$$
$$\frac{1}{2}(2r^2 + r) = \frac{1}{2}(3)$$
$$r^2 + \frac{1}{2}r = \frac{3}{2}$$
Complete the square.
$$r^2 + \frac{1}{2}r + \frac{1}{16} = \frac{3}{2} + \frac{1}{16}$$
$$\left(r + \frac{1}{4}\right)^2 = \frac{25}{16}$$
$$\sqrt{\left(r + \frac{1}{4}\right)^2} = \sqrt{\frac{25}{16}}$$
$$r + \frac{1}{4} = \pm\frac{5}{4}$$
$$r + \frac{1}{4} = \frac{5}{4} \qquad r + \frac{1}{4} = -\frac{5}{4}$$
$$r = \frac{4}{4} = 1 \qquad r = -\frac{6}{4} = -\frac{3}{2}$$
The solutions are $-\frac{3}{2}$ and 1.

43.
$$y - 2 = (y-3)(y+2)$$
$$y - 2 = y^2 - y - 6$$
$$y^2 - 2y = 4$$
Complete the square.
$$y^2 - 2y + 1 = 4 + 1$$
$$(y-1)^2 = 5$$
$$\sqrt{(y-1)^2} = \sqrt{5}$$
$$y - 1 = \pm\sqrt{5}$$
$$y - 1 = \sqrt{5} \qquad y - 1 = -\sqrt{5}$$
$$y = 1 + \sqrt{5} \qquad y = 1 - \sqrt{5}$$
The solutions are $1 - \sqrt{5}$ and $1 + \sqrt{5}$.

45.
$$6t - 2 = (2t-3)(t-1)$$
$$6t - 2 = 2t^2 - 5t + 3$$
$$2t^2 - 11t = -5$$
$$\frac{1}{2}(2t^2 - 11t) = \frac{1}{2}(-5)$$
$$t^2 - \frac{11}{2}t = -\frac{5}{2}$$
Complete the square.
$$t^2 - \frac{11}{2}t + \frac{121}{16} = -\frac{5}{2} + \frac{121}{16}$$
$$\left(t - \frac{11}{4}\right)^2 = \frac{81}{16}$$
$$\sqrt{\left(t - \frac{11}{4}\right)^2} = \sqrt{\frac{81}{16}}$$
$$t - \frac{11}{4} = \pm\frac{9}{4}$$
$$t - \frac{11}{4} = \frac{9}{4} \qquad t - \frac{11}{4} = -\frac{9}{4}$$
$$t = \frac{20}{4} = 5 \qquad t = \frac{2}{4} = \frac{1}{2}$$
The solutions are $\frac{1}{2}$ and 5.

47.
$$(x-4)(x+1) = x - 3$$
$$x^2 - 3x - 4 = x - 3$$
$$x^2 - 4x = 1$$
Complete the square.
$$x^2 - 4x + 4 = 1 + 4$$
$$(x-2)^2 = 5$$
$$\sqrt{(x-2)^2} = \sqrt{5}$$
$$x - 2 = \pm\sqrt{5}$$
$$x - 2 = \sqrt{5} \qquad x - 2 = -\sqrt{5}$$
$$x = 2 + \sqrt{5} \qquad x = 2 - \sqrt{5}$$
The solutions are $2 - \sqrt{5}$ and $2 + \sqrt{5}$.

49.
$$z^2 + 2z = 4$$
Complete the square.
$$z^2 + 2z + 1 = 4 + 1$$
$$(z+1)^2 = 5$$
$$\sqrt{(z+1)^2} = \sqrt{5}$$
$$z + 1 = \pm\sqrt{5}$$
$$z + 1 = \sqrt{5} \qquad z + 1 = -\sqrt{5}$$
$$z = \sqrt{5} - 1 \qquad z = -\sqrt{5} - 1$$
$$z \approx 1.236 \qquad z \approx -3.236$$
The solutions are −3.236 and 1.236.

51.
$$2x^2 = 4x - 1$$
$$2x^2 - 4x = -1$$
$$\frac{1}{2}(2x^2 - 4x) = \frac{1}{2}(-1)$$
$$x^2 - 2x = -\frac{1}{2}$$
Complete the square.
$$x^2 - 2x + 1 = -\frac{1}{2} + 1$$
$$(x-1)^2 = \frac{1}{2}$$
$$\sqrt{(x-1)^2} = \sqrt{\frac{1}{2}}$$
$$x - 1 = \pm\sqrt{\frac{1}{2}}$$
$$x - 1 = \sqrt{\frac{1}{2}} \qquad x - 1 = -\sqrt{\frac{1}{2}}$$
$$x = \sqrt{\frac{1}{2}} + 1 \qquad x = -\sqrt{\frac{1}{2}} + 1$$
$$x \approx 1.707 \qquad x \approx 0.293$$
The solutions are 0.293 and 1.707.

53.
$$4z^2 + 2z = 1$$
$$\frac{1}{4}(4z^2 + 2z) = \frac{1}{4}(1)$$
$$z^2 + \frac{1}{2}z = \frac{1}{4}$$
Complete the square.
$$z^2 + \frac{1}{2}z + \frac{1}{16} = \frac{1}{4} + \frac{1}{16}$$
$$\left(z + \frac{1}{4}\right)^2 = \frac{5}{16}$$
$$\sqrt{\left(z + \frac{1}{4}\right)^2} = \sqrt{\frac{5}{16}}$$
$$z + \frac{1}{4} = \pm\frac{\sqrt{5}}{4}$$
$$z + \frac{1}{4} = \frac{\sqrt{5}}{4} \qquad z + \frac{1}{4} = -\frac{\sqrt{5}}{4}$$
$$z = \frac{\sqrt{5}}{4} - \frac{1}{4} \qquad z = -\frac{\sqrt{5}}{4} - \frac{1}{4}$$
$$z \approx 0.309 \qquad z \approx -0.809$$
The solutions are −0.809 and 0.309.

Applying the Concepts

55.
$$x^2 - ax - 2a^2 = 0$$
$$x^2 - ax = 2a^2$$
Complete the square.
$$x^2 - ax + \frac{1}{4}a^2 = 2a^2 + \frac{1}{4}a^2$$
$$\left(x - \frac{1}{2}a\right)^2 = \frac{9}{4}a^2$$
$$\sqrt{\left(x - \frac{1}{2}a\right)^2} = \sqrt{\frac{9}{4}a^2}$$
$$x - \frac{1}{2}a = \pm\sqrt{\frac{9}{4}a^2} = \pm\frac{3}{2}a$$
$$x - \frac{1}{2}a = \frac{3}{2}a \qquad x - \frac{1}{2}a = -\frac{3}{2}a$$
$$x = 2a \qquad x = -a$$
The solutions are $-a$ and $2a$.

57.
$$x^2 + 3ax - 10a^2 = 0$$
$$x^2 + 3ax = 10a^2$$
Complete the square.
$$x^2 + 3ax + \frac{9}{4}a^2 = 10a^2 + \frac{9}{4}a^2$$
$$\left(x + \frac{3}{2}a\right)^2 = \frac{49}{4}a^2$$
$$\sqrt{\left(x + \frac{3}{2}a\right)^2} = \sqrt{\frac{49}{4}a^2}$$
$$x + \frac{3}{2}a = \pm\frac{7}{2}a$$
$$x + \frac{3}{2}a = \frac{7}{2}a \qquad x + \frac{3}{2}a = -\frac{7}{2}a$$
$$x = 2a \qquad x = -5a$$
The solutions are $-5a$ and $2a$.

59. **Strategy** First find the time it takes for the ball to hit the ground using $y = -16t^2 + 70t + 4$ (the answer is found in Exercise 58). Use this time to find the horizontal distance the ball travels.

Solution For $t = 4.431$ seconds, the length of time the ball is in the air is $s = 44.5t$.
$$s = 44.5(4.431)$$
$$s \approx 197.2 \text{ feet}$$

No, the ball will have gone only 197.2 ft when it hits the ground.

Section 8.3

Objective A Exercises

1. The quadratic formula is $x = \dfrac{-b \pm \sqrt{b^2 - 4ac}}{2a}$. In this formula a is the coefficient of x^2, b is the coefficient of x, and c is the constant term in the quadratic equation $ax^2 + bx + c = 0$, $a \neq 0$.

3. $x^2 - 3x - 10 = 0$

$a = 1, b = -3, c = -10$

$$x = \frac{-b \pm \sqrt{b^2 - 4ac}}{2a} = \frac{-(-3) \pm \sqrt{(-3)^2 - 4(1)(-10)}}{2(1)} = \frac{3 \pm \sqrt{9+40}}{2} = \frac{3 \pm \sqrt{49}}{2} = \frac{3 \pm 7}{2}$$

$$x = \frac{3+7}{2} = \frac{10}{2} = 5 \qquad x = \frac{3-7}{2} = -\frac{4}{2} = -2$$

The solutions are -2 and 5.

5. $y^2 + 5y - 36 = 0$

$a = 1, b = 5, c = -36$

$$y = \frac{-b \pm \sqrt{b^2 - 4ac}}{2a} = \frac{-5 \pm \sqrt{(5)^2 - 4(1)(-36)}}{2(1)} = \frac{-5 \pm \sqrt{25+144}}{2} = \frac{-5 \pm \sqrt{169}}{2} = \frac{-5 \pm 13}{2}$$

$$y = \frac{-5+13}{2} = \frac{8}{2} = 4 \qquad y = \frac{-5-13}{2} = \frac{-18}{2} = -9$$

The solutions are -9 and 4.

7. $w^2 = 8w + 72$

$w^2 - 8w - 72 = 0$

$a = 1, b = -8, c = -72$

$$w = \frac{-b \pm \sqrt{b^2 - 4ac}}{2a} = \frac{-(-8) \pm \sqrt{(-8)^2 - 4(1)(-72)}}{2(1)} = \frac{8 \pm \sqrt{64+288}}{2} = \frac{8 \pm \sqrt{352}}{2} = \frac{8 \pm 4\sqrt{22}}{2} = 4 \pm 2\sqrt{22}$$

The solutions are $4 - 2\sqrt{22}$ and $4 + 2\sqrt{22}$.

9. $v^2 = 24 - 5v$

$v^2 + 5v - 24 = 0$

$a = 1, b = 5, c = -24$

$$v = \frac{-b \pm \sqrt{b^2 - 4ac}}{2a} = \frac{-5 \pm \sqrt{(5)^2 - 4(1)(-24)}}{2(1)} = \frac{-5 \pm \sqrt{25+96}}{2} = \frac{-5 \pm \sqrt{121}}{2} = \frac{-5 \pm 11}{2}$$

$$v = \frac{-5+11}{2} = \frac{6}{2} = 3 \qquad v = \frac{-5-11}{2} = \frac{-16}{2} = -8$$

The solutions are -8 and 3.

11. $2y^2 + 5y - 1 = 0$

$a = 2, b = 5, c = -1$

$$y = \frac{-b \pm \sqrt{b^2 - 4ac}}{2a} = \frac{-5 \pm \sqrt{(5)^2 - 4(2)(-1)}}{2(2)} = \frac{-5 \pm \sqrt{25+8}}{4} = \frac{-5 \pm \sqrt{33}}{4}$$

$$y = \frac{-5+\sqrt{33}}{4} \qquad y = \frac{-5-\sqrt{33}}{4}$$

The solutions are $\dfrac{-5-\sqrt{33}}{4}$ and $\dfrac{-5+\sqrt{33}}{4}$.

13. $8s^2 = 10s + 3$

$8s^2 - 10s - 3 = 0$

$a = 8, b = -10, c = -3$

$$s = \frac{-b \pm \sqrt{b^2 - 4ac}}{2a} = \frac{-(-10) \pm \sqrt{(-10)^2 - 4(8)(-3)}}{2(8)} = \frac{10 \pm \sqrt{100+96}}{16} = \frac{10 \pm \sqrt{196}}{16} = \frac{10 \pm 14}{16}$$

$$s = \frac{10+14}{16} = \frac{24}{16} = \frac{3}{2} \qquad s = \frac{10-14}{16} = \frac{-4}{16} = -\frac{1}{4}$$

The solutions are $-\dfrac{1}{4}$ and $\dfrac{3}{2}$.

15. $x^2 = 14x - 4$

$x^2 - 14x + 4 = 0$

$a = 1, b = -14, c = 4$

$$x = \frac{-b \pm \sqrt{b^2 - 4ac}}{2a}$$

$$= \frac{-(-14) \pm \sqrt{(-14)^2 - 4(1)(4)}}{2(1)}$$

$$= \frac{14 \pm \sqrt{196 - 16}}{2} = \frac{14 \pm \sqrt{180}}{2} = \frac{14 \pm 6\sqrt{5}}{2}$$

$$= 7 \pm 3\sqrt{5}$$

$x = 7 + 3\sqrt{5} \quad x = 7 - 3\sqrt{5}$

The solutions are $7 - 3\sqrt{5}$ and $7 + 3\sqrt{5}$.

17. $2z^2 - 2z - 1 = 0$

$a = 2, b = -2, c = -1$

$$z = \frac{-b \pm \sqrt{b^2 - 4ac}}{2a}$$

$$= \frac{-(-2) \pm \sqrt{(-2)^2 - 4(2)(-1)}}{2(2)}$$

$$= \frac{2 \pm \sqrt{4 + 8}}{4} = \frac{2 \pm \sqrt{12}}{4}$$

$$= \frac{2 \pm 2\sqrt{3}}{4} = \frac{1 \pm \sqrt{3}}{2}$$

The solutions are $\frac{1 - \sqrt{3}}{2}$ and $\frac{1 + \sqrt{3}}{2}$.

19. $z^2 + 2z + 2 = 0$

$a = 1, b = 2, c = 2$

$$z = \frac{-b \pm \sqrt{b^2 - 4ac}}{2a}$$

$$= \frac{-2 \pm \sqrt{(2)^2 - 4(1)(2)}}{2(1)}$$

$$= \frac{-2 \pm \sqrt{4 - 8}}{2} = \frac{-2 \pm \sqrt{-4}}{2}$$

$$= \frac{-2 \pm 2i}{2} = -1 \pm i$$

The solutions are $-1 - i$ and $-1 + i$.

21. $y^2 - 2y + 5 = 0$

$a = 1, b = -2, c = 5$

$$y = \frac{-b \pm \sqrt{b^2 - 4ac}}{2a}$$

$$= \frac{-(-2) \pm \sqrt{(-2)^2 - 4(1)(5)}}{2(1)}$$

$$= \frac{2 \pm \sqrt{4 - 20}}{2} = \frac{2 \pm \sqrt{-16}}{2}$$

$$= \frac{2 \pm 4i}{2} = 1 \pm 2i$$

The solutions are $1 - 2i$ and $1 + 2i$.

23. $s^2 - 4s + 13 = 0$

$a = 1, b = -4, c = 13$

$$s = \frac{-b \pm \sqrt{b^2 - 4ac}}{2a}$$

$$= \frac{-(-4) \pm \sqrt{(-4)^2 - 4(1)(13)}}{2(1)}$$

$$= \frac{4 \pm \sqrt{16 - 52}}{2} = \frac{4 \pm \sqrt{-36}}{2}$$

$$= \frac{4 \pm 6i}{2} = 2 \pm 3i$$

The solutions are $2 - 3i$ and $2 + 3i$.

25. $2w^2 - 2w - 5 = 0$

$a = 2, b = -2, c = -5$

$$w = \frac{-b \pm \sqrt{b^2 - 4ac}}{2a}$$

$$= \frac{-(-2) \pm \sqrt{(-2)^2 - 4(2)(-5)}}{2(2)}$$

$$= \frac{2 \pm \sqrt{4 + 40}}{4} = \frac{2 \pm \sqrt{44}}{4}$$

$$= \frac{2 \pm 2\sqrt{11}}{4} = \frac{1 \pm \sqrt{11}}{2}$$

The solutions are $\frac{1 - \sqrt{11}}{2}$ and $\frac{1 + \sqrt{11}}{2}$.

27. $2x^2 + 6x + 5 = 0$

$a = 2, b = 6, c = 5$

$$x = \frac{-b \pm \sqrt{b^2 - 4ac}}{2a}$$

$$= \frac{-6 \pm \sqrt{(6)^2 - 4(2)(5)}}{2(2)}$$

$$= \frac{-6 \pm \sqrt{36 - 40}}{4} = \frac{-6 \pm \sqrt{-4}}{4}$$

$$= \frac{-6 \pm 2i}{4} = \frac{-3 \pm i}{2}$$

The solutions are $-\frac{3}{2} - \frac{1}{2}i$ and $-\frac{3}{2} + \frac{1}{2}i$.

29. $4t^2 - 6t + 9 = 0$

$a = 4, b = -6, c = 9$

$$t = \frac{-b \pm \sqrt{b^2 - 4ac}}{2a}$$

$$= \frac{-(-6) \pm \sqrt{(-6)^2 - 4(4)(9)}}{2(4)}$$

$$= \frac{6 \pm \sqrt{36 - 144}}{8} = \frac{6 \pm \sqrt{-108}}{8}$$

$$= \frac{6 \pm 6i\sqrt{3}}{8} = \frac{3 \pm 3i\sqrt{3}}{4}$$

The solutions are $\frac{3}{4} - \frac{3\sqrt{3}}{4}i$ and $\frac{3}{4} + \frac{3\sqrt{3}}{4}i$.

31. $p^2 - 8p + 3 = 0$
$a = 1, b = -8, c = 3$
$$p = \frac{-b \pm \sqrt{b^2 - 4ac}}{2a}$$
$$= \frac{-(-8) \pm \sqrt{(-8)^2 - 4(1)(3)}}{2(1)}$$
$$= \frac{8 \pm \sqrt{64 - 12}}{2} = \frac{8 \pm \sqrt{52}}{2}$$
$$= \frac{8 \pm 2\sqrt{13}}{2} = 4 \pm \sqrt{13}$$
The solutions are 0.394 and 7.606.

33. $w^2 + 4w = 1$
$w^2 + 4w - 1 = 0$
$a = 1, b = 4, c = -1$
$$w = \frac{-b \pm \sqrt{b^2 - 4ac}}{2a}$$
$$= \frac{-4 \pm \sqrt{(4)^2 - 4(1)(-1)}}{2(1)}$$
$$= \frac{-4 \pm \sqrt{16 + 4}}{2} = \frac{-4 \pm \sqrt{20}}{2}$$
$$= \frac{-4 \pm 2\sqrt{5}}{2} = -2 \pm \sqrt{5}$$
The solutions are −4.236 and 0.236.

35. $2y^2 = y + 5$
$2y^2 - y - 5 = 0$
$a = 2, b = -1, c = -5$
$$y = \frac{-b \pm \sqrt{b^2 - 4ac}}{2a}$$
$$= \frac{-(-1) \pm \sqrt{(-1)^2 - 4(2)(-5)}}{2(2)}$$
$$= \frac{1 \pm \sqrt{1 + 40}}{4} = \frac{1 \pm \sqrt{41}}{4}$$
The solutions are −1.351 and 1.851.

37. $3y^2 + y + 1 = 0$
$a = 3, b = 1, c = 1$
$b^2 - 4ac = 1^2 - 4(3)(1)$
$= 1 - 12 = -11$
$-11 < 0$
Since the discriminant is less than zero, the equation has two complex number solutions.

39. $4x^2 + 20x + 25 = 0$
$a = 4, b = 20, c = 25$
$b^2 - 4ac = 20^2 - 4(4)(25)$
$= 400 - 400 = 0$
Since the discriminant is equal to zero, the equation has one real number solution, a double root.

41. $3w^2 + 3w - 2 = 0$
$a = 3, b = 3, c = -2$
$b^2 - 4ac = 3^2 - 4(3)(-2)$
$= 9 + 24 = 33$
$33 > 0$
Since the discriminant is greater than zero, the equation has two real number solutions that are not equal.

Applying the Concepts

43. **Strategy** To find if the arrow reaches a height of 275 ft, use the discriminant of the equation $275 = 128t - 16t^2$ to determine if the equation has real zeros.
If $b^2 - 4ac \geq 0$, then the answer is yes.
If $b^2 - 4ac < 0$, then the answer is no.

Solution $275 = 128t - 16t^2$
$16t^2 - 128t + 275 = 0$
$a = 16, b = -128, c = 275$
$b^2 - 4ac = (-128)^2 - 4 \cdot 16 \cdot 275$
$= 16384 - 17600$
$= -1216 < 0$
No, the arrow does not reach a height of 275 ft. (The discriminant is less than zero.)

45. $x^2 - 6x + p = 0$
$x^2 - 6x = -p$
Complete the square.
$x^2 - 6x + 9 = -p + 9$
$(x - 3)^2 = -p + 9$
$\sqrt{(x - 3)^2} = \sqrt{9 - p}$
$x - 3 = \pm\sqrt{9 - p}$
$x = 3 \pm \sqrt{9 - p}$
x will be real, and there will be two nonequal solutions if $9 - p > 0$.
Solving the inequality gives $p < 9$.
The values of p are $\{p | p < 9\}$.

47. $x^2 - 2x + p = 0$
$a = 1, b = -2, c = p$
$$x = \frac{-b \pm \sqrt{b^2 - 4ac}}{2a}$$
$$x = \frac{-(-2) \pm \sqrt{(-2)^2 - 4(p)(1)}}{2(1)} = \frac{2 \pm \sqrt{4 - 4p}}{2}$$
$x = 1 \pm \sqrt{1 - p}$
x will be complex, and there will be two solutions if $1 - p < 0$.
Solving the inequality gives $p > 1$. The values of p are $(1, \infty)$.

49. $x^2 + ix + 2 = 0$

$a = 1, b = i, c = 2$

$$x = \frac{-b \pm \sqrt{b^2 - 4ac}}{2a}$$
$$= \frac{-i \pm \sqrt{(i)^2 - 4(1)(2)}}{2(1)}$$
$$= \frac{-i \pm \sqrt{-1-8}}{2} = \frac{-i \pm \sqrt{-9}}{2}$$
$$= \frac{-i \pm 3i}{2}$$
$$x = \frac{-i+3i}{2} \qquad x = \frac{-i-3i}{2}$$
$$x = \frac{2i}{2} \qquad x = \frac{-4i}{2}$$
$$x = i \qquad x = -2i$$

The solutions are $-2i$ and i.

51. $2x^2 + bx - 2 = 0$

$a = 2, b = b, c = -2$

$$x = \frac{-b \pm \sqrt{b^2 - 4ac}}{2a}$$
$$= \frac{-b \pm \sqrt{b^2 - 4(2)(-2)}}{2(2)}$$
$$= \frac{-b \pm \sqrt{b^2 + 16}}{4}$$
$$x = -\frac{b}{4} \pm \frac{1}{4}\sqrt{b^2 + 16}$$

If b is real, the quantity $\sqrt{b^2 + 16}$ can never be less than zero, and the quadratic will always have real number solutions.

Section 8.4

Objective A Exercises

1.
$$x^4 - 13x^2 + 36 = 0$$
$$(x^2)^2 - 13(x^2) + 36 = 0$$
$$u^2 - 13u + 36 = 0$$
$$(u-4)(u-9) = 0$$
$$u - 4 = 0 \qquad u - 9 = 0$$
$$u = 4 \qquad u = 9$$

Replace u by x^2.

$$x^2 = 4 \qquad x^2 = 9$$
$$\sqrt{x^2} = \sqrt{4} \qquad \sqrt{x^2} = \sqrt{9}$$
$$x = \pm 2 \qquad x = \pm 3$$

The solutions are -2, 2, -3, and 3.

3.
$$z^4 - 6z^2 + 8 = 0$$
$$(z^2)^2 - 6(z^2) + 8 = 0$$
$$u^2 - 6u + 8 = 0$$
$$(u-4)(u-2) = 0$$
$$u - 4 = 0 \qquad u - 2 = 0$$
$$u = 4 \qquad u = 2$$

Replace u by z^2.

$$z^2 = 4 \qquad z^2 = 2$$
$$\sqrt{z^2} = \sqrt{4} \qquad \sqrt{z^2} = \sqrt{2}$$
$$z = \pm 2 \qquad z = \pm\sqrt{2}$$

The solutions are -2, 2, $-\sqrt{2}$, and $\sqrt{2}$.

5.
$$p - 3p^{1/2} + 2 = 0$$
$$(p^{1/2})^2 - 3(p^{1/2}) + 2 = 0$$
$$u^2 - 3u + 2 = 0$$
$$(u-1)(u-2) = 0$$
$$u - 1 = 0 \qquad u - 2 = 0$$
$$u = 1 \qquad u = 2$$

Replace u by $p^{1/2}$.

$$p^{1/2} = 1 \qquad p^{1/2} = 2$$
$$(p^{1/2})^2 = 1^2 \qquad (p^{1/2})^2 = 2^2$$
$$p = 1 \qquad p = 4$$

The solutions are 1 and 4.

7.
$$x - x^{1/2} - 12 = 0$$
$$(x^{1/2})^2 - (x^{1/2}) - 12 = 0$$
$$u^2 - u - 12 = 0$$
$$(u+3)(u-4) = 0$$
$$u + 3 = 0 \qquad u - 4 = 0$$
$$u = -3 \qquad u = 4$$

Replace u by $x^{1/2}$.

$$x^{1/2} = -3 \qquad x^{1/2} = 4$$
$$(x^{1/2})^2 = (-3)^2 \qquad (x^{1/2})^2 = 4^2$$
$$x = 9 \qquad x = 16$$

16 checks as a solution.
9 does not check as a solution.
The solution is 16.

9.
$$z^4 + 3z^2 - 4 = 0$$
$$(z^2)^2 + 3(z^2) - 4 = 0$$
$$u^2 + 3u - 4 = 0$$
$$(u+4)(u-1) = 0$$
$$u + 4 = 0 \qquad u - 1 = 0$$
$$u = -4 \qquad u = 1$$

Replace u by z^2.

$$z^2 = -4 \qquad z^2 = 1$$
$$\sqrt{z^2} = \sqrt{-4} \qquad \sqrt{z^2} = \sqrt{1}$$
$$z = \pm 2i \qquad z = \pm 1$$

The solutions are $-2i$, $2i$, -1, and 1.

11.
$$x^4 + 12x^2 - 64 = 0$$
$$(x^2)^2 + 12(x^2) - 64 = 0$$
$$u^2 + 12u - 64 = 0$$
$$(u+16)(u-4) = 0$$
$$u + 16 = 0 \qquad u - 4 = 0$$
$$u = -16 \qquad u = 4$$

Replace u by x^2.

$$x^2 = -16 \qquad x^2 = 4$$
$$\sqrt{x^2} = \sqrt{-16} \qquad \sqrt{x^2} = \sqrt{4}$$
$$x = \pm 4i \qquad x = \pm 2$$

The solutions are $-4i$, $4i$, -2, and 2.

13.
$$p + 2p^{1/2} - 24 = 0$$
$$(p^{1/2})^2 + 2(p^{1/2}) - 24 = 0$$
$$u^2 + 2u - 24 = 0$$
$$(u+6)(u-4) = 0$$
$$u + 6 = 0 \qquad u - 4 = 0$$
$$u = -6 \qquad u = 4$$

Replace u by $p^{1/2}$.

$$p^{1/2} = -6 \qquad p^{1/2} = 4$$
$$(p^{1/2})^2 = (-6)^2 \qquad (p^{1/2})^2 = 4^2$$
$$p = 36 \qquad p = 16$$

16 checks as a solution.
36 does not check as a solution.
The solution is 16.

15.
$$y^{2/3} - 9y^{1/3} + 8 = 0$$
$$(y^{1/3})^2 - 9(y^{1/3}) + 8 = 0$$
$$u^2 - 9u + 8 = 0$$
$$(u-1)(u-8) = 0$$
$$u - 1 = 0 \qquad u - 8 = 0$$
$$u = 1 \qquad u = 8$$

Replace u by $y^{1/3}$.

$$y^{1/3} = 1 \qquad y^{1/3} = 8$$
$$(y^{1/3})^3 = 1^3 \qquad (y^{1/3})^3 = 8^3$$
$$y = 1 \qquad y = 512$$

The solutions are 1 and 512.

17.
$$9w^4 - 13w^2 + 4 = 0$$
$$9(w^2)^2 - 13(w^2) + 4 = 0$$
$$9u^2 - 13u + 4 = 0$$
$$(9u-4)(u-1) = 0$$
$$9u - 4 = 0 \qquad u - 1 = 0$$
$$9u = 4 \qquad u = 1$$
$$u = \frac{4}{9}$$

Replace u by w^2.

$$w^2 = \frac{4}{9} \qquad w^2 = 1$$
$$\sqrt{w^2} = \sqrt{\frac{4}{9}} \qquad \sqrt{w^2} = \sqrt{1}$$
$$w = \pm\frac{2}{3} \qquad w = \pm 1$$

The solutions are $-\frac{2}{3}, \frac{2}{3}, -1$, and 1.

Objective B Exercises

19. $\sqrt{x+1}+x=5$
$\sqrt{x+1}=5-x$
$(\sqrt{x+1})^2=(5-x)^2$
$x+1=25-10x+x^2$
$0=24-11x+x^2$
$0=(3-x)(8-x)$
$3-x=0 \quad 8-x=0$
$3=x \quad 8=x$
3 checks as a solution.
8 does not check as a solution.
The solution is 3.

21. $x=\sqrt{x}+6$
$x-6=\sqrt{x}$
$(x-6)^2=(\sqrt{x})^2$
$x^2-12x+36=x$
$x^2-13x+36=0$
$(x-4)(x-9)=0$
$x-4=0 \quad x-9=0$
$x=4 \quad x=9$
9 checks as a solution.
4 does not check as a solution.
The solution is 9.

23. $\sqrt{3w+3}=w+1$
$(\sqrt{3w+3})^2=(w+1)^2$
$3w+3=w^2+2w+1$
$0=w^2-w-2$
$0=(w-2)(w+1)$
$w-2=0 \quad w+1=0$
$w=2 \quad w=-1$
2 and -1 check as solutions.
The solutions are -1 and 2.

25. $\sqrt{4y+1}-y=1$
$\sqrt{4y+1}=y+1$
$(\sqrt{4y+1})^2=(y+1)^2$
$4y+1=y^2+2y+1$
$0=y^2-2y$
$0=y(y-2)$
$y=0 \quad y-2=0$
$y=2$
0 and 2 check as solutions.
The solutions are 0 and 2.

27. $\sqrt{10x+5}-2x=1$
$\sqrt{10x+5}=2x+1$
$(\sqrt{10x+5})^2=(2x+1)^2$
$10x+5=4x^2+4x+1$
$0=4x^2-6x-4$
$0=2(2x^2-3x-2)$
$0=2(2x+1)(x-2)$
$2x+1=0 \quad x-2=0$
$2x=-1 \quad x=2$
$x=-\frac{1}{2}$
$-\frac{1}{2}$ and 2 check as solutions.
The solutions are $-\frac{1}{2}$ and 2.

29. $\sqrt{p+11}=1-p$
$(\sqrt{p+11})^2=(1-p)^2$
$p+11=1-2p+p^2$
$0=-10-3p+p^2$
$0=p^2-3p-10$
$0=(p-5)(p+2)$
$p-5=0 \quad p+2=0$
$p=5 \quad p=-2$
-2 checks as a solution.
5 does not check as a solution.
The solution is -2.

31. $\sqrt{x-1}-\sqrt{x}=-1$
$\sqrt{x-1}=\sqrt{x}-1$
$(\sqrt{x-1})^2=(\sqrt{x}-1)^2$
$x-1=x-2\sqrt{x}+1$
$2\sqrt{x}=2$
$\sqrt{x}=1$
$(\sqrt{x})^2=1^2$
$x=1$
1 checks as a solution.
The solution is 1.

33. $\sqrt{2x-1}=1-\sqrt{x-1}$
$(\sqrt{2x-1})^2=(1-\sqrt{x-1})^2$
$2x-1=1-2\sqrt{x-1}+x-1$
$2\sqrt{x-1}=-x+1$
$(2\sqrt{x-1})^2=(-x+1)^2$
$4(x-1)=x^2-2x+1$
$4x-4=x^2-2x+1$
$0=x^2-6x+5$
$0=(x-5)(x-1)$
$x-5=0 \quad x-1=0$
$x=5 \quad x=1$
1 checks as a solution.
5 does not check as a solution.
The solution is 1.

35.
$$\begin{aligned}
\sqrt{t+3}+\sqrt{2t+7} &= 1\\
\sqrt{2t+7} &= 1-\sqrt{t+3}\\
(\sqrt{2t+7})^2 &= (1-\sqrt{t+3})^2\\
2t+7 &= 1-2\sqrt{t+3}+t+3\\
t+3 &= -2\sqrt{t+3}\\
(t+3)^2 &= (-2\sqrt{t+3})^2\\
t^2+6t+9 &= 4(t+3)\\
t^2+6t+9 &= 4t+12\\
t^2+2t-3 &= 0\\
(t+3)(t-1) &= 0
\end{aligned}$$
$t+3=0 \quad t-1=0$
$t=-3 \quad t=1$
-3 checks as a solution.
1 does not check as a solution.
The solution is -3.

Objective C Exercises

37.
$$\begin{aligned}
x &= \frac{10}{x-9}\\
(x-9)x &= (x-9)\frac{10}{x-9}\\
x^2-9x &= 10\\
x^2-9x-10 &= 0\\
(x-10)(x+1) &= 0
\end{aligned}$$
$x-10=0 \quad x+1=0$
$x=10 \quad x=-1$
The solutions are -1 and 10.

39.
$$\begin{aligned}
\frac{t}{t+1} &= \frac{-2}{t-1}\\
(t-1)(t+1)\frac{t}{t+1} &= (t-1)(t+1)\frac{-2}{t-1}\\
(t-1)t &= (t+1)(-2)\\
t^2-t &= -2t-2\\
t^2+t+2 &= 0\\
t &= \frac{-b\pm\sqrt{b^2-4ac}}{2a}\\
&= \frac{-1\pm\sqrt{1^2-4(1)(2)}}{2(1)}\\
&= \frac{-1\pm\sqrt{1-8}}{2}=\frac{-1\pm\sqrt{-7}}{2}=\frac{-1\pm i\sqrt{7}}{2}
\end{aligned}$$
The solutions are $-\frac{1}{2}-\frac{\sqrt{7}}{2}i$ and $-\frac{1}{2}+\frac{\sqrt{7}}{2}i$.

41.
$$\begin{aligned}
\frac{y-1}{y+2}+y &= 1\\
(y+2)\left(\frac{y-1}{y+2}+y\right) &= (y+2)1\\
(y+2)\left(\frac{y-1}{y+2}\right)+(y+2)y &= y+2\\
y-1+y^2+2y &= y+2\\
y^2+3y-1 &= y+2\\
y^2+2y-3 &= 0\\
(y+3)(y-1) &= 0
\end{aligned}$$
$y+3=0 \quad y-1=0$
$y=-3 \quad y=1$
The solutions are -3 and 1.

43.
$$\begin{aligned}
\frac{3r+2}{r+2}-2r &= 1\\
(r+2)\left(\frac{3r+2}{r+2}-2r\right) &= (r+2)1\\
(r+2)\frac{3r+2}{r+2}-(r+2)2r &= r+2\\
3r+2-2r^2-4r &= r+2\\
-2r^2-r+2 &= r+2\\
-2r^2-2r &= 0\\
-2r(r+1) &= 0
\end{aligned}$$
$-2r=0 \quad r+1=0$
$r=0 \quad r=-1$
The solutions are -1 and 0.

45.
$$\begin{aligned}
\frac{2}{2x+1}+\frac{1}{x} &= 3\\
x(2x+1)\left(\frac{2}{2x+1}+\frac{1}{x}\right) &= x(2x+1)3\\
x(2x+1)\frac{2}{2x+1}+x(2x+1)\frac{1}{x} &= 3x(2x+1)\\
2x+2x+1 &= 6x^2+3x\\
4x+1 &= 6x^2+3x\\
0 &= 6x^2-x-1\\
0 &= (2x-1)(3x+1)
\end{aligned}$$
$2x-1=0 \quad 3x+1=0$
$2x=1 \quad 3x=-1$
$x=\frac{1}{2} \quad x=-\frac{1}{3}$
The solutions are $-\frac{1}{3}$ and $\frac{1}{2}$.

47.
$$\frac{16}{z-2}+\frac{16}{z+2}=6$$
$$(z-2)(z+2)\left(\frac{16}{z-2}+\frac{16}{z+2}\right)=(z-2)(z+2)6$$
$$(z-2)(z+2)\frac{16}{z-2}+(z-2)(z+2)\frac{16}{z+2}=(z^2-4)6$$
$$(z+2)16+(z-2)16=6z^2-24$$
$$16z+32+16z-32=6z^2-24$$
$$32z=6z^2-24$$
$$0=6z^2-32z-24$$
$$0=2(3z^2-16z-12)$$
$$0=2(3z+2)(z-6)$$

$3z+2=0 \qquad z-6=0$
$3z=-2 \qquad z=6$
$z=-\frac{2}{3}$

The solutions are $-\frac{2}{3}$ and 6.

49.
$$\frac{t}{t-2}+\frac{2}{t-1}=4$$
$$(t-2)(t-1)\left(\frac{t}{t-2}+\frac{2}{t-1}\right)=(t-2)(t-1)4$$
$$(t-2)(t-1)\frac{t}{t-2}+(t-2)(t-1)\frac{2}{t-1}=(t^2-3t+2)4$$
$$(t-1)t+(t-2)2=4t^2-12t+8$$
$$t^2-t+2t-4=4t^2-12t+8$$
$$t^2+t-4=4t^2-12t+8$$
$$0=3t^2-13t+12$$
$$0=(3t-4)(t-3)$$

$3t-4=0 \qquad t-3=0$
$3t=4 \qquad t=3$
$t=\frac{4}{3}$

The solutions are $\frac{4}{3}$ and 3.

51.
$$\frac{5}{2p-1}+\frac{4}{p+1}=2$$
$$(2p-1)(p+1)\left(\frac{5}{2p-1}+\frac{4}{p+1}\right)=(2p-1)(p+1)2$$
$$(2p-1)(p+1)\frac{5}{2p-1}+(2p-1)(p+1)\frac{4}{p+1}=(2p^2+p-1)2$$
$$(p+1)5+(2p-1)4=4p^2+2p-2$$
$$5p+5+8p-4=4p^2+2p-2$$
$$13p+1=4p^2+2p-2$$
$$0=4p^2-11p-3$$
$$0=(4p+1)(p-3)$$

$4p+1=0 \qquad p-3=0$
$4p=-1 \qquad p=3$
$p=-\frac{1}{4}$

The solutions are $-\frac{1}{4}$ and 3.

Applying the Concepts

53. $(\sqrt{x}-2)^2 - 5\sqrt{x} + 14 = 0$

Let $u = \sqrt{x} - 2$.

$u^2 - 5(u+2) + 14 = 0$
$u^2 - 5u - 10 + 14 = 0$
$u^2 - 5u + 4 = 0$
$(u-4)(u-1) = 0$
$u - 4 = 0 \quad u - 1 = 0$
$u = 4 \quad u = 1$

Replace u by $\sqrt{x} - 2$.

$\sqrt{x} - 2 = 4 \quad \sqrt{x} - 2 = 1$
$\sqrt{x} = 6 \quad \sqrt{x} = 3$
$(\sqrt{x})^2 = (6)^2 \quad (\sqrt{x})^2 = (3)^2$
$x = 36 \quad x = 9$

The solutions are 9 and 36.

55. $x^4 = 10x^2 - 25$
$x^4 - 10x^2 + 25 = 0$

Let $u = x^2$.

$u^2 - 10u + 25 = 0$
$(u-5)(u-5) = 0$
$u - 5 = 0$
$u = 5$

Replace u by x^2.

$x^2 = 5$
$\sqrt{x^2} = \sqrt{5}$
$x = \pm\sqrt{5}$

The solutions are $-\sqrt{5}$ or $\sqrt{5}$.

Section 8.5

Objective A Exercises

1. It must be true that $x - 3 > 0$ and $x - 5 > 0$ or that $x - 3 < 0$ and $x - 5 < 0$. In other words, either both factors are positive or both factors are negative.

3. $(x-4)(x+2) > 0$

$\{x \mid x < -2 \text{ or } x > 4\}$

5. $x^2 - 3x + 2 \geq 0$
$(x-1)(x-2) \geq 0$

$\{x \mid x \leq 1 \text{ or } x \geq 2\}$

7. $x^2 - x - 12 < 0$
$(x+3)(x-4) < 0$

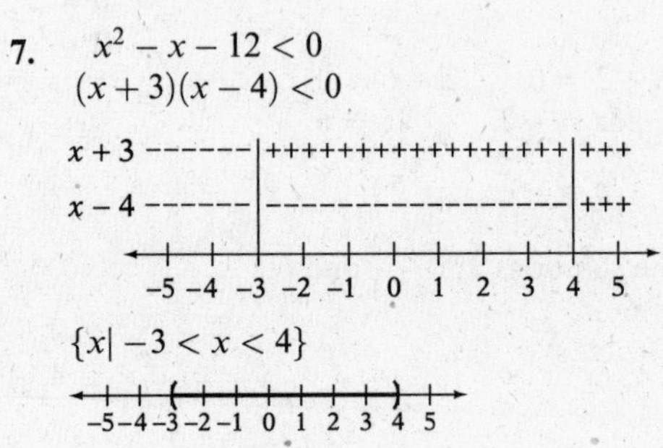

$\{x \mid -3 < x < 4\}$

9. $(x-1)(x+2)(x-3) < 0$

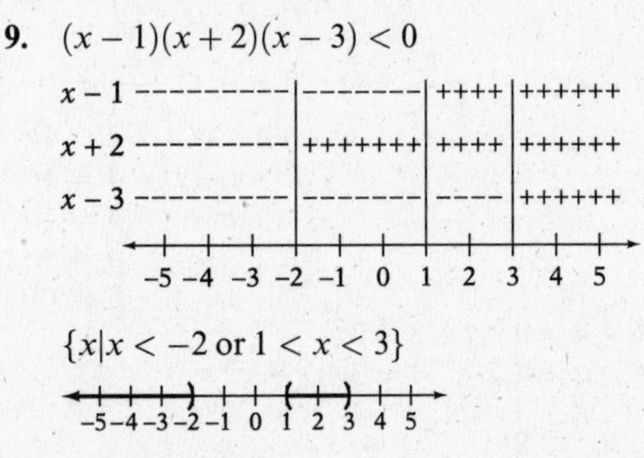

$\{x \mid x < -2 \text{ or } 1 < x < 3\}$

11. $(x+4)(x-2)(x-1) \geq 0$

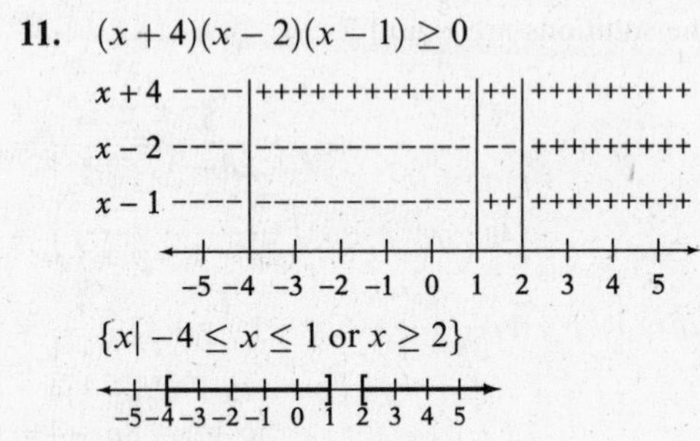

$\{x \mid -4 \leq x \leq 1 \text{ or } x \geq 2\}$

13. $\dfrac{x-4}{x+2} > 0$

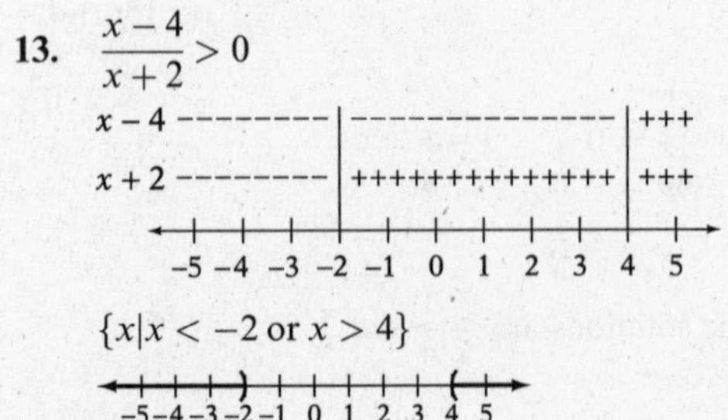

$\{x \mid x < -2 \text{ or } x > 4\}$

15. $\dfrac{x-3}{x+1} \le 0$

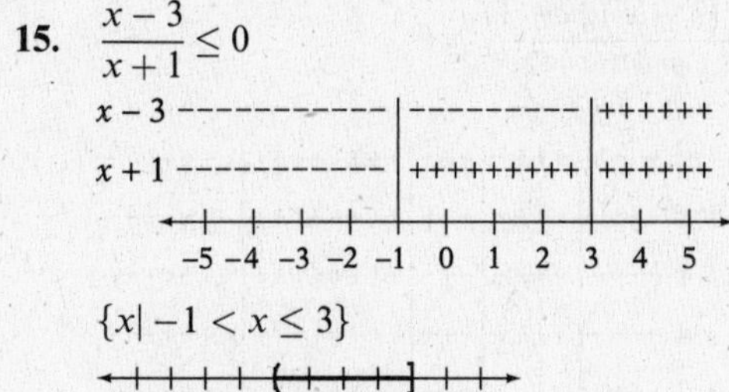

$\{x \mid -1 < x \le 3\}$

−5 −4 −3 −2 −1 0 1 2 3 4 5

17. $\dfrac{(x-1)(x+2)}{x-3} \le 0$

x − 1 --------- | ------- | ++++ | ++++++
x + 2 --------- | +++++++ | ++++ | ++++++
x − 3 --------- | ------- | ---- | ++++++
−5 −4 −3 −2 −1 0 1 2 3 4 5

$\{x \mid x \le -2 \text{ or } 1 \le x < 3\}$

−5 −4 −3 −2 −1 0 1 2 3 4 5

19.
$$x^2 - 16 > 0$$
$$(x-4)(x+4) > 0$$

x − 4 --- | ------------------ | +++
x + 4 --- | ++++++++++++++++++ | +++
−5 −4 −3 −2 −1 0 1 2 3 4 5

$\{x \mid x > 4 \text{ or } x < -4\}$

21.
$$x^2 - 9x \le 36$$
$$x^2 - 9x - 36 \le 0$$
$$(x+3)(x-12) \le 0$$

x + 3 ------- | ++++++++++++++++++ | ++
x − 12 ------- | ------------------ | ++
−8 −6 −4 −2 0 2 4 6 8 10 12

$\{x \mid -3 \le x \le 12\}$

23.
$$4x^2 - 8x + 3 < 0$$
$$(2x-1)(2x-3) < 0$$

2x − 1 --------------- | ++ | ++++++++++
2x − 3 --------------- | -- | ++++++++++
−5 −4 −3 −2 −1 0 1 2 3 4 5

$\left\{x \mid \frac{1}{2} < x < \frac{3}{2}\right\}$

25.
$$\frac{3}{x-1} < 2$$
$$\frac{3}{x-1} - 2 < 0$$
$$\frac{3}{x-1} - \frac{2x-2}{x-1} < 0$$
$$\frac{-2x+5}{x-1} < 0$$

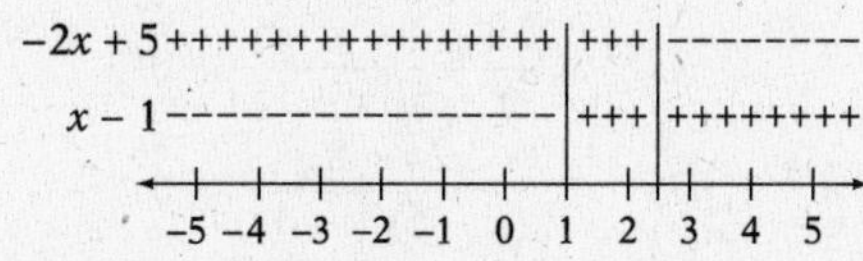

$\left\{x \mid x < 1 \text{ or } x > \frac{5}{2}\right\}$

27. $\dfrac{x-2}{(x+1)(x-1)} \le 0$

x − 2 ----------- | ---- | -- | +++++++++
x + 1 ----------- | ++++ | ++ | +++++++++
x − 1 ----------- | ---- | ++ | +++++++++
−5 −4 −3 −2 −1 0 1 2 3 4 5

$\{x \mid x < -1 \text{ or } 1 < x \le 2\}$

29.
$$\frac{x}{2x-1} \ge 1$$
$$\frac{x}{2x-1} - 1 \ge 0$$
$$\frac{x}{2x-1} - \frac{2x-1}{2x-1} \ge 0$$
$$\frac{-x+1}{2x-1} \ge 0$$

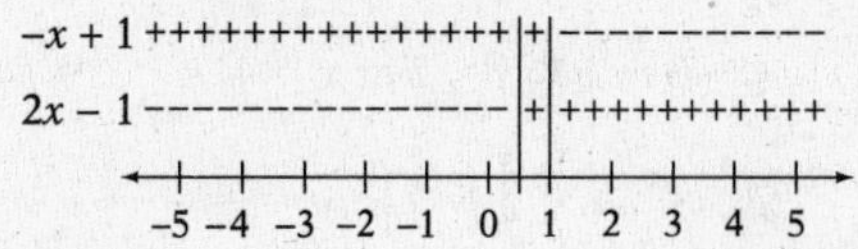

$\left\{x \mid \frac{1}{2} < x \le 1\right\}$

31.
$$\frac{x}{2-x} \le -3$$
$$\frac{x}{2-x} + 3 \le 0$$
$$\frac{x}{2-x} + \frac{3(2-x)}{2-x} \le 0$$
$$\frac{x+3(2-x)}{2-x} \le 0$$
$$\frac{x+6-3x}{2-x} \le 0$$
$$\frac{6-2x}{2-x} \le 0$$

6 − 2x +++++++++++++++++ | ++ | ------
2 − x +++++++++++++++++ | -- | ------
−5 −4 −3 −2 −1 0 1 2 3 4 5

$\{x \mid 2 < x \le 3\}$

33.

$$\frac{3}{x-5} > \frac{1}{x+1}$$
$$\frac{3}{x-5} - \frac{1}{x+1} > 0$$
$$\frac{3(x+1)}{(x-5)(x+1)} - \frac{1(x-5)}{(x-5)(x+1)} > 0$$
$$\frac{3(x+1)-1(x-5)}{(x-5)(x+1)} > 0$$
$$\frac{3x+3-x+5}{(x-5)(x+1)} > 0$$
$$\frac{2x+8}{(x-5)(x+1)} > 0$$
$$\frac{2(x+4)}{(x-5)(x+1)} > 0$$

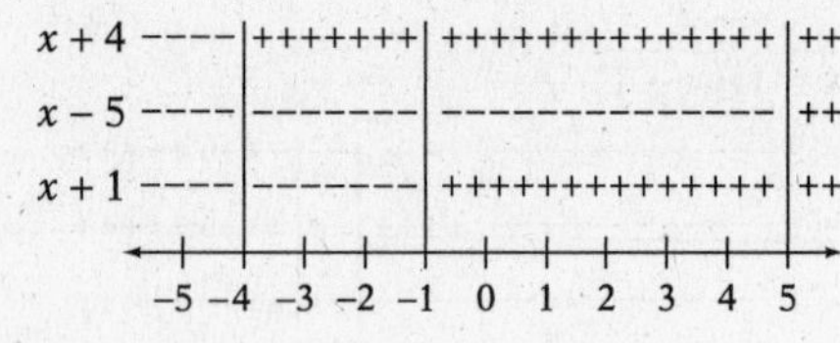

$\{x|x > 5 \text{ or } -4 < x < -1\}$

Applying the Concepts

35. $(x-1)(x+3)(x-2)(x-4) \geq 0$

```
x − 1 -------|---------|++|++++|++++
x + 3 -------|+++++++++|++|++++|++++
x − 2 -------|---------|--|++++|++++
x − 4 -------|---------|--|----|++++
      -5 -4 -3 -2 -1  0  1  2  3  4  5
```

$\{x|x \leq -3 \text{ or } 1 \leq x \leq 2 \text{ or } x \geq 4\}$

−4 −2 0 2 4

37. $(x^2+2x-3)(x^2+3x+2) \geq 0$
$(x+3)(x-1)(x+2)(x+1) \geq 0$

```
x + 3 -------|++|++|++++|+++++++++++
x − 1 -------|--|--|----|+++++++++++
x + 2 -------|--|++|++++|+++++++++++
x + 1 -------|--|--|++++|+++++++++++
      -5 -4 -3 -2 -1  0  1  2  3  4  5
```

$\{x|x \leq -3 \text{ or } -2 \leq x \leq -1 \text{ or } x \geq 1\}$

−4 −2 0 2 4

39. $\dfrac{x^2(3-x)(2x+1)}{(x+4)(x+2)} \geq 0$

```
  x²  ++++|++++|+++|++++++++|+++++++
3 − x ++++|++++|+++|++++++++|-------
2x + 1 ----|----|---|++++++++|+++++++
x + 4 ----|++++|+++|++++++++|+++++++
x + 2 ----|----|+++|++++++++|+++++++
      -5 -4 -3 -2 -1  0  1  2  3  4  5
```

$\{x| -4 < x < -2 \text{ or } -\frac{1}{2} \leq x \leq 3\}$

−4 −2 0 2 4

Section 8.6

Objective A Exercises

1. Strategy
- This is a geometry problem.
- The height of the triangle: x
  The base of the triangle: $5x - 1$
- The area of the triangle is 21cm^2.
  Use the equation for the area of a triangle.
  $\left(A = \frac{1}{2}bh\right)$

Solution

$$A = \frac{1}{2}bh$$
$$21 = \frac{1}{2}(5x-1)x$$
$$42 = 5x^2 - x$$
$$0 = 5x^2 - x - 42$$
$$0 = (5x+14)(x-3)$$
$$5x + 14 = 0 \qquad x - 3 = 0$$
$$x = -\frac{14}{5} \qquad x = 3$$

Since the height cannot be negative, $-\frac{14}{5}$ cannot be a solution.

$5x - 1 = 5(3) - 1 = 15 - 1 = 14$

The height is 3 cm.

The base is 14 cm.

3. **Strategy**
- This is a geometry problem.
- The width of the rectangle: x
  The length of the rectangle: $x + 111$
- The area of the rectangle is 104,000 sq. mi. Use the equation for the area of a rectangle. $(A = L \cdot W)$

Solution

$$A = L \cdot W$$
$$104{,}000 = (x + 111)x$$
$$104{,}000 = x^2 + 111x$$
$$0 = x^2 + 111x - 104{,}000$$
$$x = \frac{-b \pm \sqrt{b^2 - 4ac}}{2a}$$
$$x = \frac{-111 \pm \sqrt{111^2 - 4(1)(-104{,}000)}}{2(1)}$$
$$x = \frac{-111 \pm \sqrt{428{,}321}}{2}$$
$$x \approx 272 \text{ or } -383$$

Since distance cannot be a negative number, -383 cannot be a solution.

$x + 111 = 272 + 111 = 383$

The dimensions of Colorado are 272 mi by 383 mi.

5. **Strategy** To find the maximum safe speed, substitute for d in the equation $d = 0.04v^2 + 0.5v$ and solve for v.

Solution

$$d = 0.04v^2 + 0.5v$$
$$60 = 0.04v^2 + 0.5v$$
$$0 = 0.04v^2 + 0.5v - 60$$
$$v = \frac{-b \pm \sqrt{b^2 - 4ac}}{2a}$$
$$v = \frac{-0.5 \pm \sqrt{0.5^2 - 4(0.04)(-60)}}{2(0.04)}$$
$$v = \frac{-0.5 \pm \sqrt{9.85}}{0.08}$$
$$v \approx 33 \text{ or } -45$$

The speed cannot be a negative number. The maximum speed is 33 mph.

7. **Strategy** To find the time for a projectile to return to Earth, substitute the values for height ($s = 0$) and initial velocity ($v_0 = 200$ ft/s) and solve for t.

Solution

$$s = v_0t - 16t^2$$
$$0 = 200t - 16t^2$$
$$0 = 8t(25 - 2t)$$
$$8t = 0 \quad 25 - 2t = 0$$
$$t = 0 \quad t = 12.5$$

The solution $t = 0$ is not possible because the projectile has not yet left Earth. The rocket takes 12.5 s to return to Earth.

9. Strategy To find the maximum speed a driver can be going and still be able to stop within 150 m, substitute 150 in for d in the equation $d = 0.019v^2 + 0.69v$ and solve for v.

Solution

$$d = 0.019v^2 + 0.69v$$
$$150 = 0.019v^2 + 0.69v$$
$$0 = 0.019v^2 + 0.69v - 150$$
$$a = 0.019,\ b = 0.69,\ c = -150$$
$$v = \frac{-b \pm \sqrt{b^2 - 4ac}}{2a}$$
$$= \frac{-0.69 \pm \sqrt{(0.69)^2 - 4(0.019)(-150)}}{2(0.019)}$$
$$= \frac{-0.69 \pm \sqrt{0.4761 + 11.4}}{0.038} = \frac{-0.69 \pm \sqrt{11.8761}}{0.038}$$
$$t \approx -108.8 \text{ or } 72.5$$

The speed cannot be a negative number. A driver can be going approximately 72.5 km/h and still be able to stop within 150 m.

11. Strategy
- This is a work problem.
- Time for the smaller pipe to fill the tank: t.
- Time for the larger pipe to fill the tank: $t - 6$.

	Rate	Time	Part
Smaller Pipe	$\frac{1}{t}$	4	$\frac{4}{t}$
Larger Pipe	$\frac{1}{t-6}$	4	$\frac{4}{t-6}$

- The sum of the parts of the task completed must equal 1.

$$\frac{4}{t} + \frac{4}{t-6} = 1$$

Solution

$$\frac{4}{t} + \frac{4}{t-6} = 1$$
$$t(t-6)\left(\frac{4}{t} + \frac{4}{t-6}\right) = t(t-6)1$$
$$(t-6)4 + 4t = t^2 - 6t$$
$$4t - 24 + 4t = t^2 - 6t$$
$$8t - 24 = t^2 - 6t$$
$$0 = t^2 - 14t + 24$$
$$0 = (t-12)(t-2)$$
$$t - 12 = 0 \qquad t - 2 = 0$$
$$t = 12 \qquad t = 2$$
$$t - 6 = 12 - 6 = 6$$
$$t - 6 = 2 - 6 = -4$$

The solution -4 is not possible, since time cannot be a negative number.
It would take the larger pipe 6 min to fill the tank. It would take the smaller pipe 12 min to fill the tank.

13. Strategy

- This is a rate-of-wind problem.
- Rate of the wind: w.

	Distance	Rate	Time
With wind	4000	$1320 + w$	$\frac{4000}{1320+w}$
Against wind	4000	$1320 - w$	$\frac{4000}{1320-w}$

- It took 0.5 h less time to make the return trip.

Solution

$$\frac{4000}{1320 - w} - \frac{4000}{1320 + w} = 0.5$$

$$(1320 - w)(1320 + w)\left(\frac{4000}{1320 - w} - \frac{4000}{1320 + w}\right) = (1320 - w)(1320 + w)0.5$$

$$(1320 + w)4000 - (1320 - w)4000 = 0.5(1{,}742{,}400 - w^2)$$

$$5{,}280{,}000 + 4000w - 5{,}280{,}000 + 4000w = 871{,}200 - 0.5w^2$$

$$8000w = 871{,}200 - 0.5w^2$$

$$0.5w^2 + 8000w - 871{,}200 = 0$$

$$w = \frac{-b \pm \sqrt{b^2 - 4ac}}{2a}$$

$$w = \frac{-8000 \pm \sqrt{(8000)^2 - 4(0.5)(-871{,}200)}}{2(0.5)}$$

$$w = \frac{-8000 \pm \sqrt{65{,}742{,}400}}{1}$$

$w \approx -8000 \pm 8108$

$w = 108$ or $-16{,}108$

The rate cannot be a negative number.

The rate of the wind was approximately 108 mph.

15. Strategy

- This is a uniform motion problem.
- The rate of boat in calm water: r.

	Distance	Rate	Time
With current	5	$r + 4$	$\frac{5}{r+4}$
Against current	5	$r - 4$	$\frac{5}{r-4}$

- The total time for the trip is 3 h.

Solution

$$\frac{5}{r + 4} + \frac{5}{r - 4} = 3$$

$$(r + 4)(r - 4)\left(\frac{5}{r + 4} + \frac{5}{r - 4}\right) = (r + 4)(r - 4)3$$

$$(r - 4)5 + (r + 4)5 = (r^2 - 16)3$$

$$5r - 20 + 5r + 20 = 3r^2 - 48$$

$$10r = 3r^2 - 48$$

$$0 = 3r^2 - 10r - 48$$

$$0 = (3r + 8)(r - 6)$$

$3r + 8 = 0 \qquad r - 6 = 0$

$r = -\frac{8}{3} \qquad r = 6$

The rate cannot be a negative number.

The rowing rate of the guide is 6 mph.

Applying the Concepts

17. **Strategy** To find the radius of the cone, substitute 11.25π in for A and 6 in for s in the equation $A = \pi r^2 + \pi rs$ and solve for r.

Solution

$$A = \pi r^2 + \pi rs$$
$$11.25\pi = \pi r^2 + \pi r(6)$$
$$0 = \pi r^2 + 6\pi r - 11.25\pi$$
$$a = \pi,\ b = 6\pi,\ c = -11.25\pi$$
$$r = \frac{-b \pm \sqrt{b^2 - 4ac}}{2a}$$
$$= \frac{-6\pi \pm \sqrt{(6\pi)^2 - 4(\pi)(-11.25\pi)}}{2(\pi)}$$
$$= \frac{-6\pi \pm \sqrt{36\pi^2 + 45\pi^2}}{2\pi}$$
$$= \frac{-6\pi \pm \sqrt{81\pi^2}}{2\pi}$$
$$= \frac{-6\pi \pm 9\pi}{2\pi}$$
$$t = 1.5 \text{ or } -7.5$$

The radius cannot be a negative number. The radius of the cone is 1.5 in.

Chapter 8 Review Exercises

1. $$2x^2 - 3x = 0$$
$$x(2x - 3) = 0$$
$$x = 0 \quad 2x - 3 = 0$$
$$2x = 3$$
$$x = \frac{3}{2}$$

The solutions are 0 and $\frac{3}{2}$.

2. $$6x^2 + 9cx = 6c^2$$
$$6x^2 + 9cx - 6c^2 = 0$$
$$3(2x^2 + 3cx - 2c^2) = 0$$
$$3(2x - c)(x + 2c) = 0$$
$$2x - c = 0 \quad x + 2c = 0$$
$$2x = c \quad x = -2c$$
$$x = \frac{c}{2}$$

The solutions are $-2c$ and $\frac{c}{2}$.

3. $$x^2 = 48$$
$$\sqrt{x^2} = \sqrt{48}$$
$$x = \pm\sqrt{48} = \pm 4\sqrt{3}$$

The solutions are $-4\sqrt{3}$ and $4\sqrt{3}$.

4. $\left(x+\frac{1}{2}\right)^2+4=0$

$\left(x+\frac{1}{2}\right)^2=-4$

$\sqrt{\left(x+\frac{1}{2}\right)^2}=\sqrt{-4}$

$x+\frac{1}{2}=\pm\sqrt{-4}=\pm 2i$

$x+\frac{1}{2}=2i \qquad x+\frac{1}{2}=-2i$

$x=-\frac{1}{2}+2i \qquad x=-\frac{1}{2}-2i$

The solutions are $-\frac{1}{2}-2i$ and $-\frac{1}{2}+2i$.

5. $x^2+4x+3=0$

$x^2+4x=-3$

Complete the square.

$x^2+4x+4=-3+4$

$(x+2)^2=1$

$\sqrt{(x+2)^2}=\sqrt{1}$

$x+2=\pm\sqrt{1}=\pm 1$

$x+2=1 \qquad x+2=-1$

$x=-1 \qquad x=-3$

The solutions are -3 and -1.

6. $7x^2-14x+3=0$

$7x^2-14x=-3$

$\frac{1}{7}(7x^2-14x)=\frac{1}{7}(-3)$

$x^2-2x=-\frac{3}{7}$

Complete the square.

$x^2-2x+1=-\frac{3}{7}+1$

$(x-1)^2=\frac{4}{7}$

$\sqrt{(x-1)^2}=\sqrt{\frac{4}{7}}$

$x-1=\pm\sqrt{\frac{4}{7}}=\pm\frac{2\sqrt{7}}{7}$

$x-1=\frac{2\sqrt{7}}{7} \qquad x-1=-\frac{2\sqrt{7}}{7}$

$x=1+\frac{2\sqrt{7}}{7} \qquad x=1-\frac{2\sqrt{7}}{7}$

$=\frac{7+2\sqrt{7}}{7} \qquad =\frac{7-2\sqrt{7}}{7}$

The solutions are $\frac{7-2\sqrt{7}}{7}$ and $\frac{7+2\sqrt{7}}{7}$.

7. $12x^2-25x+12=0$

$a=12,\ b=-25,\ c=12$

$x=\frac{-b\pm\sqrt{b^2-4ac}}{2a}$

$=\frac{-(-25)\pm\sqrt{(-25)^2-4(12)(12)}}{2(12)}$

$=\frac{25\pm\sqrt{625-576}}{24}$

$=\frac{25\pm\sqrt{49}}{24}=\frac{25\pm 7}{24}$

$x=\frac{25+7}{24} \qquad x=\frac{25-7}{24}$

$=\frac{32}{24} \qquad =\frac{18}{24}$

$=\frac{4}{3} \qquad =\frac{3}{4}$

The solutions are $\frac{3}{4}$ and $\frac{4}{3}$.

8. $x^2-x+8=0$

$a=1,\ b=-1,\ c=8$

$x=\frac{-b\pm\sqrt{b^2-4ac}}{2a}$

$=\frac{-(-1)\pm\sqrt{(-1)^2-4(1)(8)}}{2(1)}$

$=\frac{1\pm\sqrt{1-32}}{2}$

$=\frac{1\pm\sqrt{-31}}{2}=\frac{1\pm i\sqrt{31}}{2}$

$=\frac{1}{2}\pm\frac{\sqrt{31}}{2}i$

The solutions are $\frac{1}{2}-\frac{\sqrt{31}}{2}i$ and $\frac{1}{2}+\frac{\sqrt{31}}{2}i$.

9. $(x-r_1)(x-r_2)=0$

$(x-0)[x-(-3)]=0$

$x(x+3)=0$

$x^2+3x=0$

10. $(x-r_1)(x-r_2)=0$

$\left(x-\frac{3}{4}\right)\left[x-\left(-\frac{2}{3}\right)\right]=0$

$\left(x-\frac{3}{4}\right)\left(x+\frac{2}{3}\right)=0$

$x^2-\frac{1}{12}x-\frac{1}{2}=0$

$12\left(x^2-\frac{1}{12}x-\frac{1}{2}\right)=12\cdot 0$

$12x^2-x-6=0$

11. $x^2 - 2x + 8 = 0$
$x^2 - 2x = -8$
Complete the square.
$x^2 - 2x + 1 = -8 + 1$
$(x-1)^2 = -7$
$\sqrt{(x-1)^2} = \sqrt{-7}$
$(x-1) = \pm\sqrt{-7} = \pm i\sqrt{7}$
$x - 1 = i\sqrt{7}$ $\quad x - 1 = -i\sqrt{7}$
$x = 1 + i\sqrt{7}$ $\quad x = 1 - i\sqrt{7}$
The solutions are $1 - i\sqrt{7}$ and $1 + i\sqrt{7}$.

12. $(x-2)(x+3) = x - 10$
$x^2 + x - 6 = x - 10$
$x^2 = -4$
Complete the square.
$x^2 + 0 = -4 + 0$
$x^2 = -4$
$\sqrt{x^2} = \sqrt{-4}$
$x = \pm\sqrt{-4}$
$x = \pm 2i$
The solutions are $-2i$ and $2i$.

13. $3x(x-3) = 2x - 4$
$3x^2 - 9x = 2x - 4$
$3x^2 - 11x + 4 = 0$
$a = 3, b = -11, c = 4$
$$x = \frac{-b \pm \sqrt{b^2 - 4ac}}{2a}$$
$$= \frac{-(-11) \pm \sqrt{(-11)^2 - 4(3)(4)}}{2(3)}$$
$$= \frac{11 \pm \sqrt{121 - 48}}{6}$$
$$= \frac{11 \pm \sqrt{73}}{6}$$
The solutions are $\frac{11 - \sqrt{73}}{6}$ and $\frac{11 + \sqrt{73}}{6}$.

14. $3x^2 - 5x + 1 = 0$
$a = 3, b = -5, c = 1$
$b^2 - 4ac = (-5)^2 - 4(3)(1)$
$25 - 12 = 13$
$13 > 0$
Since the discriminant is greater than zero, the equation has two real number solutions.

15. $(x+3)(2x-5) < 0$

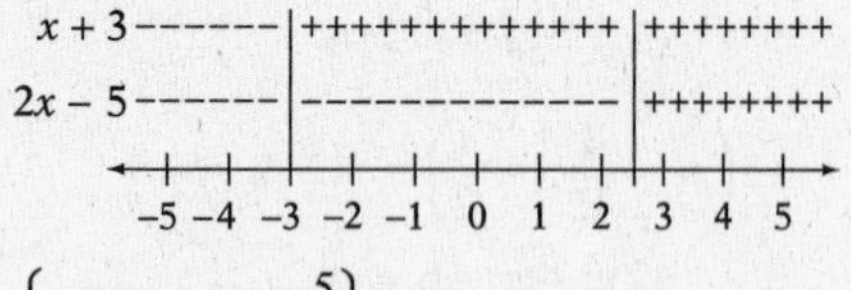

$x + 3$	------	+++++++++++++	++++++++
$2x - 5$	------	-------------	++++++++

−5 −4 −3 −2 −1 0 1 2 3 4 5

$\left\{x \mid -3 < x < \frac{5}{2}\right\}$

16. $(x-2)(x+4)(2x+3) \le 0$

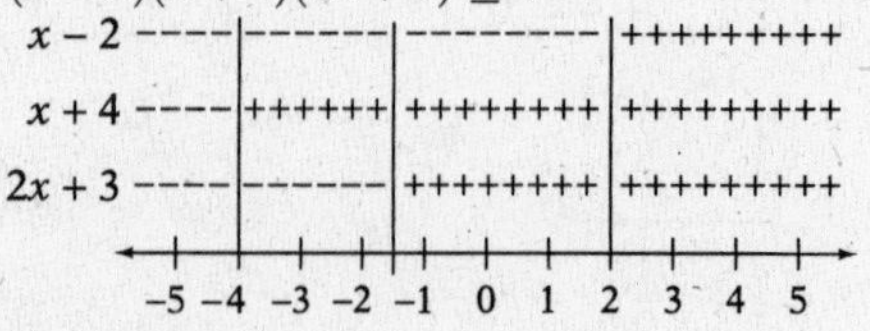

$x - 2$	----	------	--------	+++++++++
$x + 4$	----	++++++	++++++++	+++++++++
$2x + 3$	----	------	++++++++	+++++++++

−5 −4 −3 −2 −1 0 1 2 3 4 5

$\left\{x \mid x \le -4 \text{ or } -\frac{3}{2} \le x \le 2\right\}$

17. $x^{2/3} + x^{1/3} - 12 = 0$
$(x^{1/3})^2 + x^{1/3} - 12 = 0$
$u^2 + u - 12 = 0$
$(u+4)(u-3) = 0$
$u + 4 = 0$ $\quad u - 3 = 0$
$u = -4$ $\quad u = 3$
Replace u by $x^{1/3}$.
$x^{1/3} = -4$ $\quad x^{1/3} = 3$
$(x^{1/3})^3 = (-4)^3$ $\quad (x^{1/3})^3 = 3^3$
$x = -64$ $\quad x = 27$
The solutions are -64 and 27.

18. $2(x-1) + 3\sqrt{x-1} - 2 = 0$
$2(\sqrt{x-1})^2 + 3\sqrt{x-1} - 2 = 0$
$2u^2 + 3u - 2 = 0$
$(2u-1)(u+2) = 0$
$2u - 1 = 0$ $\quad u + 2 = 0$
$2u = 1$ $\quad u = -2$
$u = \frac{1}{2}$
Replace u by $\sqrt{x-1}$.
$\sqrt{x-1} = \frac{1}{2}$ $\quad \sqrt{x-1} = -2$
$(\sqrt{x-1})^2 = \left(\frac{1}{2}\right)^2$ $\quad (\sqrt{x-1})^2 = (-2)^2$
$x - 1 = \frac{1}{4}$ $\quad x - 1 = 4$
$x = \frac{5}{4}$ $\quad x = 5$
5 does not check as a solution.
The solution is $\frac{5}{4}$.

19. $3x = \frac{9}{x-2}$
$3x(x-2) = \frac{9}{x-2}(x-2)$
$3x^2 - 6x = 9$
$3x^2 - 6x - 9 = 0$
$3(x^2 - 2x - 3) = 0$
$3(x-3)(x+1) = 0$
$x - 3 = 0$ $\quad x + 1 = 0$
$x = 3$ $\quad x = -1$
The solutions are -1 and 3.

20.
$$\frac{3x+7}{x+2}+x=3$$
$$(x+2)\left(\frac{3x+7}{x+2}+x\right)=3(x+2)$$
$$3x+7+x(x+2)=3x+6$$
$$3x+7+x^2+2x=3x+6$$
$$x^2+5x+7=3x+6$$
$$x^2+2x+1=0$$
$$(x+1)^2=0$$
$$\sqrt{(x+1)^2}=\sqrt{0}$$
$$x+1=0$$
$$x=-1$$
The solution is -1.

21. $\frac{x-2}{2x-3}\geq 0$

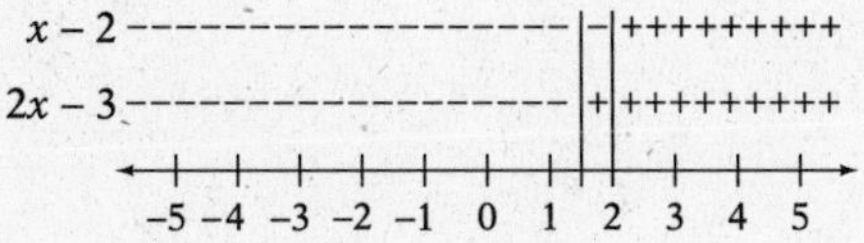

$x-2$ ------------------|-|+++++++++

$2x-3$ -----------------|+|+++++++++

$-5\ -4\ -3\ -2\ -1\ 0\ 1\ 2\ 3\ 4\ 5$

$\left\{x|x<\frac{3}{2}\text{ or }x\geq 2\right\}$

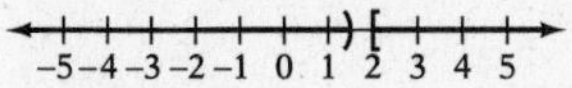

$-5\ -4\ -3\ -2\ -1\ 0\ 1\ 2\ 3\ 4\ 5$

22. $\frac{(2x-1)(x+3)}{x-4}\leq 0$

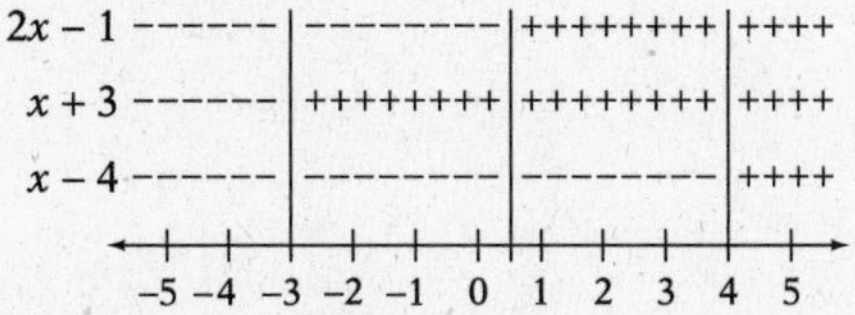

$2x-1$ ------|--------|++++++++|++++

$x+3$ ------|++++++++|++++++++|++++

$x-4$ ------|--------|--------|++++

$-5\ -4\ -3\ -2\ -1\ 0\ 1\ 2\ 3\ 4\ 5$

$\left\{x|x\leq -3\text{ or }\frac{1}{2}\leq x<4\right\}$

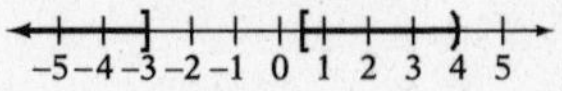

$-5\ -4\ -3\ -2\ -1\ 0\ 1\ 2\ 3\ 4\ 5$

23.
$$x=\sqrt{x}+2$$
$$x-\sqrt{x}-2=0$$
$$(\sqrt{x})^2-\sqrt{x}-2=0$$
$$u^2-u-2=0$$
$$(u-2)(u+1)=0$$
$$u-2=0\quad u+1=0$$
$$u=2\quad u=-1$$
Replace u by $\sqrt{x}$.
$$\sqrt{x}=2\qquad \sqrt{x}=-1$$
$$(\sqrt{x})^2=2^2\qquad (\sqrt{x})^2=(-1)^2$$
$$x=4\qquad x=1$$
4 checks as a solution.
1 does not check as a solution.
The solution is 4.

24.
$$2x=\sqrt{5x+24}+3$$
$$2x-3=\sqrt{5x+24}$$
$$(2x-3)^2=(\sqrt{5x+24})^2$$
$$4x^2-12x+9=5x+24$$
$$4x^2-17x-15=0$$
$$(4x+3)(x-5)=0$$
$$4x+3=0\qquad x-5=0$$
$$4x=-3\qquad x=5$$
$$x=-\frac{3}{4}$$
$-\frac{3}{4}$ does not check as a solution.
5 checks as a solution.
The solution is 5.

25.
$$\frac{x-2}{2x+3}-\frac{x-4}{x}=2$$
$$x(2x+3)\left[\frac{x-2}{2x+3}-\frac{x-4}{x}\right]=2[x(2x+3)]$$
$$x(x-2)-(2x+3)(x-4)=2x(2x+3)$$
$$x^2-2x-(2x^2-5x-12)=4x^2+6x$$
$$x^2-2x-2x^2+5x+12=4x^2+6x$$
$$-x^2+3x+12=4x^2+6x$$
$$0=5x^2+3x-12$$
$a=5,\ b=3,\ c=-12$
$$x=\frac{-b\pm\sqrt{b^2-4ac}}{2a}$$
$$=\frac{-3\pm\sqrt{3^2-4(5)(-12)}}{2(5)}$$
$$=\frac{-3\pm\sqrt{9+240}}{10}$$
$$=\frac{-3\pm\sqrt{249}}{10}$$
The solutions are $\frac{-3-\sqrt{249}}{10}$ and $\frac{-3+\sqrt{249}}{10}$.

26.
$$1-\frac{x+4}{2-x}=\frac{x-3}{x+2}$$
$$(x+2)(2-x)\left(1-\frac{x+4}{2-x}\right)=(x+2)(2-x)\frac{x-3}{x+2}$$
$$(x+2)(2-x)-(x+4)(x+2)=(x-3)(2-x)$$
$$4-x^2-(x^2+6x+8)=-x^2+5x-6$$
$$4-x^2-x^2-6x-8=-x^2+5x-6$$
$$-2x^2-6x-4=-x^2+5x-6$$
$$0=x^2+11x-2$$
$a=1,\ b=11,\ c=-2$
$$x=\frac{-b\pm\sqrt{b^2-4ac}}{2a}$$
$$=\frac{-11\pm\sqrt{11^2-4(1)(-2)}}{2(1)}$$
$$=\frac{-11\pm\sqrt{121+8}}{2}=\frac{-11\pm\sqrt{129}}{2}$$
The solutions are $\frac{-11-\sqrt{129}}{2}$ and $\frac{-11+\sqrt{129}}{2}$.

27.
$$(x - r_1)(x - r_2) = 0$$
$$\left(x - \frac{1}{3}\right)[x - (-3)] = 0$$
$$\left(x - \frac{1}{3}\right)(x + 3) = 0$$
$$x^2 + \frac{8}{3}x - 1 = 0$$
$$3\left(x^2 + \frac{8}{3}x - 1\right) = 3 \cdot 0$$
$$3x^2 + 8x - 3 = 0$$

28.
$$2x^2 + 9x = 5$$
$$2x^2 + 9x - 5 = 0$$
$$(2x - 1)(x + 5) = 0$$
$$2x - 1 = 0 \quad x + 5 = 0$$
$$2x = 1 \qquad x = -5$$
$$x = \frac{1}{2}$$
The solutions are -5 and $\frac{1}{2}$.

29.
$$2(x + 1)^2 - 36 = 0$$
$$2(x + 1)^2 = 36$$
$$(x + 1)^2 = 18$$
$$\sqrt{(x + 1)^2} = \sqrt{18}$$
$$x + 1 = \pm\sqrt{18} = \pm 3\sqrt{2}$$
$$x = -1 \pm 3\sqrt{2}$$
The solutions are $-1 + 3\sqrt{2}$ and $-1 - 3\sqrt{2}$.

30. $x^2 + 6x + 10 = 0$
$a = 1,\ b = 6,\ c = 10$
$$x = \frac{-b \pm \sqrt{b^2 - 4ac}}{2a}$$
$$x = \frac{-6 \pm \sqrt{(6)^2 - 4(1)(10)}}{2(1)}$$
$$x = \frac{-6 \pm \sqrt{-4}}{2} = \frac{-6 \pm i\sqrt{4}}{2} = \frac{-6 \pm 2i}{2}$$
$$x = -3 \pm i$$
The solutions are $-3 + i$ and $-3 - i$.

31.
$$\frac{2}{x - 4} + 3 = \frac{x}{2x - 3}$$
$$(x - 4)(2x - 3)\left[\frac{2}{x - 4} + 3\right] = (x - 4)(2x - 3) \cdot \frac{x}{2x - 3}$$
$$(2x - 3) \cdot 2 + (x - 4)(2x - 3) \cdot 3 = x(x - 4)$$
$$4x - 6 + 3(2x^2 - 11x + 12) = x^2 - 4x$$
$$4x - 6 + 6x^2 - 33x + 36 = x^2 - 4x$$
$$5x^2 - 25x + 30 = 0$$
$$x^2 - 5x + 6 = 0$$
$$(x - 2)(x - 3) = 0$$
$$x - 2 = 0 \quad x - 3 = 0$$
$$x = 2 \qquad x = 3$$
The solutions are 2 and 3.

32.
$$x^4 - 28x^2 + 75 = 0$$
$$(x^2)^2 - 28(x^2) + 75 = 0$$
$$u^2 - 28u + 75 = 0$$
$$(u - 25)(u - 3) = 0$$
$$u - 25 = 0 \quad u - 3 = 0$$
$$u = 25 \quad u = 3$$
$$x^2 = 25 \quad x^2 = 3$$
$$\sqrt{x^2} = \sqrt{25} \quad \sqrt{x^2} = \sqrt{3}$$
$$x = \pm 5 \quad x = \pm\sqrt{3}$$
The solutions are 5, -5, $\sqrt{3}$, and $-\sqrt{3}$.

33.
$$\sqrt{2x-1} + \sqrt{2x} = 3$$
$$\sqrt{2x-1} = 3 - \sqrt{2x}$$
$$(\sqrt{2x-1})^2 = (3 - \sqrt{2x})^2$$
$$2x - 1 = 9 - 6\sqrt{2x} + 2x$$
$$-10 = -6\sqrt{2x}$$
$$5 = 3\sqrt{2x}$$
$$5^2 = (3\sqrt{2x})^2$$
$$25 = 9 \cdot 2x$$
$$\frac{25}{18} = x$$
The solution is $\frac{25}{18}$.

34.
$$2x^{2/3} + 3x^{1/3} - 2 = 0$$
$$2(x^{1/3})^2 + 3(x^{1/3}) - 2 = 0$$
$$2u^2 + 3u - 2 = 0$$
$$(2u - 1)(u + 2) = 0$$
$$2u - 1 = 0 \quad u + 2 = 0$$
$$u = \frac{1}{2} \quad u = -2$$
$$x^{1/3} = \frac{1}{2} \quad x^{1/3} = -2$$
$$(x^{1/3})^3 = \left(\frac{1}{2}\right)^3 \quad (x^{1/3})^3 = (-2)^3$$
$$x = \frac{1}{8} \quad x = -8$$
The solutions are -8 and $\frac{1}{8}$.

35.
$$\sqrt{3x-2} + 4 = 3x$$
$$\sqrt{3x-2} = 3x - 4$$
$$(\sqrt{3x-2})^2 = (3x - 4)^2$$
$$3x - 2 = 9x^2 - 24x + 16$$
$$0 = 9x^2 - 27x + 18$$
$$0 = x^2 - 3x + 2$$
$$0 = (x - 2)(x - 1)$$
$$x - 2 = 0 \quad x - 1 = 0$$
$$x = 2 \quad x = 1$$
2 checks as a solution.
1 does not check as a solution.
The solution is 2.

36.
$$x^2 - 10x + 7 = 0$$
$$x^2 - 10x = -7$$
Complete the square.
$$x^2 - 10x + 25 = -7 + 25$$
$$(x - 5)^2 = 18$$
$$\sqrt{(x-5)^2} = \sqrt{18}$$
$$x - 5 = \pm 3\sqrt{2}$$
$$x = 5 \pm 3\sqrt{2}$$
The solutions are $5 + 3\sqrt{2}$ and $5 - 3\sqrt{2}$.

37.
$$\frac{2x}{x-4} + \frac{6}{x+1} = 11$$
$$(x-4)(x+1)\left[\frac{2x}{x-4} + \frac{6}{x+1}\right] = 11(x-4)(x+1)$$
$$2x(x+1) + 6(x-4) = 11x^2 - 33x - 44$$
$$2x^2 + 2x + 6x - 24 = 11x^2 - 33x - 44$$
$$0 = 9x^2 - 41x - 20$$
$$0 = (9x + 4)(x - 5)$$
$$9x + 4 = 0 \quad x - 5 = 0$$
$$x = -\frac{4}{9} \quad x = 5$$
The solutions are $-\frac{4}{9}$ and 5.

38.
$$9x^2 - 3x = 1$$
$$9x^2 - 3x - 1 = 0$$
$$a = 9,\ b = -3,\ c = -1$$
$$x = \frac{-b \pm \sqrt{b^2 - 4ac}}{2a}$$
$$x = \frac{-(-3) \pm \sqrt{(-3)^2 - 4(9)(-1)}}{2(9)}$$
$$x = \frac{3 \pm \sqrt{9 + 36}}{18} = \frac{3 \pm \sqrt{45}}{18} = \frac{3 \pm 3\sqrt{5}}{18}$$
$$x = \frac{1 \pm \sqrt{5}}{6}$$
The solutions are $\frac{1+\sqrt{5}}{6}$ and $\frac{1-\sqrt{5}}{6}$.

39.
$$2x = 4 - 3\sqrt{x-1}$$
$$2x - 4 = -3\sqrt{x-1}$$
$$(2x - 4)^2 = (-3\sqrt{x-1})^2$$
$$4x^2 - 16x + 16 = 9(x - 1)$$
$$4x^2 - 16x + 16 = 9x - 9$$
$$4x^2 - 25x + 25 = 0$$
$$(4x - 5)(x - 5) = 0$$
$$4x - 5 = 0 \quad x - 5 = 0$$
$$4x = 5 \quad x = 5$$
$$x = \frac{5}{4}$$
$\frac{5}{4}$ checks as a solution.
5 does not check as a solution.
The solution is $\frac{5}{4}$.

40.

$$1-\frac{x+3}{3-x}=\frac{x-4}{x+3}$$
$$1+\frac{x+3}{x-3}=\frac{x-4}{x+3}$$
$$(x-3)(x+3)\left[1+\frac{x+3}{x-3}\right]=(x-3)(x+3)\left(\frac{x-4}{x+3}\right)$$
$$(x-3)(x+3)+(x+3)(x+3)=(x-3)(x-4)$$
$$x^2-9+x^2+6x+9=x^2-7x+12$$
$$2x^2+6x=x^2-7x+12$$
$$x^2+13x-12=0$$

$a=1,\ b=13,\ c=-12$

$$x=\frac{-b\pm\sqrt{b^2-4ac}}{2a}$$
$$x=\frac{-13\pm\sqrt{(13)^2-4(1)(-12)}}{2(1)}$$
$$x=\frac{-13\pm\sqrt{169+48}}{2}$$
$$x=\frac{-13\pm\sqrt{217}}{2}$$

The solutions are $\frac{-13+\sqrt{217}}{2}$ and $\frac{-13-\sqrt{217}}{2}$.

41.

$$2x^2-5x=6$$
$$2x^2-5x-6=0$$

$a=2,\ b=-5,\ c=-6$

$$b^2-4ac=(-5)^2-4(2)(-6)$$
$$=25+48=73$$

Since the discriminant is greater than zero, the equation has two real number solutions.

42.

$$x^2-3x\le 10$$
$$x^2-3x-10\le 0$$
$$(x-5)(x+2)\le 0$$

The zeros are -2 and 5. The factors have opposite signs between the zeros. The solution set is $\{x|-2\le x\le 5\}$.

43. **Strategy**
- This is a uniform motion problem.
- Rate of rowing in calm water: r

	Distance	Rate	Time
With current	16	$r+2$	$\frac{16}{r+2}$
Against current	16	$r-2$	$\frac{16}{r-2}$

- The total time for the trip was 6 h.

Solution

$$\frac{16}{r+2}+\frac{16}{r-2}=6$$
$$(r+2)(r-2)\left[\frac{16}{r+2}+\frac{16}{r-2}\right]=6(r+2)(r-2)$$
$$16(r-2)+16(r+2)=6r^2-24$$
$$16r-32+16r+32=6r^2-24$$
$$0=6r^2-32r-24$$
$$0=3r^2-16r-12$$
$$0=(r-6)(3r+2)$$

$$r-6=0\qquad 3r+2=0$$
$$r=6\qquad r=-\frac{2}{3}$$

The rate cannot be a negative number. The sculling crew's rate of rowing in calm water is 6 mph.

44. **Strategy**
- This is a geometry problem.
- The width of the rectangle: x.
- The length of the rectangle: $2x + 2$.
- The area of the rectangle is 60 cm^2.

Use the equation for the area of the rectangle ($A = L \cdot W$).

Solution

$$A = L \cdot W$$
$$60 = x(2x + 2)$$
$$60 = 2x^2 + 2x$$
$$0 = 2x^2 + 2x - 60$$
$$0 = 2(x^2 + x - 30)$$
$$0 = 2(x + 6)(x - 5)$$
$$x + 6 = 0 \qquad x - 5 = 0$$
$$x = -6 \qquad x = 5$$

Since the width of the rectangle cannot be negative, -6 cannot be a solution.

$2x + 2 = 2(5) + 2 = 10 + 2 = 12$

The width of the rectangle is 5 cm.
The length of the rectangle is 12 cm.

45. **Strategy**
- This is an integer problem.
- The first integer: x.
- The second consecutive even integer: $x + 2$.
- The third consecutive even integer: $x + 4$.
- The sum of the squares of the three consecutive even integers is 56.

$x^2 + (x + 2)^2 + (x + 4)^2 = 56$

Solution

$$x^2 + (x + 2)^2 + (x + 4)^2 = 56$$
$$x^2 + x^2 + 4x + 4 + x^2 + 8x + 16 = 56$$
$$3x^2 + 12x - 36 = 0$$
$$3(x^2 + 4x - 12) = 0$$
$$3(x + 6)(x - 2) = 0$$
$$x + 6 = 0 \qquad x - 2 = 0$$
$$x = -6 \qquad x = 2$$
$$x = 2,\ x + 2 = 4,\ x + 4 = 6$$
$$x = -6,\ x + 2 = -4,\ x + 4 = -2$$

The integers are 2, 4, and 6 or -6, -4, and -2.

46. **Strategy**
- This is a work problem.
- Time for new computer to print payroll: x
- Time for older computer to print payroll: $x + 12$

	Rate	Time	Part
New Computer	$\frac{1}{x}$	8	$\frac{8}{x}$
Older Computer	$\frac{1}{x+12}$	8	$\frac{8}{x+2}$

- The sum of the parts of the task completed must be 1.

$\frac{8}{x} + \frac{8}{x + 12} = 1$

Solution

$$\frac{8}{x} + \frac{8}{x + 12} = 1$$
$$x(x + 12)\left(\frac{8}{x} + \frac{8}{x + 12}\right) = x(x + 12)(1)$$
$$8(x + 12) + 8x = x(x + 12)$$
$$8x + 96 + 8x = x^2 + 12x$$
$$16x + 96 = x^2 + 12x$$
$$0 = x^2 - 4x - 96$$
$$0 = (x - 12)(x + 8)$$
$$x - 12 = 0 \qquad x + 8 = 0$$
$$x = 12 \qquad x = -8$$

The solution -8 is not possible, since time cannot be a negative number. Working alone, the new computer can print the payroll in 12 min.

47. **Strategy**
- This is a distance-rate problem.
- Rate of the first car: r
- Rate of the second car: $r + 10$

	Distance	Rate	Time
1st car	200	r	$\frac{200}{r}$
2nd car	200	$r + 10$	$\frac{200}{r+10}$

- The second car's time is one hour less than the time of the first car.

$\frac{200}{r + 10} = \frac{200}{r} - 1$

Solution

$$\frac{200}{r + 10} = \frac{200}{r} - 1$$
$$r(r + 10)\left(\frac{200}{r + 10}\right) = \left(\frac{200}{r} - 1\right)r(r + 10)$$
$$200r = 200(r + 10) - r(r + 10)$$
$$200r = 200r + 2000 - r^2 - 10r$$
$$r^2 + 10r - 2000 = 0$$
$$(r + 50)(r - 40) = 0$$
$$r + 50 = 0 \qquad r - 40 = 0$$
$$r = -50 \qquad r = 40$$

The solution -50 is not possible, since rate cannot be a negative number.

$r + 10 = 40 + 10 = 50$

The rate of the first car is 40 mph.
The rate of the second car is 50 mph.

Chapter 8 Test

1.
$$3x^2 + 10x = 8$$
$$3x^2 + 10x - 8 = 0$$
$$(3x - 2)(x + 4) = 0$$
$$3x - 2 = 0 \qquad x + 4 = 0$$
$$3x = 2 \qquad x = -4$$
$$x = \frac{2}{3}$$

The solutions are -4 and $\frac{2}{3}$.

2. $6x^2 - 5x - 6 = 0$

$(2x-3)(3x+2) = 0$

$2x - 3 = 0 \quad 3x + 2 = 0$

$2x = 3 \quad 3x = -2$

$x = \frac{3}{2} \quad x = -\frac{2}{3}$

The solutions are $-\frac{2}{3}$ and $\frac{3}{2}$.

3. $(x - r_1)(x - r_2) = 0$

$(x-3)[x-(-3)] = 0$

$(x-3)(x+3) = 0$

$x^2 - 9 = 0$

4. $(x - r_1)(x - r_2) = 0$

$\left(x - \frac{1}{2}\right)[x - (-4)] = 0$

$\left(x - \frac{1}{2}\right)(x+4) = 0$

$x^2 + \frac{7}{2}x - 2 = 0$

$2\left(x^2 + \frac{7}{2}x - 2\right) = 2 \cdot 0$

$2x^2 + 7x - 4 = 0$

5. $3(x-2)^2 - 24 = 0$

$3(x-2)^2 = 24$

$(x-2)^2 = 8$

$\sqrt{(x-2)^2} = \sqrt{8}$

$x - 2 = \pm 2\sqrt{2}$

$x - 2 = 2\sqrt{2} \quad x - 2 = -2\sqrt{2}$

$x = 2 + 2\sqrt{2} \quad x = 2 - 2\sqrt{2}$

The solutions are $2 - 2\sqrt{2}$ and $2 + 2\sqrt{2}$.

6. $x^2 - 6x - 2 = 0$

$x^2 - 6x = 2$

Complete the square.

$x^2 - 6x + 9 = 2 + 9$

$(x-3)^2 = 11$

$\sqrt{(x-3)^2} = \sqrt{11}$

$x - 3 = \pm\sqrt{11}$

$x - 3 = \sqrt{11} \quad x - 3 = -\sqrt{11}$

$x = 3 + \sqrt{11} \quad x = 3 - \sqrt{11}$

The solutions are $3 - \sqrt{11}$ and $3 + \sqrt{11}$.

7. $3x^2 - 6x = 2$

$\frac{1}{3}(3x^2 - 6x) = \frac{1}{3}(2)$

$x^2 - 2x = \frac{2}{3}$

Complete the square.

$x^2 - 2x + 1 = \frac{2}{3} + 1$

$(x-1)^2 = \frac{5}{3}$

$\sqrt{(x-1)^2} = \sqrt{\frac{5}{3}}$

$x - 1 = \pm\frac{\sqrt{15}}{3}$

$x - 1 = \frac{\sqrt{15}}{3} \quad x - 1 = -\frac{\sqrt{15}}{3}$

$x = 1 + \frac{\sqrt{15}}{3} \quad x = 1 - \frac{\sqrt{15}}{3}$

$= \frac{3 + \sqrt{15}}{3} \quad = \frac{3 - \sqrt{15}}{3}$

The solutions are $\frac{3 - \sqrt{15}}{3}$ and $\frac{3 + \sqrt{15}}{3}$.

8. $2x^2 - 2x = 1$

$2x^2 - 2x - 1 = 0$

$a = 2,\ b = -2,\ c = -1$

$x = \frac{-b \pm \sqrt{b^2 - 4ac}}{2a}$

$= \frac{-(-2) \pm \sqrt{(-2)^2 - 4(2)(-1)}}{2(2)}$

$= \frac{2 \pm \sqrt{4+8}}{4}$

$= \frac{2 \pm \sqrt{12}}{4}$

$= \frac{2 \pm 2\sqrt{3}}{4}$

$= \frac{1 \pm \sqrt{3}}{2}$

The solutions are $\frac{1 - \sqrt{3}}{2}$ and $\frac{1 + \sqrt{3}}{2}$.

9. $x^2 + 4x + 12 = 0$

$a = 1,\ b = 4,\ c = 12$

$x = \frac{-b \pm \sqrt{b^2 - 4ac}}{2a}$

$= \frac{-4 \pm \sqrt{(4)^2 - 4(1)(12)}}{2(1)}$

$= \frac{-4 \pm \sqrt{16 - 48}}{2}$

$= \frac{-4 \pm \sqrt{-32}}{2}$

$= \frac{-4 \pm 4i\sqrt{2}}{2}$

$= -2 \pm 2i\sqrt{2}$

The solutions are $-2 - 2i\sqrt{2}$ and $-2 + 2i\sqrt{2}$.

10.
$$
\begin{aligned}
3x^2 - 4x &= 1\\
3x^2 - 4x - 1 &= 0
\end{aligned}
$$
$a = 3, b = -4, c = -1$
$$
\begin{aligned}
b^2 - 4ac &= (-4)^2 - 4(3)(-1)\\
&= 16 + 12 = 28
\end{aligned}
$$
$28 > 0$

Since the discriminant is greater than zero, the equation has two real number solutions.

11.
$$
\begin{aligned}
x^2 - 6x &= -15\\
x^2 - 6x + 15 &= 0
\end{aligned}
$$
$a = 1, b = -6, c = 15$
$$
\begin{aligned}
b^2 - 4ac &= (-6)^2 - 4(1)(15)\\
&= 36 - 60 = -24
\end{aligned}
$$
$-24 < 0$

Since the discriminant is less than zero, the equation has two complex number solutions.

12.
$$
\begin{aligned}
2x + 7x^{1/2} - 4 &= 0\\
2(x^{1/2})^2 + 7x^{1/2} - 4 &= 0\\
2u^2 + 7u - 4 &= 0\\
(2u - 1)(u + 4) &= 0
\end{aligned}
$$
$$2u - 1 = 0 \qquad u + 4 = 0$$
$$2u = 1 \qquad u = -4$$
$$u = \frac{1}{2}$$

Replace u by $x^{1/2}$.
$$x^{1/2} = \frac{1}{2} \qquad x^{1/2} = -4$$
$$(x^{1/2})^2 = \left(\frac{1}{2}\right)^2 \qquad (x^{1/2})^2 = (-4)^2$$
$$x = \frac{1}{4} \qquad x = 16$$

16 does not check as a solution.

$\frac{1}{4}$ does check as a solution.

The solution is $\frac{1}{4}$.

13.
$$
\begin{aligned}
x^4 - 4x^2 + 3 &= 0\\
(x^2)^2 - 4(x^2) + 3 &= 0\\
u^2 - 4u + 3 &= 0\\
(u - 3)(u - 1) &= 0
\end{aligned}
$$
$$u - 3 = 0 \qquad u - 1 = 0$$
$$u = 3 \qquad u = 1$$

Replace u by x^2.
$$x^2 = 3 \qquad x^2 = 1$$
$$\sqrt{x^2} = \sqrt{3} \qquad \sqrt{x^2} = \sqrt{1}$$
$$x = \pm\sqrt{3} \qquad x = \pm 1$$

The solutions are $-\sqrt{3}$, $\sqrt{3}$, -1, and 1.

14.
$$
\begin{aligned}
\sqrt{2x+1} + 5 &= 2x\\
\sqrt{2x+1} &= 2x - 5\\
(\sqrt{2x+1})^2 &= (2x - 5)^2\\
2x + 1 &= 4x^2 - 20x + 25\\
0 &= 4x^2 - 22x + 24\\
0 &= 2(2x^2 - 11x + 12)\\
0 &= 2(2x - 3)(x - 4)
\end{aligned}
$$
$$2x - 3 = 0 \qquad x - 4 = 0$$
$$2x = 3 \qquad x = 4$$
$$x = \frac{3}{2}$$

$\frac{3}{2}$ does not check as a solution.

4 does check as a solution.

The solution is 4.

15.
$$
\begin{aligned}
\sqrt{x-2} &= \sqrt{x} - 2\\
(\sqrt{x-2})^2 &= (\sqrt{x} - 2)^2\\
x - 2 &= x - 4\sqrt{x} + 4\\
-6 &= -4\sqrt{x}\\
-\frac{1}{4}(-6) &= (-4\sqrt{x})\left(-\frac{1}{4}\right)\\
\frac{3}{2} &= \sqrt{x}\\
\frac{9}{4} &= x
\end{aligned}
$$

$\frac{9}{4}$ does not check as a solution.

The equation has no solution.

16.
$$
\begin{aligned}
\frac{2x}{x-3} + \frac{5}{x-1} &= 1\\
(x-3)(x-1)\left(\frac{2x}{x-3} + \frac{5}{x-1}\right) &= (x-3)(x-1)1\\
2x(x-1) + 5(x-3) &= (x-3)(x-1)\\
2x^2 - 2x + 5x - 15 &= x^2 - 4x + 3\\
2x^2 + 3x - 15 &= x^2 - 4x + 3\\
x^2 + 7x - 18 &= 0\\
(x+9)(x-2) &= 0
\end{aligned}
$$
$$x + 9 = 0 \qquad x - 2 = 0$$
$$x = -9 \qquad x = 2$$

The solutions are -9 and 2.

17. $(x - 2)(x + 4)(x - 4) < 0$

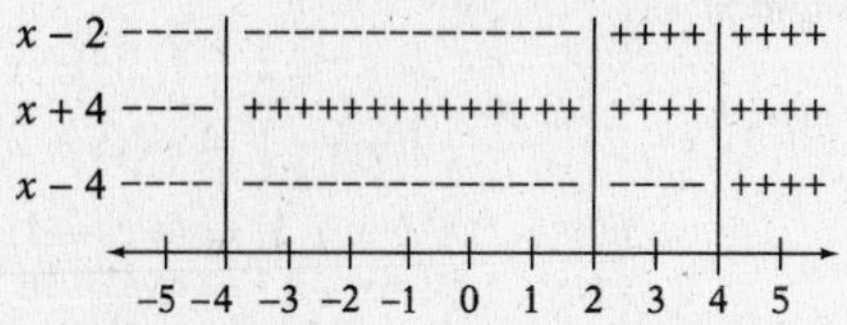

	$x < -4$	$-4 < x < 2$	$2 < x < 4$	$x > 4$
$x - 2$	----	--------------	++++	++++
$x + 4$	----	++++++++++++++	++++	++++
$x - 4$	----	--------------	----	++++

–5 –4 –3 –2 –1 0 1 2 3 4 5

$\{x | x < -4 \text{ or } 2 < x < 4\}$

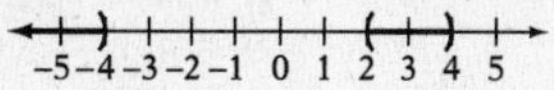

–5 –4 –3 –2 –1 0 1 2 3 4 5

18. $\frac{2x-3}{x+4} \le 0$

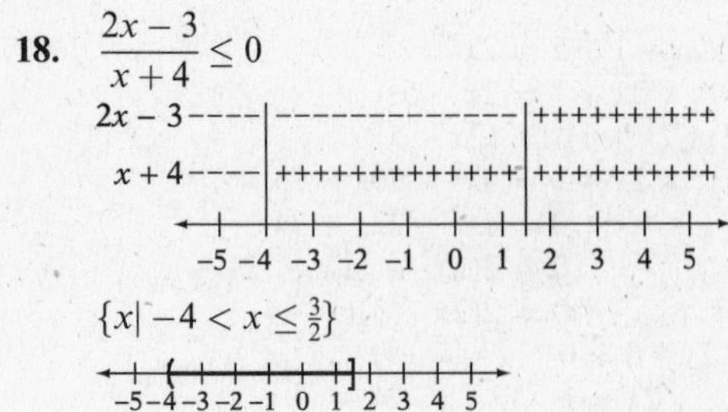

$\{x \mid -4 < x \le \frac{3}{2}\}$

19. Strategy To find the time when the ball hits the basket substitute 10 ft (the height of the basket) for h in $h = -16t^2 + 32t + 6.5$ and solve for t.

Solution

$$h = -16t^2 + 32t + 6.5$$
$$10 = -16t^2 + 32t + 6.5$$
$$0 = -16t^2 + 32t - 3.5$$
$$0 = 16t^2 + 32t + 3.5$$
$$a = 16,\ b = -32,\ c = 3.5$$
$$t = \frac{32 \pm \sqrt{(-32)^2 - 4(16)(3.5)}}{2(16)}$$
$$t = \frac{32 \pm \sqrt{1024 - 224}}{32}$$
$$t = \frac{32 \pm \sqrt{800}}{32}$$
$$t \approx 1.88 \text{ or } 0.12$$

The value 0.125 is the first time the ball is at a height of 10 ft. We need the time it takes to reach the basket after it has reached its peak.

The ball hits the basket 1.88 s after it is released.

20. Strategy
- This is a distance-rate problem.
- The rate of the canoe in calm water: x.

	Distance	Rate	Time
With current	6	$x+2$	$\frac{6}{x+2}$
Against current	6	$x-2$	$\frac{6}{x-2}$

- The total traveling time is 4 h.

$$\frac{6}{x+2} + \frac{6}{x-2} = 4$$

Solution

$$\frac{6}{x+2} + \frac{6}{x-2} = 4$$
$$(x+2)(x-2)\left(\frac{6}{x+2} + \frac{6}{x-2}\right) = (4)(x+2)(x-2)$$
$$6(x-2) + 6(x+2) = 4(x^2 - 4)$$
$$6x - 12 + 6x + 12 = 4x^2 - 16$$
$$12x = 4x^2 - 16$$
$$0 = 4x^2 - 12x - 16$$
$$0 = 4(x^2 - 3x - 4)$$
$$0 = 4(x-4)(x+1)$$
$$x - 4 = 0 \quad x + 1 = 0$$
$$x = 4 \quad x = -1$$

The solution -1 is not possible, since rate cannot be a negative number. The rate of the canoe in calm water is 4 mph.

Cumulative Review Exercises

1.
$$2a^2 - b^2 \div c = 2(3)^2 - (-4)^2 \div (-2)^2$$
$$= 2(9) - 16 \div 4$$
$$= 18 - 16 \div 4$$
$$= 18 - 4$$
$$= 14$$

2.
$$\frac{2x-3}{4} - \frac{x+4}{6} = \frac{3x-2}{8}$$
$$24\left(\frac{2x-3}{4} - \frac{x+4}{6}\right) = 24\left(\frac{3x-2}{8}\right)$$
$$6(2x-3) - 4(x+4) = 3(3x-2)$$
$$12x - 18 - 4x - 16 = 9x - 6$$
$$8x - 34 = 9x - 6$$
$$-x - 34 = -6$$
$$-x = 28$$
$$x = -28$$

The solution is -28.

3. $P_1(3, -4), P_2(-1, 2)$

$$m = \frac{y_2 - y_1}{x_2 - x_1} = \frac{2-(-4)}{-1-3} = \frac{2+4}{-4} = \frac{6}{-4} = -\frac{3}{2}$$

4.
$$x - y = 1$$
$$y = x - 1$$
$$m = 1,\ (x_1, y_1) = (1, 2)$$
$$y - y_1 = m(x - x_1)$$
$$y - 2 = 1(x - 1)$$
$$y - 2 = x - 1$$
$$y = x + 1$$

5. $-3x^3y + 6x^2y^2 - 9xy^3 = -3xy(x^2 - 2xy + 3y^2)$

6. $6x^2 - 7x - 20 = (2x-5)(3x+4)$

7.
$$a^nx + a^ny - 2x - 2y = a^n(x+y) - 2(x+y)$$
$$= (x+y)(a^n - 2)$$

8.
$$\begin{array}{r} x^2 - 3x - 4 \\ 3x-4\overline{)3x^3 - 13x^2 + 0x + 10} \\ \underline{3x^3 - 4x^2} \\ -9x^2 + 0x \\ \underline{-9x^2 + 12x} \\ -12x + 10 \\ \underline{-12x + 16} \\ -6 \end{array}$$

$$(3x^3 - 13x^2 + 10) \div (3x - 4)$$
$$= x^2 - 3x - 4 - \frac{6}{3x-4}$$

9.
$$\frac{x^2 + 2x + 1}{8x^2 + 8x} \cdot \frac{4x^3 - 4x^2}{x^2 - 1}$$
$$= \frac{(x+1)(x+1)}{8x(x+1)} \cdot \frac{4x^2(x-1)}{(x+1)(x-1)}$$
$$= \frac{(x+1)(x+1)4x^2(x-1)}{8x(x+1)(x+1)(x-1)} = \frac{x}{2}$$

10. Distance between points is
$\sqrt{(x_2 - x_1)^2 + (y_2 - y_1)^2}$.

$$\begin{aligned}\text{Distance} &= \sqrt{[2-(-2)]^2 + (5-3)^2}\\ &= \sqrt{4^2 + 2^2}\\ &= \sqrt{20}\\ &= \sqrt{4 \cdot 5}\\ &= 2\sqrt{5}\end{aligned}$$

The distance between the points is $2\sqrt{5}$.

11.
$$\begin{aligned}S &= \frac{n}{2}(a+b)\\ 2S &= n(a+b)\\ 2S &= an + bn\\ 2S - an &= bn\\ \frac{2S - an}{n} &= b\\ b &= \frac{2S - an}{n}\end{aligned}$$

12.
$$\begin{aligned}-2i(7-4i) &= -14i + 8i^2\\ &= -8 - 14i\end{aligned}$$

13.
$$\begin{aligned}a^{-1/2}(a^{1/2} - a^{3/2}) &= a^0 - a^1\\ &= 1 - a\end{aligned}$$

14.
$$\begin{aligned}\frac{\sqrt[3]{8x^4y^5}}{\sqrt[3]{16xy^6}} &= \sqrt[3]{\frac{8x^4y^5}{16xy^6}}\\ &= \sqrt[3]{\frac{x^3}{2y}}\\ &= \frac{\sqrt[3]{x^3}}{\sqrt[3]{2y}}\\ &= \frac{x}{\sqrt[3]{2y}} \cdot \frac{\sqrt[3]{4y^2}}{\sqrt[3]{4y^2}}\\ &= \frac{x\sqrt[3]{4y^2}}{\sqrt[3]{8y^3}} = \frac{x\sqrt[3]{4y^2}}{2y}\end{aligned}$$

15.
$$\begin{aligned}\frac{x}{x+2} - \frac{4x}{x+3} &= 1\\ (x+2)(x+3)\left[\frac{x}{x+2} - \frac{4x}{x+3}\right] &= (x+2)(x+3)1\\ x(x+3) - 4x(x+2) &= (x+2)(x+3)\\ x^2 + 3x - 4x^2 - 8x &= x^2 + 5x + 6\\ -3x^2 - 5x &= x^2 + 5x + 6\\ 0 &= 4x^2 + 10x + 6\\ 0 &= 2(2x^2 + 5x + 3)\\ 0 &= 2(2x+3)(x+1)\end{aligned}$$

$$\begin{aligned}2x + 3 &= 0 & x + 1 &= 0\\ 2x &= -3 & x &= -1\\ x &= -\frac{3}{2}\end{aligned}$$

The solutions are $-\frac{3}{2}$ and -1.

16.
$$\begin{aligned}\frac{x}{2x+3} - \frac{3}{4x^2-9} &= \frac{x}{2x-3}\\ \frac{x}{2x+3} - \frac{3}{(2x+3)(2x-3)} &= \frac{x}{2x-3}\\ (2x+3)(2x-3)\left[\frac{x}{2x+3} - \frac{3}{(2x+3)(2x-3)}\right] &= \frac{x}{2x-3}(2x+3)(2x-3)\\ (2x-3)x - 3 &= (2x+3)x\\ 2x^2 - 3x - 3 &= 2x^2 + 3x\\ -3 &= 6x\\ x &= -\frac{1}{2}\end{aligned}$$

Check:
$$\begin{aligned}\frac{-\frac{1}{2}}{2(-\frac{1}{2}) + 3} - \frac{3}{4(-\frac{1}{2})^2 - 9} &= \frac{-\frac{1}{2}}{2(-\frac{1}{2}) - 3}\\ \frac{-\frac{1}{2}}{2} - \frac{3}{-8} &= \frac{-\frac{1}{2}}{-4}\\ \frac{1}{8} &= \frac{1}{8}\end{aligned}$$

The solution is $-\frac{1}{2}$.

17.
$$x^4 - 6x^2 + 8 = 0$$
$$(x^2)^2 - 6(x^2) + 8 = 0$$
$$u^2 - 6y + 8 = 0$$
$$(u-4)(u-2) = 0$$
$$u - 4 = 0 \quad u - 2 = 0$$
$$u = 4 \quad u = 2$$
Replace u by x^2.
$$x^2 = 4 \qquad x^2 = 2$$
$$\sqrt{x^2} = \sqrt{4} \qquad \sqrt{x^2} = \sqrt{2}$$
$$x = \pm 2 \qquad x = \pm\sqrt{2}$$
The solutions are -2, 2, $-\sqrt{2}$, and $\sqrt{2}$.

18.
$$\sqrt{3x+1} - 1 = x$$
$$\sqrt{3x+1} = x + 1$$
$$(\sqrt{3x+1})^2 = (x+1)^2$$
$$3x + 1 = x^2 + 2x + 1$$
$$0 = x^2 - x$$
$$0 = x(x-1)$$
$$x = 0 \quad x - 1 = 0$$
$$x = 1$$
0 and 1 both check as solutions.
The solutions are 0 and 1.

19.
$$|3x - 2| < 8$$
$$-8 < 3x - 2 < 8$$
$$-8 + 2 < 3x - 2 + 2 < 8 + 2$$
$$-6 < 3x < 10$$
$$\frac{1}{3}\cdot(-6) < \frac{1}{3}\cdot(3x) < \frac{1}{3}\cdot 10$$
$$\left\{x \middle| -2 < x < \frac{10}{3}\right\}$$

20.
$$6x - 5y = 15$$
$$6x - 5(0) = 15$$
$$6x = 15$$
$$x = \frac{15}{6} = \frac{5}{2}$$
The x-intercept is $\left(\frac{5}{2}, 0\right)$.
$$6(0) - 5y = 15$$
$$-5y = 15$$
$$y = -3$$
The y-intercept is (0, -3).

21. Solve each inequality.
$$x + y \le 3 \qquad 2x - y < 4$$
$$y \le 3 - x \qquad -y < 4 - 2x$$
$$y > -4 + 2x$$

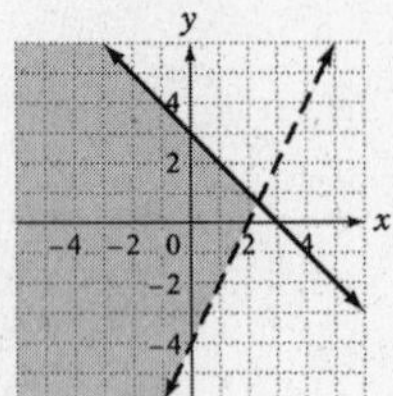

22.
$$x + y + z = 2$$
$$-x + 2y - 3z = -9$$
$$x - 2y - 2z = -1$$
$$D = \begin{vmatrix} 1 & 1 & 1 \\ -1 & 2 & -3 \\ 1 & -2 & -2 \end{vmatrix}$$
$$= \begin{vmatrix} 2 & -3 \\ -2 & -2 \end{vmatrix} - \begin{vmatrix} -1 & -3 \\ 1 & -2 \end{vmatrix} + \begin{vmatrix} -1 & 2 \\ 1 & -2 \end{vmatrix} = -15$$
$$D_x = \begin{vmatrix} 2 & 1 & 1 \\ -9 & 2 & -3 \\ -1 & -2 & -2 \end{vmatrix}$$
$$= 2\begin{vmatrix} 2 & -3 \\ -2 & -2 \end{vmatrix} - \begin{vmatrix} -9 & -3 \\ -1 & -2 \end{vmatrix} + \begin{vmatrix} -9 & 2 \\ -1 & -2 \end{vmatrix} = -15$$
$$D_y = \begin{vmatrix} 1 & 2 & 1 \\ -1 & -9 & -3 \\ 1 & -1 & -2 \end{vmatrix}$$
$$= \begin{vmatrix} -9 & -3 \\ -1 & -2 \end{vmatrix} - 2\begin{vmatrix} -1 & -3 \\ 1 & -2 \end{vmatrix} + \begin{vmatrix} -1 & -9 \\ 1 & -1 \end{vmatrix} = 15$$
$$D_z = \begin{vmatrix} 1 & 1 & 2 \\ -1 & 2 & -9 \\ 1 & -2 & -1 \end{vmatrix}$$
$$= \begin{vmatrix} 2 & -9 \\ -2 & -1 \end{vmatrix} - \begin{vmatrix} -1 & -9 \\ 1 & -1 \end{vmatrix} + 2\begin{vmatrix} -1 & 2 \\ 1 & -2 \end{vmatrix} = -30$$
$$x = \frac{D_x}{D} = \frac{-15}{-15} = 1$$
$$y = \frac{D_y}{D} = \frac{15}{-15} = -1$$
$$z = \frac{D_z}{D} = \frac{-30}{-15} = 2$$
The solution is (-1, 1, 2).

23. $f(-2) = \dfrac{2(-2) - 3}{(-2)^2 - 1} = \dfrac{-4 - 3}{4 - 1} = \dfrac{-7}{3} = -\dfrac{7}{3}$

24.
$$f(x) = \frac{x - 2}{x^2 - 2x - 15}$$
$$f(x) = \frac{x - 2}{(x-5)(x+3)}$$
The domain of $f(x)$ is all real numbers except $x = -3$ and $x = 5$, where the function is undefined.
$\{x | x \ne -3, 5\}$

25.
$$x^3 + x^2 - 6x < 0$$
$$x(x^2 + x - 6) < 0$$
$$x(x+3)(x-2) < 0$$

	$x < -3$	$-3 < x < 0$	$0 < x < 2$	$x > 2$
x	------	--------	++++	+++++++++
$x+3$	------	+++++++	++++	+++++++++
$x-2$	------	--------	----	+++++++++

−5 −4 −3 −2 −1 0 1 2 3 4 5

$\{x | x < -3 \text{ or } 0 < x < 2\}$

−5 −4 −3 −2 −1 0 1 2 3 4 5

26. $\dfrac{(x-1)(x-5)}{x+3} \geq 0$

$x-1$	------	---------	+++++++++	++
$x-5$	------	---------	---------	++
$x+3$	------	+++++++++	+++++++++	++

−5 −4 −3 −2 −1 0 1 2 3 4 5

$\{x \mid -3 < x \leq 1 \text{ or } x \geq 5\}$

−5 −4 −3 −2 −1 0 1 2 3 4 5

27. Strategy Let p represent the length of the piston rod, T the tolerance, and m the given length. Solve the absolute value inequality $|m-p| \leq T$ for m.

Solution

$$|m-p| \leq T$$
$$\left|m - 9\frac{3}{8}\right| \leq \frac{1}{64}$$
$$-\frac{1}{64} \leq m - 9\frac{3}{8} \leq \frac{1}{64}$$
$$-\frac{1}{64} + 9\frac{3}{8} \leq m \leq \frac{1}{64} + 9\frac{3}{8}$$
$$9\frac{23}{64} \leq m \leq 9\frac{25}{64}$$

The lower limit is $9\frac{23}{64}$ in.

The upper limit is $9\frac{25}{64}$ in.

28.
$$A = \frac{1}{2}b \cdot h$$
$$= \frac{1}{2}(x+8)(2x-4)$$
$$= \frac{1}{2}(2x^2 + 12x - 32)$$
$$= (x^2 + 6x - 16)\text{ ft}^2$$

29. $2x^2 + 4x + 3 = 0$

$a = 2,\ b = 4,\ c = 3$

$b^2 - 4ac = 4^2 - 4(2)(3)$

$= 16 - 24 = -8$

$-8 < 0$

Since the discriminant is less than zero, the equation has two complex number solutions.

30.
$$m = \frac{y_2 - y_1}{x_2 - x_1}$$
$$= \frac{0 - 250{,}000}{30 - 0}$$
$$= \frac{-250{,}000}{30} = -\frac{25{,}000}{3}$$

The building depreciates $\frac{\$25{,}000}{3}$, or about \$8333, each year.

Chapter 9: Functions and Relations

Prep Test

1. $-\dfrac{(-4)}{2(2)} = -\left(\dfrac{-4}{4}\right) = -(-1) = 1$

2. $y = -(-2)^2 + 2(-2) + 1 = -4 - 4 + 1 = -7$

3. $f(-4) = (-4)^2 - 3(-4) + 2 = 16 + 12 + 2 = 30$

4. $$\begin{aligned} p(2+h) &= (2+h)^2 - 5 \\ &= 4 + 4h + h^2 - 5 \\ &= h^2 + 4h - 1 \end{aligned}$$

5. $$0 = (3x+2)(x-3)$$
$$0 = 3x + 2 \qquad 0 = x - 3$$
$$\left(-\frac{1}{3}\right) - 3x = 2\left(-\frac{1}{3}\right) \qquad x = 3$$
$$x = -\frac{2}{3}$$
The solutions are $-\dfrac{2}{3}$ and 3.

6. $$x = \frac{-b \pm \sqrt{b^2 - 4ac}}{2a} = \frac{-(-4) \pm \sqrt{(-4)^2 - 4(1)(1)}}{2(1)} = \frac{4 \pm \sqrt{16-4}}{2}$$
$$= \frac{4 \pm \sqrt{12}}{2} = \frac{4 \pm 2\sqrt{3}}{2} = \frac{4}{2} \pm \frac{2\sqrt{3}}{2} = 2 \pm \sqrt{3}$$
The solutions are $2 - \sqrt{3}$ and $2 + \sqrt{3}$.

7. $$\begin{aligned} x &= 2y + 4 \\ 2y + 4 &= x \\ \left(\frac{1}{2}\right)2y &= (x - 4)\left(\frac{1}{2}\right) \\ y &= \frac{1}{2}x - 2 \end{aligned}$$

8. The domain of the function is the set of the first components in the ordered pair.
D: {−2, 3, 4, 6}
The range of the function is the set of the second components in the ordered pair.
R: {4, 5, 6}
The relation is a function because each number in the domain is paired with only one number in the range.

9. $x = 8$

10.
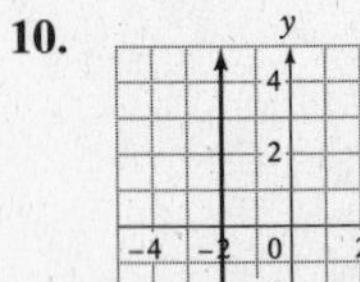

Go Figure

Answer: 22 chimes

Section 9.1

Objective A Exercises

1. A quadratic function is a function of the form $f(x) = ax^2 + bx + c,\ a \neq 0$.

3. The vertex of a parabola is the point with the smallest y-coordinate or the largest y-coordinate. When $a > 0$, the parabola opens up and the vertex of the parabola is the point with the smallest y-coordinate. When $a < 0$, the parabola opens down and the vertex is the point with the largest y-coordinate.

5. −5

7. $x = 7$

9. $$y = x^2 - 2x - 4$$
$$-\frac{b}{2a} = -\frac{-2}{2(1)} = 1$$
$$y = 1^2 - 2(1) - 4 = -5$$
Vertex: (1, −5)
Axis of symmetry: $x = 1$

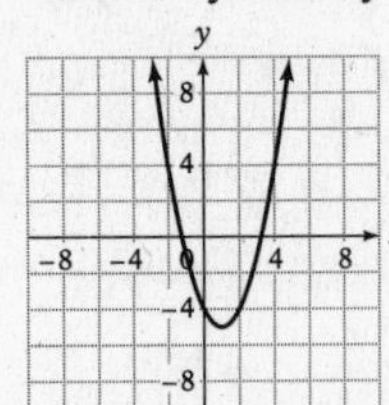

11. $$y = -x^2 + 2x - 3$$
$$-\frac{b}{2a} = -\frac{2}{2(-1)} = 1$$
$$y = -(1)^2 + 2(1) - 3 = -2$$
Vertex: (1, −2)
Axis of symmetry: $x = 1$

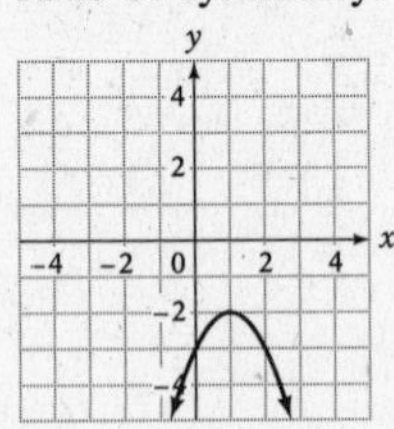

13. $f(x) = x^2 - x - 6$

$$-\frac{b}{2a} = \frac{-(-1)}{2 \cdot 1} = \frac{1}{2}$$

$$f(x) = \left(\frac{1}{2}\right)^2 - \left(\frac{1}{2}\right) - 6 = -\frac{25}{4}$$

Vertex: $\left(\frac{1}{2}, -\frac{25}{4}\right)$

Axis of symmetry: $x = \frac{1}{2}$

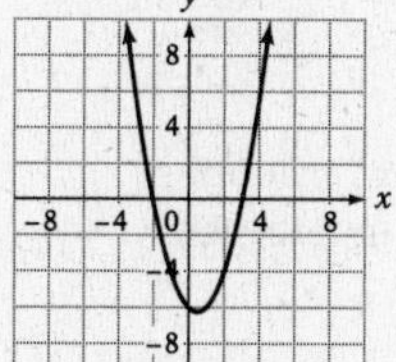

15. $F(x) = x^2 - 3x + 2$

$$-\frac{b}{2a} = -\frac{-3}{2(1)} = \frac{3}{2}$$

$$F(x) = \left(\frac{3}{2}\right)^2 - 3\left(\frac{3}{2}\right) + 2 = -\frac{1}{4}$$

Vertex: $\left(\frac{3}{2}, -\frac{1}{4}\right)$

Axis of symmetry: $x = \frac{3}{2}$

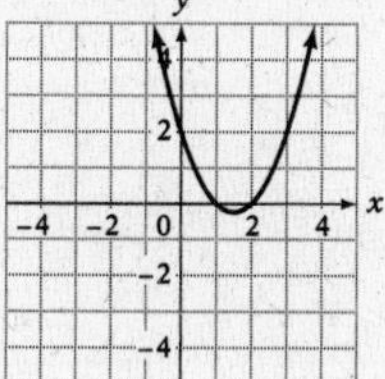

17. $y = -2x^2 + 6x$

$$-\frac{b}{2a} = -\frac{6}{2(-2)} = \frac{3}{2}$$

$$y = -2\left(\frac{3}{2}\right)^2 + 6\left(\frac{3}{2}\right) = \frac{9}{2}$$

Vertex: $\left(\frac{3}{2}, \frac{9}{2}\right)$

Axis of symmetry: $x = \frac{3}{2}$

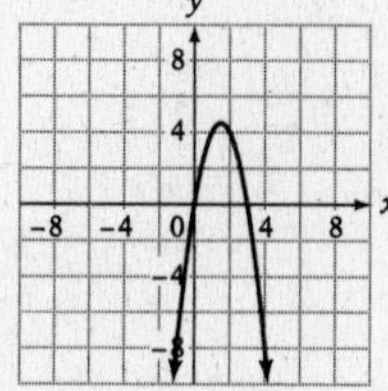

19. $y = -\frac{1}{4}x^2 - 1$

$$-\frac{b}{2a} = -\frac{0}{2\left(-\frac{1}{4}\right)} = 0$$

$$y = -\frac{1}{4}(0)^2 - 1 = -1$$

Vertex: $(0, -1)$

Axis of symmetry: $x = 0$

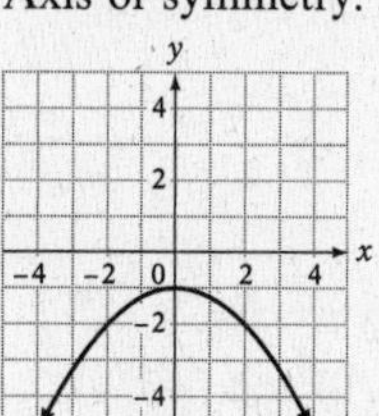

21. $P(x) = -\frac{1}{2}x^2 + 2x - 3$

$$-\frac{b}{2a} = -\frac{2}{2\left(-\frac{1}{2}\right)} = 2$$

$$P(x) = -\frac{1}{2}(2)^2 + 2(2) - 3 = -1$$

Vertex: $(2, -1)$

Axis of symmetry: $x = 2$

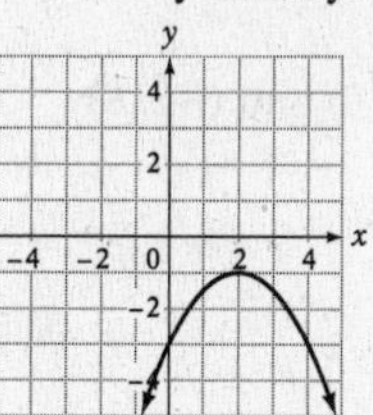

23. $y = -\frac{1}{2}x^2 + x - 3$

$$-\frac{b}{2a} = -\frac{1}{2\left(-\frac{1}{2}\right)} = 1$$

$$y = -\frac{1}{2}(1)^2 + 1 - 3 = -\frac{5}{2}$$

Vertex: $\left(1, -\frac{5}{2}\right)$

Axis of symmetry: $x = 1$

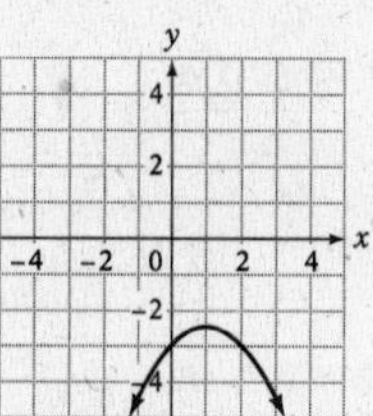

25. Domain: $\{x \mid x \in \text{real numbers}\}$
Range: $\{y \mid y \geq -5\}$

27. Domain: $\{x \mid x \in \text{real numbers}\}$
Range: $\{y \mid y \leq 0\}$

29. Domain: $\{x | x \in \text{real numbers}\}$
Range: $\{y | y \geq -7\}$

Objective B Exercises

31. **a.** A y-intercept of the graph of a parabola is a point at which the graph crosses the y-axis. It is a point at which $x = 0$.

b. The graph of a parabola has one y-intercept.

33. $y = x^2 - 9$
$0 = x^2 - 9$
$0 = (x - 3)(x + 3)$
$x - 3 = 0 \quad x + 3 = 0$
$x = 3 \quad x = -3$
The x-intercepts are (3, 0) and (−3, 0).

35. $y = 3x^2 + 6x$
$0 = 3x^2 + 6x$
$0 = 3x(x + 2)$
$3x = 0 \quad x + 2 = 0$
$x = 0 \quad x = -2$
The x-intercepts are (0, 0) and (−2, 0).

37. $y = x^2 - 2x - 8$
$0 = x^2 - 2x - 8$
$0 = (x - 4)(x + 2)$
$x - 4 = 0 \quad x + 2 = 0$
$x = 4 \quad x = -2$
The x-intercepts are (4, 0) and (−2, 0).

39. $y = 2x^2 - 5x - 3$
$0 = 2x^2 - 5x - 3$
$0 = (2x + 1)(x - 3)$
$2x + 1 = 0 \quad x - 3 = 0$
$2x = -1 \quad x = 3$
$x = -\frac{1}{2}$
The x-intercepts are $\left(-\frac{1}{2}, 0\right)$ and (3, 0).

41. $y = x^2 + 4x - 3$
$0 = x^2 + 4x - 3$
$a = 1, b = 4, c = -3$

$$x = \frac{-b \pm \sqrt{b^2 - 4ac}}{2a}$$
$$= \frac{-4 \pm \sqrt{(4)^2 - 4(1)(-3)}}{2(1)}$$
$$= \frac{-4 \pm \sqrt{16 + 12}}{2} = \frac{-4 \pm \sqrt{28}}{2}$$
$$= \frac{-4 \pm 2\sqrt{7}}{2} = -2 \pm \sqrt{7}$$

The x-intercepts are
$(-2 + \sqrt{7}, 0)$ and $(-2 - \sqrt{7}, 0)$.

43. $y = -x^2 - 4x - 5$
$0 = -1x^2 - 4x - 5$
$a = -1, b = -4, c = -5$

$$x = \frac{-b \pm \sqrt{b^2 - 4ac}}{2a}$$
$$= \frac{-(-4) \pm \sqrt{(-4)^2 - 4(-1)(-5)}}{2(-1)}$$
$$= \frac{4 \pm \sqrt{16 - 20}}{-2} = \frac{4 \pm \sqrt{-4}}{-2}$$
$$= \frac{4 \pm 2i}{-2} = -2 \pm i$$

The equation has no real solutions.
The parabola has no x-intercepts.

45. $f(x) = x^2 - 6x + 9$
$x^2 - 6x + 9 = 0$
$(x - 3)(x - 3) = 0$
$x - 3 = 0 \quad x - 3 = 0$
$x = 3 \quad x = 3$
The zero is 3.

47. $f(x) = -x^2 + 3x + 8$
$-x^2 + 3x + 8 = 0$

$$x = \frac{-b \pm \sqrt{b^2 - 4ac}}{2a}$$
$$x = \frac{-3 \pm \sqrt{3^2 - 4(-1)(8)}}{2(-1)} = \frac{-3 \pm \sqrt{41}}{-2}$$
$$x = \frac{3}{2} \pm \frac{\sqrt{41}}{2}$$

The zeros are $\frac{3 + \sqrt{41}}{2}$ and $\frac{3 - \sqrt{41}}{2}$.

49. $f(x) = -3x^2 + 4x$
$-3x^2 + 4x = 0$
$x(-3x + 4) = 0$
$x = 0 \quad -3x + 4 = 0$
$-3x = -4$
$x = \frac{4}{3}$

The zeros are 0 and $\frac{4}{3}$.

51. $f(x) = 3x^2 + 6$
$3x^2 + 6 = 0$
$3(x^2 + 2) = 0$
$x^2 + 2 = 0$
$x^2 = -2$
$x = \pm\sqrt{-2} = \pm i\sqrt{2}$
The zeros are $i\sqrt{2}$ and $-i\sqrt{2}$.

53. $f(x) = 3x^2 - x + 4$
$3x^2 - x + 4 = 0$
$$x = \frac{-b \pm \sqrt{b^2 - 4ac}}{2a}$$
$$x = \frac{-(-1) \pm \sqrt{(-1)^2 - 4(3)(4)}}{2(3)} = \frac{1 \pm \sqrt{-47}}{6}$$
$$x = \frac{1 \pm \sqrt{47}i}{6} = \frac{1}{6} \pm \frac{\sqrt{47}}{6}i$$
The zeros are $\frac{1}{6} + \frac{\sqrt{47}}{6}i$ and $\frac{1}{6} - \frac{\sqrt{47}}{6}i$.

55. $f(x) = -2x^2 + x + 5$
$-2x^2 + x + 5 = 0$
$$x = \frac{-b \pm \sqrt{b^2 - 4ac}}{2a}$$
$$x = \frac{-1 \pm \sqrt{1^2 - 4(-2)(5)}}{2(-2)} = \frac{-1 \pm \sqrt{41}}{-4}$$
$$x = \frac{1}{4} \pm \frac{\sqrt{41}}{4}$$
The zeros are $\frac{1 + \sqrt{41}}{4}$ and $\frac{1 - \sqrt{41}}{4}$.

57. $y = -x^2 - x + 3$
$a = -1, b = -1, c = 3$
$b^2 - 4ac$
$(-1)^2 - 4(-1)(3) = 1 + 12 = 13$
$13 > 0$
Since the discriminant is greater than zero, the parabola has two x-intercepts.

59. $y = x^2 - 10x + 25$
$a = 1, b = -10, c = 25$
$b^2 - 4ac$
$(-10)^2 - 4(1)(25) = 100 - 100 = 0$
Since the discriminant is equal to zero, the parabola has one x-intercept.

61. $y = -2x^2 + x - 1$
$a = -2, b = 1, c = -1$
$b^2 - 4ac$
$(1)^2 - 4(-2)(-1) = 1 - 8 = -7$
$-7 < 0$
Since the discriminant is less than zero, the parabola has no x-intercepts.

63. $y = 4x^2 - x - 2$
$a = 4, b = -1, c = -2$
$b^2 - 4ac$
$(-1)^2 - 4(4)(-2) = 1 + 32 = 33$
$33 > 0$
Since the discriminant is greater than zero, the parabola has two x-intercepts.

65. $y = 2x^2 + x + 4$
$a = 2, b = 1, c = 4$
$b^2 - 4ac$
$(1)^2 - 4(2)(4) = 1 - 32 = -31$
$-31 < 0$
Since the discriminant is less than zero, the parabola has no x-intercepts.

67. $y = 4x^2 + 2x - 5$
$a = 4, b = 2, c = -5$
$b^2 - 4ac$
$(2)^2 - 4(4)(-5) = 4 + 80 = 84$
$84 > 0$
Since the discriminant is greater than zero, the parabola has two x-intercepts.

69. The zeros of a function describe the x-intercepts of the graph of a function, therefore, the x-intercepts are (−4, 0) and (5, 0).

Objective C Exercises

71. To find the minimum or maximum value of a quadratic function, find the x-coordinate of the vertex. Then evaluate the function at the value of this x-coordinate.

73. **a.** Because a is positive, the parabola opens up. Therefore, the function has a minimum.

b. Because a is negative, the parabola opens down. Therefore, the function has a maximum.

c. Because a is positive, the parabola opens up. Therefore, the function has a minimum.

75. $f(x) = 2x^2 + 4x$
$$x = -\frac{b}{2a} = -\frac{4}{2(2)} = -1$$
$f(1) = 2(-1)^2 + 4(-1) = 2 - 4 = -2$
Since a is positive, the function has a minimum value. The minimum value of the function is −2.

77. $f(x) = -2x^2 + 4x - 5$
$$x = -\frac{b}{2a} = -\frac{4}{2(-2)} = 1$$
$f(1) = -2(1)^2 + 4(1) - 5 = -3$
Since a is negative, the function has a maximum value. The maximum value of the function is −3.

79. $f(x) = -2x^2 - 3x$
$$x = -\frac{b}{2a} = -\frac{-3}{2(-2)} = -\frac{3}{4}$$
$$f\left(-\frac{3}{4}\right) = -2\left(-\frac{3}{4}\right)^2 - 3\left(-\frac{3}{4}\right) = \frac{9}{8}$$
Since a is negative, the function has a maximum value. The maximum value of the function is $\frac{9}{8}$.

81. $f(x) = 3x^2 + 3x - 2$

$x = -\frac{b}{2a} = -\frac{3}{2(3)} = -\frac{1}{2}$

$f(x) = 3x^2 + 3x - 2$

$f\left(-\frac{1}{2}\right) = 3\left(-\frac{1}{2}\right)^2 + 3\left(-\frac{1}{2}\right) - 2$

$= \frac{3}{4} - \frac{3}{2} - 2 = -\frac{11}{4}$

Since a is positive, the function has a minimum value. The minimum value of the function is $-\frac{11}{4}$.

83. $f(x) = -x^2 - x + 2$

$x = -\frac{b}{2a} = -\frac{-1}{2(-1)} = -\frac{1}{2}$

$f(x) = -x^2 - x + 2$

$f\left(-\frac{1}{2}\right) = -\left(-\frac{1}{2}\right)^2 - \left(-\frac{1}{2}\right) + 2$

$= -\frac{1}{4} + \frac{1}{2} + 2 = \frac{9}{4}$

Since a is negative, the function has a maximum value. The maximum value of the function is $\frac{9}{4}$.

85. $f(x) = 3x^2 + 5x + 2$

$x = -\frac{b}{2a} = -\frac{5}{2(3)} = -\frac{5}{6}$

$f(x) = 3x^2 + 5x + 2$

$f\left(-\frac{5}{6}\right) = 3\left(-\frac{5}{6}\right)^2 + 5\left(-\frac{5}{6}\right) + 2$

$= 3\left(\frac{25}{36}\right) + \left(\frac{-25}{6}\right) + 2$

$= \frac{25}{12} - \frac{25}{6} + 2 = -\frac{1}{12}$

Since a is positive, the function has a minimum value. The minimum value of the function is $-\frac{1}{12}$.

87. **a.** $f(x) = -2x^2 + 2x - 1$

$x = -\frac{b}{2a} = -\frac{2}{2(-2)} = \frac{1}{2}$

$f(x) = -2x^2 + 2x - 1$

$f\left(\frac{1}{2}\right) = -2\left(\frac{1}{2}\right)^2 + 2\left(\frac{1}{2}\right) - 1$

$= -\frac{1}{2} + 1 - 1 = -\frac{1}{2}$

Since a is negative, the function has a maximum value. The maximum value of the function is $-\frac{1}{2}$.

b. $f(x) = -x^2 + 8x - 2$

$x = -\frac{b}{2a} = -\frac{8}{2(-1)} = 4$

$f(x) = -x^2 + 8x - 2$

$f(4) = -(4)^2 + 8(4) - 2$

$= -16 + 32 - 2 = 14$

Since a is negative, the function has a maximum value. The maximum value of the function is 14.

c. $f(x) = -4x^2 + 0x + 3$

$x = -\frac{b}{2a} = -\frac{0}{2(-4)} = 0$

$f(x) = -4x^2 + 0x + 3$

$f(0) = -4(0)^2 + 3$

$= 0 + 3 = 3$

Since a is negative, the function has a maximum value. The maximum value of the function is 3.

Parabola **b** has the greatest maximum value.

89. Since the parabola opens down, a is negative and the function has a maximum value. The maximum value of the function is -5.

Objective D Exercises

91. **Strategy** To find the price that will give the maximum revenue, find the P-coordinate of the vertex.

Solution $P = -\frac{b}{2a} = -\frac{125}{2\left(-\frac{1}{4}\right)} = 250$

A price of $250 will give the maximum revenue.

93. **Strategy** To find the time it takes for the diver to reach the maximum height, find the t-coordinate of the vertex.

To find the maximum height, evaluate the function at the t-coordinate of the vertex.

Solution $t = -\frac{b}{2a} = -\frac{7.8}{2(-4.9)} = \frac{78}{98}$

The diver reaches the maximum height in $\frac{78}{98}$ s.

$s(t) = -4.9t^2 + 7.8t + 10$

$s\left(\frac{78}{98}\right) = -4.9\left(\frac{78}{98}\right)^2 + 7.8\left(\frac{78}{98}\right) + 10$

$\approx -3.104 + 6.2082 + 10$

≈ 13.1

The diver reaches a height of 13.1 m above the water.

95. Strategy To find the distance from one end of the bridge where the cable is at its minimum height, find the x-coordinate of the vertex.
To find the minimum height, evaluate the function at the x-coordinate of the vertex.

Solution $x = -\frac{b}{2a} = -\frac{-0.8}{2(0.25)} = 1.6$
The cable is at its minimum height 1.6 ft from one end of the bridge.
$h(x) = 0.25x^2 - 0.8x + 25$
$h(1.6) = 0.25(1.6)^2 - 0.8(1.6) + 25$
$= 0.64 - 1.28 + 25 = 24.36$
The minimum height is 24.36 ft.

97. Strategy To find the height at which the water will land, evaluate the function at $x = 40$.

Solution $s(x) = -\frac{1}{30}x^2 + 2x + 5$
$s(40) = -\frac{1}{30}(40)^2 + 2(40) + 5$
$s(40) = -\frac{160}{3} + \frac{240}{3} + \frac{15}{3}$
$s(40) = \frac{95}{3} = 31\frac{2}{3}$
The water will land at a height of $31\frac{2}{3}$ ft.

99. Strategy To find the speed a car can travel and still stop 44 ft away, solve the equation for v when $s(v) = 44$.

Solution $s(v) = 0.055v^2 + 1.1v$
$44 = 0.055v^2 + 1.1v$
$0 = 0.055v^2 + 1.1v - 44$
$v = \frac{-b \pm \sqrt{b^2 - 4ac}}{2a}$
$= \frac{-1.1 \pm \sqrt{(1.1)^2 - 4(0.055)(-44)}}{2(0.055)}$
$= \frac{-1.1 \pm \sqrt{1.21 + 9.68}}{0.11}$
$= \frac{-1.1 \pm 9.68}{0.11}$
$= 20 \text{ or } -40$
Since -40 cannot be a speed, the car can be traveling 20 mph and still stop at a stop sign 44 ft away.

101. Strategy Let x represent one number. Because the sum of the 2 numbers is 20, let $20 - x$ represent the other number. Then their product is represented by $x(20 - x) = 20x - x^2 = -x^2 + 20x$.
• To find one of the numbers, find the x-coordinate of the vertex.
$f(x) = -x^2 + 20x$.
• To find the other number, replace x in $20 - x$ by the x-coordinate of the vertex and evaluate.

Solution $x = -\frac{b}{2a} = -\frac{20}{2(-1)} = 10$
$20 - x = 20 - 10 = 10$
The two numbers are 10 and 10.

103. Strategy Let x represent the width of the rectangular corral. The length will then be represented by $200 - 2x$. The area will be
$f(x) = x(200 - 2x)$
$= 200x - 2x^2 = -2x^2 + 200x$.
• To determine the width of the corral that will maximize the enclosed area, find the x-coordinate of the vertex $f(x) = -2x^2 + 200x$.
• To find the length of the corral, replace x in $f(x) = -2x^2 + 200x$ by the x-coordinate of the vertex and evaluate.

Solution $x = -\frac{b}{2a} = -\frac{200}{2(-2)} = 50$
$f(20) = 200 - 2(50) = 200 - 100 = 100$
The length is 100 ft and the width is 50 ft.

Applying the Concepts

105. Strategy Find the x-coordinate of the vertex point. The y-coordinate will be 0 since the vertex is said to be on the x-axis.
• Plug x and y into the given equation of the parabola and solve for k.

Solution $x = -\frac{b}{2a} = -\frac{-8}{2(1)} = 4$
Plug $x = 4$ and $y = 0$ into $y = x^2 - 8x + k$.
$0 = 4^2 - 8(4) + k$
$0 = 16 - 32 + k$
$0 = -16 + k$
$k = 16$

Section 9.2

Objective A Exercises

1. A vertical line intersects the function no more than once. Yes. The graph is a function.

3. A vertical line intersects the relation more than once. No. The graph is not a function.

5. A vertical line intersects the function no more than once. Yes. The graph is a function.

7. $f(x) = 3|x - 2|$
Domain: $\{x|x \in \text{real numbers}\}$
Range: $\{y|y \geq 0\}$

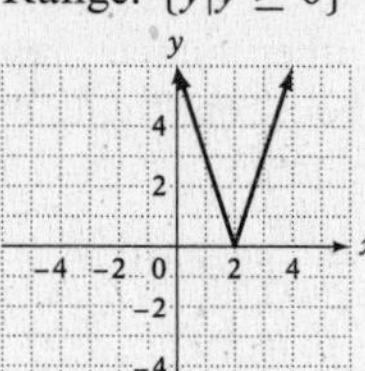

9. $f(x) = 1 - x^3$
Domain: $\{x|x \in \text{real numbers}\}$
Range: $\{y|y \in \text{real numbers}\}$

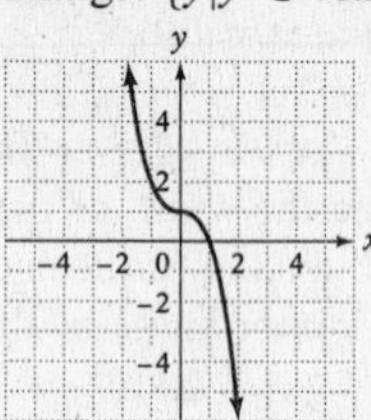

11. $f(x) = \sqrt{4 - x}$
Domain: $\{x|x \leq 4\}$
Range: $\{y|y \geq 0\}$

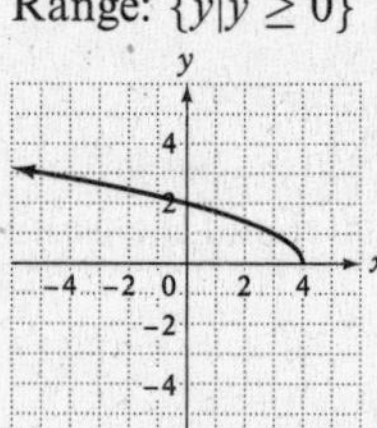

13. $f(x) = x^3 + 4x^2 + 4x$
Domain: $\{x|x \in \text{real numbers}\}$
Range: $\{y|y \in \text{real numbers}\}$

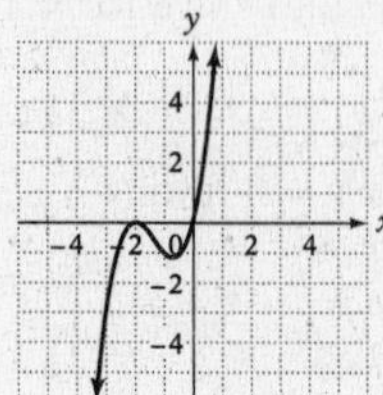

15. $f(x) = -\sqrt{x + 2}$
Domain: $\{x|x \geq -2\}$
Range: $\{y|y \leq 0\}$

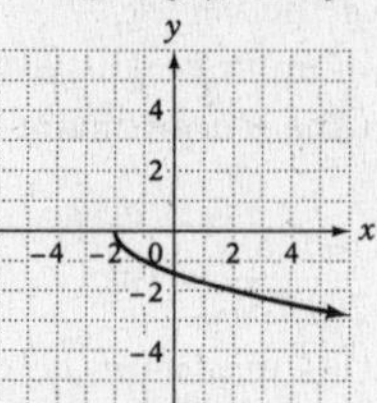

17. $f(x) = |2x + 2|$
Domain: $\{x|x \in \text{real numbers}\}$
Range: $\{y|y \geq 0\}$

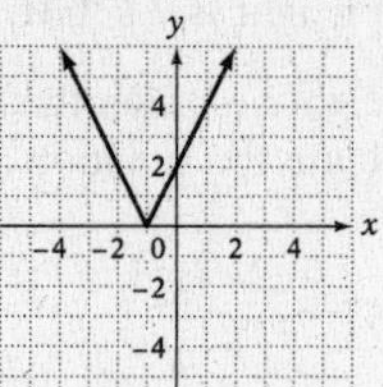

Applying the Concepts

19. $f(x) = \sqrt{x - 2}$
$f(a) = 4 = \sqrt{a - 2}$
$4^2 = (\sqrt{a - 2})^2$
$16 = a - 2$
$a = 18$

21. $f(a, b) = a + b$
$g(a, b) = a \cdot b$
$f(2, 5) = 2 + 5 = 7$
$g(2, 5) = 2 \cdot 5 = 10$
$f(2, 5) + g(2, 5) = 7 + 10 = 17$

23. $f(14) = 8$

25.
x − 2 ---------- | ---------- | +++++++++
x + 2 ---------- | +++++++++ | +++++++++
−5 −4 −3 −2 −1 0 1 2 3 4 5

$\{x| -2 < x < 2\}$

27. $f(x) = |2x - 2|$
$f(x)$ is smallest when $2x - 2 = 0$
$2x = 2$
$x = 1$

Section 9.3

Objective A Exercises

1. $f(2) - g(2) = (2 \cdot 2^2 - 3) - (-2 \cdot 2 + 4)$
$= (2 \cdot 4 - 3) - (-4 + 4)$
$= (8 - 3) - (0)$
$= 5$

3. $f(0)+g(0)=(2\cdot 0^2-3)-(-2\cdot 0+4)$
$=(0-3)+(0+4)$
$=-3+4$
$=1$

5. $(f\cdot g)(2)=f(2)\cdot g(2)$
$=(2\cdot 2^2-3)\cdot(-2\cdot 2+4)$
$=(2\cdot 4-3)\cdot(-4+4)$
$=(8-3)\cdot 0$
$=5\cdot 0$
$=0$

7. $\left(\frac{f}{g}\right)4=\frac{f(4)}{g(4)}$
$=\frac{2(4)^2-3}{-2\cdot 4+4}$
$=\frac{2\cdot 16-3}{-8+4}$
$=\frac{32-3}{-4}$
$=\frac{29}{-4}$
$=-\frac{29}{4}$

9. $\left(\frac{g}{f}\right)(-3)=\frac{g(-3)}{f(-3)}$
$=\frac{-2(-3)+4}{2(-3)^2-3}$
$=\frac{6+4}{2(9)-3}$
$=\frac{10}{18-3}$
$=\frac{10}{15}$
$=\frac{2}{3}$

11. $f(1)+g(1)=[2(1)^2+3(1)-1]+[2(1)-4]$
$=[2(1)+3-1]+[2-4]$
$=[2+3-1]+[-2]$
$=[4]+[-2]$
$=2$

13. $f(4)-g(4)=[2(4)^2+3(4)-1]-[2(4)-4]$
$=[2(16)+3(4)-1]-[8-4]$
$=[32+12-1]-[4]$
$=[43]-[4]$
$=39$

15. $(f\cdot g)(1)=[2(1)^2+3(1)-1]\cdot[2(1)-4]$
$=[2(1)+3(1)-1]\cdot[2-4]$
$=[2+3-1]\cdot[-2]$
$=[4]\cdot[-2]$
$=-8$

17. $\left(\frac{f}{g}\right)(-3)=\frac{f(-3)}{g(-3)}$
$=\frac{2(-3)^2+3(-3)-1}{2(-3)-4}$
$=\frac{2(9)-9-1}{-6-4}$
$=\frac{18-9-1}{-10}$
$=\frac{8}{-10}$
$=-\frac{4}{5}$

19. $f(2)-g(2)=[2^2+3(2)-5]-[2^3-2(2)-3]$
$=[4+6-5]-[8-4+3]$
$=5-7$
$=-2$

21. $\left(\frac{f}{g}\right)(-2)=\frac{f(-2)}{g(-2)}$
$=\frac{(-2)^2+3(-2)-5}{(-2)^3-2(-2)+3}$
$=\frac{4-6-5}{-8+4+3}$
$=\frac{-7}{-1}$
$=7$

23. The expression $(f\circ g)(-2)$ means to evaluate the function f at $g(-2)$.

Objective B Exercises

25. $f(x)=2x-3$
$f(0)=2(0)-3$
$=0-3=-3$
$g(x)=4x-1$
$g(-3)=4(-3)-1$
$=-12-1=-13$
$g[f(0)]=-13$

27. $f(x)=2x-3$
$f(-2)=2(-2)-3$
$=-4-3=-7$
$g(x)=4x-1$
$g(-7)=4(-7)-1$
$=-28-1=-29$
$f[g(-2)]=-29$

29. $f(x)=2x-3$
$g(x)=4x-1$
$g(2x-3)=4(2x-3)-1$
$=8x-12-1$
$=8x-13$
$g[f(x)]=8x-13$

31.
$$\begin{aligned} h(x) &= 2x+4 \\ h(0) &= 2(0)+4 \\ &= 0+4=4 \\ f(x) &= \frac{1}{2}x+2 \\ f(4) &= \frac{1}{2}(4)+2 \\ &= 2+2=4 \\ f[h(0)] &= 4 \end{aligned}$$

33.
$$\begin{aligned} h(x) &= 2x+4 \\ h(-1) &= 2(-1)+4 \\ &= -2+4=2 \\ f(x) &= \frac{1}{2}x+2 \\ f(2) &= \frac{1}{2}(2)+2 \\ f(2) &= \frac{1}{2}(2)+2 \\ &= 1+2=3 \\ f[h(-1)] &= 3 \end{aligned}$$

35.
$$\begin{aligned} h(x) &= 2x+4 \\ f(x) &= \frac{1}{2}x+2 \\ f(2x+4) &= \frac{1}{2}(2x+4)+2 \\ &= x+2+2 \\ &= x+4 \\ f[h(x)] &= x+4 \end{aligned}$$

37.
$$\begin{aligned} g(x) &= x^2+3 \\ g(0) &= 0^2+3=3 \\ h(x) &= x-2 \\ h(3) &= 3-2=1 \\ h[g(0)] &= 1 \end{aligned}$$

39.
$$\begin{aligned} g(x) &= x^2+3 \\ g(-2) &= (-2)^2+3 \\ &= 4+3=7 \\ h(x) &= x-2 \\ h(7) &= 7-2=5 \\ h[g(-2)] &= 5 \end{aligned}$$

41.
$$\begin{aligned} g(x) &= x^2+3 \\ h(x) &= x-2 \\ h(x^2+3) &= x^2+3-2 \\ &= x^2+1 \\ h[g(x)] &= x^2+1 \end{aligned}$$

43.
$$\begin{aligned} f(x) &= x^2+x+1 \\ f(0) &= 0^2+0+1 \\ &= 0+0+1=1 \\ h(x) &= 3x+2 \\ h(1) &= 3(1)+2 \\ &= 3+2=5 \\ h[f(0)] &= 5 \end{aligned}$$

45.
$$\begin{aligned} f(x) &= x^2+x+1 \\ f(-2) &= (-2)^2-2+1 \\ &= 4-2+1=3 \\ h(x) &= 3x+2 \\ h(3) &= 3(3)+2 \\ &= 9+2=11 \\ h[f(-2)] &= 11 \end{aligned}$$

47.
$$\begin{aligned} f(x) &= x^2+x+1 \\ h(x) &= 3x+2 \\ h(x^2+x+1) &= 3(x^2+x+1)+2 \\ &= 3x^2+3x+3+2 \\ &= 3x^2+3x+5 \\ h[f(x)] &= 3x^2+3x+5 \end{aligned}$$

49.
$$\begin{aligned} g(x) &= x^3 \\ g(-1) &= (-1)^3=-1 \\ f(x) &= x-2 \\ f(-1) &= -1-2=-3 \\ f[g(-1)] &= -3 \end{aligned}$$

51.
$$\begin{aligned} f(x) &= x-2 \\ f(-1) &= -1-2=-3 \\ g(x) &= x^3 \\ g(-3) &= (-3)^3=-27 \\ g[f(2)] &= -27 \end{aligned}$$

53.
$$\begin{aligned} f(x) &= x-2 \\ g(x-2) &= (x-2)^3 \\ g[f(x)] &= x^3-6x^2+12x-8 \end{aligned}$$

55. **a. Strategy** Selling price equals the cost plus mark-up. If the cost is x and the mark-up is 60%, then $S = x + 0.60x$. If $M(x) = \dfrac{50x + 10{,}000}{x}$ is the cost per camera, then $(S \circ M)(x)$ is the selling price per camera. Find $(S \circ M)(x)$.

Solution
$$\begin{aligned} (S \circ M)(x) &= (S(M(x)) \\ &= M(x) + 0.60(M(x)) \\ &= 1.60(M(x)) \\ &= 1.60\left(\frac{50x+10{,}000}{x}\right) \\ &= \left(\frac{80x+16{,}000}{x}\right) \\ S(M(x)) &= 80 + \frac{16{,}000}{x} \end{aligned}$$

b. $(S \circ M)(5000) = 80 + \dfrac{16{,}000}{5000} = 80 + 3.2 = \$83.20$

c. When 5000 digital cameras are manufactured, the camera store sells each camera for \$83.20.

57. rebate: $r(p) = p - 1500$
discounted price: $d(p) = 0.90p$

a. If the dealer takes the rebate first and then the discount, we are finding
$d(r(p)) = 0.90(p - 1500) = 0.90p - 1350$

b. If the dealer takes the discount first and then the rebate, we are finding
$r(d(p)) = 0.90p - 1500$

c. As a buyer, you would prefer the dealer use $r(d(p))$ (The cost is less.)

Applying the Concepts

59. Reading from the graphs,
$f(1) = 2$ and $g(2) = 0$.
$g[f(1)] = g(2) = 0$

61. $(f \circ g)(3) = f(g(3))$
Reading from the graphs,
$g(3) = 5$ and $f(5) = -2$
$(f \circ g)(3) = f(g(3)) = f(5) = -2$

63. Reading from the graphs,
$g(0) = -4$ and $f(-4) = 7$
$f[g(0)] = f(-4) = 7$

65. $$\begin{aligned} g(3+h) - g(3) &= (3+h)^2 - 1 - ((3)^2 - 1) \\ &= 9 + 6h + h^2 - 1 - 8 \\ g(3+h) - g(3) &= h^2 + 6h \end{aligned}$$

67. $$\begin{aligned} \frac{g(1+h) - g(1)}{h} &= \frac{[(1+h)^2 - 1] - [(1)^2 - 1]}{h} \\ &= \frac{1 + 2h + h^2 - 1 - 0}{h} \\ &= \frac{2h + h^2}{h} \\ \frac{g(1+h) - g(1)}{h} &= 2 + h \end{aligned}$$

69. $$\begin{aligned} \frac{g(a+h) - g(a)}{h} &= \frac{[(a+h)^2 - 1] - (a^2 - 1)}{h} \\ &= \frac{a^2 + 2ah + h^2 - 1 - a^2 + 1}{h} \\ &= \frac{2ah + h^2}{h} \\ \frac{g(a+h) - g(a)}{h} &= 2a + h \end{aligned}$$

71. $$\begin{aligned} f(x) &= 2x \\ f(1) &= 2 \cdot 1 = 2 \\ h(x) &= x - 2 \\ h(2) &= 2 - 2 = 0 \\ g(x) &= 3x - 1 \\ g(0) &= 3 \cdot 0 - 1 = -1 \\ g(h[f(1)]) &= -1 \end{aligned}$$

73. $$\begin{aligned} g(x) &= 3x - 1 \\ g(0) &= 3 \cdot 0 - 1 = -1 \\ h(x) &= x - 2 \\ h(-1) &= -1 - 2 = -3 \\ f(x) &= 2x \\ f(-3) &= 2(-3) = -6 \\ f(h[f(0)]) &= -6 \end{aligned}$$

75. $$\begin{aligned} h(x) &= x - 2 \\ f(x-2) &= 2(x-2) = 2x - 4 \\ g(2x-4) &= 3(2x-4) - 1 = 6x - 12 - 1 = 6x - 13 \\ g(f[h(x)]) &= 6x - 13 \end{aligned}$$

Section 9.4

Objective A Exercises

1. A function is a 1–1 function if, for any a and b in the domain of f, $f(a) = f(b)$ implies $a = b$. Here is an alternative definition: A function is a set of ordered pairs in which no two ordered pairs that have the same first coordinate have different second coordinates. This means that given an x, there is only one y that can be paired with that x. A 1–1 function satisfies the additional condition that given any y, there is only one x that can be paired with that y.

3. Yes. The graph represents a 1–1 function.

5. No. The graph is not a 1–1 function. It fails the horizontal-line test.

7. Yes. The graph is a 1–1 function.

9. No. The graph is not a 1–1 function. It fails the horizontal- and vertical-line tests.

11. No. The graph is not a 1–1 function. It fails the horizontal-line test.

13. No. The graph is not a 1–1 function. It fails the horizontal-line test.

Objective B Exercises

15. The coordinates of each ordered pair of the inverse of a function are in the reverse order of the coordinates of the ordered pairs of the original function. For example, if $(-1, 5)$ is an ordered pair of the original function, then $(5, -1)$ is an ordered pair of the inverse of a function.

17. The inverse of $\{(1, 0), (2, 3), (3, 8), (4, 15)\}$ is $\{(0, 1), (3, 2), (8, 3), (15, 4)\}$.

19. $\{(3, 5), (-3, -5), (2, 5), (-2, -5)\}$ has no inverse because the numbers 5 and -5 would be paired with different members of the range.

21. The inverse of $\{(0, -2), (-1, 5), (3, 3), (-4, 6)\}$ is $\{(-2, 0), (5, -1), (3, 3), (6, -4)\}$.

23. No inverse.

25.
$$f(x) = 4x - 8$$
$$y = 4x - 8$$
$$x = 4y - 8$$
$$x + 8 = 4y$$
$$\frac{1}{4}x + 2 = y$$
The inverse function is $f^{-1}(x) = \frac{1}{4}x + 2$.

27.
$$f(x) = 2x + 4$$
$$y = 2x + 4$$
$$x = 2y + 4$$
$$x - 4 = 2y$$
$$\frac{1}{2}x - 2 = y$$
The inverse function is $f^{-1}(x) = \frac{1}{2}x - 2$.

29.
$$f(x) = \frac{1}{2}x - 1$$
$$y = \frac{1}{2}x - 1$$
$$x = \frac{1}{2}y - 1$$
$$x + 1 = \frac{1}{2}y$$
$$2x + 2 = y$$
The inverse function is $f^{-1}(x) = 2x + 2$.

31.
$$f(x) = -2x + 2$$
$$y = -2x + 2$$
$$x = -2y + 2$$
$$2y = -x + 2$$
$$y = -\frac{1}{2}x + 1$$
The inverse function is $f^{-1}(x) = -\frac{1}{2}x + 1$.

33.
$$f(x) = \frac{2}{3}x + 4$$
$$y = \frac{2}{3}x + 4$$
$$x = \frac{2}{3}y + 4$$
$$x - 4 = \frac{2}{3}y$$
$$\frac{3}{2}(x - 4) = y$$
$$\frac{3}{2}x - 6 = y$$
The inverse function is $f^{-1}(x) = \frac{3}{2}x - 6$.

35.
$$f(x) = -\frac{1}{3}x + 1$$
$$y = -\frac{1}{3}x + 1$$
$$x = -\frac{1}{3}y + 1$$
$$x - 1 = -\frac{1}{3}y$$
$$-3(x - 1) = y$$
$$-3x + 3 = y$$
The inverse function is $f^{-1}(x) = -3x + 3$.

37.
$$f(x) = 2x - 5$$
$$y = 2x - 5$$
$$x = 2y - 5$$
$$x + 5 = 2y$$
$$\frac{1}{2}x + \frac{5}{2} = y$$
The inverse function is $f^{-1}(x) = \frac{1}{2}x + \frac{5}{2}$.

39.
$$f(x) = 5x - 2$$
$$y = 5x - 2$$
$$x = 5y - 2$$
$$x + 2 = 5y$$
$$\frac{1}{5}x + \frac{2}{5} = y$$
The inverse function is $f^{-1}(x) = \frac{1}{5}x + \frac{2}{5}$.

41.
$$f(x) = 6x - 3$$
$$y = 6x - 3$$
$$x = 6y - 3$$
$$x + 3 = 6y$$
$$\frac{1}{6}x + \frac{1}{2} = y$$
The inverse function is $f^{-1}(x) = \frac{1}{6}x + \frac{1}{2}$.

43.
$$f(x) = 3x - 5$$
$$y = 3x - 5$$
$$x = 3y - 5$$
$$x + 5 = 3y$$
$$\frac{1}{3}x + \frac{5}{3} = y$$
$$f^{-1}(x) = \frac{1}{3}x + \frac{5}{3}$$
$$f^{-1}(0) = \frac{1}{3}(0) + \frac{5}{3}$$
$$= \frac{5}{3}$$

45.
$$f(x) = 3x - 5$$
$$y = 3x - 5$$
$$x = 3y - 5$$
$$x + 5 = 3y$$
$$\frac{1}{3}x + \frac{5}{3} = y$$
$$f^{-1}(x) = \frac{1}{3}x + \frac{5}{3}$$
$$f^{-1}(4) = \frac{1}{3}(4) + \frac{5}{3}$$
$$= \frac{4}{3} + \frac{5}{3} = 3$$

47. Using the vertical-line test, the graph is a function. Using the horizontal-line test, the graph is 1–1 and therefore does have an inverse.

49.
$$f(g(x)) = f\left(\frac{x}{4}\right)$$
$$= 4\left(\frac{x}{4}\right)$$
$$= x$$
$$g(f(x)) = g(4x)$$
$$= \frac{4x}{4}$$
$$= x$$
Yes. The functions are inverses of each other.

51. $f(h(x)) = f\left(\frac{1}{3x}\right)$
$= 3\left(\frac{1}{3x}\right)$
$= \frac{1}{x}$
$h(f(x)) = h(3x)$
$= \frac{1}{3(3x)}$
$= \frac{1}{9x}$
No. The functions are not inverses of each other.

53. $g(f(x)) = g\left(\frac{1}{3}x - \frac{2}{3}\right)$
$= 3\left(\frac{1}{3}x - \frac{2}{3}\right) + 2$
$= x - 2 + 2 = x$
$f(g(x)) = f(3x + 2)$
$= \frac{1}{3}(3x + 2) - \frac{2}{3}$
$= x + \frac{2}{3} - \frac{2}{3} = x$
Yes. The functions are inverses of each other.

55. $f(g(x)) = f(2x + 3)$
$= \frac{1}{2}(2x + 3) - \frac{3}{2}$
$= x + \frac{3}{2} - \frac{3}{2} = x$
$g(f(x)) = g\left(\frac{1}{2}x - \frac{3}{2}\right)$
$= 2\left(\frac{1}{2}x - \frac{3}{2}\right) + 3$
$= x - 3 + 3 = x$
Yes. The functions are inverses of each other.

Applying the Concepts

57.

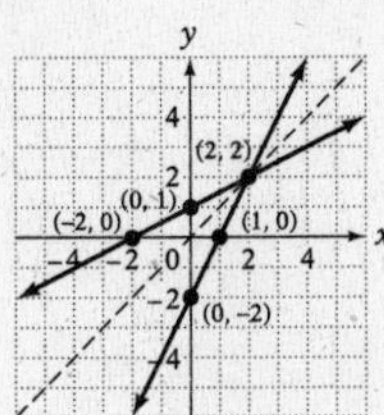

59.

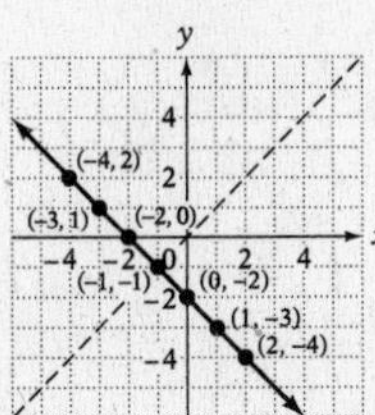

The inverse is the same graph.

61.

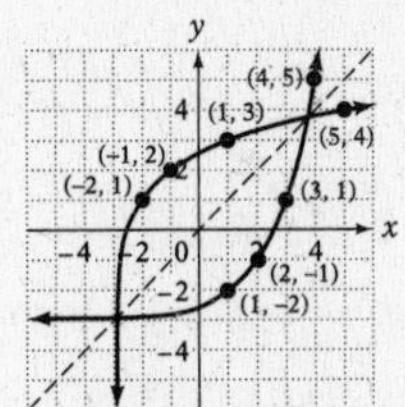

63. Inverse of the Function

Grade	Score
A	90–100
B	80–89
C	70–79
D	60–69
F	0–59

No, the inverse of the grading scale is not a function because each grade is paired with more than one score.

65. $-3 = f(5)$
$f^{-1}[f(5)] = 5$

67. $1 = f(0)$
$f^{-1}[f(0)] = 0$

69. $7 = f(-4)$
$f^{-1}[f(-4)] = -4$

71. A constant function is defined as $y = b$, where b is a constant. The inverse of this function would be $x = a$, where a is a constant. This is not a function.

Chapter 9 Review Exercises

1. Yes, the graph is that of a function. It passes the vertical-line test.

2. Yes. The graph is a 1–1 function. It passes the horizontal- and vertical-line tests.

3. $f(x) = 3x^3 - 2$
Domain: $\{x | x \in \text{real numbers}\}$
Range: $\{y | y \in \text{real numbers}\}$

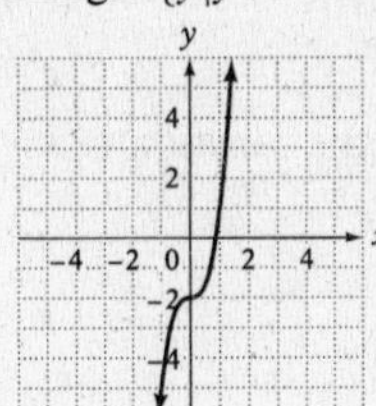

4. $f(x) = \sqrt{x+4}$
Domain: $\{x|x \geq -4\}$
Range: $\{y|y \geq 0\}$

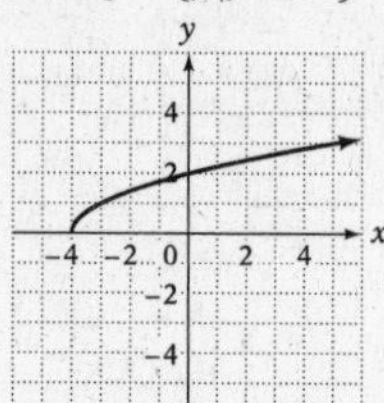

5. $y = -3x^2 + 4x + 6$
$a = -3,\ b = 4,\ c = 6,$
$b^2 - 4ac$
$(4)^2 - 4(-3)(6) = 16 + 72 = 88$
$88 > 0$
Since the discriminant is greater than zero, the function has two x-intercepts.

6. $y = 3x^2 + 9x$
$0 = 3x^2 + 9x$
$0 = 3x(x+3)$
$3x = 0 \quad x + 3 = 0$
$x = 0 \quad\quad x = -3$
The x-intercepts are (0, 0) and (−3, 0).

7. $f(x) = 3x^2 + 2x + 2$
$0 = 3x^2 + 2x + 2$
$$x = \frac{-b \pm \sqrt{b^2 - 4ac}}{2a}$$
$$x = \frac{-2 \pm \sqrt{2^2 - 4(3)(2)}}{2 \cdot 3} = \frac{-2 \pm \sqrt{-20}}{6}$$
$$x = -\frac{1}{3} \pm \frac{2i\sqrt{5}}{6} = -\frac{1}{3} \pm \frac{i\sqrt{5}}{3}$$
The zeros are $-\frac{1}{3} + \frac{\sqrt{5}}{3}i$ and $-\frac{1}{3} - \frac{\sqrt{5}}{3}i$.

8. $f(x) = -2x^2 + 4x + 1$
$$x = \frac{-b}{2a} = \frac{-4}{2 \cdot (-2)} = 1$$
$f(1) = -2(1)^2 + 4(1) + 1$
$= -2 + 4 + 1$
$= 3$
The maximum value of the function is 3.

9. $f(x) = x^2 - 7x + 8$
$$x = \frac{-b}{2a} = \frac{-(-7)}{2 \cdot 1} = \frac{7}{2}$$
$$f\left(\frac{7}{2}\right) = \left(\frac{7}{2}\right)^2 - 7\left(\frac{7}{2}\right) + 8$$
$$= \frac{49}{4} - \frac{49}{2} + 8$$
$$= -\frac{17}{4}$$
The minimum value of the function is $-\frac{17}{4}$.

10. $f(x) = x^2 + 4,\ g(x) = 4x - 1$
$g(0) = 4(0) - 1 = 0 - 1 = -1$
$f(-1) = (-1)^2 + 4 = 1 + 4 = 5$
$f[g(0)] = 5$

11. $f(x) = 6x + 8,\ g(x) = 4x + 2$
$f(-1) = 6(-1) + 8 = -6 + 8 = 2$
$g(2) = 4(2) + 2 = 8 + 2 = 10$
$g[f(-1)] = 10$

12. $f(x) = 3x^2 - 4,\ g(x) = 2x + 1$
$f[g(x)] = f(2x+1)$
$= 3(2x+1)^2 - 4$
$= 3(4x^2 + 4x + 1) - 4$
$= 12x^2 + 12x + 3 - 4$
$f[g(x)] = 12x^2 + 12x - 1$

13. $f(g(x)) = f(-4x+5)$
$$= -\frac{1}{4}(-4x+5) + \frac{5}{4}$$
$$= x - \frac{5}{4} + \frac{5}{4} = x$$
$$g(f(x)) = g\left(-\frac{1}{4}x + \frac{5}{4}\right)$$
$$= -4\left(-\frac{1}{4}x + \frac{5}{4}\right) + 5$$
$$= x - 5 + 5 = x$$
Yes. The functions are inverses of each other.

14. $f(x) = x^2 + 2x - 4$
Domain: $\{x|x \in \text{real numbers}\}$
Range: $\{y|y \geq -5\}$

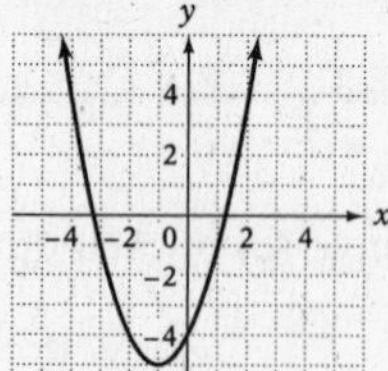

15. $y = x^2 - 2x + 3$
$y = (x-1)^2 - 1 + 3$
$y = (x-1)^2 + 2$
Vertex: (1, 2)
Axis of symmetry: $x = 1$

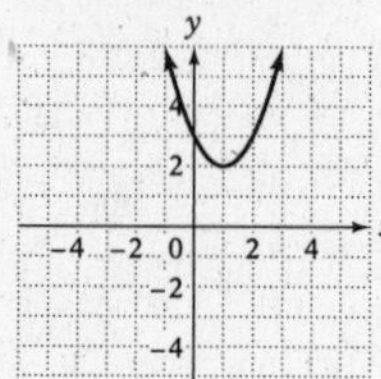

16. $f(x) = |x| - 3$
Domain: $\{x|x \in \text{real numbers}\}$
Range: $\{y|y \geq -3\}$

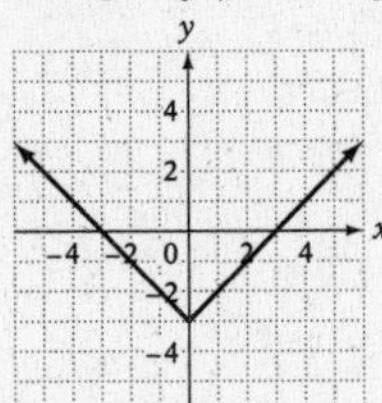

17. No. The graph is not a 1–1 function. It fails the horizontal-line test.

18.
$$\begin{aligned}(f+g)(2) &= f(2) + g(2)\\ &= [2^2 + 2(2) - 3] + [2^2 - 2]\\ &= [4 + 4 - 3] + [4 - 2]\\ &= [5] + [2]\\ &= 7\end{aligned}$$

19.
$$\begin{aligned}(f-g)(-4) &= f(-4) - g(-4)\\ &= [(-4)^2 + 2(-4) - 3] - [(-4)^2 - 2]\\ &= [16 - 8 - 3] - [16 - 2]\\ &= [5] - [14]\\ &= -9\end{aligned}$$

20.
$$\begin{aligned}(f \cdot g)(-4) &= f(-4) \cdot g(-4)\\ &= [(-4)^2 + 2(-4) - 3] \cdot [(-4)^2 - 2]\\ &= [16 - 8 - 3] \cdot [16 - 2]\\ &= [5] \cdot [14]\\ &= 70\end{aligned}$$

21. $\left(\dfrac{f}{g}\right)(3) = \dfrac{f(3)}{g(3)} = \dfrac{(3)^2 + 2(3) - 3}{3^2 - 2} = \dfrac{9 + 6 - 3}{9 - 2} = \dfrac{12}{7}$

22. $f(x) = 2x^2 + x - 5$, $g(x) = 3x - 1$
$$\begin{aligned}g[f(x)] &= g(2x^2 + x - 5)\\ &= 3(2x^2 + x - 5) - 1\\ &= 6x^2 + 3x - 15 - 1\\ g[f(x)] &= 6x^2 + 3x - 16\end{aligned}$$

23.
$$\begin{aligned}f(x) &= -6x + 4\\ y &= -6x + 4\\ x &= -6y + 4\\ x - 4 &= -6y\\ -\frac{1}{6}x + \frac{2}{3} &= y\end{aligned}$$
The inverse function is $f^{-1}(x) = -\frac{1}{6}x + \frac{2}{3}$.

24.
$$\begin{aligned}f(x) &= \frac{2}{3}x - 12\\ y &= \frac{2}{3}x - 12\\ x &= \frac{2}{3}y - 12\\ x + 12 &= \frac{2}{3}y\\ \frac{3}{2}(x + 12) &= y\\ \frac{3}{2}x + 18 &= y\end{aligned}$$
The inverse function is $f^{-1}(x) = \frac{3}{2}x + 18$.

25.
$$\begin{aligned}f(x) &= \frac{1}{2}x + 8\\ y &= \frac{1}{2}x + 8\\ x &= \frac{1}{2}y + 8\\ x - 8 &= \frac{1}{2}y\\ 2(x - 8) &= 2 \cdot \frac{1}{2}y\\ 2x - 16 &= y\end{aligned}$$
The inverse function is $f^{-1}(x) = 2x - 16$.

26. **Strategy** The perimeter is 28 ft.
$$\begin{aligned}28 &= 2L + 2W\\ 14 &= L + W\\ 14 - L &= W\end{aligned}$$
The area is
$$\begin{aligned}L \cdot W &= L \cdot (14 - L)\\ &= 14L - L^2\end{aligned}$$
- To find the length, find the L-coordinate of the vertex of the function $f(L) = -L^2 + 14L$.
- To find the width, replace L in $14 - L$ by the L-coordinate of the vertex and evaluate.
- To find the maximum area, multiply the length by the width.

Solution $L = \dfrac{-b}{2a} = \dfrac{-14}{2 \cdot (-1)} = 7$
The length is 7 ft.
$14 - L = 14 - 7 = 7$
The width is 7 ft.
$L \cdot W = 7 \cdot 7 = 49$
The dimensions are 7 ft by 7 ft.
The maximum area is 49 ft^2.

Chapter 9 Test

1. $y = 2x^2 - 3x + 4$
$0 = 2x^2 - 3x + 4$
$x = \dfrac{-b \pm \sqrt{b^2 - 4ac}}{2a}$
$x = \dfrac{-(-3) \pm \sqrt{(-3)^2 - 4(2)(4)}}{2(2)}$
$= \dfrac{3 \pm \sqrt{-23}}{4}$
$x = \dfrac{3}{4} \pm \dfrac{i\sqrt{23}}{4}$
The zeros of the function are
$\dfrac{3}{4} + \dfrac{\sqrt{23}}{4}i$ and $\dfrac{3}{4} - \dfrac{i\sqrt{23}}{4}$.

2. $g(x) = x^2 + 3x - 8$
$0 = x^2 + 3x - 8$
$x = \dfrac{-b \pm \sqrt{b^2 - 4ac}}{2a}$
$x = \dfrac{-3 \pm \sqrt{3^2 - 4(1)(-8)}}{2 \cdot 1}$
$= \dfrac{-3 \pm \sqrt{41}}{2}$
The x-intercepts of the function are
$\dfrac{-3 + \sqrt{41}}{2}$ and $\dfrac{-3 - \sqrt{41}}{2}$.

3. $y = 3x^2 + 2x - 4$
$a = 3,\ b = 2,\ c = -4$
$b^2 - 4ac$
$2^2 - 4(3)(-4) = 4 + 48 = 52$
$52 > 0$
Since the discriminant is greater than zero, the parabola has two real zeros.

4. $D(x) = f(x) - g(x)$
$= x^2 + 2x - 3 - (x^3 - 1)$
$D(x) = -x^3 + x^2 + 2x - 2$
$D(2) = -(2)^3 + (2)^2 + 2(2) - 2$
$= -8 + 4 + 4 - 2 = -2$
$D(2) = -2 = (f - g)(2)$

5. $P(x) = f(x) \cdot g(x)$
$= (x^3 + 1) \cdot (2x - 3)$
$P(x) = 2x^4 - 3x^3 + 2x - 3$
$P(-3) = 2(-3)^4 - 3(-3)^3 + 2(-3) - 3$
$= 162 + 81 - 6 - 3 = 234$
$P(-3) = 234 = (f \cdot g)(-3)$

6. $\left(\dfrac{f}{g}\right)(x) = \dfrac{4x - 5}{x^2 + 3x + 4}$
$\left(\dfrac{f}{g}\right)(-2) = \dfrac{4(-2) - 5}{(-2)^2 + 3(-2) + 4}$
$= \dfrac{-13}{4 - 6 + 4} = \dfrac{-13}{2} = -\dfrac{13}{2}$

7. $D(x) = f(x) - g(x)$
$= x^2 + 4 - (2x^2 + 2x + 1)$
$D(x) = -x^2 - 2x + 3$
$D(-4) = -(-4)^2 - 2(-4) + 3$
$= -16 + 8 + 3 = -5$
$D(-4) = -5 = (f - g)(-4)$

8. $g(x) = \dfrac{x}{x + 1}$
$g(3) = \dfrac{3}{3 + 1} = \dfrac{3}{4}$
$f(x) = 4x + 2$
$f\left(\dfrac{3}{4}\right) = 4 \cdot \dfrac{3}{4} + 2 = 5$
$f[g(3)] = 5$

9. $g(x) = x - 1$
$f[g(x)] = 2(x - 1)^2 - 7$
$= 2(x^2 - 2x + 1) - 7$
$= 2x^2 - 4x + 2 - 7$
$f[g(x)] = 2x^2 - 4x - 5$

10. $f(x) = -x^2 + 8x - 7$
$x = -\dfrac{b}{2a} = \dfrac{-8}{2 \cdot (-1)} = 4$
$f(4) = -(4)^2 + 8(4) - 7$
$= -16 + 32 - 7 = 9$
The maximum value of the function is 9.

11. $f(x) = 4x - 2$
$y = 4x - 2$
$x = 4y - 2$
$x + 2 = 4y$
$\dfrac{1}{4}x + \dfrac{1}{2} = y$
The inverse function is $f^{-1}(x) = \dfrac{1}{4}x + \dfrac{1}{2}$.

12. $f(x) = \dfrac{1}{4}x - 4$
$y = \dfrac{1}{4}x - 4$
$x = \dfrac{1}{4}y - 4$
$x + 4 = \dfrac{1}{4}y$
$4x + 16 = y$
The inverse of the function is $f^{-1}(x) = 4x + 16$.

13. The inverse of the function of $\{(2, 6), (3, 5), (4, 4), (5, 3)\}$ is $\{(6, 2), (5, 3), (4, 4), (3, 5)\}$.

14. $f[g(x)] = \dfrac{1}{2}(2x - 4) + 2$
$= x - 2 + 2$
$= x$
$g[f(x)] = 2\left(\dfrac{1}{2}x + 2\right) - 4$
$= x + 4 - 4$
$= x$
Yes. The functions are inverses of each other.

15. Strategy
- Let x represent one number. Since the sum of the two numbers is 20, the other number is $20 - x$. Their product is $20x - x^2$.
- To find the first number, find the x-coordinate of the vertex of the function $f(x) = -x^2 + 20x$.
- To find the other number, replace x in $20 - x$ with the x-coordinate of the vertex and evaluate.

Solution $x = -\dfrac{b}{2a} = \dfrac{-20}{2 \cdot (-1)} = 10$

$20 - x = 20 - 10 = 10$

The numbers are 10 and 10, so their product is 100.

16. $f(x) = -\sqrt{3 - x}$
Domain: $\{x|x \le 3\}$
Range: $\{y|y \le 0\}$.

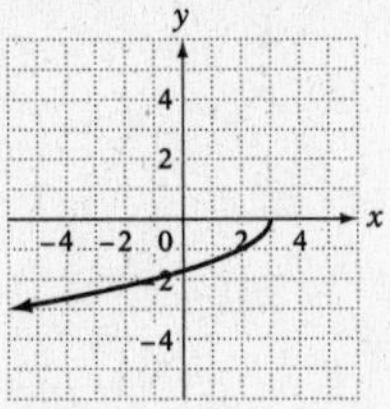

17. $f(x) = |\frac{1}{2}x| - 2$
Domain: $\{x|x \in \text{real numbers}\}$
Range: $\{y|y \ge -2\}$.

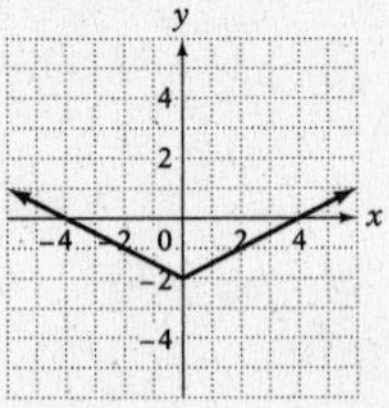

18. $f(x) = x^3 - 3x + 2$
Domain: $\{x|x \in \text{real numbers}\}$
Range: $\{y|y \in \text{real numbers}\}$

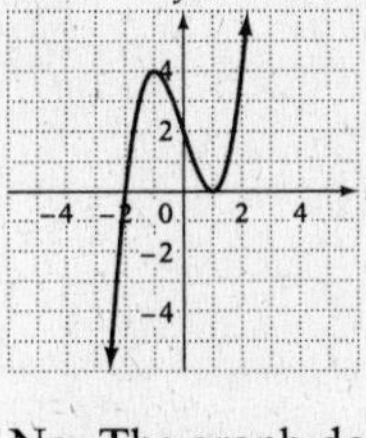

19. No. The graph does not represent a 1–1 function. It fails the horizontal- and vertical-line tests.

20. Strategy The perimeter is 200 cm.

$200 = 2L + 2W$
$100 = L + W$
$100 - L = W$

The area is $L \cdot W = L(100 - L)$
$= 100L - L^2$

- To find the length, find the L-coordinate of the vertex of the function $f(L) = -L^2 + 100L$.
- To find the width, replace L in $100 - L$ by the L-coordinate of the vertex and evaluate.
- To find the maximum area, multiply the length by the width.

Solution $L = -\dfrac{b}{2a} = -\dfrac{100}{2(-1)} = 50$

The length is 50 cm.
$100 - L = 100 - 50 = 50$
The width is 50 cm.
$L \cdot W = 50 \cdot 50 = 2500$
Dimensions of 50 cm by 50 cm would give the maximum area as 2500 cm^2.

Cumulative Review Exercises

1. $-3a + \left|\dfrac{3b - ab}{3b - c}\right| = -3(2) + \left|\dfrac{3(2) - 2(2)}{3(2) - (-2)}\right|$

$= -6 + \left|\dfrac{6 - 4}{6 + 2}\right| = -6 + \left|\dfrac{2}{8}\right|$

$= -6 + \left|\dfrac{1}{4}\right| = -6 + \dfrac{1}{4} = -\dfrac{23}{4}$

2. 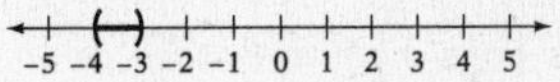

3. $\dfrac{3x - 1}{6} - \dfrac{5 - x}{4} = \dfrac{5}{6}$

$12\left(\dfrac{3x - 1}{6} - \dfrac{5 - x}{4}\right) = 12\left(\dfrac{5}{6}\right)$

$2(3x - 1) - 3(5 - x) = 2(5)$
$6x - 2 - 15 + 3x = 10$
$9x - 17 = 10$
$9x = 27$
$x = 3$

The solution is 3.

4. $4x - 2 < -10$ or $3x - 1 > 8$

$4x - 2 < -10$ $\quad$ $3x - 1 > 8$
$4x < -8$ $\quad$ $3x > 9$
$x < -2$ $\quad$ $x > 3$

$\{x|x < -2\}$ or $\{x|x > 3\}$
$\{x|x < -2\} \cup \{x|x > 3\} = \{x|x < -2 \text{ or } x > 3\}$

5. $|8 - 2x| \ge 0$

$8 - 2x \le 0$ $\quad$ $8 - 2x \ge 0$
$8 \le 2x$ $\quad$ $8 \ge 2x$
$4 \le x$ $\quad$ $4 \ge x$

$\{x|x \ge 4\}$ or $\{x|x \le 4\}$
$\{x|x \ge 4\} \cup \{x|x \le 4\} = \{x|x \in \text{real numbers}\}$

6. $\left(\frac{3a^3b}{2a}\right)^2\left(\frac{a^2}{-3b^2}\right)^3 = \left(\frac{3a^2b}{2}\right)^2\left(\frac{a^2}{-3b^2}\right)^3$

$= \left(\frac{3^2a^4b^2}{2^2}\right)\left(\frac{a^6}{(-3)^3b^6}\right)$

$= \left(\frac{9a^4b^2}{4}\right)\left(\frac{a^6}{-27b^6}\right)$

$= \frac{9a^4b^2a^6}{4(-27)b^6}$

$= \frac{9a^{10}b^2}{-108b^6}$

$= -\frac{a^{10}}{12b^4}$

7. $(x-4)(2x^2+4x-1)$

$= 2x^2(x-4)+4x(x-4)-1(x-4)$

$= 2x^3-8x^2+4x^2-16x-x+4$

$= 2x^3-4x^2-17x+4$

8. $6x-2y=-3$

$4x+y=5$

$6x-2y=-3$

$8x+2y=10$

$14x = 7$

$x=\frac{1}{2}$

$6x-2y=-3$

$6\left(\frac{1}{2}\right)-2y=-3$

$3-2y=-3$

$-2y=-6$

$y=3$

The solution is $\left(\frac{1}{2}, 3\right)$.

9. $x^3y+x^2y^2-6xy^3 = xy(x^2+xy-6y^2)$

$= xy(x+3y)(x-2y)$

10. $(b+2)(b-5)=2b+14$

$b^2-3b-10=2b+14$

$b^2-5b-24=0$

$(b-8)(b+3)=0$

$b-8=0 \quad b+3=0$

$b=8 \quad b=-3$

The solutions are -3 and 8.

11. $x^2-2x>15$

$x^2-2x-15>0$

$(x-5)(x+3)>0$

$x-5$: − − − − − − − − (from −5 to 5) ++

$x+3$: − − − − − − ++++++++++++ (from −3) ++

−5 −4 −3 −2 −1 0 1 2 3 4 5

$\{x \mid x<-3 \text{ or } x>5\}$

12. $\frac{x^2+4x-5}{2x^2-3x+1}-\frac{x}{2x-1}$

$= \frac{(x+5)(x-1)}{(2x-1)(x-1)}-\frac{x}{2x-1}$

$= \frac{x+5}{2x-1}-\frac{x}{2x-1}$

$= \frac{x+5-x}{2x-1} = \frac{5}{2x-1}$

13. $\frac{5}{x^2+7x+12} = \frac{9}{x+4}-\frac{2}{x+3}$

$\frac{5}{(x+4)(x+3)} = \frac{9}{x+4}\cdot\frac{x+3}{x+3}-\frac{2}{x+3}\cdot\frac{x+4}{x+4}$

$\frac{5}{(x+4)(x+3)} = \frac{9x+27}{(x+4)(x+3)}-\frac{2x+8}{(x+4)(x+3)}$

$5=(9x+27)-(2x+8)$

$5=9x+27-2x-8$

$5=7x+19$

$-14=7x$

$-2=x$

The solution is -2.

14. $\frac{4-6i}{2i} = \frac{4-6i}{2i}\cdot\frac{i}{i}$

$= \frac{4i-6i^2}{2i^2}$

$= \frac{4i+6}{-2}$

$= -3-2i$

15. Vertex: (0, 0)

Axis of symmetry: $x=0$

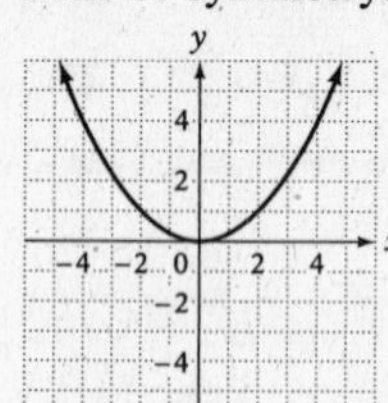

16. $3x-4y \geq 8$

$-4y \geq -3x+8$

$y \leq \frac{3}{4}x-2$

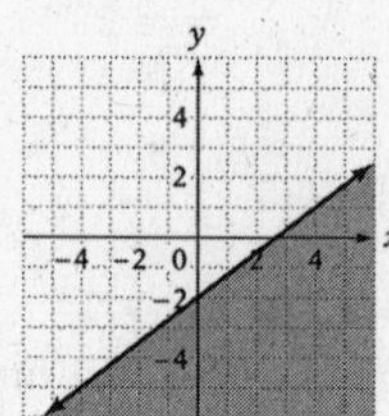

17. $m = \frac{y_2 - y_1}{x_2 - x_1} = \frac{-6-4}{2-(-3)} = \frac{-10}{5} = -2$

$y - y_1 = m(x - x_1)$
$y - 4 = -2[x - (-3)]$
$y - 4 = -2(x + 3)$
$y - 4 = -2x - 6$
$y = -2x - 2$

18. The product of the slopes of perpendicular lines is -1.

$2x - 3y = 6$ $\quad m_1 \cdot m_2 = -1$
$-3y = -2x + 6$ $\quad \frac{2}{3} \cdot m_2 = -1$
$y = \frac{2}{3}x - 2$ $\quad m_2 = -\frac{3}{2}$

$y - y_1 = m(x - x_1)$
$y - 1 = -\frac{3}{2}[x - (-3)]$
$y - 1 = -\frac{3}{2}(x + 3)$
$y - 1 = -\frac{3}{2}x - \frac{9}{2}$
$y = -\frac{3}{2}x - \frac{7}{2}$

19. $3x^2 = 3x - 1$
$3x^2 - 3x + 1 = 0$
$a = 3,\ b = -3,\ c = 1$

$$x = \frac{-b \pm \sqrt{b^2 - 4ac}}{2a}$$
$$= \frac{-(-3) \pm \sqrt{(-3)^2 - 4(3)(1)}}{2(3)}$$
$$= \frac{3 \pm \sqrt{9 - 12}}{6}$$
$$= \frac{3 \pm \sqrt{-3}}{6} = \frac{3 \pm i\sqrt{-3}}{6} = \frac{1}{2} \pm \frac{\sqrt{3}}{6}i$$

The solutions are $\frac{1}{2} + \frac{\sqrt{3}}{6}i$ and $\frac{1}{2} - \frac{\sqrt{3}}{6}i$.

20. $\sqrt{8x+1} = 2x - 1$
$(\sqrt{8x+1})^2 = (2x-1)^2$
$8x + 1 = 4x^2 - 4x + 1$
$0 = 4x^2 - 12x$
$0 = 4x(x - 3)$
$4x = 0 \quad x - 3 = 0$
$x = 0 \quad x = 3$

Check:

$\sqrt{8x+1}$	$= 2x - 1$
$\sqrt{8(0)+1}$	$2(0) - 1$
$\sqrt{1}$	-1

$1 \neq -1$

Check:

$\sqrt{8x+1}$	$= 2x - 1$
$\sqrt{8(3)+1}$	$2(3) - 1$
$\sqrt{24+1}$	$6 - 1$
$\sqrt{25}$	5

$5 = 5$

The solution is 3.

21. $f(x) = 2x^2 - 3$
$a = 2,\ b = 0,\ c = -3$
$x = \frac{-b}{2a} = \frac{-0}{2 \cdot 2} = 0$
$f(0) = 2(0)^2 - 3 = -3$

The minimum value of the function is -3.

22. $f(x) = |3x - 4|$;
domain $= \{0, 1, 2, 3\}$
$f(x) = |3x - 4|$
$f(0) = |3(0) - 4| = |0 - 4| = |-4| = 4$
$f(1) = |3(1) - 4| = |3 - 4| = |-1| = 1$
$f(2) = |3(2) - 4| = |6 - 4| = |2| = 2$
$f(3) = |3(3) - 4| = |9 - 4| = |5| = 5$

The range is $= \{1, 2, 4, 5\}$.

23. $\{(-3, 0), (-2, 0), (-1, 1), (0, 1)\}$

Yes. Each member of the domain is paired with only one member of the range. The set of ordered pairs is a function.

24. $\sqrt[3]{5x - 2} = 2$
$(\sqrt[3]{5x - 2})^3 = 2^3$
$5x - 2 = 8$
$5x = 10$
$x = 2$

The solution is 2.

25. $h(x) = \frac{1}{2}x + 4$
$h(2) = \frac{1}{2}(2) + 4 = 1 + 4 = 5$
$g(x) = 3x - 5$
$g(5) = 3(5) - 5 = 15 - 5 = 10$

$g(h(2)) = g(5) = 10$

26. $f(x) = -3x + 9$
$$y = -3x + 9$$
$$x = -3y + 9$$
$$3y = -x + 9$$
$$y = -\frac{1}{3}x + 3$$
The inverse function is $f^{-1}(x) = -\frac{1}{3}x + 3$.

27. Strategy • Cost per pound of the mixture: x

	Amount	Cost	Value
\$4.50 tea	30	4.50	4.50(3)
\$3.60 tea	45	3.60	3.60(45)
Mixture	75	x	$75x$

• The sum of the values before mixing equals the value after mixing.

Solution
$$4.50(30) + 3.60(45) = 75x$$
$$135 + 162 = 75x$$
$$297 = 75x$$
$$3.96 = x$$
The cost per pound of the mixture is \$3.96.

28. Strategy • Pounds of 80% copper alloy: x

	Amount	Percent	Quantity
80%	x	0.80	$0.80x$
20%	50	0.20	0.20(50)
40%	$50 + x$	0.40	$0.40(50 + x)$

• The sum of the quantities before mixing is equal to the quantity after mixing.

Solution
$$0.80x + 0.20(50) = 0.40(50 + x)$$
$$0.80x + 10 = 20 + 0.40x$$
$$0.40x + 10 = 20$$
$$0.40x = 10$$
$$x = 25$$
25 lb of the 80% copper alloy must be used.

29. Strategy To find the additional amount of insecticide, write and solve a proportion using x to represent the additional amount of insecticide. Then, $x + 6$ is the total amount of insecticide.

Solution
$$\frac{6}{16} = \frac{x+6}{28}$$
$$\frac{3}{8} = \frac{x+6}{28}$$
$$\frac{3}{8} \cdot 56 = \frac{x+6}{28} \cdot 56$$
$$21 = (x + 6)2$$
$$21 = 2x + 12$$
$$9 = 2x$$
$$4.5 = x$$
An additional 4.5 oz of insecticide are required.

30. Strategy • This is a work problem.
• Time for the smaller pipe to fill the tank: t
Time for the larger pipe to fill the tank: $t - 8$

	Rate	Time	Part
Smaller pipe	$\frac{1}{t}$	3	$\frac{3}{t}$
Larger pipe	$\frac{1}{t-8}$	3	$\frac{3}{t-8}$

• The sum of the parts of the task completed must equal 1.

Solution
$$\frac{3}{t} + \frac{3}{t-8} = 1$$
$$t(t-8)\left(\frac{3}{t} + \frac{3}{t-8}\right) = t(t-8)$$
$$(t-8)3 + 3t = t^2 - 8t$$
$$3t - 24 + 3t = t^2 - 8t$$
$$6t - 24 = t^2 - 8t$$
$$0 = t^2 - 14t + 24$$
$$= (t-2)(t-12)$$
$t - 2 = 0$ $\quad t - 12 = 0$
$t = 2$ $\quad t = 12$
The solution 2 is not possible since the time for the larger pipe would then be a negative number.
$t - 8 = 12 - 8 = 4$
It would take the larger pipe 4 min to fill the tank.

31. Strategy To find the distance:
• Write the basic direct variation equation, replace the variables by the given values, and solve for k.
• Write the direct variation equation, replacing k by its value. Substitute 40 for f and solve for d.

Solution
$d = kf$ $\quad d = \frac{3}{5}f$
$30 = k(50)$ $\quad = \frac{3}{5}(40)$
$\frac{3}{5} = k$ $\quad = 24$
A force of 40 lb will stretch the string 24 in.

32. Strategy To find the frequency:
• Write the basic indirect variation equation, replace the variables by the given values, and solve for k.
• Write the inverse variation equation, replacing k by its value. Substitute 1.5 for L and solve for f.

Solution
$f = \frac{k}{L}$ $\quad f = \frac{120}{L}$
$60 = \frac{k}{2}$ $\quad = \frac{120}{1.5}$
$120 = k$ $\quad = 80$
The frequency is 80 vibrations/min.

Chapter 10: Exponential and Logarithmic Functions

Prep Test

1. $3^{-2} = \frac{1}{3^2} = \frac{1}{9}$

2. $\left(\frac{1}{2}\right)^{-4} = \left(\frac{2}{1}\right)^4 = 2^4 = 16$

3. $\frac{1}{8} = \frac{1}{2^3} = 2^{-3}; ? = -3$

4. $f(-1) = (-1)^4 + (-1)^3 = 1 + (-1) = 0$
   $f(3) = (3)^4 + (3)^3 = 81 + 27 = 108$

5. $3x + 7 = x - 5$
   $2x = -12$
   $x = -6$

6. $16 = x^2 - 6x$
   $x^2 - 6x - 16 = 0$
   $(x+2)(x-8) = 0$
   $x + 2 = 0 \qquad x - 8 = 0$
   $x = -2 \qquad x = 8$

   The solutions are −2 and 8.

7. $A(1+i)^n = 5000(1+0.04)^6$
   $= 5000(1.04)^6$
   $= 6326.60$

8. 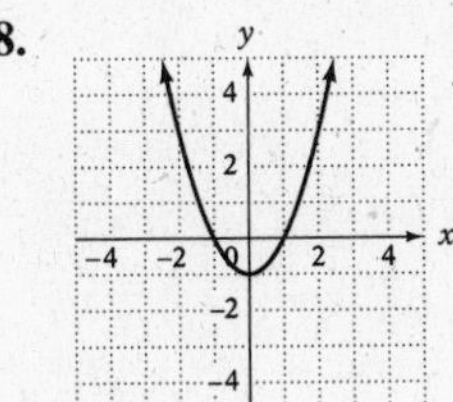

Go Figure

By observation, one can see that the odd powers of 9 end in 9. The even powers of 9 end in 1, and the first term (9^0) also is 1. Therefore, when beginning with an even power of 9 and ending with an odd power of 9, the resulting sum will always be a multiple of 10, with a ones digit of 0.

Answer: 0.

Section 10.1

Objective A Exercises

1. An exponential function with base b is defined by $f(x) = b^x$, $b > 0$, $b \neq 1$, and x is any real number.

3. When $f(x) = b^x$, $b > 0$, $b \neq 1$. Therefore **c** cannot be the base of an exponential function because it is negative.

5. $f(x) = 3^x$

   a. $f(2) = 3^2 = 9$

   b. $f(0) = 3^0 = 1$

   c. $f(-2) = 3^{-2} = \frac{1}{3^2} = \frac{1}{9}$

7. $g(x) = 2^{x+1}$

   a. $g(3) = 2^{3+1} = 2^4 = 16$

   b. $g(1) = 2^{1+1} = 2^2 = 4$

   c. $g(-3) = 2^{-3+1} = 2^{-2} = \frac{1}{2^2} = \frac{1}{4}$

9. $P(x) = \left(\frac{1}{2}\right)^{2x}$

   a. $P(0) = \left(\frac{1}{2}\right)^{2\cdot 0} = \left(\frac{1}{2}\right)^0 = 1$

   b. $P\left(\frac{3}{2}\right) = \left(\frac{1}{2}\right)^{2\cdot\frac{3}{2}} = \left(\frac{1}{2}\right)^3 = \frac{1}{8}$

   c. $P(-2) = \left(\frac{1}{2}\right)^{2(-2)} = \left(\frac{1}{2}\right)^{-4} = 2^4 = 16$

11. $G(x) = e^{x/2}$

   a. $G(4) = e^{4/2} = e^2 = 7.3891$

   b. $G(-2) = e^{-2/2} = e^{-1} = \frac{1}{e} \approx 0.3679$

   c. $G\left(\frac{1}{2}\right) = e^{\frac{1}{2}/2} = e^{1/4} = e^{0.25} \approx 1.2840$

13. $H(r) = e^{-r+3}$

   a. $H(-1) = e^{-(-1)+3} = e^4 \approx 54.5982$

   b. $H(3) = e^{-3+3} = e^0 = 1$

   c. $H(5) = e^{-5+3} = e^{-2} = \frac{1}{e^2} \approx 0.1353$

15. $F(x) = 2^{x^2}$

   a. $F(2) = 2^{2^2} = 2^4 = 16$

   b. $F(-2) = 2^{(-2)^2} = 2^4 = 16$

   c. $F\left(\frac{3}{4}\right) = 2^{\left(\frac{3}{4}\right)^2} = 2^{\frac{9}{16}} = \sqrt[16]{2^9} = \sqrt[16]{512} \approx 1.4768$

17. $f(x) = e^{-x^2/2}$

   a. $f(-2) = e^{-(-2)^2/2}$
   $= e^{-4/2}$
   $= e^{-2} = \frac{1}{e^2} \approx 0.1353$

   b. $f(2) = e^{-(2)^2/2} = e^{-4/2} = e^{-2} = \frac{1}{e^2} \approx 0.1353$

   c. $f(-3) = e^{-(-3)^2/2} = e^{-9/2} = \frac{1}{e^{9/2}} \approx 0.0111$

Objective B Exercises

19.

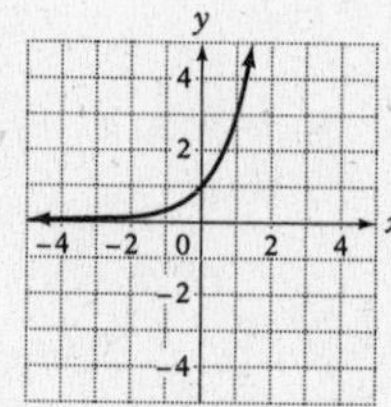

21.

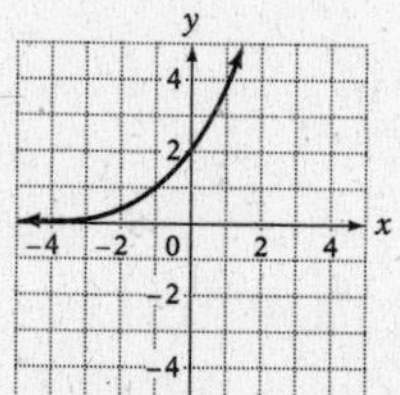

23.

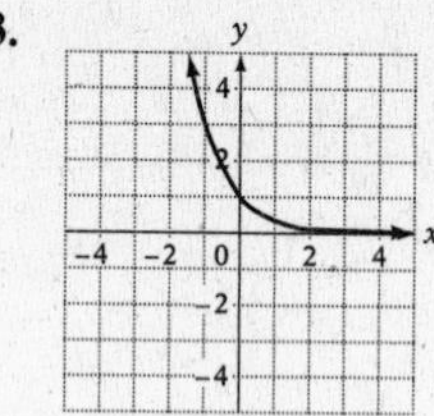

25.

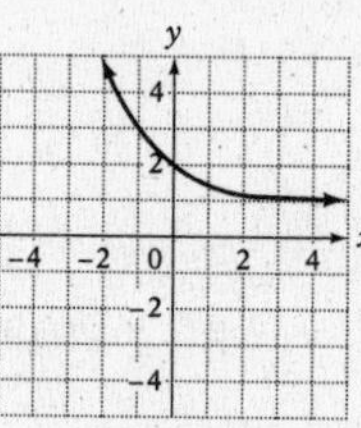

27.

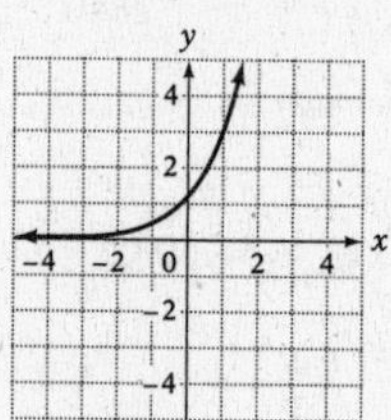

29.

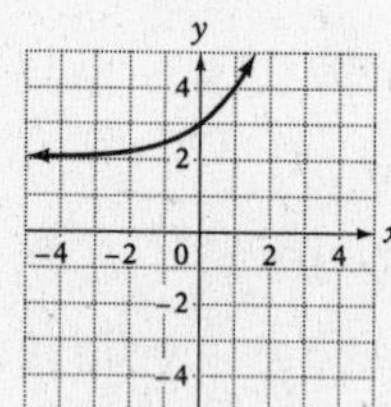

31. $\left(\frac{1}{3}\right)^x = \frac{1^x}{3^x} = \frac{1}{3^x} = 3^{-x}$; therefore, **b** and **d** have the same graph.

33.

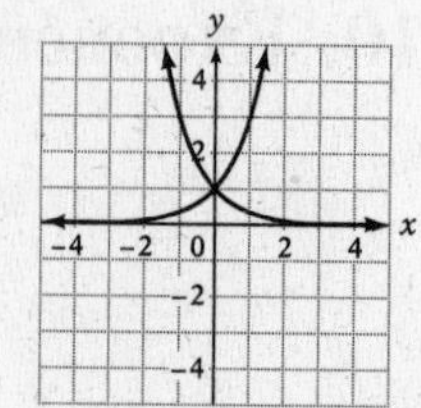

The intersection of the graphs is (0, 1).

35.

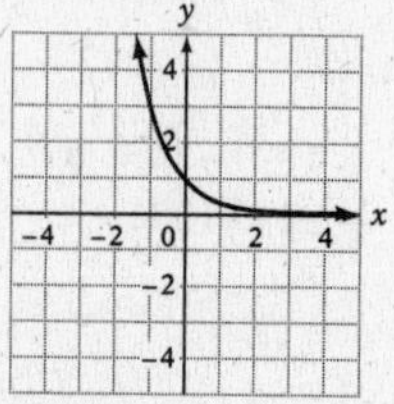

There is no x-intercept and the y-intercept is (0, 1).

Applying the Concepts

37. $P(x) = (\sqrt{3})^x$

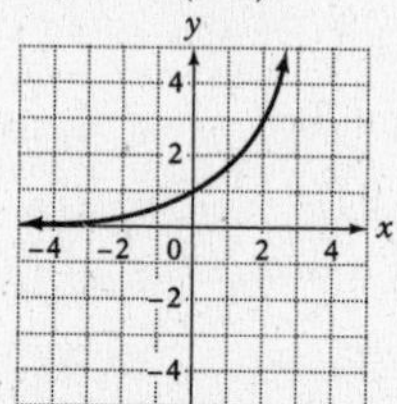

39. $f(x) = \pi^x$

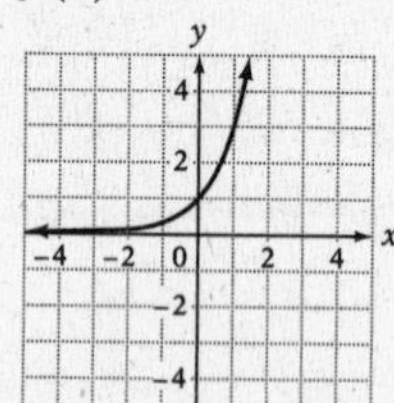

41. **a.**

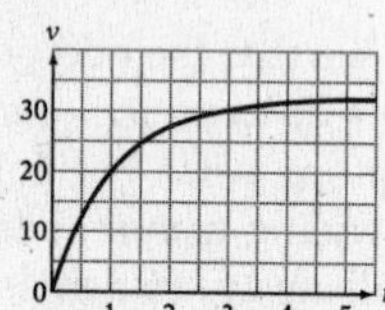

b. The point (2, 27.7) means that after 2 s, the object is falling at a speed of 27.7 ft/s.

Section 10.2

Objective A Exercises

1. **a.** A common logarithm is a logarithm with base 10.

   b. $\log 4z$

3. $5^2 = 25$ is equivalent to $\log_5 25 = 2$.

5. $4^{-2} = \frac{1}{16}$ is equivalent to $\log_4 \frac{1}{16} = -2$.

7. $10^y = x$ is equivalent to $\log_{10} x = y$.

9. $a^x = w$ is equivalent to $\log_a w = x$.

11. $\log_3 9 = 2$ is equivalent to $3^2 = 9$.

13. $\log 0.01 = -2$ is equivalent to $10^{-2} = 0.01$.

15. $\ln x = y$ is equivalent to $e^y = x$.

17. $\log_b u = v$ is equivalent to $b^v = u$.

19. $\log_3 81 = x$
$81 = 3^x$
$x = 4$
$\log_3 81 = 4$

21. $\log_2 128 = x$
$2^x = 128$
$x = 7$
$\log_2 128 = 7$

23. $\log 100 = x$
$10^x = 100$
$x = 2$
$\log 100 = 2$

25. $\ln e^3 = x$
$3\ln e = x$
$3(1) = x$
$x = 3$
$\ln e^3 = 3$

27. $\log_8 1 = x$
$8^x = 1$
$x = 0$
$\log_8 1 = 0$

29. $\log_5 625 = x$
$5^x = 625$
$x = 4$
$\log_5 625 = 4$

31. $\log_3 x = 2$
$3^2 = x$
$9 = x$

33. $\log_4 x = 3$
$4^3 = x$
$64 = x$

35. $\log_7 x = -1$
$7^{-1} = x$
$\frac{1}{7} = x$

37. $\log_6 x = 0$
$6^0 = x$
$1 = x$

39. $\log x = 2.5$
$10^{2.5} = x$
$316.23 \approx x$

41. $\log x = -1.75$
$10^{-1.75} = x$
$0.02 \approx x$

43. $\ln x = 2$
$e^2 = x$
$7.39 \approx x$

45. $\ln x = -\frac{1}{2}$
$e^{-1/2} = x$
$0.61 \approx x$

Objective B Exercises

47. Answers will vary. For example, the log of a product is equal to the sum of the logs: $\log_b xy = \log_b x + \log_b y$.

49. $\log_3 x^3 + \log_3 y^2 = \log_3(x^3y^2)$

51. $\ln x^4 - \ln y^2 = \ln \frac{x^4}{y^2}$

53. $3\log_7 x = \log_7 x^3$

55. $3\ln x + 4\ln y = \ln x^3 + \ln y^4 = \ln(x^3y^4)$

57. $2(\log_4 x + \log_4 y) = 2\log_4(xy)$
$= \log_4(xy)^2$
$= \log_4(x^2y^2)$

59. $2\log_3 x - \log_3 y + 2\log_3 z$
$= \log_3 x^2 - \log_3 y + \log_3 z^2$
$= \log_3 \frac{x^2}{y} + \log_3 z^2$
$= \log_3 \frac{x^2z^2}{y}$

61. $\ln x - (2\ln y + \ln z) = \ln x - (\ln y^2 + \ln z)$
$= \ln x - \ln(y^2 \cdot z)$
$= \ln \frac{x}{y^2z}$

63. $\frac{1}{2}(\log_6 x - \log_6 y) = \frac{1}{2}\log_6 \frac{x}{y} = \log_6 \left(\frac{x}{y}\right)^{1/2}$
$= \log_6 \sqrt{\frac{x}{y}}$

65. $2(\log_4 s - 2\log_4 t + \log_4 r)$
$= 2(\log_4 s - \log_4 t^2 + \log_4 r)$
$= 2\left(\log_4 \frac{s}{t^2} + \log_4 r\right)$
$= 2\log_4 \frac{sr}{t^2}$
$= \log_4 \left(\frac{sr}{t^2}\right)^2$
$= \log_4 \frac{s^2r^2}{t^4}$

67. $\ln x - 2(\ln y + \ln z) = \ln x - 2[\ln(yz)]$
$= \ln x - \ln(yz)^2$
$= \ln x - \ln(y^2z^2) = \ln \frac{x}{y^2z^2}$

69. $3\log_2 t - 2(\log_2 r - \log_2 v)$
$= \log_2 t^3 - 2\log_2 \frac{r}{v}$
$= \log_2 t^3 - \log_2 \left(\frac{r}{v}\right)^2$
$= \log_2 t^3 - \log_2 \frac{r^2}{v^2}$
$= \log_2 \frac{t^3 v^2}{r^2}$

71. $\frac{1}{2}(3\log_4 x - 2\log_4 y + \log_4 z)$
$= \frac{1}{2}(\log_4 x^3 - \log_4 y^2 + \log_4 z)$
$= \frac{1}{2}\left(\log_4 \frac{x^3}{y^2} + \log_4 z\right)$
$= \frac{1}{2}\log_4 \frac{x^3 z}{y^2}$
$= \log_4 \left(\frac{x^3 z}{y^2}\right)^{1/2}$
$= \log_4 \sqrt{\frac{x^3 z}{y^2}}$

73. $\frac{1}{2}(\ln x - 3\ln y)$
$= \frac{1}{2}(\ln x - \ln y^3)$
$= \frac{1}{2}\left[\ln \frac{x}{y^3}\right]$
$= \ln \left(\frac{x}{y^3}\right)^{1/2}$
$= \ln \sqrt{\frac{x}{y^3}}$

75. $\frac{1}{2}\log_2 x - \frac{2}{3}\log_2 y + \frac{1}{2}\log_2 z$
$= \log_2 x^{1/2} - \log_2 y^{2/3} + \log_2 z^{1/2}$
$= \log_2 \frac{x^{1/2}}{y^{2/3}} + \log_2 z^{1/2}$
$= \log_2 \frac{x^{1/2} z^{1/2}}{y^{2/3}}$
$= \log_2 \frac{\sqrt{xz}}{\sqrt[3]{y^2}}$

77. $\log_8 (xz) = \log_8 x + \log_8 z$

79. $\log_3 x^5 = 5\log_3 x$

81. $\log_b \frac{r}{s} = \log_b r - \log_b s$

83. $\log_3(x^2 y^6) = \log_3 x^2 + \log_3 y^6$
$= 2\log_3 x + 6\log_3 y$

85. $\log_7 \frac{u^3}{v^4} = \log_7 u^3 - \log_7 v^4$
$= 3\log_7 u - 4\log_7 v$

87. $\log_2 (rs)^2 = 2\log_2 (rs)$
$= 2(\log_2 r + \log_2 s)$
$= 2\log_2 r + 2\log_2 s$

89. $\ln x^2 yz = \ln x^2 + \ln y + \ln z$
$= 2\ln x + \ln y + \ln z$

91. $\log_5 \frac{xy^2}{z^4} = \log_5 (xy^2) - \log_5 z^4$
$= \log_5 x + \log_5 y^2 - \log_5 z^4$
$= \log_5 x + 2\log_5 y - 4\log_5 z$

93. $\log_8 \frac{x^2}{yz^2} = \log_8 x^2 - \log_8 (yz^2)$
$= 2\log_8 x - (\log_8 y + \log_8 z^2)$
$= 2\log_8 x - \log_8 y - \log_8 z^2$
$= 2\log_8 x - \log_8 y - 2\log_8 z$

95. $\log_4 \sqrt{x^3 y} = \log_4 (x^3 y)^{1/2}$
$= \frac{1}{2}[\log_4 (x^3 y)]$
$= \frac{1}{2}(\log_4 x^3 + \log_4 y)$
$= \frac{1}{2}(3\log_4 x + \log_4 y)$
$= \frac{3}{2}\log_4 x + \frac{1}{2}\log_4 y$

97. $\log_7 \sqrt{\frac{x^3}{y}} = \log_7 \left(\frac{x^3}{y}\right)^{1/2}$
$= \frac{1}{2}\log_7 \frac{x^3}{y}$
$= \frac{1}{2}(\log_7 x^3 - \log_7 y)$
$= \frac{1}{2}(3\log_7 x - \log_7 y)$
$= \frac{3}{2}\log_7 x - \frac{1}{2}\log_7 y$

99. $\log_3 \frac{t}{\sqrt{x}} = \log_3 t - \log_3 \sqrt{x}$
$= \log_3 t - \log_3 x^{1/2}$
$= \log_3 t - \frac{1}{2}\log_3 x$

Objective C Exercises

101. $\log_{10} 7 \approx 0.8451$

103. $\log_{10} \frac{3}{5} = \log_{10} 3 - \log_{10} 5 \approx -0.2218$

105. $\ln 4 \approx 1.3863$

107. $\ln \frac{17}{6} = \ln 17 - \ln 6 \approx 1.0415$

109. $\log_8 6 = \frac{\log_{10} 6}{\log_{10} 8} \approx 0.8617$

111. $\log_5 30 = \frac{\log_{10} 30}{\log_{10} 5} \approx 2.1133$

113. $\log_3 0.5 = \frac{\log_{10}(0.5)}{\log_{10} 3} \approx -0.6309$

115. $\log_7 1.7 = \dfrac{\log_{10} 1.7}{\log_{10} 7} \approx 0.2727$

117. $\log_5 15 = \dfrac{\log_{10} 15}{\log_{10} 5} \approx 1.6826$

119. $\log_{12} 120 = \dfrac{\log_{10} 120}{\log_{10} 12} \approx 1.9266$

121. $\log_4 2.55 = \dfrac{\log_{10} 2.55}{\log_{10} 4} \approx 0.6752$

123. $\log_5 67 = \dfrac{\log_{10} 67}{\log_{10} 5} \approx 2.6125$

Applying the Concepts

125. **a.** $\log_3(-9) = -2$

$3^{-2} = \dfrac{1}{3^2} = \dfrac{1}{9}$

$\dfrac{1}{9} \neq -9$ False

b. $x^y = z$ is $\log_x z = y$ True

c. $\log(x^{-1}) = \dfrac{1}{\log x}$

$\log(x^{-1}) = -\log x$

$-\log x \neq \dfrac{1}{\log x}$ False

d. $\log \dfrac{x}{y} = \log x - \log y$ True

e. $\log(x \cdot y) = \log x \cdot \log y$ False

f. If $\log x = \log y$, then $x = y$. True

Section 10.3

Objective A Exercises

1. Yes. Answers will vary. For example, the function passes both the vertical-line test and the horizontal-line test.

3. They are the same graph.

5. $f(x) = \log_4 x$

$y = \log_4 x$

$y = \log_4 x$ is equivalent to $x = 4^y$

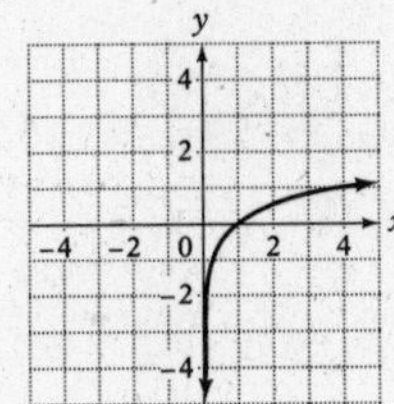

7. $f(x) = \log_3(2x - 1)$

$y = \log_3(2x - 1)$

$y = \log_3(2x - 1)$ is equivalent to $(2x - 1) = 3^y$, $2x = 3^y + 1$, or $x = \dfrac{1}{2}(3^y + 1)$.

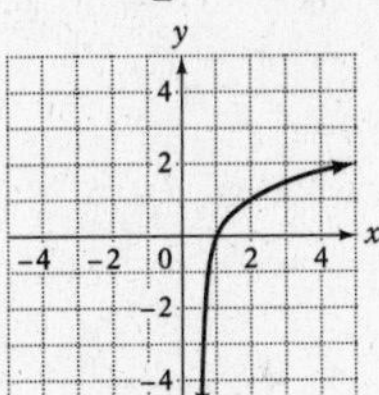

9. $f(x) = 3\log_2 x$

$y = 3\log_2 x$

$\dfrac{y}{3} = \log_2 x$

$\dfrac{y}{3} = \log_2 x$ is equivalent to $x = 2^{\frac{y}{3}}$.

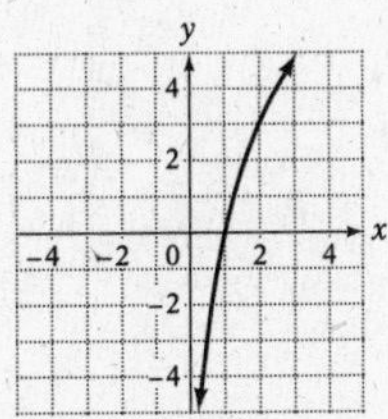

11. $f(x) = -\log_2 x$

$y = -\log_2 x$

$-y = \log_2 x$

$-y = \log_2 x$ is equivalent to $x = 2^{-y}$.

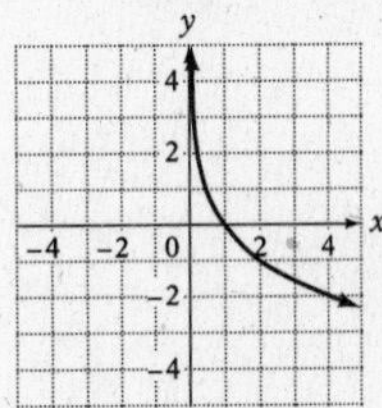

13. $f(x) = \log_2(x - 1)$

$y = \log_2(x - 1)$

$y = \log_2(x - 1)$ is equivalent to $(x - 1) = 2^y$, or $x = 2^y + 1$.

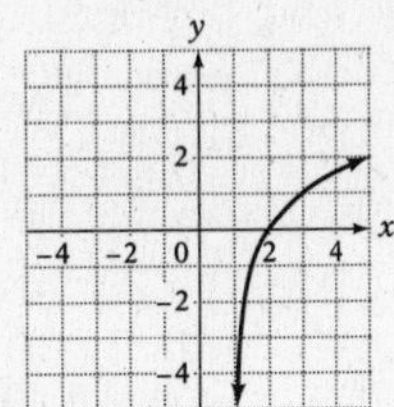

15. $f(x) = -\log_2(x-1)$
$y = -\log_2(x-1)$
$-y = \log_2(x-1)$
$-y = \log_2(x-1)$ is equivalent to $(x-1) = 2^{-y}$, or $x = 2^{-y} + 1$.

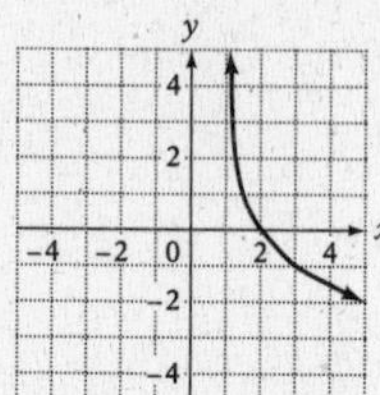

Applying the Concepts

17. $f(x) = x - \log_2(1-x)$
$$y = x - \log_2(1-x)$$
$$y = x - \frac{\log(1-x)}{\log 2}$$
$$y \approx x - \frac{\log(1-x)}{0.3010}$$

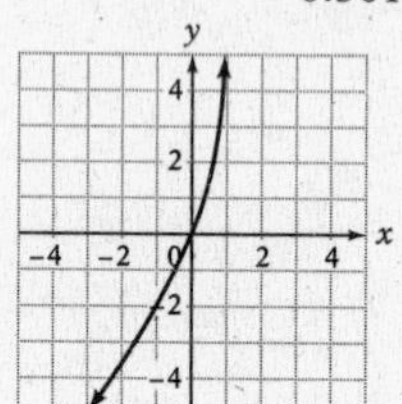

19. $f(x) = \frac{x}{2} - 2\log_2(x+1)$
$$y = \frac{x}{2} - \log_2(x+1)^2$$
$$= \frac{x}{2} - \frac{\log(x+1)^2}{\log 2}$$
$$\approx \frac{x}{2} - \frac{\log(x+1)^2}{0.3010}$$

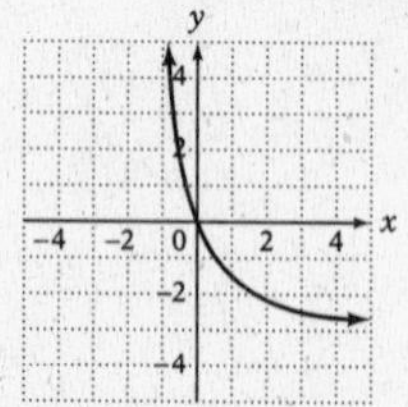

21. $f(x) = x^2 - 10\ln(x-1)$
$y = x^2 - 10\ln(x-1)$

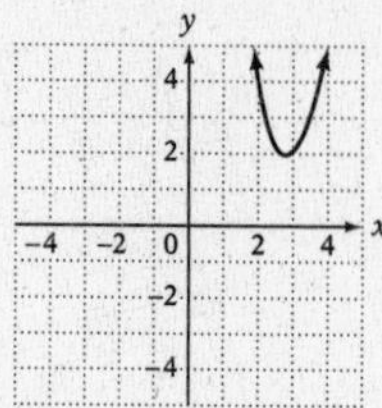

23. a. $M = 5\log s - 5$

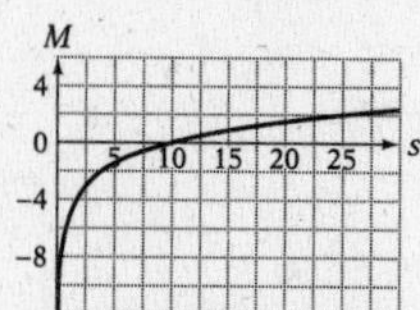

b. The point (25.1, 2) means that a star that is 25.1 parsecs from Earth has a distance modulus of 2.

Section 10.4

Objective A Exercises

1. An exponential equation is one in which a variable occurs in an exponent.

3. $$5^{4x-1} = 5^{x-2}$$
$$4x - 1 = x - 2$$
$$3x - 1 = -2$$
$$3x = -1$$
$$x = -\frac{1}{3}$$
The solution is $-\frac{1}{3}$.

5. $$8^{x-4} = 8^{5x+8}$$
$$x - 4 = 5x + 8$$
$$-4x - 4 = 8$$
$$-4x = 12$$
$$x = -3$$
The solution is -3.

7. $$9^x = 3^{x+1}$$
$$3^{2x} = 3^{x+1}$$
$$2x = x + 1$$
$$x = 1$$
The solution is 1.

9. $$8^{x+2} = 16^x$$
$$(2^3)^{x+2} = 2^{4x}$$
$$2^{3x+6} = 2^{4x}$$
$$3x + 6 = 4x$$
$$6 = x$$
The solution is 6.

11. $$16^{2-x} = 32^{2x}$$
$$(2^4)^{2-x} = (2^5)^{2x}$$
$$2^{8-4x} = 2^{10x}$$
$$8 - 4x = 10x$$
$$8 = 14x$$
$$\frac{4}{7} = x$$
The solution is $\frac{4}{7}$.

13. $$\begin{aligned} 25^{3-x} &= 125^{2x-1} \\ (5^2)^{3-x} &= (5^3)^{2x-1} \\ 5^{6-2x} &= 5^{6x-3} \\ 6-2x &= 6x-3 \\ 9 &= 8x \\ \frac{9}{8} &= x \end{aligned}$$
The solution is $\frac{9}{8}$.

15. $$\begin{aligned} 5^x &= 6 \\ \log 5^x &= \log 6 \\ x\log 5 &= \log 6 \\ x &= \frac{\log 6}{\log 5} \\ x &\approx 1.1133 \end{aligned}$$
The solution is 1.1133.

17. $$\begin{aligned} e^x &= 3 \\ \ln e^x &= \ln 3 \\ x\ln e &= \ln 3 \\ x &= \frac{\ln 3}{\ln e} \\ x &\approx 1.0986 \end{aligned}$$
The solution is 1.0986.

19. $$\begin{aligned} 10^x &= 21 \\ \log 10^x &= \log 21 \\ x\log 10 &= \log 21 \\ x &= \frac{\log 21}{\log 10} \\ x &\approx 1.3222 \end{aligned}$$
The solution is 1.3222.

21. $$\begin{aligned} 2^{-x} &= 7 \\ \log 2^{-x} &= \log 7 \\ -x\log 2 &= \log 7 \\ -x &= \frac{\log 7}{\log 2} \\ -x &\approx 2.8074 \\ x &\approx -2.8074 \end{aligned}$$
The solution is −2.8074.

23. $$\begin{aligned} 2^{x-1} &= 6 \\ \log 2^{x-1} &= \log 6 \\ (x-1)\log 2 &= \log 6 \\ x-1 &= \frac{\log 6}{\log 2} \\ x-1 &\approx 2.5850 \\ x &\approx 3.5850 \end{aligned}$$
The solution is 3.5850.

25. $$\begin{aligned} 3^{2x-1} &= 4 \\ \log 3^{2x-1} &= \log 4 \\ (2x-1)\log 3 &= \log 4 \\ 2x-1 &= \frac{\log 4}{\log 3} \\ 2x-1 &\approx 1.26186 \\ 2x &\approx 2.26186 \\ x &\approx 1.1309 \end{aligned}$$
The solution is 1.1309.

Objective B Exercises

27. A logarithmic equation is an equation in which one or more of the terms is a logarithmic expression.

29. $\log_2(2x-3) = 3$
Rewrite in exponential form.
$$\begin{aligned} 2^3 &= 2x-3 \\ 8 &= 2x-3 \\ 11 &= 2x \\ \frac{11}{2} &= x \end{aligned}$$
The solution is $\frac{11}{2}$.

31. $\log_2(x^2+2x) = 3$
Rewrite in exponential form.
$$\begin{aligned} 2^3 &= x^2+2x \\ 8 &= x^2+2x \\ 0 &= x^2+2x-8 \\ 0 &= (x+4)(x-2) \end{aligned}$$
$$\begin{aligned} x+4 &= 0 & x-2 &= 0 \\ x &= -4 & x &= 2 \end{aligned}$$
The solutions are −4 and 2.

33. $\log_5 \frac{2x}{x-1} = 1$
Rewrite in exponential form.
$$\begin{aligned} 5^1 &= \frac{2x}{x-1} \\ (x-1)5 &= (x-1)\frac{2x}{x-1} \\ 5x-5 &= 2x \\ 3x-5 &= 0 \\ 3x &= 5 \\ x &= \frac{5}{3} \end{aligned}$$
The solution is $\frac{5}{3}$.

35. $\log x = \log(1-x)$
Use the fact that if $\log_b u = \log_b v$, then $u = v$.
$$\begin{aligned} x &= 1-x \\ 2x &= 1 \\ x &= \frac{1}{2} \end{aligned}$$
The solution is $\frac{1}{2}$.

37. $\ln 5 = \ln(4x - 13)$
Use the fact that if $\ln u = \ln v$, then $u = v$

$$5 = 4x - 13$$
$$18 = 4x$$
$$\frac{18}{4} = x$$
$$\frac{9}{2} = x$$

The solution is $\frac{9}{2}$.

39. $\ln(3x + 2) = 4$
This equation is equivalent to

$$e^4 = 3x + 2$$
$$e^4 - 2 = 3x$$
$$\frac{e^4 - 2}{3} = x$$
$$17.5327 \approx x$$

The solution is 17.5327.

41. $\log_2 8x - \log_2(x^2 - 1) = \log_2 3$

$$\log_2 \frac{8x}{x^2 - 1} = \log_2 3$$

Use the fact that if $\log_b u = \log_b v$, then $u = v$.

$$\frac{8x}{x^2 - 1} = 3$$
$$(x^2 - 1)\frac{8x}{x^2 - 1} = (x^2 - 1)3$$
$$8x = 3x^2 - 3$$
$$0 = 3x^2 - 8x - 3$$
$$0 = (3x + 1)(x - 3)$$

$3x + 1 = 0 \qquad x - 3 = 0$
$3x = -1 \qquad x = 3$
$x = -\frac{1}{3}$

$-\frac{1}{3}$ does not check as a solution.
The solution is 3.

43. $\log_9 x + \log_9(2x - 3) = \log_9 2$

$$\log_9 x(2x - 3) = \log_9 2$$

Use the fact that if $\log_b u = \log_b v$, then $u = v$.

$$x(2x - 3) = 2$$
$$2x^2 - 3x = 2$$
$$2x^2 - 3x - 2 = 0$$
$$(2x + 1)(x - 2) = 0$$

$2x + 1 = 0 \qquad x - 2 = 0$
$2x = -1 \qquad x = 2$
$x = -\frac{1}{2}$

$-\frac{1}{2}$ does not check as a solution.
The solution is 2.

45. $\log_8 6x = \log_8 2 + \log_8(x - 4)$

$$\log_8 6x = \log_8 2(x - 4)$$

Use the fact that if $\log_b u = \log_b v$, then $u = v$.

$$6x = 2(x - 4)$$
$$6x = 2x - 8$$
$$4x = -8$$
$$x = -2$$

-2 does not check as a solution. The equation has no solution.

Applying the Concepts

47.

$$8^{\frac{x}{2}} = 6$$
$$\log 8^{\frac{x}{2}} = \log 6$$
$$\frac{x}{2}\log 8 = \log 6$$
$$\frac{x}{2} = \frac{\log 6}{\log 8}$$
$$x = 2\left(\frac{\log 6}{\log 8}\right)$$
$$x \approx 1.7233$$

49.

$$5^{\frac{3x}{2}} = 7$$
$$\log 5^{\frac{3x}{2}} = \log 7$$
$$\frac{3x}{2}\log 5 = \log 7$$
$$\frac{3x}{2} = \frac{\log 7}{\log 5}$$
$$x = \frac{2}{3}\left(\frac{\log 7}{\log 5}\right)$$
$$x \approx 0.8060$$

51.

$$1.2^{\frac{x}{2} - 1} = 1.4$$
$$\log 1.2^{\frac{x}{2} - 1} = \log 1.4$$
$$\left(\frac{x}{2} - 1\right)\log 1.2 = \log 1.4$$
$$\frac{x}{2} - 1 = \frac{\log 1.4}{\log 1.2}$$
$$\frac{x}{2} - 1 \approx 1.84549$$
$$\frac{x}{2} \approx 2.84549$$
$$x \approx 5.6910$$

53. **a.** $s = 312.5\ln\frac{e^{0.32t} + e^{-0.32t}}{2}$

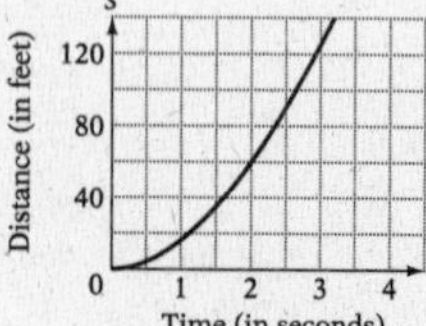

b. Use graphing calculator to find value of t when $s = 100$.
$t \approx 2.64$
It will take 2.64 s for the object to travel 100 ft.

55. The error is in the second step because the value of log 0.5 is less than zero. Multiplying each side of the inequality by this negative quantity changes the direction of the inequality symbol.

Section 10.5

Objective A Exercises

1. **Strategy** To find the value of the investment, use the compound interest formula. $P = 1000$, $n = 8$, and $i = \frac{8\%}{4} = \frac{0.08}{4} = 0.02$

   Solution
   $$A = P(1+i)^n$$
   $$A = 1000(1+0.02)^8$$
   $$A = 1000(1.02)^8$$
   $$A \approx 1171.66$$
   The value of the investment after 2 years is \$1172.

3. **Strategy** To find out how many years it will take for the investment to be worth \$15,000, solve the compound interest formula for n. Use $A = 15{,}000$, $P = 5000$, and $i = \frac{6\%}{12} = \frac{0.06}{12} = 0.005$.

   Solution
   $$P = A(1+i)^n$$
   $$15{,}000 = 5000(1+0.005)^n$$
   $$3 = (1.005)^n$$
   $$\log 3 = \log(1.005)^n$$
   $$\log 3 = n \log 1.005$$
   $$\frac{\log 3}{\log 1.005} = n$$
   $$220 \approx n$$
   $$\frac{n}{12} \approx \frac{220}{12} \approx 18$$
   The investment will be worth \$15,000 in approximately 18 years.

5. a. **Strategy** To find the technetium level after 3 hours, use the exponential decay equation. $k = 6$, $A_0 = 30$, and $t = 3$.

   Solution
   $$A = A_0\left(\frac{1}{2}\right)^{t/k}$$
   $$= 30\left(\frac{1}{2}\right)^{3/6}$$
   $$\approx 21.2$$
   After 3 hours, the technetium level will be 21.2 mg.

   b. **Strategy** To find how long it will take for the technetium level to reach 20 mg, solve the equation for t. $k = 6$, $A_0 = 30$, and $A = 20$.

   Solution
   $$A = A_0\left(\frac{1}{2}\right)^{t/k}$$
   $$20 = 30\left(\frac{1}{2}\right)^{t/6}$$
   $$\frac{2}{3} = \left(\frac{1}{2}\right)^{t/6}$$
   $$\log\frac{2}{3} = \log\left(\frac{1}{2}\right)^{\frac{t}{6}}$$
   $$\frac{6\log\frac{2}{3}}{\log\frac{1}{2}} = t$$
   $$t \approx 3.5$$
   The technetium level will reach 20 mg after 3.5 h.

7. **Strategy** To find the half-life, solve for k in the exponential decay equation. Use $A_0 = 25$, $A = 18.95$, and $t = 1$.

   Solution
   $$A = A_0\left(\frac{1}{2}\right)^{t/k}$$
   $$18.95 = 25\left(\frac{1}{2}\right)^{1/k}$$
   $$0.758 = \left(\frac{1}{2}\right)^{1/k}$$
   $$0.758 = (0.5)^{1/k}$$
   $$\log 0.758 = \log(0.5)^{1/k}$$
   $$\log 0.758 = \frac{1}{k}\log 0.5$$
   $$k \log 0.758 = \log 0.5$$
   $$k = \frac{\log(0.5)}{\log(0.76)}$$
   $$k \approx 2.5$$
   The half-life is 2.5 years.

9. a. **Strategy** To find the pressure at 40 km above Earth's surface, use the atmospheric pressure equation. $h = 40$.

   Solution
   $$P(h) = 10.13e^{-0.116h}$$
   $$= 10.13e^{-0.116(40)}$$
   $$= 10.13e^{-4.64}$$
   $$\approx 0.098$$
   At 40 km above Earth's surface, the atmospheric pressure is approximately 0.098 newtons/cm^2.

b. Strategy To find the pressure on the Earth's surface, use the atmospheric pressure equation. $h = 0$.

Solution
$$P(h) = 10.13e^{-0.116h}$$
$$= 10.13e^{-0.116(0)}$$
$$= 10.13e^{0}$$
$$= 10.13$$

On Earth's surface, the atmospheric pressure is approximately 10.13 newtons/cm^2.

c. Since the atmospheric pressure is approximately 10.13 newtons/cm^2 on Earth's surface and 0.098 newtons/cm^2 40 km above Earth's surface, the atmospheric pressure decreases as you rise above Earth's surface.

11. Strategy To find the pH, replace H^+ with its given value and solve for pH.

Solution
$$\text{pH} = -\log(\text{H}^+)$$
$$= -\log(3.97 \times 10^{-7})$$
$$\approx 6.4$$

The pH of milk is 6.4.

13. Strategy To find the thickness, solve the equation for d. Use $P = 75\% = 0.75$ and $k = 0.05$.

Solution
$$\log P = -kd$$
$$\log 0.75 = -0.05d$$
$$\frac{\log 0.75}{-0.05} = d$$
$$2.499 \approx d$$

The depth is 2.5 m.

15. Strategy To find the number of decibels, replace I with its given value in the equation and solve for D.

Solution
$$D = 10(\log I + 16)$$
$$= 10[\log(3.2 \times 10^{-10}) + 16]$$
$$\approx 10(6.5051)$$
$$= 65.051$$

Normal conversation emits 65 decibels.

17. Strategy To find the time in minutes that a Major League baseball game increased from 1981 to 1999, find $x = 19$ and $x = 1$.

Solution
$$T(x) = 149.57 + 7.63\ln x$$
$$x = 19$$
$$T(x) = 149.57 + 7.63\ln 19$$
$$\approx 172.04$$
$$x = 1$$
$$T(x) = 149.57 + 7.63\ln 1$$
$$= 149.57$$

$172.04 - 149.57 = 22.47$

The average time of a Major League baseball game increased by 22.5 min.

19. Strategy To find the thickness of copper needed so that the intensity of x-ray passing through copper is 25% of its original intensity, replace I_0 and I with values and solve for x. Let $I_0 = 1$, then $I = 0.25$.

Solution
$$I = I_0e^{-kx}$$
$$0.25 = (1)e^{-3.2x}$$
$$0.25 = e^{-3.2x}$$
$$\ln 0.25 = \ln e^{-3.2x}$$
$$-1.39 \approx -3.2x$$
$$x \approx 0.43$$

The thickness of the copper is 0.4 cm.

21. Strategy To determine the magnitude of the earthquake, solve for M in the given equation. Use 6,309,573 for I_0.

Solution
$$M = \log\frac{I}{I_0}$$
$$= \log\frac{6{,}309{,}573I_0}{I_0}$$
$$= \log 6{,}309{,}573$$
$$\approx 6.8$$

The Richter scale magnitude of the earthquake was 6.8.

23. Strategy To determine the intensity of the earthquake, solve for I in the given equation. Use 8.9 for M.

Solution
$$M = \log\frac{I}{I_0}$$
$$8.9 = \log\frac{I}{I_0}$$
$$10^{8.9} = 10^{\log\frac{I}{I_0}}$$
$$10^{8.9} = \frac{I}{I_0}$$
$$I = I_0 10^{8.9}$$
$$I = 794{,}328{,}235I_0$$

The intensity of the earthquake was $794{,}328{,}235I_0$.

25. **Strategy** To determine the magnitude of the earthquake for the seismogram given, solve for M in the given equation. Use $A = 23$ and $t = 24$.

Solution
$$\begin{aligned} M &= \log A + 3\log 8t - 2.92 \\ &= \log 23 + 3\log 8(24) - 2.92 \\ &\approx 1.36 + 6.85 - 2.92 \\ &= 5.29 \end{aligned}$$
The magnitude of the earthquake for the seismogram given is 5.3.

27. **Strategy** To determine the magnitude of the earthquake for the seismogram given, solve for M in the given equation. Use $A = 28$ and $t = 28$.

Solution
$$\begin{aligned} M &= \log A + 3\log 8t - 2.92 \\ &= \log 28 + 3\log 8(28) - 2.92 \\ &\approx 1.45 + 7.05 - 2.92 \\ &= 5.58 \end{aligned}$$
The magnitude of the earthquake for the seismogram given is 5.6.

Applying the Concepts

29. **Strategy** To find the value of the investment, solve the continuous compounding formula for P. Use $A = 2500$, $t = 5$, and $r = 0.05$.

Solution
$$\begin{aligned} P &= Ae^{rt} \\ P &= 2500e^{0.05(5)} \\ P &= 2500e^{0.25} \\ P &= 2500(1.284) \\ P &\approx 3210.06 \end{aligned}$$
The investment has a value of \$3210.06 after 5 years.

Chapter 10 Review Exercises

1. $f(x) = e^{x-2}$
$f(2) = e^{2-2} = e^0 = 1$

2. $\log_5 25 = 2$ is equivalent to $5^2 = 25$.

3. $f(x) = 3^{-x} + 2$

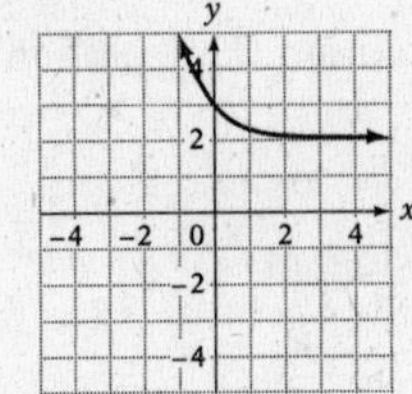

4. $f(x) = \log_3(x-1)$
$y = \log_3(x-1)$
$y = \log_3(x-1)$ is equivalent to $3^y = x - 1$ or $3^y + 1 = x$

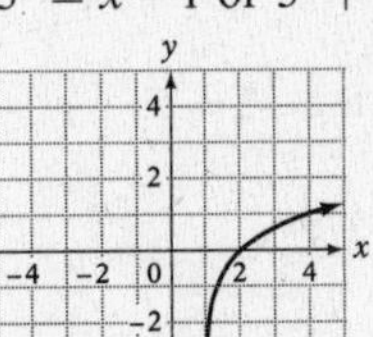

5. $$\begin{aligned} \log_3 \sqrt[5]{x^2y^4} &= \frac{1}{5}(\log_3 x^2 + \log_3 y^4) \\ &= \frac{1}{5}(2\log_3 x + 4\log_3 y) \\ &= \frac{2}{5}\log_3 x + \frac{4}{5}\log_3 y \end{aligned}$$

6. $$\begin{aligned} 2\log_3 x - 5\log_3 y &= \log_3 x^2 - \log_3 y^5 \\ &= \log_3 \frac{x^2}{y^5} \end{aligned}$$

7. $$\begin{aligned} 27^{2x+4} &= 81^{x-3} \\ 3^{3(2x+4)} &= 3^{4(x-3)} \\ 6x + 12 &= 4x - 12 \\ 2x &= -24 \\ x &= -12 \end{aligned}$$

8. $\log_5 \frac{7x+2}{3x} = 1$
Rewrite in exponential form.
$$\begin{aligned} 5^1 &= \frac{7x+2}{3x} \\ 15x &= 7x + 2 \\ 8x &= 2 \\ x &= \frac{1}{4} \end{aligned}$$
The solution is $\frac{1}{4}$.

9. $\log_6 22 = \frac{\log_{10} 22}{\log_{10} 6} \approx 1.7251$

10. $$\begin{aligned} \log_2 x &= 5 \\ 2^5 &= x \\ 32 &= x \end{aligned}$$

11. $\log_3(x+2) = 4$
Rewrite in exponential form.
$$\begin{aligned} 3^4 &= x + 2 \\ 81 &= x + 2 \\ 79 &= x \end{aligned}$$
The solution is 79.

12. $$\begin{aligned} \log_{10} x &= 3 \\ 10^3 &= x \\ 1000 &= x \end{aligned}$$

13. $$\begin{aligned} \frac{1}{3}(\log_7 x + 4\log_7 y) &= \frac{1}{3}(\log_7 xy^4) \\ &= \log_7 \sqrt[3]{xy^4} \end{aligned}$$

14. $\log_8\sqrt{\dfrac{x^5}{y^3}} = \dfrac{1}{2}(\log_8 x^5 - \log_8 y^3)$

$= \dfrac{1}{2}(5\log_8 x - 3\log_8 y)$

15. $2^5 = 32$ is equivalent to $\log_2 32 = 5$.

16. $\log_3 1.6 = \dfrac{\log_{10}1.6}{\log_{10}3} \approx 0.4278$

17.

$$\begin{aligned} 3^{x+2} &= 5 \\ \log 3^{x+2} &= \log 5 \\ (x+2)\log 3 &= \log 5 \\ x+2 &= \frac{\log 5}{\log 3} \\ x &= \frac{\log 5}{\log 3} - 2 \\ &\approx -0.535 \end{aligned}$$

The solution is -0.535.

18. $f(x) = \left(\dfrac{2}{3}\right)^{x+2}$

$f(-3) = \left(\dfrac{2}{3}\right)^{-3+2} = \left(\dfrac{2}{3}\right)^{-1} = \dfrac{3}{2}$

19. $\log_8(x+2) - \log_8 x = \log_8 4$

$\log_8\left(\dfrac{x+2}{x}\right) = \log_8 4$

Use the fact that if $\log_b u = \log_b v$, then $u = v$.

$$\begin{aligned} \frac{x+2}{x} &= 4 \\ x+2 &= 4x \\ 2 &= 3x \\ \frac{2}{3} &= x \end{aligned}$$

The solution is $\dfrac{2}{3}$.

20. $\log_6 2x = \log_6 2 + \log_6(3x-4)$

$\log_6 2x = \log_6 2(3x-4)$

Use the fact that if $\log_b u = \log_b v$, then $u = v$.

$$\begin{aligned} 2x &= 2(3x-4) \\ 2x &= 6x - 8 \\ -4x &= -8 \\ x &= 2 \end{aligned}$$

The solution is 2.

21. $f(x) = \left(\dfrac{2}{3}\right)^{x+1}$

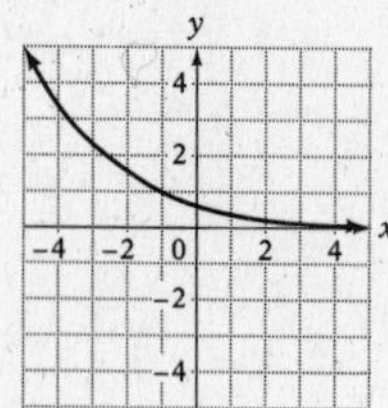

22. $f(x) = \log_2(2x-1)$

$y = \log_2(2x-1)$

is equivalent to $2^y = 2x - 1$ or $\dfrac{1}{2}\cdot 2^y + \dfrac{1}{2} = x$.

23. $\log_6 36 = x$

$6^x = 36 = 6^2$

$x = 2$

24. $\dfrac{1}{3}(\log_2 x - \log_2 y) = \log_2\sqrt[3]{\dfrac{x}{y}}$

25.

$$\begin{aligned} 9^{2x} &= 3^{x+3} \\ (3^2)^{2x} &= 3^{x+3} \\ 3^{4x} &= 3^{x+3} \\ 4x &= x+3 \\ 3x &= 3 \\ x &= 1 \end{aligned}$$

26. $f(x) = \left(\dfrac{3}{5}\right)^x$

$f(0) = \left(\dfrac{3}{5}\right)^0$

$f(0) = 1$

27. $\log_5 x = -1$

$5^{-1} = x$

$\dfrac{1}{5} = x$

28. $3^4 = 81$ is equivalent to $\log_3 81 = 4$

29. $\log x + \log(x-2) = \log 15$

$\log x(x-2) = \log 15$

Use the fact that if $\log_b u = \log_b v$, then $u = v$.

$$\begin{aligned} x(x-2) &= 15 \\ x^2 - 2x - 15 &= 0 \\ (x-5)(x+3) &= 0 \end{aligned}$$

$x - 5 = 0 \quad x + 3 = 0$

$x = 5 \quad x = -3$

-3 does not check as a solution. The solution is 5.

30.

$$\begin{aligned} \log_5\sqrt[3]{x^2y} &= \frac{1}{3}\log_5(x^2y) \\ &= \frac{1}{3}(\log_5 x^2 + \log_5 y) \\ &= \frac{1}{3}(2\log_5 x + \log_5 y) \\ &= \frac{2}{3}\log_5 x + \frac{1}{3}\log_5 y \end{aligned}$$

31.

$$\begin{aligned} 9^{2x-5} &= 9^{x-3} \\ 2x - 5 &= x - 3 \\ x - 5 &= -3 \\ x &= 2 \end{aligned}$$

32. $f(x) = 7^{x+2}$
$f(-3) = 7^{-3+2} = 7^{-1}$
$f(-3) = \frac{1}{7}$

33. $\log_2 16 = x$
$2^x = 16 = 2^4$
$x = 4$

34. $\log_6 x = \log_6 2 + \log_6(2x - 3)$
$\log_6 x = \log_6 2(2x - 3)$
Use the fact that if $\log_b u = \log_b v$, then $u = v$.
$x = 2(2x - 3)$
$x = 4x - 6$
$-3x = -6$
$x = 2$

35. $\log_2 5 = \dfrac{\log_{10} 5}{\log_{10} 2} \approx 2.3219$

36. $4^x = 8^{x-1}$
$(2^2)^x = (2^3)^{x-1}$
$2^{2x} = 2^{3x-3}$
$2x = 3x - 3$
$-x = -3$
$x = 3$

37. $\log_5 x = 4$
$5^4 = x$
$625 = x$

38. $3\log_b x - 7\log_b y = \log_b x^3 - \log_b y^7$
$= \log_b \dfrac{x^3}{y^7}$

39. $f(x) = 5^{-x-1}$
$f(-2) = 5^{-(-2)-1} = 5^{2-1}$
$= 5$

40. $5^{x-2} = 7$
$(x - 2)\log 5 = \log 7$
$x - 2 = \dfrac{\log 7}{\log 5}$
$x = \dfrac{\log 7}{\log 5} + 2$
$x \approx 3.2091$

41. Strategy To find the value of the investment, solve the compound interest formula for A. Use $P = 4000$, $n = 24$, and $i = \dfrac{8\%}{12} = \dfrac{0.08}{12} = 0.00\overline{6}$

Solution $A = P(1 + r)^n$
$A = 4000(1 + 0.0\overline{6})^{24}$
$= 4000(1.00\overline{6})^{24}$
≈ 4691.55
The value of the investment in 2 years is \$4692.

42. Strategy To find the Richter scale magnitude of the earthquake, use the Richter equation $M = \log \dfrac{I}{I_0}$.

Solution Using $I = 199{,}526{,}232 I_0$
$M = \log \dfrac{I}{I_0}$
$M = \log \dfrac{199{,}526{,}232 I_0}{I_0}$
$M \approx 8.3$
The Richter scale magnitude of the earthquake is 8.3.

43. Strategy To find the half-life, solve for k in the exponential decay equation. Use $A_0 = 25$, $A = 15$, and $t = 20$.

Solution $A = A_0\left(\dfrac{1}{2}\right)^{t/k}$
$15 = 25\left(\dfrac{1}{2}\right)^{20/k}$
$0.6 = \left(\dfrac{1}{2}\right)^{20/k}$
$\log 0.6 = \log 0.5^{20/k}$
$\log 0.6 = \dfrac{20}{k}\log 0.5$
$k = \dfrac{20 \log 0.5}{\log 0.6}$
$k \approx 27.14$
The half-life is 27 days.

44. Strategy To find the number of decibels, replace I with its given value in the equation and solve for D.

Solution $D = 10(\log I + 16)$
$= 10[\log(5 \times 10^{-6}) + 16]$
$= 10(10.6990)$
$= 106.99$
The sound emitted from a busy street corner is 107 decibels.

Chapter 10 Test

1. $f(x) = \left(\dfrac{2}{3}\right)^x$
$f(0) = \left(\dfrac{2}{3}\right)^0$
$f(0) = 1$

2. $f(x) = 3^{x+1}$
$f(-2) = 3^{-2+1} = 3^{-1}$
$f(2) = \frac{1}{3}$

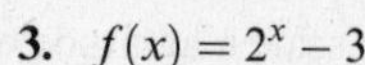

3. $f(x) = 2^x - 3$

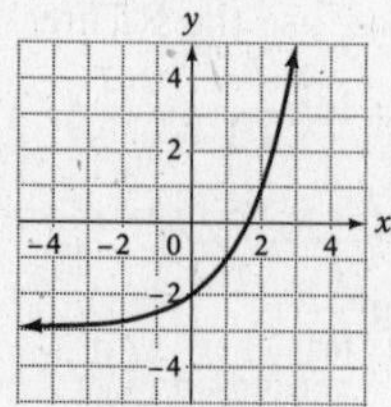

4. $f(x) = 2^x + 2$

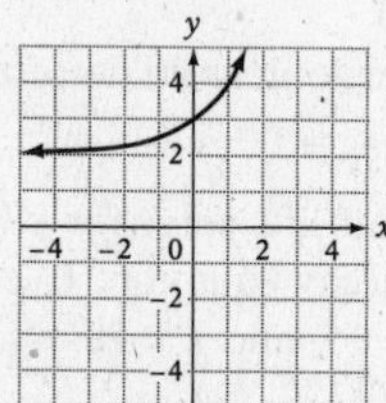

5. $\log_4 16 = x$
$$4^x = 16$$
$$x = 2$$

6. $\log_3 x = -2$
$$3^{-2} = x$$
$$\frac{1}{9} = x$$

7. $f(x) = \log_2(2x)$
$$y = \log_2(2x)$$
$$2^y = 2x$$
$$\frac{1}{2} \cdot 2^y = x$$

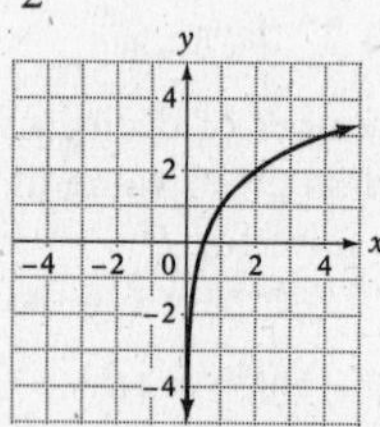

8. $f(x) = \log_3(x+1)$
$$y = \log_3(x+1)$$
$$3^y = x+1$$
$$-1 + 3^y = x$$

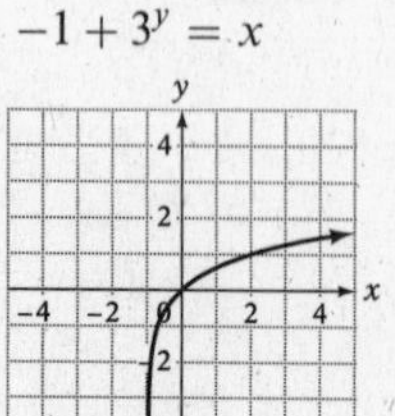

9. $\log_6 \sqrt{xy^3} = \frac{1}{2}\log_6(xy^3)$
$$= \frac{1}{2}(\log_6 x + \log_6 y^3)$$
$$= \frac{1}{2}(\log_6 x + 3\log_6 y)$$

10. $\frac{1}{2}(\log_3 x - \log_3 y) = \frac{1}{2}\left(\log_3 \frac{x}{y}\right) = \log_3 \sqrt{\frac{x}{y}}$

11. $\ln \frac{x}{\sqrt{z}} = \ln x - \ln \sqrt{z}$
$$= \ln x - \frac{1}{2}\ln z$$

12. $3\ln x - \ln y - \frac{1}{2}\ln z = \ln x^3 - \ln y - \ln z^{1/2}$
$$= \ln \frac{x^3}{y} - \ln z^{1/2}$$
$$= \ln \frac{x^3}{yz^{1/2}} = \ln \frac{x^3}{y\sqrt{z}}$$

13. $3^{7x+1} = 3^{4x-5}$
$$7x + 1 = 4x - 5$$
$$3x = -6$$
$$x = -2$$
The solution is -2.

14. $8^x = 2^{x-6}$
$$(2^3)^x = 2^{x-6}$$
$$3x = x - 6$$
$$2x = -6$$
$$x = -3$$
The solution is -3.

15. $3^x = 17$
$$\log 3^x = \log 17$$
$$x\log 3 = \log 17$$
$$x = \frac{\log 17}{\log 3}$$
$$x \approx 2.5789$$
The solution is 2.5789.

16. $\log x + \log(x-4) = \log 12$
$$\log x(x-4) = \log 12$$
Use the fact that if $\log_b u = \log_b v$, then $u = v$.
$$x(x-4) = 12$$
$$x^2 - 4x - 12 = 0$$
$$(x-6)(x+2) = 0$$
$$x - 6 = 0 \quad x + 2 = 0$$
$$x = 6 \quad x = -2$$
-2 does not check as a solution. The solution is 6.

17. $\log_6 x + \log_6(x-1) = 1$
$$\log_6 x(x-1) = 1$$
$$x(x-1) = 6^1$$
$$x^2 - x - 6 = 0$$
$$(x-3)(x+2) = 0$$
$$x - 3 = 0 \quad x + 2 = 0$$
$$x = 3 \quad x = -2$$
-2 does not check as a solution. The solution is 3.

18. $\log_5 9 = \frac{\log_{10} 9}{\log_{10} 5} \approx 1.3652$

19. $\log_3 19 = \frac{\log 19}{\log 3} \approx 2.6801$

20. Strategy To find the half-life, solve for k in the exponential decay equation. Use $A_0 = 10$, $A = 9$, and $t = 5$.

Solution

$$A = A_0\left(\frac{1}{2}\right)^{t/k}$$
$$9 = 10\left(\frac{1}{2}\right)^{5/k}$$
$$0.9 = \left(\frac{1}{2}\right)^{5/k}$$
$$\log 0.9 = \log(0.5)^{5/k}$$
$$\log 0.9 = \frac{5}{k}\log 0.5$$
$$k = \frac{5\log 0.5}{\log 0.9}$$
$$k \approx 32.9$$

The half-life is 33 h.

Cumulative Review Exercises

1.
$$\begin{aligned} 4 - 2[x - 3(2 - 3x) - 4x] &= 2x \\ 4 - 2[x - 6 + 9x - 4x] &= 2x \\ 4 - 2[6x - 6] &= 2x \\ 4 - 12x + 12 &= 2x \\ 16 &= 14x \\ \frac{8}{7} &= x \end{aligned}$$

2.
$$\begin{aligned} 2x - y &= 5 \\ -y &= -2x + 5 \\ y &= 2x - 5 \end{aligned}$$
$$m = 2$$
$$\begin{aligned} y - y_1 &= m(x - x_1) \\ y - (-2) &= 2(x - 2) \\ y + 2 &= 2x - 4 \\ y &= 2x - 6 \end{aligned}$$

3. $4x^{2n} + 7x^n + 3 = (4x^n + 3)(x^n + 1)$

4.
$$\begin{aligned} \frac{1 - \frac{5}{x} + \frac{6}{x^2}}{1 + \frac{1}{x} - \frac{6}{x^2}} \cdot \frac{x^2}{x^2} &= \frac{x^2 - 5x + 6}{x^2 + x - 6} \\ &= \frac{(x - 2)(x - 3)}{(x + 3)(x - 2)} \\ &= \frac{x - 3}{x + 3} \end{aligned}$$

5.
$$\begin{aligned} \frac{\sqrt{xy} \cdot (\sqrt{x} + \sqrt{y})}{(\sqrt{x} - \sqrt{y}) \cdot (\sqrt{x} + \sqrt{y})} &= \frac{\sqrt{x^2y} + \sqrt{xy^2}}{\sqrt{x^2} - \sqrt{y^2}} \\ &= \frac{x\sqrt{y} + y\sqrt{x}}{x - y} \end{aligned}$$

6.
$$\begin{aligned} x^2 - 4x - 6 &= 0 \\ x^2 - 4x + 4 &= 6 + 4 \\ (x - 2)^2 &= 10 \\ \sqrt{(x - 2)^2} &= \sqrt{10} \\ x - 2 &= \pm\sqrt{10} \\ x &= 2 \pm \sqrt{10} \end{aligned}$$

The solutions are $2 + \sqrt{10}$ and $2 - \sqrt{10}$.

7.
$$\begin{aligned} (x - r_1)(x - r_2) &= 0 \\ \left(x - \frac{1}{3}\right)(x - (-3)) &= 0 \\ \left(x - \frac{1}{3}\right)(x + 3) &= 0 \\ x^2 + \frac{8}{3}x - 1 &= 0 \\ 3x^2 + 8x - 3 &= 0 \end{aligned}$$

8.
$$\begin{aligned} 2x - y &< 3 \\ -y &< 3 - 2x \\ y &> -3 + 2x \end{aligned} \qquad \begin{aligned} x + y &< 1 \\ y &< 1 - x \end{aligned}$$

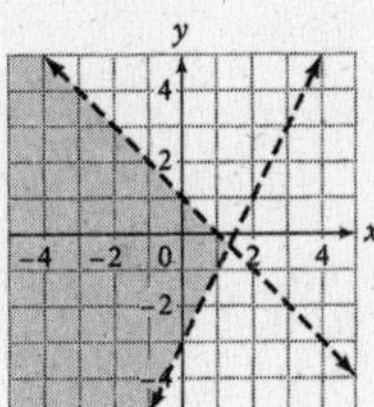

9. (1) $3x - y + z = 3$
(2) $x + y + 4z = 7$
(3) $3x - 2y + 3z = 8$

Eliminate y. Add Equation (1) to Equation (2).

$3x - y + z = 3$
$x + y + 4z = 7$

(4) $4x + 5z = 10$

Multiply Equation (2) by 2 and add to Equation (3).

$2(x + y + 4z) = 2(7)$
$3x - 2y + 3z = 8$

$2x + 2y + 8z = 14$
$3x - 2y + 3z = 8$

(5) $5x + 11z = 22$

Eliminate x. Multiply Equation (4) by -5 and Equation (5) by 4, and add.

$-5(4x + 5z) = -5(10)$
$4(5x + 11z) = 4(22)$

$-20x - 25z = -50$
$20x + 44z = 88$

$19z = 38$
$z = 2$

Replace z by 2 in Equation (4).

$4x + 5(2) = 10$
$4x + 10 = 10$
$4x = 0$
$x = 0$

Replace x by 0 and z by 2 in Equation (2).

$0 + y + 4(2) = 7$
$y + 8 = 7$
$y = -1$

The solution is $(0, -1, 2)$.

10. $\dfrac{x-4}{2-x} - \dfrac{1-6x}{2x^2-7x+6}$

$= \dfrac{x-4}{2-x} - \dfrac{1-6x}{(2x-3)(x-2)}$

$= \dfrac{x-4}{2-x} + \dfrac{1-6x}{(2x-3)(2-x)}$

$= \dfrac{(x-4)}{(2-x)} \cdot \dfrac{(2x-3)}{(2x-3)} + \dfrac{1-6x}{(2x-3)(2-x)}$

$= \dfrac{2x^2 - 11x + 12 + 1 - 6x}{(2-x)(2x-3)}$

$= \dfrac{2x^2 - 17x + 13}{(2-x)(2x-3)} = -\dfrac{2x^2 - 17x + 13}{(x-2)(2x-3)}$

11. $x^2 + 4x - 5 \le 0$
$(x+5)(x-1) \le 0$

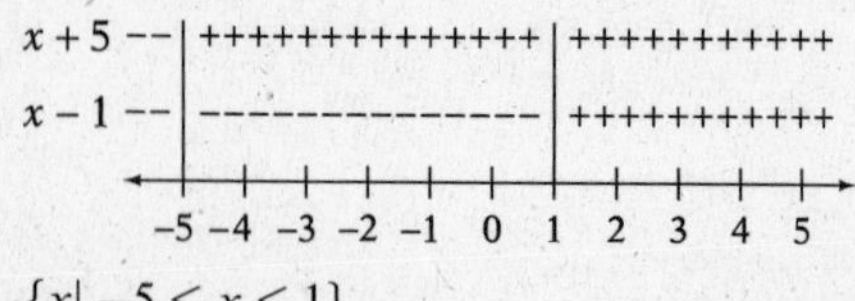

$\{x \mid -5 \le x \le 1\}$

12. $|2x - 5| \le 3$
$-3 \le 2x - 5 \le 3$
$-3 + 5 \le 2x \le 3 + 5$
$2 \le 2x \le 8$
$1 \le x \le 4$
$\{x \mid 1 \le x \le 4\}$

13. $f(x) = \left(\dfrac{1}{2}\right)^x + 1$

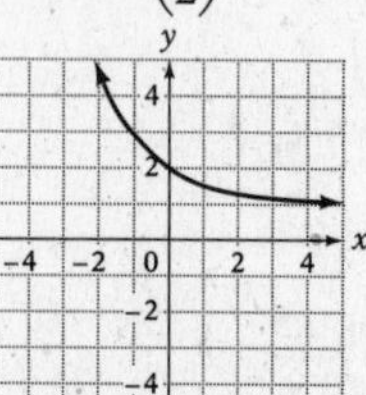

14. $f(x) = \log_2 x - 1$
$y = \log_2 x - 1$
$y + 1 = \log_2 x$
$2^{y+1} = x$

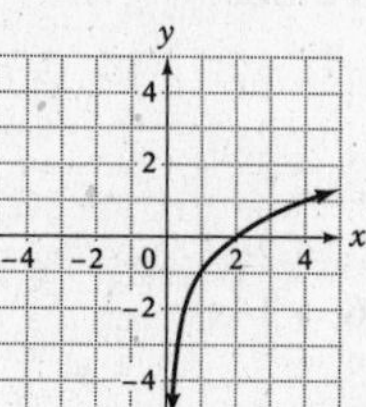

15. $f(x) = 2^{-x-1}$
$f(-3) = 2^{-(-3)-1} = 2^{3-1}$
$= 2^2 = 4$

16. $\log_5 x = 3$
$5^3 = x$
$125 = x$

17. $3\log_b x - 5\log_b y = \log_b x^3 - \log_b y^5$
$= \log_b \dfrac{x^3}{y^5}$

18. $\log_3 7 = \dfrac{\log 7}{\log 3} \approx 1.7712$

19. $4^{5x-2} = 4^{3x+2}$
$5x - 2 = 3x + 2$
$2x - 2 = 2$
$2x = 4$
$x = 2$

The solution is 2.

20. $\log x + \log(2x + 3) = \log 2$
$\log x(2x + 3) = \log 2$

Use the fact that if $\log_b u = \log_b v$, then $u = v$.

$x(2x + 3) = 2$
$2x^2 + 3x - 2 = 0$
$(2x - 1)(x + 2) = 0$
$2x - 1 = 0 \quad x + 2 = 0$
$x = \dfrac{1}{2} \quad\quad x = -2$

-2 does not check as a solution. The solution is $\dfrac{1}{2}$.

21. Strategy To find the number of checks, write and solve an inequality using c to represent the number of checks.

Solution

$$\begin{aligned} 5.00 + 0.02c &> 2.00 + 0.08c \\ 5 - 0.06c &> 2 \\ -0.06c &> -3 \\ \frac{-0.06c}{-0.06} &< \frac{-3}{-0.06} \\ c &< 50 \end{aligned}$$

The customer can write at most 49 checks.

22. Strategy • Cost per pound of mixture: x

	Amount	Cost	Value
4.00 chocolate	16	4.00	16(4.00)
2.50 chocolate	24	2.50	24(2.50)
Mixture	40	x	$40x$

• The sum of the values before mixing equals the value after mixing.

Solution

$$\begin{aligned} 16(4.00) + 24(2.50) &= 40x \\ 64 + 60 &= 40x \\ 124 &= 40x \\ 3.1 &= x \end{aligned}$$

The cost per pound of the mixture is \$3.10.

23. Strategy • Rate of the wind: r

	Distance	Rate	Time
With wind	1000	$225 + r$	$\frac{1000}{225+r}$
Against wind	800	$225 - r$	$\frac{1000}{225-r}$

• The time flying with the wind equals the time flying against the wind.

Solution

$$\begin{aligned} \frac{1000}{225+r} &= \frac{800}{225-r} \\ (225+r)(225-r)\frac{1000}{225+r} &= (225+r)(225-r)\frac{800}{225-r} \\ (225-r)(1000) &= (225+r)(800) \\ 225000 - 1000r &= 180000 + 800r \\ 45000 &= 1800r \\ 25 &= r \end{aligned}$$

The rate of the wind is 25 mph.

24. Strategy To find how far the force will stretch the spring:

• Write the basic direct variation equation, replace the variable by the given values and solve for k.

• Write the direct variation equation, replacing k by its value. Substitute 34 for f and solve for d.

Solution

$$\begin{aligned} d &= kf \\ 6 &= k(20) \\ 0.3 &= k \end{aligned}$$

$$\begin{aligned} d &= 0.3f \\ d &= 0.3(34) \\ d &= 10.2 \end{aligned}$$

The spring will stretch 10.2 in.

25. Strategy • Cost of redwood: x
Cost of fir: y
First purchase:

	Amount	Cost	Value
Redwood	80	x	$80x$
Fir	140	y	$140y$

Second purchase:

	Amount	Cost	Value
Redwood	140	x	$140x$
Fir	100	y	$100y$

• The total cost of the first purchase is \$67. The total cost of the second purchase is \$81.

$$80x + 140y = 67$$
$$140x + 100y = 81$$

Solution

$$80x + 140y = 67$$
$$140x + 100y = 81$$

$$-5(80x + 140y) = -5(67)$$
$$7(140x + 100y) = 7(81)$$

$$-400x - 700y = -335$$
$$980x + 700y = 567$$

$$580x = 232$$
$$x = 0.40$$

$$80x + 140y = 67$$
$$80(0.40) + 140y = 67$$
$$32 + 140y = 67$$
$$140y = 35$$
$$y = 0.25$$

The cost of redwood is \$0.40 per foot.
The cost of fir is \$0.25 per foot.

26. Strategy To find the time, solve the compound interest formula for n. Use $A = 10{,}000$, $P = 5000$, and $i = \frac{7\%}{2} = \frac{0.07}{2} = 0.035$.

Solution

$$A = P(1+i)^n$$
$$10{,}000 = 5000(1+0.035)^n$$
$$10{,}000 = 5000(1.035)^n$$
$$2 = (1.035)^n$$
$$\log 2 = \log(1.035)^n$$
$$\log 2 = n\log(1.035)$$
$$\frac{\log 2}{\log 1.035} = n$$
$$20 \approx n$$

20 compounding periods $\div 2$ compounding periods per years = 10 years

The investment will double in 10 years.

Chapter 11: Conic Sections

Prep Test

1. $d = \sqrt{(-2-4)^2 + (3-(-1))^2}$
$= \sqrt{(-6)^2 + (4)^2}$
$= \sqrt{36+16}$
$= \sqrt{52} = 7.21$

2. $x^2 - 8x + \left[\frac{1}{2}(-8)\right]^2$
$= x^2 - 8x + (-4)^2$
$= x^2 - 8x + 16$
$= (x-4)^2$

3. $\frac{x^2}{16} + \frac{3^2}{9} = 1$
$\frac{x^2}{16} + 1 = 1$
$\frac{x^2}{16} = 0$
$x^2 = 0$

$\frac{x^2}{16} + \frac{0^2}{9} = 1$
$\frac{x^2}{16} + 0 = 1$
$\frac{x^2}{16} = 1$
$x^2 = 16$
$\sqrt{x^2} = \sqrt{16}$
$x = \pm\sqrt{16}$
$x = \pm 4$
The solutions are 0, −4, and 4.

4. $7x + 4(x-2) = 3$
$7x + 4x - 8 = 3$
$11x - 8 = 3$
$\left(\frac{1}{11}\right)11x = 11\left(\frac{1}{11}\right)$
$x = 1$

$y = x - 2$
$y = 1 - 2$
$y = -1$
$(1, -1)$

5. (1) $4x - y = 9$
(2) $2x + 3y = -13$
Eliminate x.
$4x - y = 9$
$-2(2x + 3y) = -2(-13)$

$4x - y = 9$
$-4x - 6y = 26$
Add the equations.
$\left(-\frac{1}{7}\right) - 7y = 35\left(-\frac{1}{7}\right)$
$y = -5$
Replace y in equation (1).
$4x - (-5) = 9$
$4x + 5 = 9$
$\left(\frac{1}{4}\right)4x = 4\left(\frac{1}{4}\right)$
$x = 1$
$(1, -5)$

6. $y = x^2 - 4x + 2$
axis of symmetry:
$-\frac{b}{2a} = -\frac{(-4)}{2(1)} = -(-2) = 2;\ x = 2$
To find the vertex, replace x by 2 in the original function.
$y = (2)^2 - 4(2) + 2 = 4 - 8 + 2 = -2;\ y = -2$
Vertex: $(2, -2)$

7.
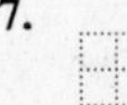
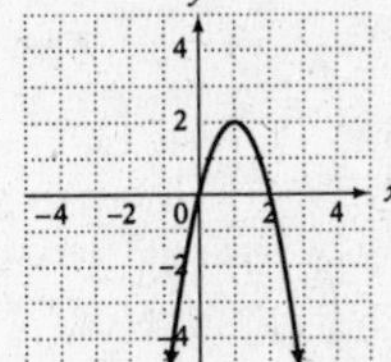

8.
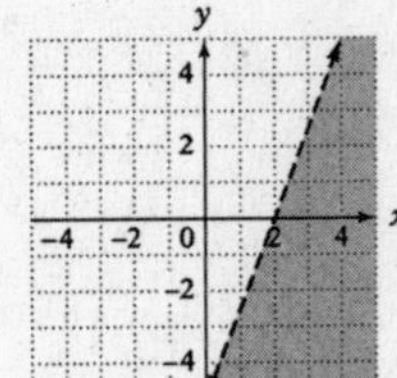

9.

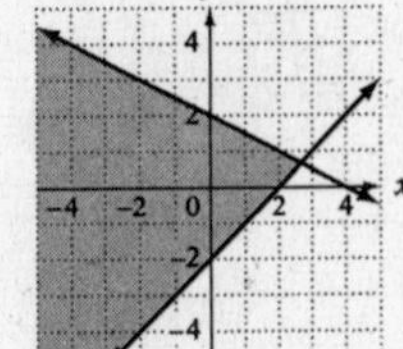

Go Figure

The height of the Empire State Building is 1250 ft or 15,000 in.

Let x represent the number of times the paper is folded. Then the thickness is $(0.01)2^x$. This thickness must be at least 15,000.

$(0.01)2^x \geq 15{,}000$

$2^x \geq 1{,}500{,}000$

This happens when $x = 21$, or when the paper has been folded 21 times.

Section 11.1

Objective A Exercises

1. a. The graph of a parabola with form $y = ax^2 + bx + c$, $a \neq 0$ has a vertical line of symmetry.

   b. Since $a > 0$, the parabola opens up.

3. a. The graph of a parabola with form $x = ay^2 + by + c$, $a \neq 0$ has a horizontal line of symmetry.

   b. Since $a > 0$, the parabola opens right.

5. a. The graph of a parabola with form $x = ay^2 + by + c$, $a \neq 0$ has a horizontal line of symmetry.

   b. Since $a < 0$, the parabola opens left.

7. $x = y^2 - 3y - 4$

   $-\dfrac{b}{2a} = -\dfrac{-3}{2(1)} = \dfrac{3}{2}$

   Axis of symmetry: $y = \dfrac{3}{2}$

   $x = \left(\dfrac{3}{2}\right)^2 - \left(\dfrac{3}{2}\right) - 4 = -\dfrac{25}{4}$

   Vertex: $\left(-\dfrac{25}{4}, \dfrac{3}{2}\right)$

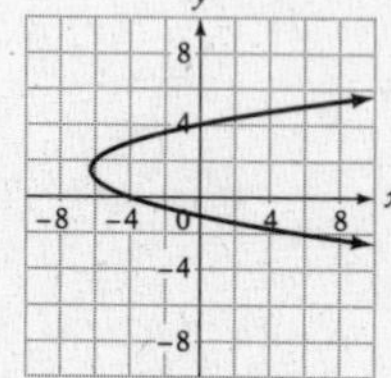

9. $y = x^2 + 2$

   $-\dfrac{b}{2a} = -\dfrac{0}{2(1)} = 0$

   Axis of symmetry: $x = 0$

   $y = (0)^2 + 2 = 2$

   Vertex: (0, 2)

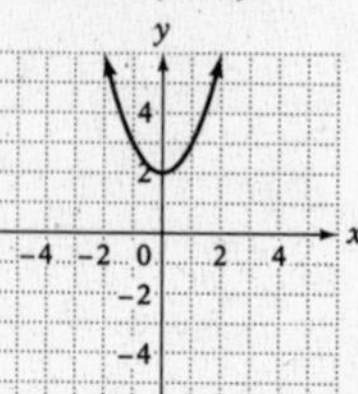

11. $x = -\dfrac{1}{4}y^2 - 1$

   $-\dfrac{b}{2a} = -\dfrac{0}{2\left(-\frac{1}{4}\right)} = 0$

   Axis of symmetry: $y = 0$

   $x = -\dfrac{1}{4}(0)^2 - 1 = -1$

   Vertex: (−1, 0)

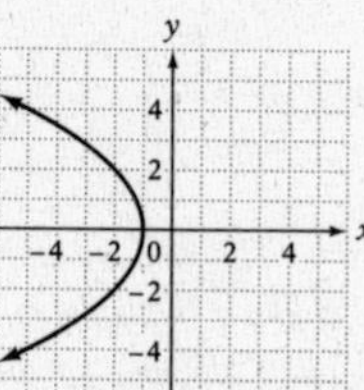

13. $x = -\dfrac{1}{2}y^2 + 2y - 3$

   $-\dfrac{b}{2a} = -\dfrac{2}{2\left(-\frac{1}{2}\right)} = 2$

   Axis of symmetry: $y = 2$

   $x = -\dfrac{1}{2}(2)^2 + 2(2) - 3 = -1$

   Vertex: (−1, 2)

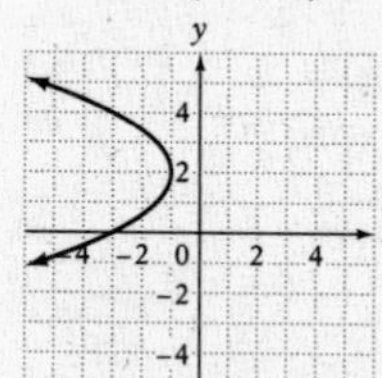

15. $y = \frac{1}{2}x^2 + x - 3$

$-\frac{b}{2a} = -\frac{1}{2\left(\frac{1}{2}\right)} = -1$

Axis of symmetry: $x = -1$

$y = -\frac{1}{2}(-1)^2 + (-1) - 3 = -\frac{7}{2}$

Vertex: $\left(-1, -\frac{7}{2}\right)$

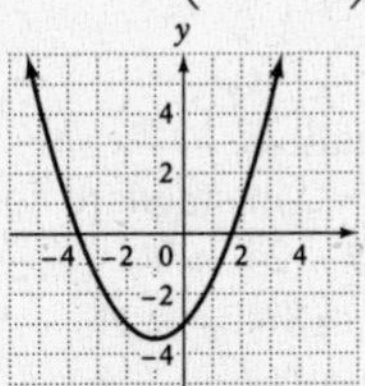

17. $x = y^2 - y - 6$

$-\frac{b}{2a} = -\frac{-1}{2(1)} = \frac{1}{2}$

Axis of symmetry: $y = \frac{1}{2}$

$x = \left(\frac{1}{2}\right)^2 - \left(\frac{1}{2}\right) - 6 = -\frac{25}{4}$

Vertex: $\left(-\frac{25}{4}, \frac{1}{2}\right)$

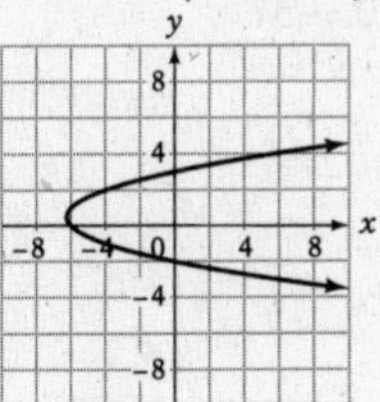

19. $y = 2x^2 + 4x - 5$

$-\frac{b}{2a} = -\frac{4}{2(2)} = -1$

Axis of symmetry: $x = -1$

$y = 2(-1)^2 + 4(-1) - 5 = -7$

Vertex: $(-1, -7)$

21. $y = 2x^2 - x - 3$

$-\frac{b}{2a} = -\frac{-1}{2(2)} = \frac{1}{4}$

Axis of symmetry: $x = \frac{1}{4}$

$y = 2\left(\frac{1}{4}\right)^2 - \frac{1}{4} - 3 = -\frac{25}{8}$

Vertex: $\left(\frac{1}{4}, -\frac{25}{8}\right)$

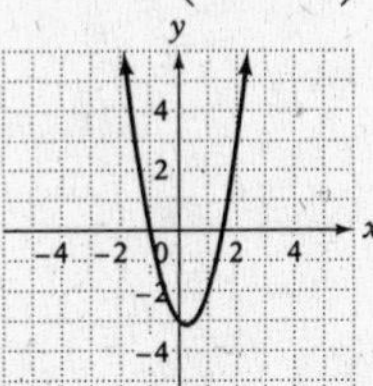

23. $y = x^2 + 5x + 6$

$-\frac{b}{2a} = -\frac{5}{2(1)} = -\frac{5}{2}$

Axis of symmetry: $x = -\frac{5}{2}$

$y = \left(-\frac{5}{2}\right)^2 + 5\left(-\frac{5}{2}\right) + 6 = -\frac{1}{4}$

Vertex: $\left(-\frac{5}{2}, -\frac{1}{4}\right)$

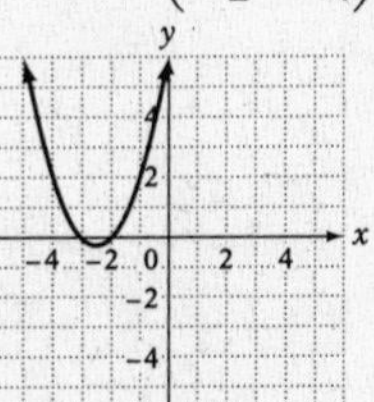

Applying the Concepts

25. $p = \frac{1}{4(a)}$; $a = 2$

$p = \frac{1}{4(2)} = \frac{1}{8}$

The focus is $\left(0, \frac{1}{8}\right)$.

Section 11.2

Objective A Exercises

1. Center: $(2, -2)$
Radius: $\sqrt{9} = 3$

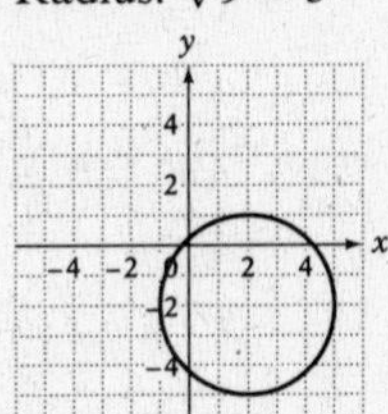

3. Center: $(-3, 1)$
Radius: $\sqrt{25} = 5$

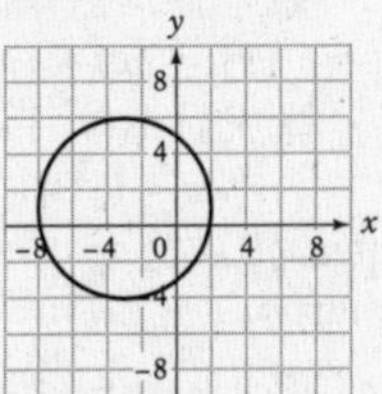

5. Center: $(-2, -2)$
Radius: $\sqrt{4} = 2$

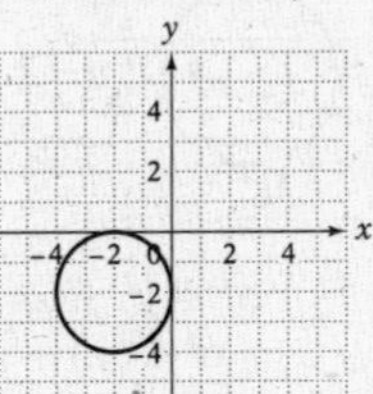

7. $(x - h)^2 + (y - k)^2 = r^2$

$(x - 2)^2 + [y - (-1)]^2 = 2^2$

$(x - 2)^2 + (y + 1)^2 = 4$

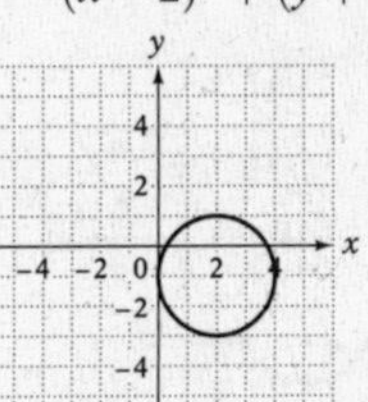

9. $(x_1, y_1) = (1, 2)\,(x_2, y_2) = (-1, 1)$

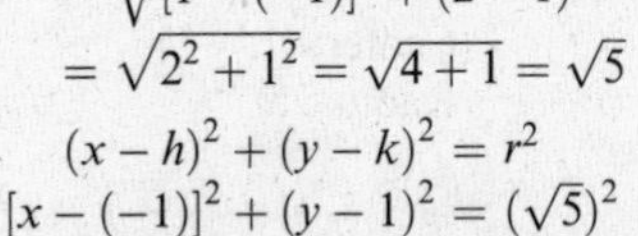

$d = \sqrt{(x_1 - x_2)^2 + (y_1 - y_2)^2}$

$= \sqrt{[1 - (-1)]^2 + (2 - 1)^2}$

$= \sqrt{2^2 + 1^2} = \sqrt{4 + 1} = \sqrt{5}$

$(x - h)^2 + (y - k)^2 = r^2$

$[x - (-1)]^2 + (y - 1)^2 = (\sqrt{5})^2$

$(x + 1)^2 + (y - 1)^2 = 5$

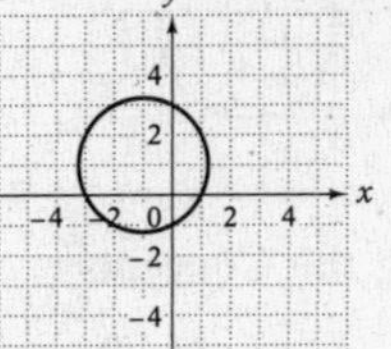

Objective B Exercises

11.
$$\begin{aligned} x^2 + y^2 - 2x + 4y - 20 &= 0 \\ (x^2 - 2x) + (y^2 + 4y) &= 20 \\ (x^2 - 2x + 1) + (y^2 + 4y + 4) &= 20 + 1 + 4 \\ (x - 1)^2 + (y + 2)^2 &= 25 \end{aligned}$$

Center: (1, −2)
Radius: 5

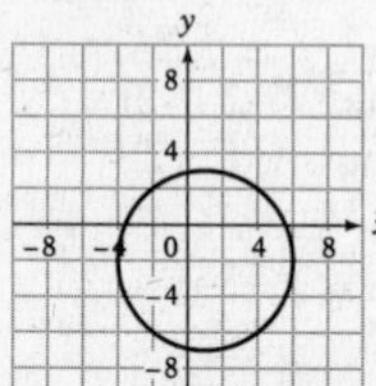

13.
$$\begin{aligned} x^2 + y^2 + 6x + 8y + 9 &= 0 \\ (x^2 + 6x) + (y^2 + 8y) &= -9 \\ (x^2 + 6x + 9) + (y^2 + 8y + 16) &= -9 + 9 + 16 \\ (x + 3)^2 + (y + 4)^2 &= 16 \end{aligned}$$

Center: (−3, −4)
Radius: 4

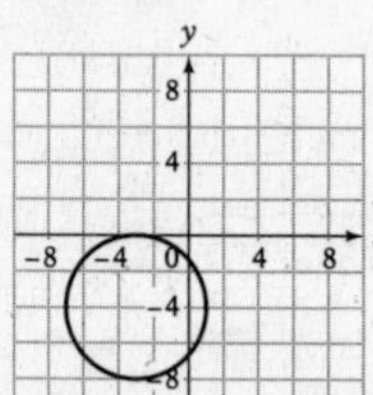

15.
$$\begin{aligned} x^2 + y^2 - 2x + 2y - 23 &= 0 \\ (x^2 - 2x) + (y^2 + 2y) &= 23 \\ (x^2 - 2x + 1) + (y^2 + 2y + 1) &= 23 + 1 + 1 \\ (x - 1)^2 + (y + 1)^2 &= 25 \end{aligned}$$

Center: (1, −1)
Radius: 5

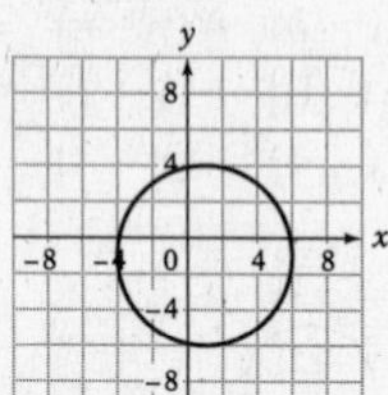

Applying the Concepts

17. $(x_1, y_1) = (4, 0)\,(x_2, y_2) = (0, 0)$

$$\begin{aligned} r &= \sqrt{(x_1 - x_2)^2 + (y_1 - y_2)^2} \\ &= \sqrt{(4 - 0)^2 + (0 - 0)^2} \\ &= \sqrt{4^2 + 0^2} = 4 \end{aligned}$$

$$\begin{aligned} (x - h)^2 + (y - k)^2 &= r^2 \\ (x - 4)^2 + (y - 0)^2 &= 4^2 \\ (x - 4)^2 + y^2 &= 16 \end{aligned}$$

19. The radius of the circle will be $6\sqrt{3}$ inches.

$$\begin{aligned} r^2 + 6^2 &= 12^2 \\ r^2 + 36 &= 144 \\ r^2 &= 108 \\ r &= \sqrt{108} = 6\sqrt{3}\text{ in.} \end{aligned}$$

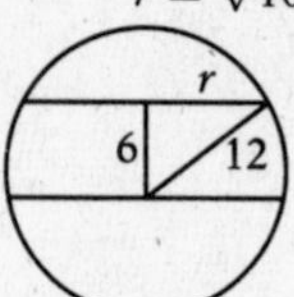

21. Attempt to write the equation $x^2 + y^2 + 4x + 8y + 24 = 0$ in standard form.

$$\begin{aligned} x^2 + y^2 + 4x + 8y + 24 &= 0 \\ (x^2 + 4x) + (y^2 + 8y) &= -24 \\ (x^2 + 4x + 4) + (y^2 + 8y + 16) &= -24 + 4 + 16 \\ (x + 2)^2 + (y + 4)^2 &= -4 \end{aligned}$$

This is not the equation of a circle because r^2 is negative ($r^2 = -4$) and the square of a real number cannot be negative.

Section 11.3

Objective A Exercises

1. x-intercepts: (2, 0) and (−2, 0)
y-intercepts: (0, 3) and (0, −3).

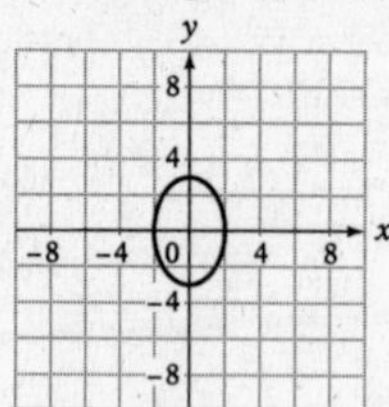

3. x-intercepts: (5, 0) and (−5, 0)
y-intercepts: (0, 3) and (0, −3).

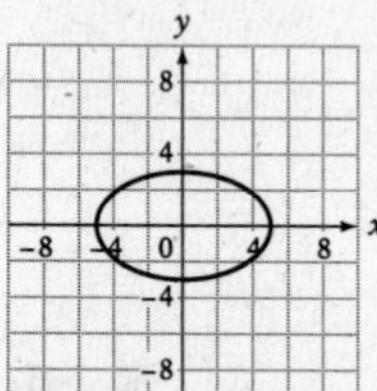

5. x-intercepts: (6, 0) and (−6, 0)
y-intercepts: (0, 4) and (0, −4).

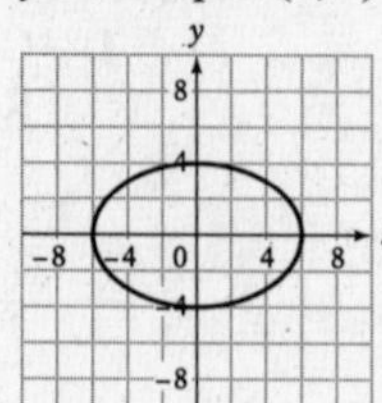

7. x-intercepts: (4, 0) and (−4, 0)
y-intercepts: (0, 7) and (0, −7).

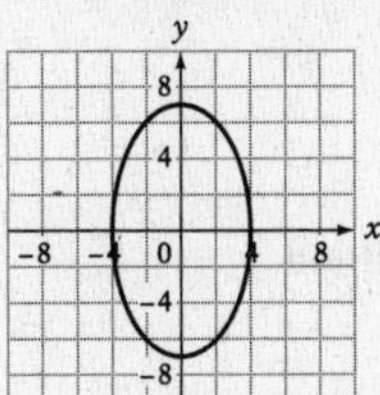

9. x-intercepts: (2, 0) and (−2, 0)
y-intercepts: (0, 5) and (0, −5).

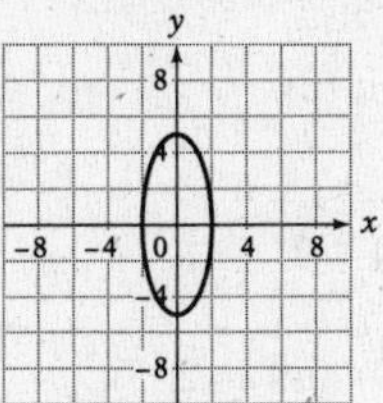

Objective B Exercises

11. Axis of symmetry: x-axis
Vertices: (5, 0) and (−5, 0)
Asymptotes: $y = \frac{2}{5}x$ and
$y = -\frac{2}{5}x$

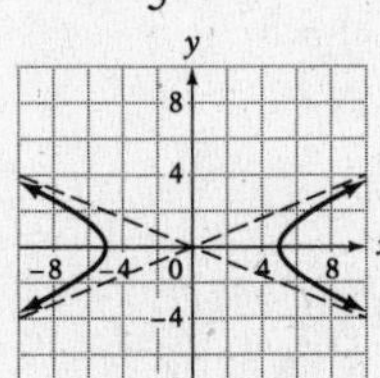

13. Axis of symmetry: y-axis
Vertices: (0, 4) and (0, −4)
Asymptotes: $y = \frac{4}{5}x$ and
$y = -\frac{4}{5}x$

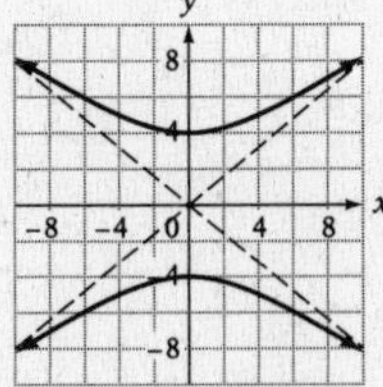

15. Axis of symmetry: x-axis
Vertices: (3, 0) and (−3, 0)
Asymptotes: $y = \frac{7}{3}x$ and
$y = -\frac{7}{3}x$

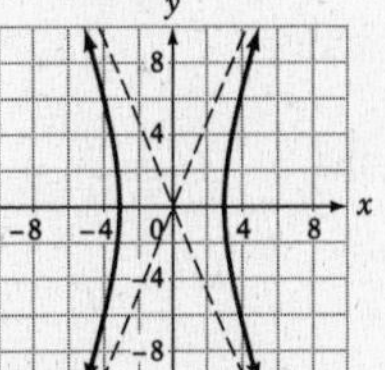

17. Axis of symmetry: y-axis
Vertices: (0, 2) and (0, −2)
Asymptotes: $y = \frac{1}{2}x$ and
$y = -\frac{1}{2}x$

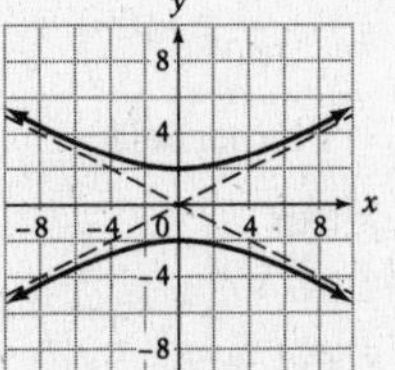

19. Axis of symmetry: x-axis
Vertices: (6, 0) and (−6, 0)
Asymptotes: $y = \frac{1}{2}x$ and
$y = -\frac{1}{2}x$

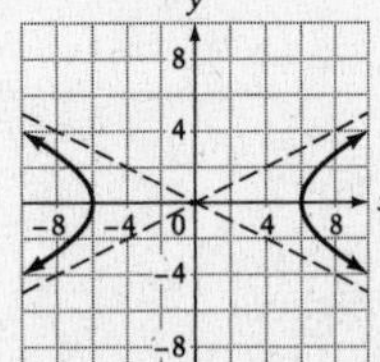

21. Axis of symmetry: y-axis
Vertices: (0, 5) and (0, −5)
Asymptotes: $y = \frac{5}{2}x$ and
$y = -\frac{5}{2}x$

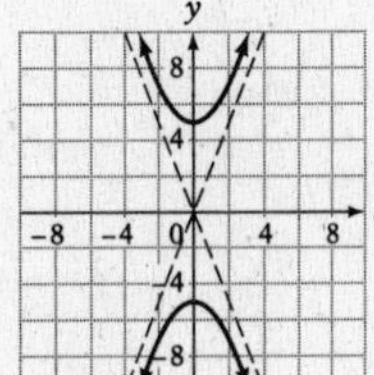

Applying the Concepts

23. a.
$$16x^2 + 25y^2 = 400$$
$$\frac{1}{400}(16x^2 + 25y^2) = \frac{1}{400} \cdot 400$$
$$\frac{x^2}{25} + \frac{y^2}{16} = 1 \text{ ellipse}$$

b. x-intercepts: (5, 0) and (−5, 0)
y-intercepts: (0, 4) and (0, −4)

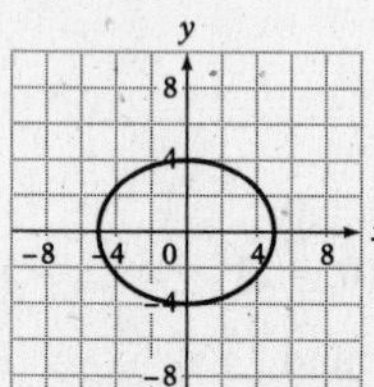

25. a.
$$25y^2 - 4x^2 = -100$$
$$-\frac{1}{100}(25y^2 - 4x^2) = -\frac{1}{100}(-100)$$
$$\frac{x^2}{25} - \frac{y^2}{4} = 1 \text{ hyperbola}$$

b. Axis of symmetry: x-axis
Vertices: (5, 0) and (−5, 0)
Asymptotes: $y = \frac{2}{5}x$ and $y = -\frac{2}{5}x$

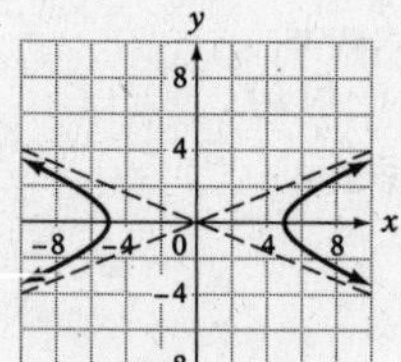

27. a.
$$4y^2 - x^2 = 36$$
$$\frac{1}{36}(4y^2 - x^2) = \frac{1}{36}(36)$$
$$\frac{y^2}{9} - \frac{x^2}{36} = 1 \text{ hyperbola}$$

b. Axis of symmetry: y-axis
Vertices: (0, 3) and (0, −3)
Asymptotes: $y = \frac{1}{2}x$ and $y = -\frac{1}{2}x$

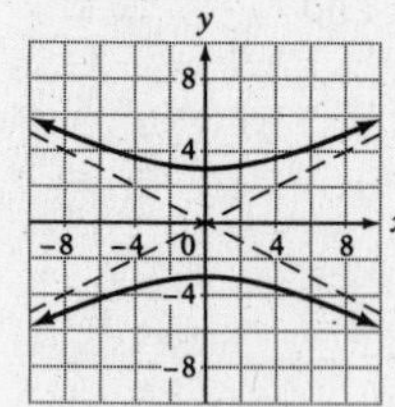

Section 11.4

Objective A Exercises

1. (1) $y = x^2 - x - 1$
(2) $y = 2x + 9$
Use the substitution method.
$$y = x^2 - x - 1$$
$$2x + 9 = x^2 - x - 1$$
$$0 = x^2 - 3x - 10$$
$$0 = (x - 5)(x + 2)$$
$$x - 5 = 0 \quad x + 2 = 0$$
$$x = 5 \quad x = -2$$
Substitute into Equation (2).
$$y = 2x + 9 \quad y = 2x + 9$$
$$y = 2(5) + 9 \quad y = 2(-2) + 9$$
$$y = 10 + 9 \quad y = -4 + 9$$
$$y = 19 \quad y = 5$$
The solutions are (5, 19) and (−2, 5).

3. (1) $y^2 = -x + 3$
(2) $x - y = 1$

Solve Equation (2) for x.
$$x - y = 1$$
$$x = y + 1$$
Use the substitution method.
$$y^2 = -x + 3$$
$$y^2 = -(y + 1) + 3$$
$$y^2 = -y - 1 + 3$$
$$y^2 = -y + 2$$
$$y^2 + y - 2 = 0$$
$$(y + 2)(y - 1) = 0$$
$$y + 2 = 0 \quad y - 1 = 0$$
$$y = -2 \quad y = 1$$
Substitute into Equation (2).
$$x - y = 1 \quad x - y = 1$$
$$x - (-2) = 1 \quad x - 1 = 1$$
$$x + 2 = 1 \quad x = 2$$
$$x = -1$$
The solutions are (−1, −2) and (2, 1).

5. (1) $y^2 = 2x$
(2) $x + 2y = -2$
Solve Equation (2) for x.
$x + 2y = -2$
$x = -2y - 2$
Use the substitution method.

$$\begin{aligned} y^2 &= 2x \\ y^2 &= 2(-2y - 2) \\ y^2 &= -4y - 4 \\ y^2 + 4y + 4 &= 0 \\ (y + 2)(y + 2) &= 0 \end{aligned}$$

$y + 2 = 0 \quad y + 2 = 0$
$y = -2 \quad y = -2$
Substitute into Equation (2).

$$\begin{aligned} x + 2y &= -2 \\ x + 2(-2) &= -2 \\ x - 4 &= -2 \\ x &= 2 \end{aligned}$$

The solution is a double root.
The solution is $(2, -2)$

7. (1) $x^2 + 2y^2 = 12$
(2) $2x - y = 2$
Solve Equation (2) for y.
$2x - y = 2$
$-y = -2x + 2$
$y = 2x - 2$
Use the substitution method.

$$\begin{aligned} x^2 + 2y^2 &= 12 \\ x^2 + 2(2x - 2)^2 &= 12 \\ x^2 + 2(4x^2 - 8x + 4) &= 12 \\ x^2 + 8x^2 - 16x + 8 &= 12 \\ 9x^2 - 16x - 4 &= 0 \\ (x - 2)(9x + 2) &= 0 \end{aligned}$$

$x - 2 = 0 \quad 9x + 2 = 0$
$x = 2 \quad 9x = -2$
$x = -\frac{2}{9}$

Substitute into Equation (2).
$2x - y = 2 \qquad 2x - y = 2$
$2(2) - y = 2 \qquad 2\left(-\frac{2}{9}\right) - y = 2$
$4 - y = 2 \qquad -\frac{4}{9} - y = 2$
$-y = -2 \qquad -y = \frac{22}{9}$
$y = 2 \qquad y = -\frac{22}{9}$

The solutions are $(2, 2)$ and $\left(-\frac{2}{9}, -\frac{22}{9}\right)$.

9. (1) $x^2 + y^2 = 13$
(2) $x + y = 5$
Solve Equation (2) for y.
$x + y = 5$
$y = -x + 5$
Use the substitution method.

$$\begin{aligned} x^2 + y^2 &= 13 \\ x^2 + (-x + 5)^2 &= 13 \\ x^2 + x^2 - 10x + 25 &= 13 \\ 2x^2 - 10x + 12 &= 0 \\ 2(x^2 - 5x + 6) &= 0 \\ 2(x - 3)(x - 2) &= 0 \end{aligned}$$

$x - 3 = 0 \quad x - 2 = 0$
$x = 3 \quad x = 2$

Substitute into Equation (2).
$x + y = 5 \quad x + y = 5$
$3 + y = 5 \quad 2 + y = 5$
$y = 2 \quad y = 3$

The solutions are $(3, 2)$ and $(2, 3)$.

11. (1) $4x^2 + y^2 = 12$
(2) $y = 4x^2$
Use the substitution method.

$$\begin{aligned} 4x^2 + y^2 &= 12 \\ 4x^2 + (4x^2)^2 &= 12 \\ 4x^2 + 16x^4 &= 12 \\ 16x^4 + 4x^2 - 12 &= 0 \\ 4(4x^4 + x^2 - 3) &= 0 \\ 4(4x^2 - 3)(x^2 + 1) &= 0 \end{aligned}$$

$4x^2 - 3 = 0 \qquad x^2 + 1 = 0$
$4x^2 = 3 \qquad x^2 = -1$
$x^2 = \frac{3}{4} \qquad x = \pm\sqrt{-1}$
$x = \pm\frac{\sqrt{3}}{2}$

Substitute the real number solutions into Equation (2).
$y = 4x^2 \qquad y = 4x^2$
$y = 4\left(\frac{\sqrt{3}}{2}\right)^2 \qquad y = 4\left(-\frac{\sqrt{3}}{2}\right)^2$
$y = 4\left(\frac{3}{4}\right) \qquad y = 4\left(\frac{3}{4}\right)$
$y = 3 \qquad y = 3$

The solutions are $\left(\frac{\sqrt{3}}{2}, 3\right)$ and $\left(-\frac{\sqrt{3}}{2}, 3\right)$.

13. (1) $y = x^2 - 2x - 3$
(2) $y = x - 6$
Use the substitution method.

$$y = x^2 - 2x - 3$$
$$x - 6 = x^2 - 2x - 3$$
$$0 = x^2 - 3x + 3$$
$$x = \frac{-b \pm \sqrt{b^2 - 4ac}}{2a}$$
$$= \frac{-(-3) \pm \sqrt{(-3)^2 - 4(1)(3)}}{2(1)}$$
$$= \frac{3 \pm \sqrt{9 - 12}}{2}$$
$$= \frac{3 \pm \sqrt{-3}}{2}$$

The system of equations has no real number solution.

15. (1) $3x^2 - y^2 = -1$
(2) $x^2 + 4y^2 = 17$
Use the addition method.
Multiply Equation (1) by 4.

$$12x^2 - 4y^2 = -4$$
$$x^2 + 4y^2 = 17$$
$$13x^2 = 13$$
$$x^2 = 1$$
$$x = \pm\sqrt{1} = \pm 1$$

Substitute into Equation (2).

$$x^2 + 4y^2 = 17 \qquad x^2 + 4y^2 = 17$$
$$1^2 + 4y^2 = 17 \qquad (-1)^2 + 4y^2 = 17$$
$$1 + 4y^2 = 17 \qquad 1 + 4y^2 = 17$$
$$4y^2 = 16 \qquad 4y^2 = 16$$
$$y^2 = 4 \qquad y^2 = 4$$
$$y = \pm\sqrt{4} \qquad y = \pm\sqrt{4}$$
$$y = \pm 2 \qquad y = \pm 2$$

The solutions are (1, 2), (1, −2), (−1, 2), and (−1, −2).

17. (1) $2x^2 + 3y^2 = 30$
(2) $x^2 + y^2 = 13$
Use the addition method.
Multiply Equation (2) by −2.

$$2x^2 + 3y^2 = 30$$
$$-2x^2 - 2y^2 = -26$$
$$y^2 = 4$$
$$y = \pm\sqrt{4} = \pm 2$$

Substitute into Equation (2).

$$x^2 + y^2 = 13 \qquad x^2 + y^2 = 13$$
$$x^2 + 2^2 = 13 \qquad x^2 + (-2)^2 = 13$$
$$x^2 + 4 = 13 \qquad x^2 + 4 = 13$$
$$x^2 = 9 \qquad x^2 = 9$$
$$x = \pm\sqrt{9} \qquad x = \pm\sqrt{9}$$
$$x = \pm 3 \qquad x = \pm 3$$

The solutions are (3, 2), (3, −2), (−3, 2), and (−3, −2).

19. (1) $y = 2x^2 - x + 1$
(2) $y = x^2 - x + 5$
Use the substitution method.

$$y = 2x^2 - x + 1$$
$$x^2 - x + 5 = 2x^2 - x + 1$$
$$0 = x^2 - 4$$
$$0 = (x + 2)(x - 2)$$
$$x + 2 = 0 \qquad x - 2 = 0$$
$$x = -2 \qquad x = 2$$

Substitute into Equation (2).

$$y = x^2 - x + 5 \qquad y = x^2 - x + 5$$
$$y = (-2)^2 - (-2) + 5 \qquad y = 2^2 - 2 + 5$$
$$y = 4 + 2 + 5 \qquad y = 4 - 2 + 5$$
$$y = 11 \qquad y = 7$$

The solutions are (2, 7) and (−2, 11).

21. (1) $2x^2 + 3y^2 = 24$
(2) $x^2 - y^2 = 7$
Use the addition method.
Multiply Equation (2) by 3.

$$2x^2 + 3y^2 = 24$$
$$3x^2 - 3y^2 = 21$$
$$5x^2 = 45$$
$$x^2 = 9$$
$$x = \pm\sqrt{9} = \pm 3$$

Substitute into Equation (2).

$$x^2 - y^2 = 7 \qquad x^2 - y^2 = 7$$
$$3^2 - y^2 = 7 \qquad (-3)^2 - y^2 = 7$$
$$9 - y^2 = 7 \qquad 9 - y^2 = 7$$
$$-y^2 = -2 \qquad -y^2 = -2$$
$$y^2 = 2 \qquad y^2 = 2$$
$$y = \pm\sqrt{2} \qquad y = \pm\sqrt{2}$$

The solutions are $(3, \sqrt{2})$, $(3, -\sqrt{2})$, $(-3, \sqrt{2})$, and $(-3, -\sqrt{2})$.

23. (1) $x^2 + y^2 = 36$
(2) $4x^2 + 9y^2 = 36$
Use the addition method.
Multiply Equation (1) by −4.

$$-4x^2 - 4y^2 = -144$$
$$4x^2 + 9y^2 = 36$$
$$5y^2 = -108$$
$$y^2 = -\frac{108}{5}$$
$$y = \pm\sqrt{-\frac{108}{5}}$$

The system of equations has no real number solution.

25. (1) $11x^2 - 2y^2 = 4$
(2) $3x^2 + y^2 = 15$
Use the addition method.
Multiply Equation (2) by 2.

$$11x^2 - 2y^2 = 4$$
$$6x^2 + 2y^2 = 30$$
$$17x^2 = 34$$
$$x^2 = 2$$
$$x = \pm\sqrt{2}$$

Substitute into Equation (2).

$$3x^2 + y^2 = 15 \qquad 3x^2 + y^2 = 15$$
$$3(\sqrt{2})^2 + y^2 = 15 \qquad 3(-\sqrt{2})^2 + y^2 = 15$$
$$3(2) + y^2 = 15 \qquad 3(2) + y^2 = 15$$
$$6 + y^2 = 15 \qquad 6 + y^2 = 15$$
$$y^2 = 9 \qquad y^2 = 9$$
$$y = \pm\sqrt{9} \qquad y = \pm\sqrt{9}$$
$$y = \pm 3 \qquad y = \pm 3$$

The solutions are $(\sqrt{2}, 3)$, $(\sqrt{2}, -3)$, $(-\sqrt{2}, 3)$, and $(-\sqrt{2}, -3)$.

27. (1) $2x^2 - y^2 = 7$
(2) $2x - y = 5$
Solve Equation (2) for y.

$$2x - y = 5$$
$$-y = -2x + 5$$
$$y = 2x - 5$$

Use the substitution method.

$$2x^2 - y^2 = 7$$
$$2x^2 - (2x - 5)^2 = 7$$
$$2x^2 - (4x^2 - 20x + 25) = 7$$
$$2x^2 - 4x^2 + 20x - 25 = 7$$
$$-2x^2 + 20x - 32 = 0$$
$$-2(x^2 - 10x + 16) = 0$$
$$-2(x - 2)(x - 8) = 0$$
$$x - 2 = 0 \qquad x - 8 = 0$$
$$x = 2 \qquad x = 8$$

Substitution into Equation (2).

$$2x - y = 5 \qquad 2x - y = 5$$
$$2(2) - y = 5 \qquad 2(8) - y = 5$$
$$4 - y = 5 \qquad 16 - y = 5$$
$$-y = 1 \qquad -y = -11$$
$$y = -1 \qquad y = 11$$

The solutions are $(2, -1)$ and $(8, 11)$.

29. (1) $y = 3x^2 + x - 4$
(2) $y = 3x^2 - 8x + 5$
Use the substitution method.

$$y = 3x^2 + x - 4$$
$$3x^2 - 8x + 5 = 3x^2 + x - 4$$
$$0 = 9x - 9$$
$$0 = 9(x - 1)$$
$$x - 1 = 0$$
$$x = 1$$

Substitute into Equation (1).

$$y = 3x^2 + x - 4$$
$$y = 3(1)^2 + 1 - 4$$
$$y = 3 + 1 - 4$$
$$y = 0$$

The solution is $(1, 0)$.

31. (1) $x = y + 3$
(2) $x^2 + y^2 = 5$
Use the substitution method.

$$x^2 + y^2 = 5$$
$$(y + 3)^2 + y^2 = 5$$
$$y^2 + 6y + 9 + y^2 = 5$$
$$2y^2 + 6y + 9 = 5$$
$$2y^2 + 6y + 4 = 0$$
$$2(y^2 + 3y + 2) = 0$$
$$2(y + 1)(y + 2) = 0$$
$$y + 1 = 0 \qquad y + 2 = 0$$
$$y = -1 \qquad y = -2$$

Substitute into Equation (1).

$$x = y + 3 \qquad x = y + 3$$
$$x = -1 + 3 \qquad x = -2 + 3$$
$$x = 2 \qquad x = 1$$

The solutions are $(2, -1)$ and $(1, -2)$.

33. (1) $y = x^2 + 4x + 4$
(2) $x + 2y = 4$
Use the substitution method.

$$x + 2y = 4$$
$$x + 2(x^2 + 4x + 4) = 4$$
$$x + 2x^2 + 8x + 8 = 4$$
$$2x^2 + 9x + 8 = 4$$
$$2x^2 + 9x + 4 = 0$$
$$(2x + 1)(x + 4) = 0$$
$$2x + 1 = 0 \qquad x + 4 = 0$$
$$2x = -1 \qquad x = -4$$
$$x = -\frac{1}{2}$$

Substitute into Equation (2).

$$x + 2y = 4 \qquad x + 2y = 4$$
$$-\frac{1}{2} + 2y = 4 \qquad -4 + 2y = 4$$
$$2y = 8$$
$$2y = 4 + \frac{1}{2} \qquad y = 4$$
$$2y = \frac{9}{2}$$
$$y = \frac{9}{4}$$

The solutions are $\left(-\frac{1}{2}, \frac{9}{4}\right)$ and $(-4, 4)$.

Applying the Concepts

35. $y = 2^x$

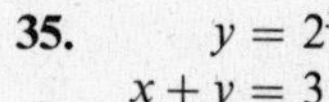

$x + y = 3$

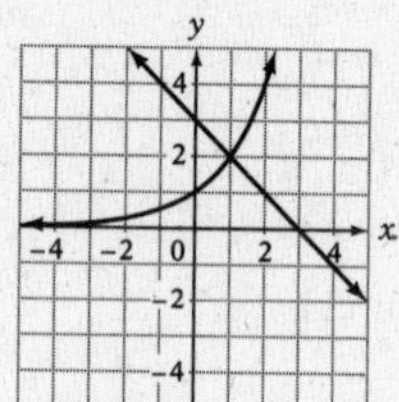

The solution is (1.000, 2.000).

37. $y = \log_2 x$

$\frac{x^2}{9} + \frac{y^2}{1} = 1$

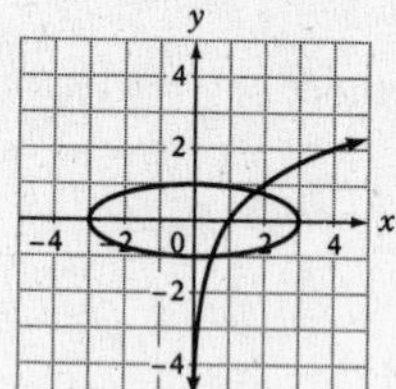

The approximate solutions are (1.755, 0.811) and (0.505, −0.986).

39. $y = -\log_3 x$

$x + y = 4$

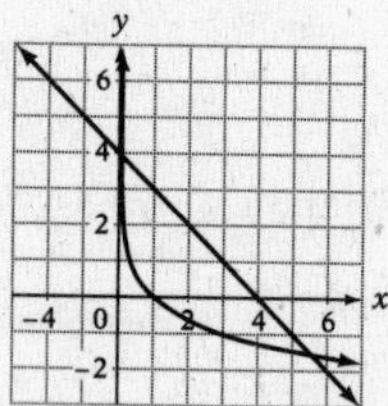

The approximate solutions are (5.562, −1.562) and (0.013, 3.987).

Section 11.5

Objective A Exercises

1. A solid curve is used for the boundaries of inequalities that use $\le$ or $\ge$, and a dashed curve is used for the boundaries of inequalities that use $<$ or $>$.

3. $y \le x^2 - 4x + 3$

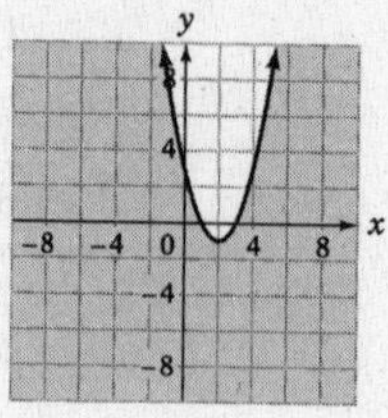

5. $(x - 1)^2 + (y + 2)^2 \le 9$

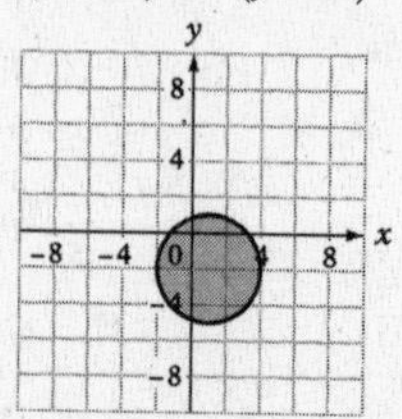

7. $(x + 3)^2 + (x - 2)^2 \ge 9$

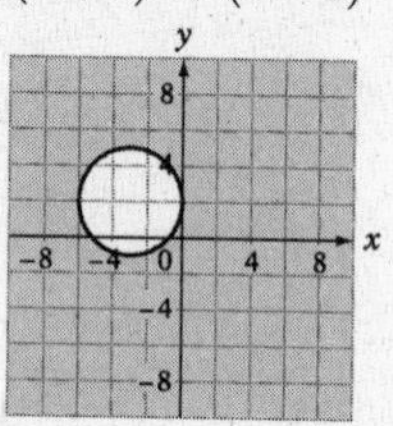

9. $\frac{x^2}{16} + \frac{y^2}{25} < 1$

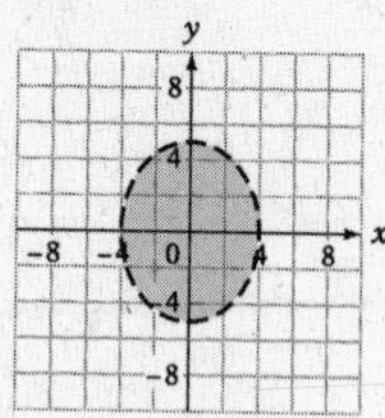

11. $\frac{x^2}{25} - \frac{y^2}{9} \le 1$

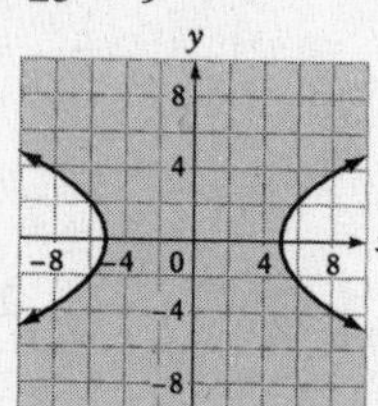

13. $\frac{x^2}{4} + \frac{y^2}{16} \ge 1$

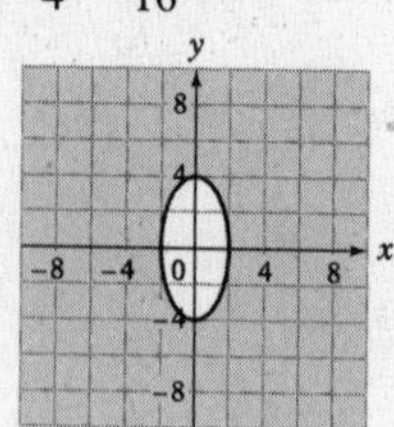

15. $y \le x^2 - 2x + 3$

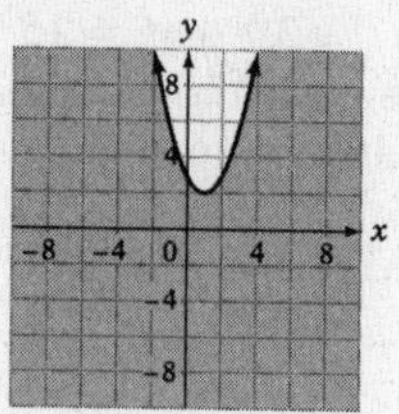

17. $\frac{y^2}{9} - \frac{x^2}{16} \le 1$

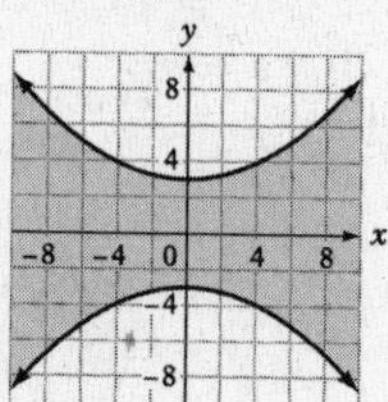

19. $\frac{x^2}{9} + \frac{y^2}{1} \le 1$

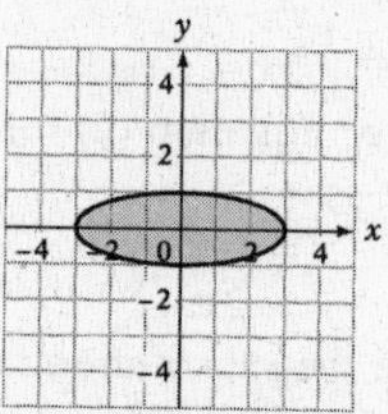

21. $(x - 1)^2 + (y + 3)^2 \le 25$

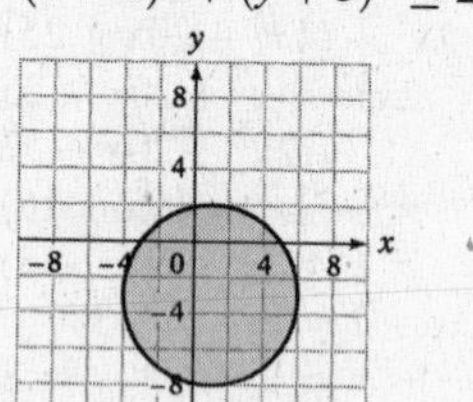

23. $\frac{y^2}{25} - \frac{x^2}{4} \le 1$

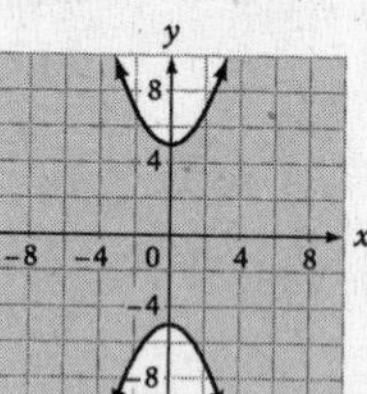

25. $\frac{x^2}{25} + \frac{y^2}{9} \le 1$

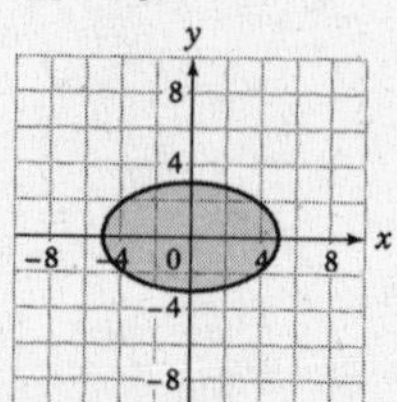

Objective B Exercises

27. $y \le (x-2)^2$
$y + x > 4$

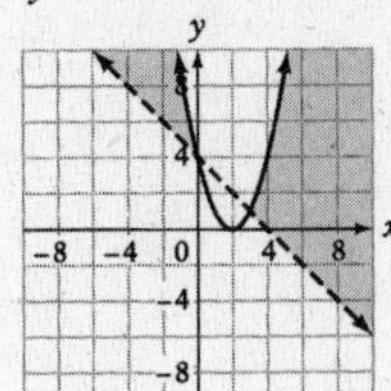

29. $x^2 + y^2 < 16$
$y > x + 1$

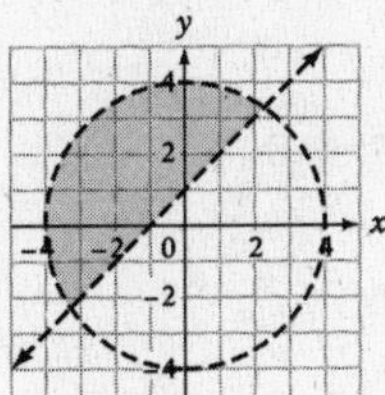

31. $\dfrac{x^2}{4} + \dfrac{y^2}{16} \le 1$
$y \le -\dfrac{1}{2}x + 2$

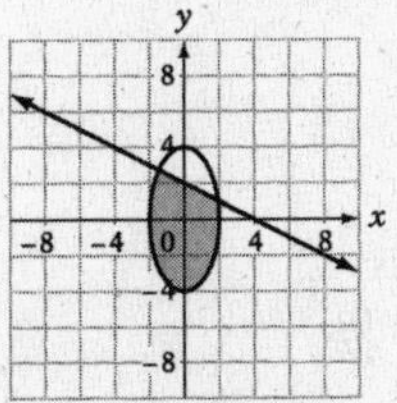

33. $x \ge y^2 - 3y + 2$
$y \ge 2x - 2$

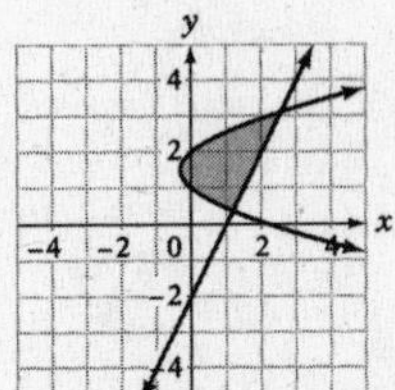

35. $x^2 + y^2 < 25$
$\dfrac{x^2}{9} + \dfrac{y^2}{36} < 1$

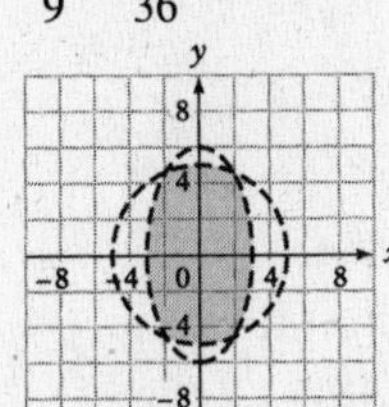

37. $x^2 + y^2 > 4$
$x^2 + y^2 < 25$

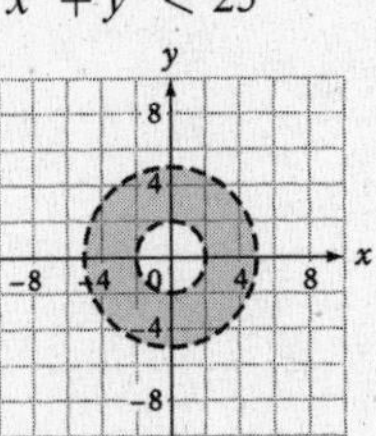

Applying the Concepts

39. $y > x^2 - 3$
$y < x + 3$
$x \le 0$

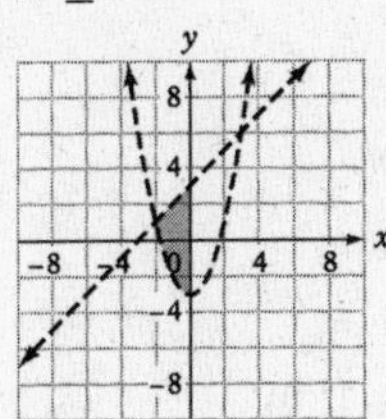

41. $x^2 + y^2 < 3$
$x > y^2 - 1$
$y \ge 0$

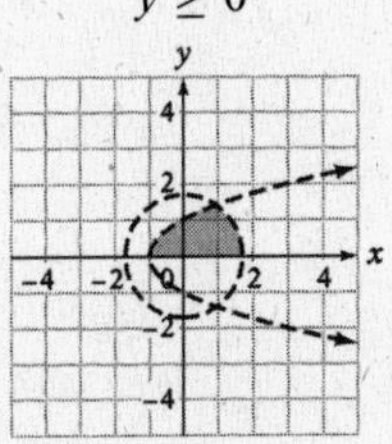

43. $\dfrac{x^2}{16} + \dfrac{y^2}{4} \le 1$
$x^2 + y^2 \le 4$
$x \ge 0$
$y \le 0$

45. $y > 2^x$
$x + y < 4$

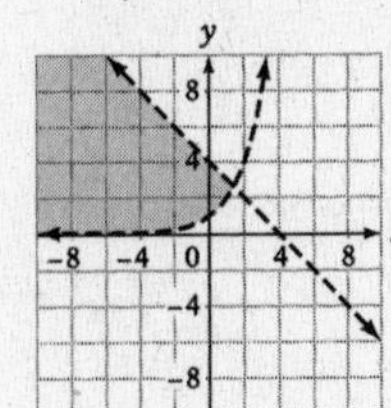

47. $y \geq \log_2 x$
$x^2 + y^2 < 9$

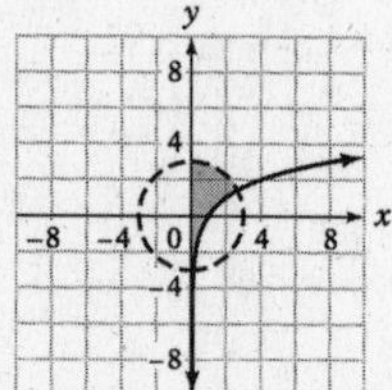

Chapter 11 Review Exercises

1. $y = -2x^2 + x - 2$

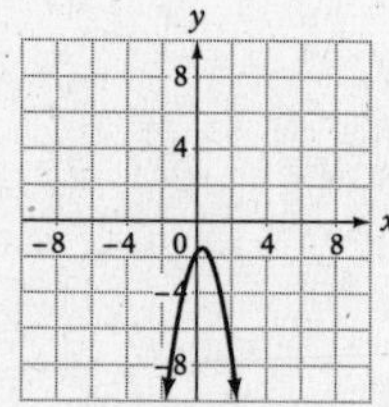

2. x-intercepts: (1, 0) and (−1, 0)
y-intercepts: (0, 3) and (0, −3)

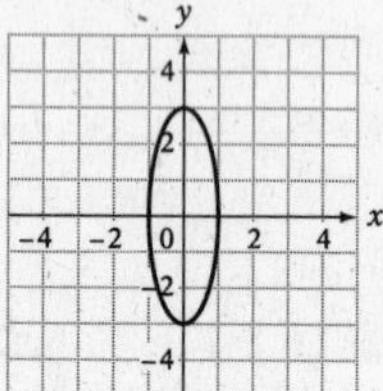

3. $\frac{x^2}{9} - \frac{y^2}{16} < 1$

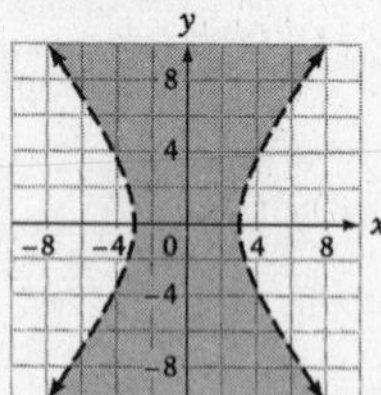

4. $(x+3)^2 + (y+1)^2 = 1$

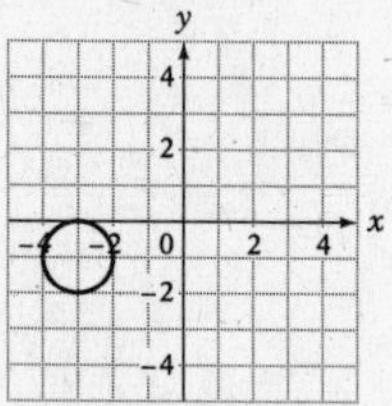

Center: (−3, −1)
Radius: 1

5. $y = x^2 - 4x + 8$
$-\frac{b}{2a} = -\frac{-4}{2(1)} = \frac{4}{2} = 2$
Axis of symmetry: $x = 2$
$y = 2^2 - 4(2) + 8 = 4$
Vertex: (2, 4)

6. (1) $y^2 = 2x^2 - 3x + 6$
(2) $y^2 = 2x^2 + 5x - 2$
Use the addition method. Multiply equation (2) by −1.
$y^2 = 2x^2 - 3x + 6$
$-y^2 = -2x^2 - 5x + 2$
$0 = -8x + 8$
$8x = 8$
$x = 1$
Substitute into Equation (1).
$y^2 = 2x^2 - 3x + 6$
$y^2 = 2(1)^2 - 3(1) + 6$
$y^2 = 2 - 3 + 6$
$y^2 = 5$
$y = \pm\sqrt{5}$
The solutions are $(1, \sqrt{5})$ and $(1, -\sqrt{5})$.

7. $\frac{x^2}{25} + \frac{y^2}{16} \leq 1$
$\frac{y^2}{4} - \frac{x^2}{4} \geq 1$

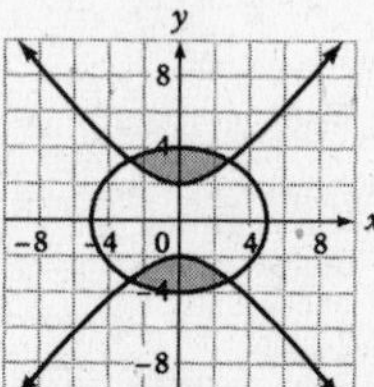

8. Axis of symmetry: x-axis
Vertices: (5, 0) and (−5, 0)
Asymptotes: $y = \frac{1}{5}x$ and $y = -\frac{1}{5}x$

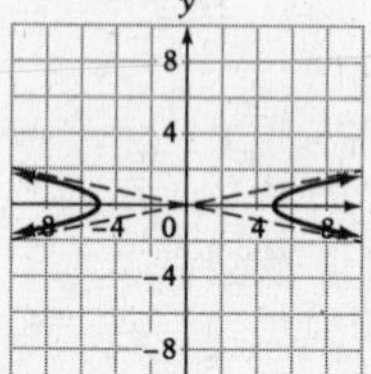

9. $(x_1, y_1) = (2, -1)$ $(x_2, y_2) = (-1, 2)$
$r = \sqrt{(x_1 - x_2)^2 + (y_1 - y_2)^2}$
$= \sqrt{[2-(-1)]^2 + (-1-2)^2}$
$= \sqrt{3^2 + (-3)^2}$
$= \sqrt{9+9} = \sqrt{18}$

$(x-h)^2 + (y-k)^2 = r^2$
$[x-(-1)]^2 + (y-2)^2 = (\sqrt{18})^2$
$(x+1)^2 + (y-2)^2 = 18$

10. $y = -x^2 + 7x - 8$

$-\frac{b}{2a} = \frac{-7}{2(-1)} = \frac{7}{2}$

Axis of symmetry: $x = \frac{7}{2}$

$y = -\left(\frac{7}{2}\right)^2 + 7\left(\frac{7}{2}\right) - 8 = \frac{17}{4}$

Vertex: $\left(\frac{7}{2}, \frac{17}{4}\right)$

11. (1) $x = 2y^2 - 3y + 1$

(2) $3x - 2y = 0$

Solve Equation (2) for x.

$3x - 2y = 0$

$3x = 2y$

$x = \frac{2}{3}y$

Use the substitution method.

$\frac{2}{3}y = 2y^2 - 3y + 1$

$2y = 6y^2 - 9y + 3$

$0 = 6y^2 - 11y + 3$

$0 = (2y - 3)(3y - 1)$

$0 = 2y - 3 \qquad 0 = 3y - 1$

$2y = 3 \qquad 3y = 1$

$y = \frac{3}{2} \qquad y = \frac{1}{3}$

Substitute into Equation (2).

$3x - 2\left(\frac{3}{2}\right) = 0 \qquad 3x - 2\left(\frac{1}{3}\right) = 0$

$3x = 3 \qquad 3x = \frac{2}{3}$

$x = 1 \qquad x = \frac{2}{9}$

The solutions are $\left(1, \frac{3}{2}\right)$ and $\left(\frac{2}{9}, \frac{1}{3}\right)$.

12. $(x_1, y_1) = (4, 6)$ $(x_2, y_2) = (0, -3)$

$d = \sqrt{(x_1 - x_2)^2 + (y_1 - y_2)^2}$

$= \sqrt{(4-0)^2 + (6-(-3))^2}$

$= \sqrt{4^2 + 9^2}$

$= \sqrt{16 + 81} = \sqrt{97}$

$(x - h)^2 + (y - k)^2 = r^2$

$(x - 0)^2 + (y - (-3))^2 = (\sqrt{97})^2$

$x^2 + (y + 3)^2 = 97$

13. $(x - h)^2 + (y - k)^2 = r^2$

$(x - (-1))^2 + (y - 5)^2 = 6^2$

$(x + 1)^2 + (y - 5)^2 = 36$

14. (1) $2x^2 + y^2 = 19$

(2) $3x^2 - y^2 = 6$

Use the addition method.

$2x^2 + y^2 = 19$

$3x^2 - y^2 = 6$

$5x^2 = 25$

$x^2 = 5$

$x = \pm\sqrt{5}$

Substitute the Equation (1).

$2x^2 + y^2 = 19 \qquad 2x^2 + y^2 = 19$

$2(\sqrt{5})^2 + y^2 = 19 \qquad 2(-\sqrt{5})^2 + y^2 = 19$

$10 + y^2 = 19 \qquad 10 + y^2 = 19$

$y^2 = 9 \qquad y^2 = 9$

$y = \pm\sqrt{9} \qquad y = \pm\sqrt{9}$

$y = \pm 3 \qquad y = \pm 3$

The solutions are $(\sqrt{5}, 3)$, $(\sqrt{5}, -3)$, $(-\sqrt{5}, 3)$, and $(-\sqrt{5}, -3)$.

15. $x^2 + y^2 + 4x - 2y = 4$

$(x^2 + 4x) + (y^2 - 2y) = 4$

$(x^2 + 4x + 4) + (y^2 - 2y + 1) = 4 + 4 + 1$

$(x + 2)^2 + (y - 1)^2 = 9$

16. (1) $y = x^2 + 5x - 6$

(2) $y = x - 10$

Use the substitution method.

$x - 10 = x^2 + 5x - 6$

$0 = x^2 + 4x + 4$

$0 = (x + 2)^2$

$x + 2 = 0 \qquad x + 2 = 0$

$x = -2 \qquad x = -2$

Substitute into Equation (2).

$y = -2 - 10 \qquad y = -2 - 10$

$y = -12 \qquad y = -12$

The solution is $(-2, -12)$.

17. $\frac{x^2}{16} + \frac{y^2}{4} > 1$

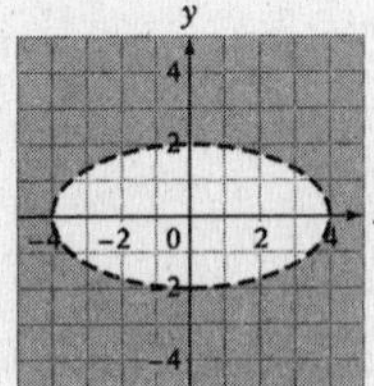

18. Axis of symmetry: y-axis

Vertices: $(0, 4)$ and $(0, -4)$

Asymptotes: $y = \frac{4}{3}x$ and $y = -\frac{4}{3}x$.

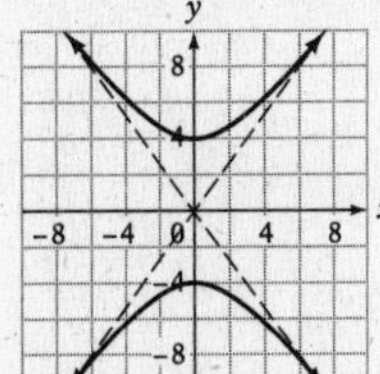

19. $(x-2)^2+(y+1)^2 \le 16$

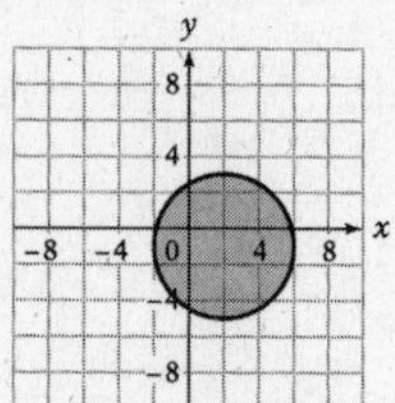

20. $x^2+(y-2)^2=9$

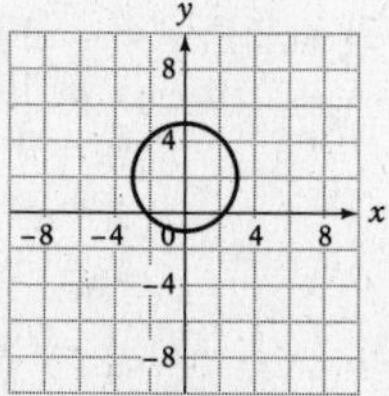

Center: (0, 2)
Radius: 3

21. x-intercepts: (5, 0) and (−5, 0)
y-intercepts: (0, 3) and (0, −3).

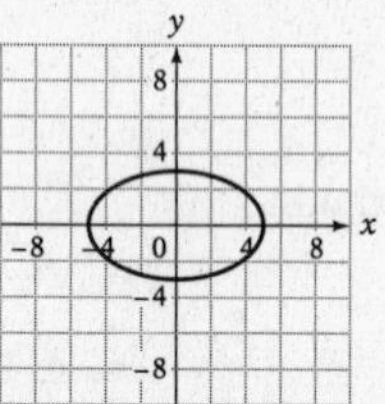

22. $y \ge -x^2-2x+3$

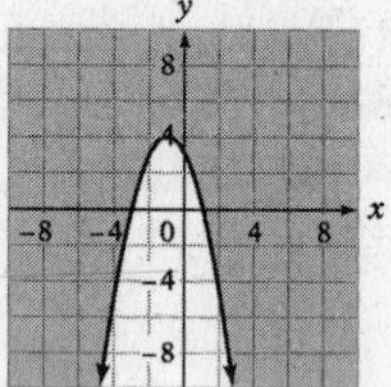

23. $\frac{x^2}{16}+\frac{y^2}{4}<1$
$x^2+y^2>9$

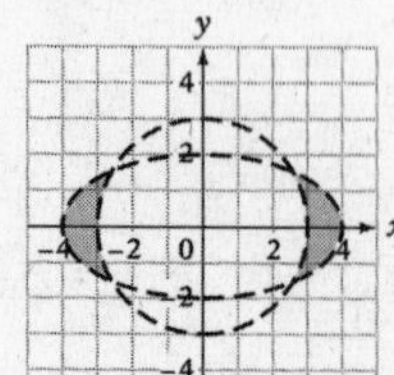

24. (1) $y \ge x^2-4x+2$
(2) $y \le \frac{1}{3}x-1$
Write Equation (1) in standard form.
$y \ge (x^2-4x+4)-4+2$
$y \ge (x-2)^2-2$

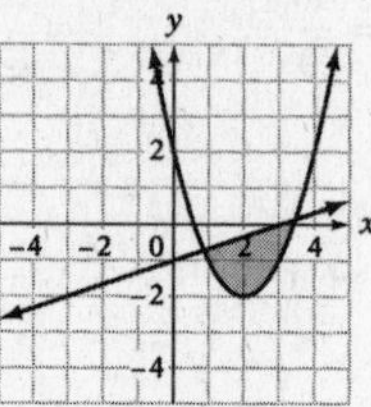

25. $x=2y^2-6y+5$

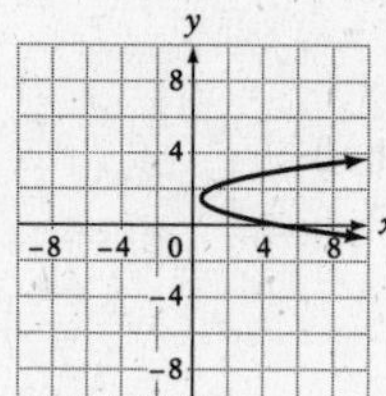

26. $\frac{x^2}{9}+\frac{y^2}{1}\ge 1$
$\frac{x^2}{4}-\frac{y^2}{1}\le 1$

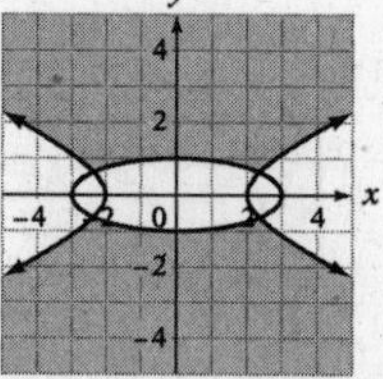

27. $y=x^2-4x-1$

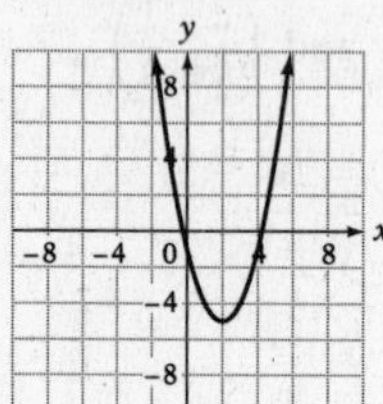

28. $x=y^2-1$

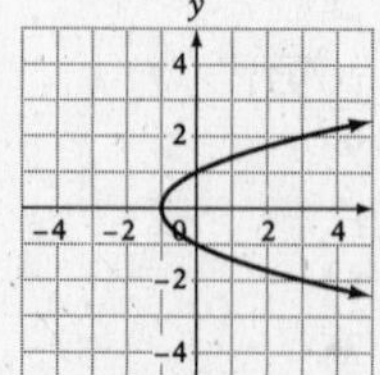

29.
$$(x-h)^2+(y-k)^2=r^2$$
$$(x-3)^2+[y-(-4)]^2=5^2$$
$$(x-3)^2+(y+4)^2=25$$

30.
$$x^2 + y^2 - 6x + 4y - 23 = 0$$
$$(x^2 - 6x) + (y^2 + 4y) = 23$$
$$(x^2 - 6x + 9) + (y^2 + 4y + 4) = 23 + 9 + 4$$
$$(x - 3)^2 + (y + 2)^2 = 36$$

31. $(x + 1)^2 + (y - 4)^2 = 36$

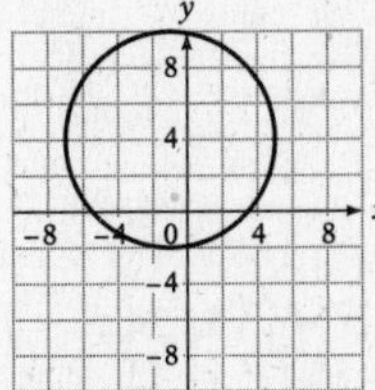

Center: $(-1, 4)$
Radius: $\sqrt{36} = 6$

32.
$$x^2 + y^2 - 8x + 4y + 16 = 0$$
$$(x^2 - 8x) + (y^2 + 4y) = -16$$
$$(x^2 - 8x + 16) + (y^2 + 4y + 4) = -16 + 16 + 4$$
$$(x - 4)^2 + (y + 2)^2 = 4$$

Center: $(4, -2)$; $r = 2$

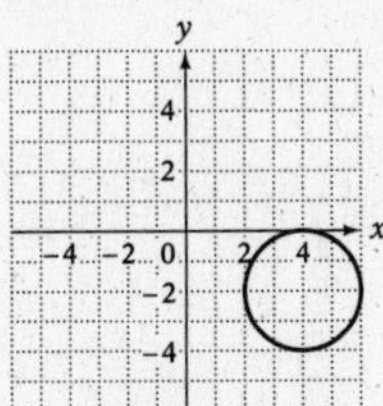

33. $(6, 0)$ and $(-6, 0)$
$(0, 4)$ and $(0, -4)$

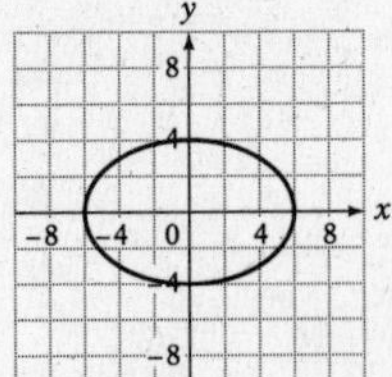

34. $(1, 0)$ and $(-1, 0)$
$(0, 5)$ and $(0, -5)$

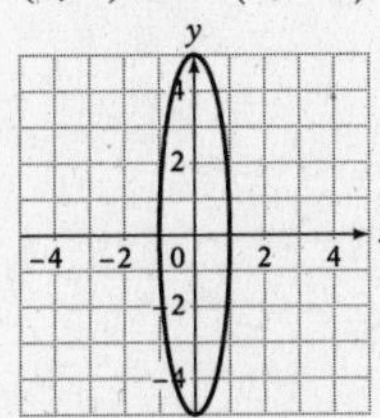

35. Axis of symmetry: x-axis
Vertices: $(3, 0)$ and $(-3, 0)$
Asymptotes: $y = 2x$ and $y = -2x$

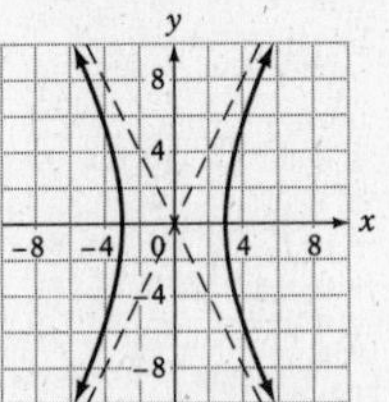

36. Axis of symmetry: y-axis
Vertices: $(0, 6)$ and $(0, -6)$
Asymptotes: $y = 3x$ and $y = -3x$

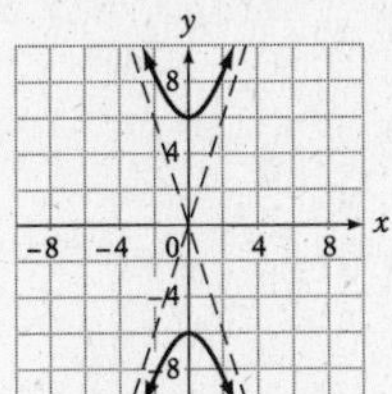

37. (1) $y^2 = x^2 - 3x - 4$
(2) $y = x + 1$

Use the substitution method.
$$y = x^2 - 3x - 4$$
$$x + 1 = x^2 - 3x - 4$$
$$0 = x^2 - 4x - 5$$
$$0 = (x + 1)(x - 5)$$
$$x + 1 = 0 \qquad x - 5 = 0$$
$$x = -1 \qquad x = 5$$

Substitute into Equation (2).
$$y = x + 1 \qquad y = x + 1$$
$$y = -1 + 1 \qquad y = 5 + 1$$
$$y = 0 \qquad y = 6$$

The solutions are $(-1, 0)$ and $(5, 6)$.

38. $x \le y^2 + 2y - 3$

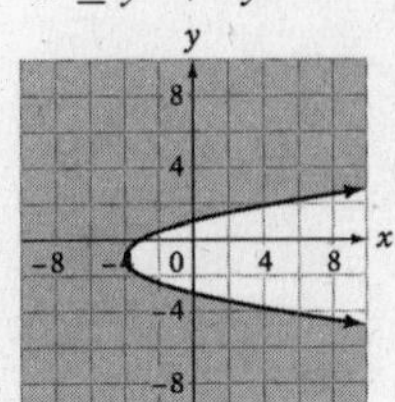

Chapter 11 Test

1. $(x-h)^2+(y-k)^2=r^2$
$[x-(-3)]^2+[y-(-3)]^2=4^2$
$(x+3)^2+(y+3)^2=16$

2. $(x_1, y_1)=(2, 5)\ (x_2, y_2)=(-2, 1)$
$d=\sqrt{(x_1-x_2)^2+(y_1-y_2)^2}$
$=\sqrt{(2-(-2))^2+(5-1)^2}$
$=\sqrt{4^2+4^2}=\sqrt{16+16}$
$=\sqrt{32}=4\sqrt{2}$
$(x-h)^2+(y-k)^2=r^2$
$(x-(-2))^2+(y-1)^2=(4\sqrt{2})^2$
$(x+2)^2+(y-1)^2=32$

3. Axis of symmetry: y-axis
Vertices: (0, 5) and (0, −5)
Asymptotes: $y=\frac{5}{4}x$ and $y=-\frac{5}{4}x$

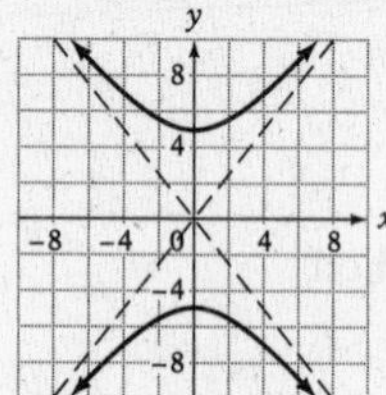

4. $x^2+y^2<36$
$x+y>4$

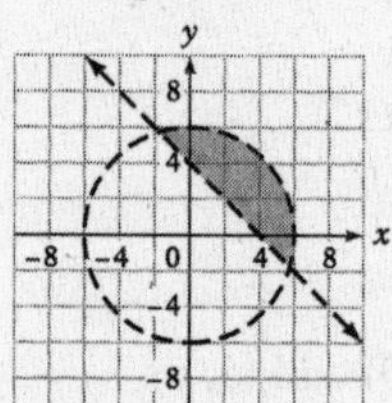

5. $y=-x^2+6x-5$
$-\frac{b}{2a}=-\frac{6}{2(-1)}=3$
Axis of symmetry: $x=3$

6. (1) $x^2-y^2=24$
(2) $2x^2+5y^2=55$
Use the addition method.
Multiply Equation (1) by −2.
$-2x^2+2y^2=-48$
$2x^2+5y^2=55$
$7y^2=7$
$y^2=1$
$y=\pm\sqrt{1}$
$y=\pm 1$
Substitute into Equation (1).

$x^2-y^2=24$	$x^2-y^2=24$
$x^2-(1)^2=24$	$x^2-(-1)^2=24$
$x^2-1=24$	$x^2-1=24$
$x^2=25$	$x^2=25$
$x=\pm\sqrt{25}$	$x=\pm\sqrt{25}$
$x=\pm 5$	$x=\pm 5$

The solutions are (5, 1), (−5, 1), (5, −1), and (−5, −1).

7. $y=-x^2+3x-2$
$-\frac{b}{2a}=-\frac{3}{2(-1)}=\frac{3}{2}$
$y=-\left(\frac{3}{2}\right)^2+3\left(\frac{3}{2}\right)-2=\frac{1}{4}$
Vertex: $\left(\frac{3}{2}, \frac{1}{4}\right)$

8. $x=y^2-y-2$

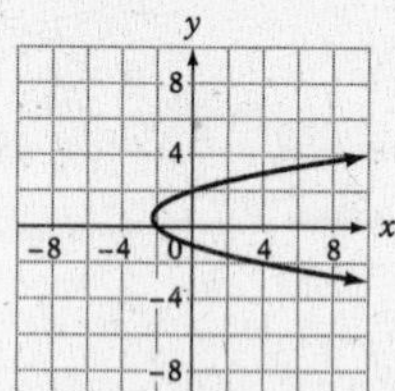

9. x-intercepts: (4, 0) and (−4, 0)
y-intercepts: (0, 2) and (0, −2)

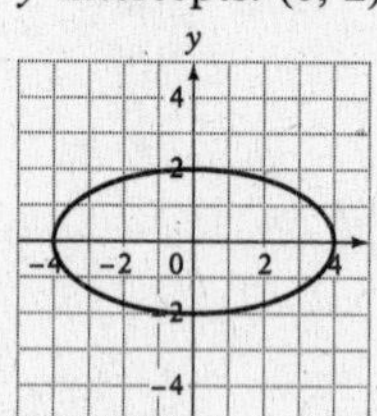

10. $\frac{x^2}{25}+\frac{y^2}{4}\le 1$

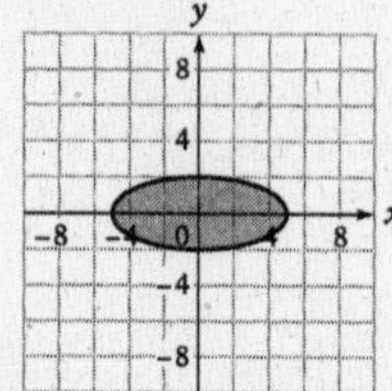

11. $(x-h)^2+(y-k)^2=r^2$
$[(x-(-2)]^2+(y-4)^2=3^2$
$(x+2)^2+(y-4)^2=9$

12. (1) $x=3y^2+2y-4$
(2) $x=y^2-5y$
Use the addition method.
Multiply Equation (2) by -1.
$x=3y^2+2y-4$
$-x=-y^2+5y$
$0=2y^2+7y-4$
$0=(2y-1)(y+4)$
$2y-1=0 \quad y+4=0$
$y=\frac{1}{2} \quad y=-4$

Substitute into Equation (2).
$x=y^2-5y \quad x=y^2-5y$
$x=\left(\frac{1}{2}\right)^2-5\left(\frac{1}{2}\right) \quad x=(-4)^2-5(-4)$
$x=16+20$
$x=36$
$x=\frac{1}{4}-\frac{5}{2}$
$x=-\frac{9}{4}$
The solutions are $\left(-\frac{9}{4},\frac{1}{2}\right)$ and $(36,-4)$.

13. (1) $x^2+2y^2=4$
(2) $x+y=2$
Solve Equation (2) for y.
$x+y=2$
$y=2-x$
Use the substitution method.
$x^2+2(2-x)^2=4$
$x^2+2(4-4x+x^2)=4$
$x^2+8-8x+2x^2=4$
$3x^2-8x+4=0$
$(3x-2)(x-2)=0$
$3x-2=0 \quad x-2=0$
$x=\frac{2}{3} \quad x=2$
Substitute into Equation (2).
$x+y=2 \quad x+y=2$
$\frac{2}{3}+y=2 \quad 2+y=2$
$y=0$
$y=\frac{4}{3}$
The solutions are $\left(\frac{2}{3},\frac{4}{3}\right)$ and $(2, 0)$.

14. $(x_1, y_1)=(2, 4)\ (x_2, y_2)=(-1,-3)$
$r=\sqrt{(x_1-x_2)^2+(y_1-y_2)^2}$
$=\sqrt{[2-(-1)]^2+[4-(-3)]^2}$
$=\sqrt{3^2+7^2}$
$=\sqrt{9+49}=\sqrt{58}$
$(x-h)^2+(y-k)^2=r^2$
$[x-(-1)]^2+[y-(-3)]^2=(\sqrt{58})^2$
$(x+1)^2+(y+3)^2=58$

15. $\frac{x^2}{25}-\frac{y^2}{16}>1$
$x^2+y^2<3$

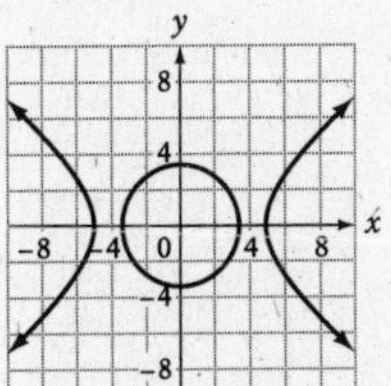

∅ The solution sets of these inequalities do not intersect, so the system has no real number solution.

16. a. $x^2+y^2-4x+2y+1=0$
$(x^2-4x)+(y^2+2y)=-1$
$(x^2-4x+4)+(y^2+2y+1)=-1+4+1$
$(x-2)^2+(y+1)^2=4$

b. Center: $(2,-1)$
Radius: 2

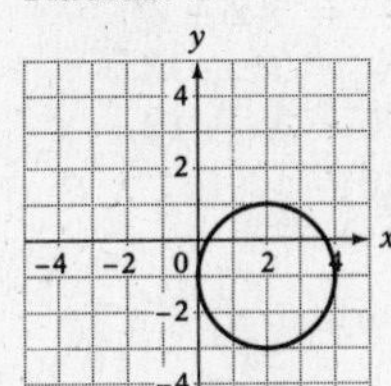

17. $y=\frac{1}{2}x^2+x-4$

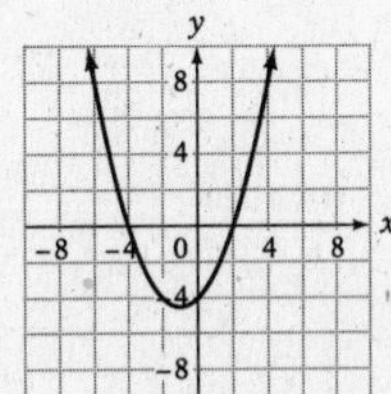

18. $(x-2)^2+(y+1)^2=9$

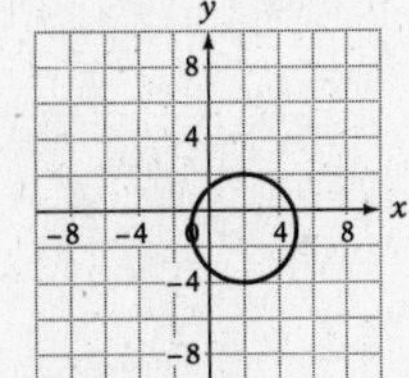

19. Axis of symmetry: x-axis
Vertices: (3, 0) and (−3, 0)
Asymptotes: $y = \frac{2}{3}x$ and $y = -\frac{2}{3}x$

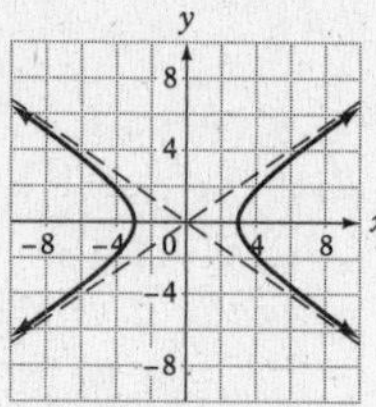

20. $\frac{x^2}{16} - \frac{y^2}{25} < 1$

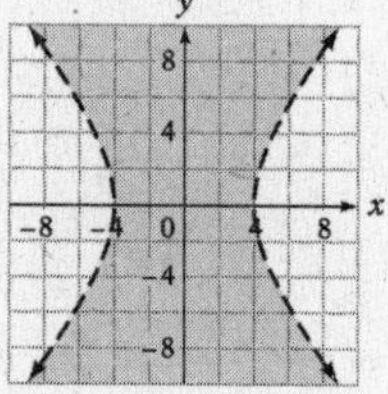

Cumulative Review Exercises

1. $\{x|x < 4\} \cap \{x|x > 2\} = \{x|2 < x < 4\}$

2.
$$\frac{5x-2}{3} - \frac{1-x}{5} = \frac{x+4}{10}$$
$$30\left(\frac{5x-2}{3} - \frac{1-x}{5}\right) = 30\left(\frac{x+4}{10}\right)$$
$$10(5x-2) - 6(1-x) = 3(x+4)$$
$$50x - 20 - 6 + 6x = 3x + 12$$
$$56x - 26 = 3x + 12$$
$$53x - 26 = 12$$
$$53x = 38$$
$$x = \frac{38}{53}$$

The solution is $\frac{38}{53}$.

3.
$$4 + |3x+2| < 6$$
$$|3x+2| < 2$$
$$-2 < 3x + 2 < 2$$
$$-4 < 3x < 0$$
$$-\frac{4}{3} < x < 0$$
$$\left\{x \middle| -\frac{4}{3} < x < 0\right\}$$

4. $(x_1, y_1) = (2, -3)$ $m = -\frac{3}{2}$
$$y - y_1 = m(x - x_1)$$
$$y - (-3) = -\frac{3}{2}(x-2)$$
$$y + 3 = -\frac{3}{2}x + 3$$
$$y = -\frac{3}{2}x$$

5. The product of the slopes of two perpendicular lines is −1.
$$m_1 \cdot m_2 = -1 \qquad y - y_1 = m(x - x_1)$$
$$m_1 \cdot -1 = -1 \qquad y - (-2) = 1(x-4)$$
$$m_1 = 1 \qquad y + 2 = x - 4$$
$$y = x - 6$$

6. $x^{2n}(x^{2n} + 2x^n - 3x) = x^{4n} + 2x^{3n} - 3x^{2n+1}$

7.
$$(x-1)^3 - y^3$$
$$= [(x-1) - y][(x-1)^2 + (x-1)y + y^2]$$
$$= (x - 1 - y)(x^2 - 2x + 1 + xy - y + y^2)$$
$$= (x - y - 1)(x^2 - 2x + 1 + xy - y + y^2)$$

8.
$$\frac{3x-2}{x+4} \le 1$$
$$\frac{3x-2}{x+4} - 1 \le 0$$
$$\frac{3x-2}{x+4} - \frac{x+4}{x+4} \le 0$$
$$\frac{2x-6}{x+4} \le 0$$

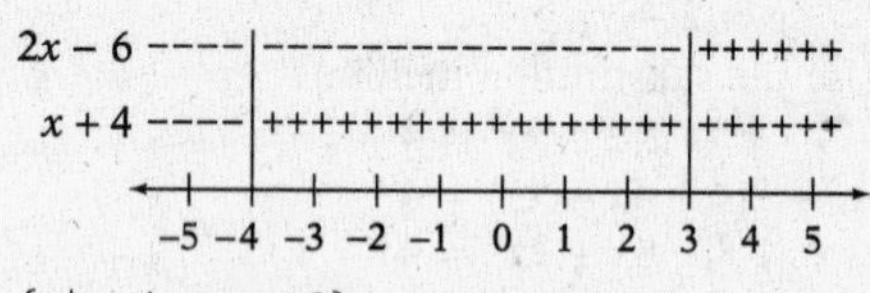

$\{x| -4 < x \le 3\}$

9.
$$\frac{ax - bx}{ax + ay - bx - by} = \frac{x(a-b)}{(ax+ay) + (-bx - by)}$$
$$= \frac{x(a-b)}{a(x+y) - b(x+y)}$$
$$= \frac{x(a-b)}{(x+y)(a-b)} = \frac{x}{x+y}$$

10. $$\frac{x-4}{3x-2}-\frac{1+x}{3x^2+x-2}=\frac{x-4}{3x-2}-\frac{1+x}{(3x-2)(x+1)}$$
$$=\frac{x-4}{3x-2}\cdot\frac{x+1}{x+1}-\frac{1+x}{(3x-2)(x+1)}$$
$$=\frac{x^2-3x-4}{(3x-2)(x+1)}-\frac{1+x}{(3x-2)(x+1)}$$
$$=\frac{(x^2-3x-4)-(1+x)}{(3x-2)(x+1)}$$
$$=\frac{x^2-3x-4-1-x}{(3x-2)(x+1)}$$
$$=\frac{x^2-4x-5}{(3x-2)(x+1)}$$
$$=\frac{(x-5)(x+1)}{(3x-2)(x+1)}=\frac{x-5}{3x-2}$$

11. $$\frac{6x}{2x-3}-\frac{1}{2x-3}=7$$
$$(2x-3)\left(\frac{6x}{2x-3}-\frac{1}{2x-3}\right)=(2x-3)(7)$$
$$6x-1=14x-21$$
$$-8x-1=-21$$
$$-8x=-20$$
$$x=\frac{5}{2}$$

The solution is $\frac{5}{2}$.

12. $$5x+2y>10$$
$$2y>-5x+10$$
$$y>-\frac{5}{2}x+5$$

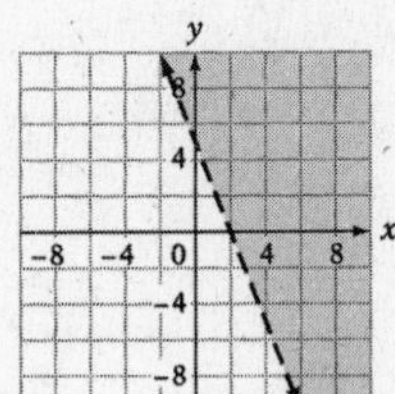

13. $$\left(\frac{12a^2b^2}{a^{-3}b^{-4}}\right)^{-1}\left(\frac{ab}{4^{-1}a^{-2}b^4}\right)^2=(12a^5b^6)^{-1}\left(\frac{4a^3}{b^3}\right)^2$$
$$=(12^{-1}a^{-5}b^{-6})\left(\frac{4^2a^6}{b^6}\right)$$
$$=\left(\frac{1}{12a^5b^6}\right)\left(\frac{4^2a^6}{b^6}\right)=\frac{16a^6}{12a^5b^{12}}=\frac{4a}{3b^{12}}$$

14. $2\sqrt[4]{x^3}=2x^{3/4}$

15. $$\sqrt{18}-\sqrt{-25}=\sqrt{18}-i\sqrt{25}$$
$$=3\sqrt{2}-5i$$

16. $$2x^2+2x-3=0$$
$$x=\frac{-b\pm\sqrt{b^2-4ac}}{2a}=\frac{-2\pm\sqrt{2^2-4(2)(-3)}}{2(2)}$$
$$=\frac{-2\pm\sqrt{4+24}}{4}=\frac{-2\pm\sqrt{28}}{4}$$
$$=\frac{-2\pm2\sqrt{7}}{4}=\frac{-1\pm\sqrt{7}}{2}$$

The solutions are $\frac{-1+\sqrt{7}}{2}$ and $\frac{-1-\sqrt{7}}{2}$.

17. (1) $x^2+y^2=20$
(2) $x^2-y^2=12$

Use the addition method.
$$x^2+y^2=20$$
$$x^2-y^2=12$$
$$2x^2=32$$
$$x^2=16$$
$$x=\pm\sqrt{16}$$
$$x=\pm4$$

Substitute into Equation (1).

$x^2+y^2=20$	$x^2+y^2=20$
$(4)^2+y^2=20$	$(-4)^2+y^2=20$
$16+y^2=20$	$16+y^2=20$
$y^2=4$	$y^2=4$
$y=\pm\sqrt{4}$	$y=\pm\sqrt{4}$
$y=\pm2$	$y=\pm2$

The solutions are (4, 2), (4, −2), (−4, 2), and (−4, −2).

18. $$x-\sqrt{2x-3}=3$$
$$-\sqrt{2x-3}=-x+3$$
$$\sqrt{2x-3}=x-3$$
$$(\sqrt{2x-3})^2=(x-3)^2$$
$$2x-3=x^2-6x+9$$
$$0=x^2-8x+12$$
$$0=(x-6)(x-2)$$

$x-6=0 \quad x-2=0$
$x=6 \quad x=2$

Check:

$x-\sqrt{2x-3}=3$		$x-\sqrt{2x-3}=3$	
$6-\sqrt{2(6)-3}$	3	$2-\sqrt{2(2)-3}$	3
$6-\sqrt{9}$		$2-\sqrt{1}$	
$6-3$		$2-1$	
$3=3$		$1\neq3$	

The solution is 6.

19. $f(x) = -x^2 + 3x - 2$
$f(-3) = -(-3)^2 + 3(-3) - 2$
$= -9 - 9 - 2$
$= -20$

20. $f(x) = 4x + 8$
$y = 4x + 8$
$x = 4y + 8$
$x - 8 = 4y$
$\frac{1}{4}x - 2 = y$

The inverse of the function is $f^{-1}(x) = \frac{1}{4}x - 2$.

21. $f(x) = -2x^2 + 4x - 2$
$x = -\frac{b}{2a} = -\frac{4}{2(-2)} = \frac{-4}{-4} = 1$
$f(x) = -2x^2 + 4x - 2$
$f(1) = -2(1)^2 + 4(1) - 2$
$= -2 + 4 - 2$
$= 0$
The maximum value of the function is 0.

22. Strategy
- Let x represent one number. Since the sum of the two numbers is 40, $40 - x$ represents the other number. Then their product is represented by $-x^2 + 40x$.
- To find the first of the two numbers, find the x-coordinate of the vertex of the function $f(x) = -x^2 + 40x$.
- To find the other number, replace x in $40 - x$ by the x-coordinate of the vertex and evaluate.

Solution $x = -\frac{b}{2a} = -\frac{40}{2(-1)} = \frac{-40}{-2} = 20$
$40 - x = 40 - 20 = 20$
$20 \cdot 20 = 400$
The maximum product of the two numbers whose sum is 40 is 400.

23. $(x_1, y_1) = (2, 4)\ (x_2, y_2) = (-1, 0)$
$d = \sqrt{(x_1 - x_2)^2 + (y_1 - y_2)^2}$
$= \sqrt{[2 - (-1)^2] + (4 - 0)^2}$
$= \sqrt{3^2 + 4^2}$
$= \sqrt{9 + 16} = \sqrt{25} = 5$

24. $(x_1, y_1) = (3, 1)(x_2, y_2) = (-1, 2)$
$r = \sqrt{(x_1 - x_2)^2 + (y_1 - y_2)^2}$
$= \sqrt{[3 - (-1)]^2 + (1 - 2)^2}$
$= \sqrt{4^2 + (-1)^2}$
$= \sqrt{16 + 1} = \sqrt{17}$
$(x - h)^2 + (y - k)^2 = r^2$
$[x - (-1)]^2 + (y - 2)^2 = (\sqrt{17})^2$
$(x + 1)^2 + (y - 2)^2 = 17$

25. $x = y^2 - 2y + 3$
$-\frac{b}{2a} = -\frac{-2}{2(1)} = \frac{2}{2} = 1$
$x = (1)^2 - 2(1) + 3 = 2$
Vertex: (2, 1)
Axis of symmetry: $y = 1$

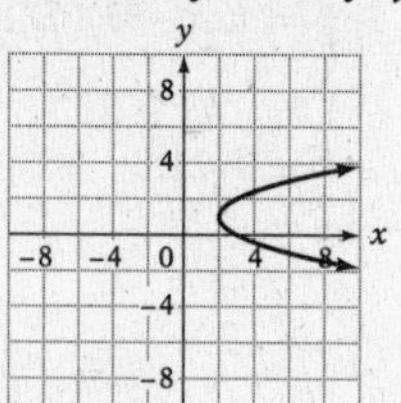

26. $\frac{x^2}{25} + \frac{y^2}{4} = 1$
x-intercepts: (5, 0) and (−5, 0)
y-intercepts: (0, 2) and (0, −2).

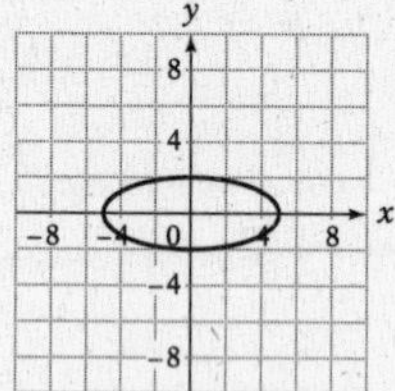

27. $\frac{y^2}{4} - \frac{x^2}{25} < 1$
Axis of symmetry: y-axis
Vertices: (0, 2) and (0, −2)
Asymptotes: $y = \frac{2}{5}x$ and $y = -\frac{2}{5}x$

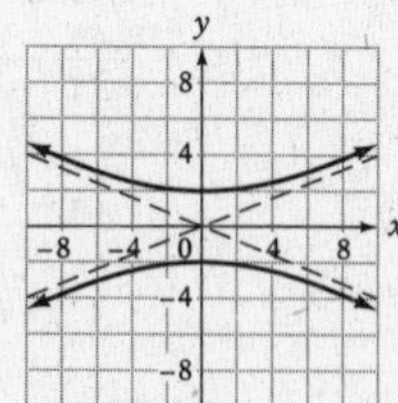

28. $(x-1)^2 + y^2 \le 25$
$y^2 < x$

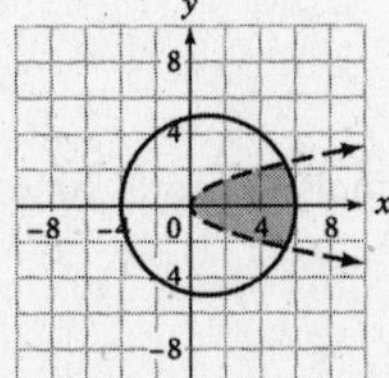

29. Strategy • Number of adult tickets: x
Number of child tickets: $192 - x$

	Amount	Cost	Value
Adult tickets	x	12.00	$12x$
Child tickets	$192 - x$	4.50	$4.5(192 - x)$

• The sum of the values of each type of ticket sold equals the total value of all the tickets sold ($1479).

Solution

$$12x + 4.5(192 - x) = 1479$$
$$12x + 864 - 4.5x = 1479$$
$$7.5x + 864 = 1479$$
$$7.5x = 615$$
$$x = 82$$

There were 82 adult tickets sold.

30. Strategy • The unknown rate of the motorcycle: r
The unknown rate of the car: $r - 12$

	Distance	Rate	Time
Motorcycle	180	r	$\frac{180}{r}$
Car	144	$r - 12$	$\frac{144}{r-12}$

• The time traveled by the motorcycle is the same as the time traveled by the car.

Solution

$$\frac{180}{r} = \frac{144}{r-12}$$
$$r(r-12)\left(\frac{180}{r}\right) = r(r-12)\left(\frac{144}{r-12}\right)$$
$$(r-12)(180) = r(144)$$
$$180r - 2160 = 144r$$
$$36r - 2160 = 0$$
$$36r = 2160$$
$$r = 60$$

The rate of the motorcycle is 60 mph.

31. Strategy • The unknown rowing rate of the crew in calm water: r

	Distance	Rate	Time
With current	12	$r + 1.5$	$\frac{12}{r+1.5}$
Against current	12	$r - 1.5$	$\frac{12}{r-1.5}$

• The sum of the time traveling upriver and the time traveling downriver equals the total time traveled (6 h).

Solution

$$\frac{12}{r+1.5} + \frac{12}{r-1.5} = 6$$
$$(r+1.5)(r-1.5)\left(\frac{12}{r+1.5} + \frac{12}{r-1.5}\right) = (r+1.5)(r-1.5)(6)$$
$$(r-1.5)(12) + (r+1.5)(12) = (r^2 - 2.25)(6)$$
$$12r - 18 + 12r + 18 = 6r^2 - 13.5$$
$$24r = 6r^2 - 13.5$$
$$0 = 6r^2 - 24r - 13.5$$
$$0 = 12r^2 - 48r - 27$$
$$0 = 3(4r^2 - 16r - 9)$$
$$0 = 4r^2 - 16r - 9$$
$$0 = (2r+1)(2r-9)$$

$$2r + 1 = 0 \qquad 2r - 9 = 0$$
$$2r = -1 \qquad 2r = 9$$
$$r = -\frac{1}{2} \qquad r = \frac{9}{2}$$

Since rate cannot be a negative number, the solution $-\frac{1}{2}$ is not possible.
The rowing rate of the crew is 4.5 mph.

32. Strategy To find the speed:
• Write the basic inverse variation equation, replace the variables by the given values, and solve for k.
• Write the inverse variation equation, replacing k by its value. Substitute 60 for t and solve for v.

Solution

$$v = \frac{k}{t} \qquad v = \frac{1080}{t}$$
$$30 = \frac{k}{36} \qquad = \frac{1080}{60}$$
$$1080 = k \qquad = 18$$

The gear will make 18 revolutions/min.

Chapter 12: Sequences and Series

Prep Test

1. $[3(1)-2]+[3(2)-2]+[3(3)-2]$
$=[3-2]+[6-2]+[9-2]$
$=1+4+7$
$=12$

2. $f(6)=\dfrac{6}{6+2}=\dfrac{6}{8}=\dfrac{\overset{1}{\cancel{2}}\cdot 3}{\underset{1}{\cancel{2}}\cdot 2\cdot 2}=\dfrac{3}{4}$

3. $2+(5-1)(4)$
$=2+4(4)$
$=2+16$
$=18$

4. $(-3)(-2)^{6-1}$
$=(-3)(-2)^5$
$=(-3)(-32)$
$=96$

5. $\dfrac{-2[1-(-4)^5]}{1-(-4)}$
$=\dfrac{-2[1-(-1024)]}{1+4}$
$=\dfrac{-2(1025)}{5}=\dfrac{-2(205\cdot \overset{1}{\cancel{5}})}{\underset{1}{\cancel{5}}}$
$=-2(205)=-410$

6. $\dfrac{\frac{4}{10}}{\frac{10}{10}-\frac{1}{10}}=\dfrac{\frac{4}{10}}{\frac{9}{10}}=\dfrac{4}{10}\div\dfrac{9}{10}=\dfrac{4}{\underset{1}{\cancel{10}}}\cdot\dfrac{\overset{1}{\cancel{10}}}{9}=\dfrac{4}{9}$

7. $(x+y)(x+y)$
$=x^2+xy+xy+y^2$
$=x^2+2xy+y^2$

8. $(x+y)^2(x+y)$
$=(x^2+xy+xy+y^2)(x+y)$

$$\begin{array}{r} x^2+2xy+y^2 \\ x+y \\ \hline x^2y+2xy^2+y^3 \\ x^3+2x^2y+xy^2 \\ \hline x^3+3x^2y+3xy^2+y^3 \end{array}$$

Go Figure

$1=1^6$, $32=2^5$, $81=3^4$, $64=4^3$, $25=5^2$, $6=6^1$
Next in the sequence would be 7^0 which equals 1.
Answer: 1

Section 12.1

Objective A Exercises

1. A sequence is an ordered list of numbers.

3. 8

5. $a_n=n+1$
$a_1=1+1=2$ The first term is 2.
$a_2=2+1=3$ The second term is 3.
$a_3=3+1=4$ The third term is 4.
$a_4=4+1=5$ The fourth term is 5.

7. $a_n=2n+1$
$a_1=2(1)+1=3$ The first term is 3.
$a_2=2(2)+1=5$ The second term is 5.
$a_3=2(3)+1=7$ The third term is 7.
$a_4=2(4)+1=9$ The fourth term is 9.

9. $a_n=2-2n$
$a_1=2-2(1)=0$ The first term is 0.
$a_2=2-2(2)=-2$ The second term is -2.
$a_3=2-2(3)=-4$ The third term is -4.
$a_4=2-2(4)=-6$ The fourth term is -6.

11. $a_n=2^n$
$a_1=2^1=2$ The first term is 2.
$a_2=2^2=4$ The second term is 4.
$a_3=2^3=8$ The third term is 8.
$a_4=2^4=16$ The fourth term is 16.

13. $a_n=n^2+1$
$a_1=1^2+1=2$ The first term is 2.
$a_2=2^2+1=5$ The second term is 5.
$a_3=3^2+1=10$ The third term is 10.
$a_4=4^2+1=17$ The fourth term is 17.

15. $a_n=n^2-\dfrac{1}{n}$
$a_1=1^2-\dfrac{1}{1}$ The first term is 0.
$a_2=2^2-\dfrac{1}{2}=\dfrac{7}{2}$ The second term is $\dfrac{7}{2}$.
$a_3=3^2-\dfrac{1}{3}=\dfrac{26}{3}$ The third term is $\dfrac{26}{3}$.
$a_4=4^2-\dfrac{1}{4}=\dfrac{63}{4}$ The fourth term is $\dfrac{63}{4}$.

17. $a_n=3n+4$
$a_{12}=3(12)+4=40$
The twelfth term is 40.

19. $a_n=n(n-1)$
$a_{11}=11(11-1)=110$
The eleventh term is 110.

21. $a_n = (-1)^{n-1}n^2$
$a_{15} = (-1)^{14}(15)^2 = 225$
The fifteenth term is 225.

23. $a_n = \left(\frac{1}{2}\right)^n$
$a_8 = \left(\frac{1}{2}\right)^8 = \frac{1}{256}$
The eighth term is $\frac{1}{256}$.

25. $a_n = (n+2)(n+3)$
$a_{17} = (17+2)(17+3) = (19)(20) = 380$
The seventeenth term is 380.

27. $a_n = \frac{(-1)^{2n-1}}{n^2}$
$a_6 = \frac{(-1)^{11}}{(6)^2} = -\frac{1}{36}$
The sixth term is $-\frac{1}{36}$.

Objective B Exercises

29. $\sum_{n=1}^{5}(2n+3) = (2\cdot1+3)+(2\cdot2+3)+(2\cdot3+3)+(2\cdot4+3)+(2\cdot5+3)$
$= 5+7+9+11+13$
$= 45$

31. $\sum_{i=1}^{4} 2i = 2(1)+2(2)+2(3)+2(4) = 2+4+6+8 = 20$

33. $\sum_{i=1}^{6} i^2 = 1^2+2^2+3^2+4^2+5^2+6^2 = 1+4+9+16+25+36 = 91$

35. $\sum_{n=1}^{6}(-1)^n = (-1)^1+(-1)^2+(-1)^3+(-1)^4+(-1)^5+(-1)^6$
$= -1+1-1+1-1+1 = 0$

37. $\sum_{i=3}^{6} i^3 = 3^3+4^3+5^3+6^3 = 27+64+125+216 = 432$

39. $\sum_{n=3}^{5}\frac{(-1)^{n-1}}{n-2} = \frac{(-1)^2}{3-2}+\frac{(-1)^3}{4-2}+\frac{(-1)^4}{5-2} = 1-\frac{1}{2}+\frac{1}{3} = \frac{5}{6}$

41. $\sum_{n=1}^{5} 2x^n = 2x+2x^2+2x^3+2x^4+2x^5$

43. $\sum_{i=1}^{5}\frac{x^i}{i} = x+\frac{x^2}{2}+\frac{x^3}{3}+\frac{x^4}{4}+\frac{x^5}{5}$

45. $\sum_{n=1}^{5} x^{2n} = x^{2(1)}+x^{2(2)}+x^{2(3)}+x^{2(4)}+x^{2(5)} = x^2+x^4+x^6+x^8+x^{10}$

47. $\sum_{i=1}^{4}\frac{x^i}{i^2} = \frac{x^1}{1^2}+\frac{x^2}{2^2}+\frac{x^3}{3^2}+\frac{x^4}{4^2} = x+\frac{x^2}{4}+\frac{x^3}{9}+\frac{x^4}{16}$

49. $\sum_{i=1}^{5} x^{-i} = x^{-1}+x^{-2}+x^{-3}+x^{-4}+x^{-5} = \frac{1}{x}+\frac{1}{x^2}+\frac{1}{x^3}+\frac{1}{x^4}+\frac{1}{x^5}$

Applying the Concepts

51. $\frac{1}{1}+\frac{1}{2}+\frac{1}{3}+\cdots+\frac{1}{n}=\sum_{i=1}^{n}\frac{1}{i}$

53. Sigma notation was first used to denote summation in 1755, in a differential calculus book by the famous mathematician Leonhard Euler. Another famous mathematician, Lagrange, also used sigma notation in the 18th century, but it was not widely utilized until the 19th century when mathematicians such as Fourier, Cauchy, and Jacobi used sigma notation to denote sums and infinite series.

Section 12.2

Objective A Exercises

1. $d = a_2 - a_1 = 11 - 1 = 10$
$a_n = a_1 + (n-1)d$
$a_{15} = 1 + (15-1)(10)$
$= 1 + 14(10) = 1 + 140$
$a_{15} = 141$

3. $d = a_2 - a_1 = -2 - (-6) = 4$
$a_n = a_1 + (n-1)d$
$a_{15} = -6 + (15-1)4$
$= -6 + 14(4) = -6 + 56$
$a_{15} = 50$

5. $d = a_2 - a_1 = \frac{5}{2} - 2 = \frac{1}{2}$
$a_n = a_1 + (n-1)d$
$a_{31} = 2 + (31-1)\frac{1}{2} = 2 + 30\left(\frac{1}{2}\right) = 2 + 15$
$a_{31} = 17$

7. $d = a_2 - a_1 = -\frac{5}{2} - (-4) = \frac{3}{2}$
$a_n = a_1 + (n-1)d$
$a_{12} = -4 + (12-1)\frac{3}{2} = -4 + 11\left(\frac{3}{2}\right) = -4 + \frac{33}{2}$
$a_{12} = \frac{25}{2} = 12\frac{1}{2}$

9. $d = a_2 - a_1 = 5 - 8 = -3$
$a_n = a_1 + (n-1)d$
$a_{40} = 8 + (40-1)(-3) = 8 + 39(-3) = 8 - 117$
$a_{40} = -109$

11. $d = a_2 - a_1 = 4 - 1 = 3$
$a_n = a_1 + (n-1)d$
$a_n = 1 + (n-1)3$
$a_n = 1 + 3n - 3$
$a_n = 3n - 2$

13. $d = a_2 - a_1 = 0 - 3 = -3$
$a_n = a_1 + (n-1)d$
$a_n = 3 + (n-1)(-3)$
$a_n = 3 - 3n + 3$
$a_n = -3n + 6$

15. $d = a_2 - a_1 = 4.5 - 7 = -2.5$
$a_n = a_1 + (n-1)d$
$a_n = 7 + (n-1)(-2.5)$
$a_n = 7 - 2.5n + 2.5$
$a_n = -2.5n + 9.5$

17. $d = a_2 - a_1 = 8 - 3 = 5$
$a_n = a_1 + (n-1)d$
$98 = 3 + (n-1)5$
$98 = 3 + 5n - 5$
$98 = 5n - 2$
$100 = 5n$
$20 = n$
There are 20 terms in the sequence.

19. $d = a_2 - a_1 = -3 - 1 = -4$
$a_n = a_1 + (n-1)d$
$-75 = 1 + (n-1)(-4)$
$-75 = 1 - 4n + 4$
$-75 = 5 - 4n$
$-80 = -4n$
$20 = n$
There are 20 terms in the sequence.

21. $d = a_2 - a_1 = \frac{13}{3} - \frac{7}{3} = 2$
$a_n = a_1 + (n-1)d$
$\frac{79}{3} = \frac{7}{3} + (n-1)2$
$\frac{79}{3} = \frac{7}{3} + 2n - 2$
$\frac{79}{3} = 2n + \frac{1}{3}$
$26 = 2n$
$13 = n$
There are 13 terms in the sequence.

23. $d = a_2 - a_1 = 2 - 3.5 = -1.5$
$a_n = a_1 + (n-1)d$
$-25 = 3.5 + (n-1)(-1.5)$
$-25 = 3.5 - 1.5n + 1.5$
$-25 = 5 - 1.5n$
$-30 = -1.5n$
$20 = n$
There are 20 terms in the sequence.

Objective B Exercises

25. $d = a_2 - a_1 = 3 - 1 = 2$
$a_n = a_1 + (n-1)d$
$a_n = 1 + (50-1)2 = 1 + 49(2) = 99$
$S_n = \frac{n}{2}(a_1 + a_2)$
$S_n = \frac{50}{2}(1 + 99) = 25(100) = 2500$

27. $d = a_2 - a_1 = 18 - 20 = -2$
$a_n = a_1 + (n-1)d$
$a_n = 20 + (40-1)(-2) = 20 + 39(-2)$
$= 20 - 78 = -58$
$S_n = \frac{n}{2}(a_1 + a_n)$
$S_n = \frac{40}{2}[20 + (-58)] = \frac{40}{2}(-38) = -760$

29. $d = a_2 - a_1 = 1 - \frac{1}{2} = \frac{1}{2}$
$a_n = a_1 + (n-1)d$
$a_n = \frac{1}{2} + (27-1)\frac{1}{2} = \frac{1}{2} + (26)\frac{1}{2} = \frac{27}{2}$
$S_n = \frac{n}{2}(a_1 + a_n)$
$S_n = \frac{27}{2}\left(\frac{1}{2} + \frac{27}{2}\right) = \frac{27}{2}(14) = 189$

31. $a_i = 3i - 1$
$a_1 = 3(1) - 1 = 2$
$a_{15} = 3(15) - 1 = 44$
$S_i = \frac{i}{2}(a_1 + a_i)$
$S_i = \frac{15}{2}(2 + 44) = \frac{15}{2}(46) = 345$

33. $a_n = \frac{1}{2}n + 1$
$a_1 = \frac{1}{2}(1) + 1 = \frac{3}{2}$
$a_{17} = \frac{1}{2}(17) + 1 = \frac{19}{2}$
$S_n = \frac{n}{2}(a_1 + a_n)$
$S_n = \frac{17}{2}\left(\frac{3}{2} + \frac{19}{2}\right) = \frac{17}{2}(11) = \frac{187}{2}$

35. $a_i = 4 - 2i$
$a_1 = 4 - 2(1) = 2$
$a_{15} = 4 - 2(15) = -26$
$S_i = \frac{i}{2}(a_1 + a_i)$
$S_i = \frac{15}{2}(2 - 26) = \frac{15}{2}(-24) = -180$

Objective C Exercises

37. Strategy To find the number of weeks:
- Write the arithmetic sequence.
- Find the common difference of the arithmetic sequence.
- Use the Formula for the *n*th Term of an Arithmetic Sequence to find the number of terms in the sequence.

Solution 10, 15, 20, ..., 60
$d = a_2 - a_1 = 15 - 10 = 5$
$a_n = a_1 + (n-1)d$
$60 = 10 + (n-1)5$
$60 = 10 + 5n - 5$
$60 = 5 + 5n$
$55 = 5n$
$11 = n$
In 11 weeks, the person will walk 60 min per day.

39. Strategy To find the distance the object will fall during the fifth second:
- Write the common difference of the arithmetic sequence.
- Find the common difference of the arithmetic sequence.
- Use the Formula for the *n*th Term of an Arithmetic Sequence to find the 5th term.

Solution 16, 48, 80, ...
$d = a_2 - a_1 = 48 - 16 = 32$
$a_n = a_1 + (n-1)d$
$a_5 = 16 + (5-1)32$
$a_5 = 16 + 4(32)$
$a_5 = 16 + 128 = 144$
The object will drop 144 ft during the 5th second.

41. Strategy To find the monthly salary during the eighth month and to find the total salary for the eight-month period:

- Write the arithmetic sequence.
- Find the common difference of the arithmetic sequence.
- Use the Formula for the nth term of an Arithmetic Sequence to find the salary during the eighth month.
- Use the Formula for the Sum of n Terms of an Arithmetic Sequence to find the total salary for the eight months.

Solution 1500, 1800, 2100, ...

$$d = a_1 - a_2 = 1800 - 1500 = 300$$
$$a_8 = a_1 + (n-1)d$$
$$a_8 = 1500 + (8-1)(300)$$
$$a_8 = 1500 + 7(300) = 1500 + 2100$$
$$a_8 = 3600$$
$$S_n = \frac{n}{2}(a_1 + a_n)$$
$$S_8 = \frac{8}{2}(1500 + 3600)$$
$$S_8 = 4(5100) = 20{,}400$$

The salary for the eighth month is \$3600. The total salary for eight months is \$20,400.

Applying the Concepts

43. $\sum_{i=1}^{2} \log 2i = \log 2(1) + \log 2(2)$

$$= \log 2 + \log 4$$
$$= \log(2 \cdot 4)$$
$$= \log 8$$

Section 12.3

Objective A Exercises

1. An arithmetic sequence is one in which the *difference* between any two consecutive terms is a constant. A geometric sequence is one in which each successive term of the sequence is the same *nonzero constant multiple* of the preceding term.

3. $r = \frac{a_2}{a_1} = \frac{8}{2} = 4$

$$a_n = a_1 r^{n-1}$$
$$a_9 = 2(4)^{9-1}$$
$$= 2(4)^8$$
$$= 2(65{,}536) = 131{,}072$$

5. $r = \frac{a_2}{a_1} = \frac{-4}{6} = -\frac{2}{3}$

$$a_n = a_1 r^{n-1}$$
$$a_7 = 6\left(-\frac{2}{3}\right)^{7-1}$$
$$= 6\left(-\frac{2}{3}\right)^6$$
$$= 6\left(\frac{64}{729}\right) = \frac{128}{243}$$

7. $r = \frac{a_2}{a_1} = \frac{\frac{1}{8}}{-\frac{1}{16}} = -2$

$$a_n = a_1 r^{n-1}$$
$$a_{10} = -\frac{1}{16}(-2)^{10-1}$$
$$= -\frac{1}{16}(-2)^9$$
$$= -\frac{1}{16}(-512) = 32$$

9. $a_n = a_1 r^{n-1}$

$$a_4 = 9r^{4-1}$$
$$\frac{8}{3} = 9r^{4-1}$$
$$\frac{8}{3} = 9r^3$$
$$\frac{8}{27} = r^3$$
$$\frac{2}{3} = r$$
$$a_n = a_1 r^{n-1}$$
$$a_2 = 9\left(\frac{2}{3}\right)^{2-1} = 9\left(\frac{2}{3}\right) = 6$$
$$a_3 = 9\left(\frac{2}{3}\right)^{3-1} = 9\left(\frac{2}{3}\right)^2 = 9\left(\frac{4}{9}\right) = 4$$

11. $a_n = a_1 r^{n-1}$

$$a_4 = 3r^{4-1}$$
$$-\frac{8}{9} = 3r^{4-1}$$
$$-\frac{8}{9} = 3r^3$$
$$-\frac{8}{27} = r^3$$
$$-\frac{2}{3} = r$$
$$a_n = a_1 r^{n-1}$$
$$a_2 = 3\left(-\frac{2}{3}\right)^{2-1} = 3\left(-\frac{2}{3}\right) = -2$$
$$a_3 = 3\left(-\frac{2}{3}\right)^{3-1} = 3\left(-\frac{2}{3}\right)^2 = 3\left(\frac{4}{9}\right) = \frac{4}{3}$$

13. $a_n = a_1 r^{n-1}$
$a_4 = (-3)r^{4-1}$
$192 = (-3)r^{4-1}$
$192 = (-3)r^3$
$-64 = r^3$
$-4 = r$
$a_n = a_1 r^{n-1}$
$a_2 = -3(-4)^{2-1} = -3(-4) = 12$
$a_3 = -3(-4)^{3-1} = -3(-4)^2 = -3(16) = -48$

Objective B Exercises

15. $r = \frac{a_2}{a_1} = \frac{6}{2} = 3$

$S_n = \frac{a_1(1-r^n)}{1-r}$

$S_7 = \frac{2(1-3^7)}{1-3} = \frac{2(1-2187)}{-2}$

$= \frac{2(-2186)}{-2} = 2186$

17. $r = \frac{a_2}{a_1} = \frac{9}{12} = \frac{3}{4}$

$S_n = \frac{a_1(1-r^n)}{1-r}$

$S_5 = \frac{12\left[1-\left(\frac{3}{4}\right)^5\right]}{1-\frac{3}{4}}$

$= \frac{12\left(1-\frac{243}{1024}\right)}{\frac{1}{4}}$

$= 48\left(\frac{781}{1024}\right) = \frac{2343}{64}$

19. $a_n = (2)^i$
$a_1 = (2)^1 = 2$
$a_2 = (2)^2 = 4$
$r = \frac{a_2}{a_1} = \frac{4}{2} = 2$

$S_n = \frac{a_1(1-r^n)}{1-r}$

$S_5 = \frac{2(1-2^5)}{1-2} = \frac{2(1-32)}{-1}$

$= -2(-31) = 62$

21. $a_n = \left(\frac{1}{3}\right)^i$

$a_1 = \left(\frac{1}{3}\right)^1 = \frac{1}{3}$

$a_2 = \left(\frac{1}{3}\right)^2 = \frac{1}{9}$

$r = \frac{a_2}{a_1} = \frac{1}{9} \div \frac{1}{3} = \frac{1}{9} \cdot \frac{3}{1} = \frac{1}{3}$

$S_n = \frac{a_1(1-r^n)}{1-r}$

$S_5 = \frac{\frac{1}{3}\left[1-\left(\frac{1}{3}\right)^5\right]}{1-\frac{1}{3}} = \frac{\frac{1}{3}\left(1-\frac{1}{243}\right)}{\frac{2}{3}}$

$= \frac{\frac{1}{3}\left(\frac{242}{243}\right)}{\frac{2}{3}} = \frac{\frac{242}{729}}{\frac{2}{3}} = \frac{121}{243}$

Objective B Exercises

23. $r = \frac{a_2}{a_1} = \frac{2}{3}$

$S = \frac{a_1}{1-r} = \frac{3}{1-\frac{2}{3}} = \frac{3}{\frac{1}{3}} = 9$

25. $r = \frac{a_2}{a_1} = \frac{\frac{7}{100}}{\frac{7}{10}} = \frac{1}{10}$

$S = \frac{a_1}{1-r} = \frac{\frac{7}{10}}{1-\frac{1}{10}} = \frac{\frac{7}{10}}{\frac{9}{10}} = \frac{7}{9}$

27. $0.8\overline{88} = 0.8 + 0.08 + 0.008 + \cdots$

$= \frac{8}{10} + \frac{8}{100} + \frac{8}{1000} + \cdots$

$S = \frac{a_1}{1-r} = \frac{\frac{8}{10}}{1-\frac{1}{10}} = \frac{\frac{8}{10}}{\frac{9}{10}} = \frac{8}{9}$

An equivalent fraction is $\frac{8}{9}$.

29. $0.2\overline{22} = 0.2 + 0.02 + 0.002 + \cdots$

$= \frac{2}{10} + \frac{2}{100} + \frac{2}{1000} + \cdots$

$S = \frac{a_1}{1-r} = \frac{\frac{2}{10}}{1-\frac{1}{10}} = \frac{\frac{2}{10}}{\frac{9}{10}} = \frac{2}{9}$

An equivalent fraction is $\frac{2}{9}$.

31. $0.4\overline{545} = 0.45 + 0.0045 + 0.000045 + \cdots$

$$= \frac{45}{100} + \frac{45}{10{,}000} + \frac{45}{1{,}000{,}000} + \cdots$$

$$S = \frac{a_1}{1-r} = \frac{\frac{45}{100}}{1 - \frac{1}{100}} = \frac{\frac{45}{100}}{\frac{99}{100}} = \frac{45}{99} = \frac{5}{11}$$

An equivalent fraction is $\frac{5}{11}$.

33. $0.16\overline{66} = 0.1 + 0.06 + 0.006 + 0.0006 + \cdots$

$$= \frac{1}{10} + \frac{6}{100} + \frac{6}{1000} + \frac{6}{10{,}000} + \cdots$$

$$S = \frac{a_1}{1-r} = \frac{\frac{6}{100}}{1 - \frac{1}{10}} = \frac{\frac{6}{100}}{\frac{9}{10}} = \frac{6}{90} = \frac{1}{15}$$

$$0.16\overline{66} = \frac{1}{10} + \frac{1}{15} = \frac{5}{30} = \frac{1}{6}$$

An equivalent fraction is $\frac{1}{6}$.

Objective D Exercises

35. **Strategy** To find the amount of radioactive material at the beginning of the fourth hour, use the Formula for the nth Term of a Geometric Sequence.

Solution $n = 4$, $a_1 = 400$, $r = \frac{1}{2}$

$$a_n = a_1 r^{n-1}$$

$$a_4 = 400\left(\frac{1}{2}\right)^{4-1}$$

$$= 400\left(\frac{1}{2}\right)^3 = 400\left(\frac{1}{8}\right) = 50$$

There will be 50 mg of radioactive material in the sample at the beginning of the fourth hour.

37. **Strategy** To find the height of the ball on the sixth bounce, use the Formula for the nth Term of a Geometric Sequence.

Solution $n = 6$, $a_1 = 75\%$ of $10 = 7.5$, $r = 75\% = \frac{3}{4}$

$$a_n = a_1 r^{n-1}$$

$$a_6 = 7.5\left(\frac{3}{4}\right)^{6-1}$$

$$= 7.5\left(\frac{3}{4}\right)^5 = 7.5\left(\frac{243}{1024}\right) \approx 1.8$$

The ball bounces to a height of 1.8 ft on the sixth bounce.

39. **Strategy** To find the total amount of money earned in 30 days, use the Formula for the Sum of n Terms of a Finite Geometric Series.

Solution $n = 30$, $a_1 = 1$, $r = 2$

$$S_n = \frac{a_1(1 - r^n)}{1 - r}$$

$$S_{30} = \frac{1(1 - 2^{30})}{1 - 2}$$

$$= \frac{1(1 - 1{,}073{,}741{,}824)}{-1}$$

$$= \frac{-1{,}073{,}741{,}823}{-1} = 1{,}073{,}741{,}823$$

The total amount earned in 30 days is 1,073,741,823 cents, or \$10,737,418.23.

Applying the Concepts

41. a_1, a_2, 3 a_4, a_5, $\frac{1}{9}$

Use the two known terms in the Formula for the nth Term of a Geometric Sequence, and solve for r.

$$a_3 = a_1 r^{3-1} \qquad a_6 = a_1 r^{6-1}$$

$$3 = a_1 r^2 \qquad (2)\ \frac{1}{9} = a_1 r^5$$

$$a_1 = \frac{3}{r^2}$$

$$(1)\ a_1 = 3r^{-2}$$

Substitute Equation (1) into Equation (2).

$$\frac{1}{9} = a_1 r^5$$

$$\frac{1}{9} = (3r^{-2})r^5$$

$$\frac{1}{9} = 3r^3$$

$$\frac{1}{27} = r^3$$

$$\frac{1}{3} = r$$

Substitute r into Equation (1) to find a_1.

$$a_1 = 3r^{-2}$$

$$a_1 = 3\left(\frac{1}{3}\right)^{-2}$$

$$a_1 = 27$$

The first term is 27.

Section 12.4

Objective A Exercises

1. $3! = 3 \cdot 2 \cdot 1 = 6$

3. $8! = 8 \cdot 7 \cdot 6 \cdot 5 \cdot 4 \cdot 3 \cdot 2 \cdot 1 = 40{,}320$

5. $0! = 1$

7. $\frac{5!}{2!3!} = \frac{5 \cdot 4 \cdot 3 \cdot 2 \cdot 1}{(2 \cdot 1)(3 \cdot 2 \cdot 1)} = 10$

9. $\frac{6!}{6!0!} = \frac{6 \cdot 5 \cdot 4 \cdot 3 \cdot 2 \cdot 1}{(6 \cdot 5 \cdot 4 \cdot 3 \cdot 2 \cdot 1)(1)} = 1$

11. $\frac{9!}{6!3!} = \frac{9 \cdot 8 \cdot 7 \cdot 6 \cdot 5 \cdot 4 \cdot 3 \cdot 2 \cdot 1}{(6 \cdot 5 \cdot 4 \cdot 3 \cdot 2 \cdot 1)(3 \cdot 2 \cdot 1)} = 84$

13. $\binom{7}{2} = \frac{7!}{(7-2)!2!} = \frac{7!}{5!2!} = \frac{7 \cdot 6 \cdot 5 \cdot 4 \cdot 3 \cdot 2 \cdot 1}{(5 \cdot 4 \cdot 3 \cdot 2 \cdot 1)(2 \cdot 1)} = 21$

15. $\binom{10}{2} = \frac{10!}{(10-2)!2!} = \frac{10!}{8!2!} = \frac{10 \cdot 9 \cdot 8 \cdot 7 \cdot 6 \cdot 5 \cdot 4 \cdot 3 \cdot 2 \cdot 1}{(8 \cdot 7 \cdot 6 \cdot 5 \cdot 4 \cdot 3 \cdot 2 \cdot 1)(2 \cdot 1)} = 45$

17. $\binom{9}{0} = \frac{9!}{(9-0)!0!} = \frac{9!}{9!0!} = \frac{9 \cdot 8 \cdot 7 \cdot 6 \cdot 5 \cdot 4 \cdot 3 \cdot 2 \cdot 1}{(9 \cdot 8 \cdot 7 \cdot 6 \cdot 5 \cdot 4 \cdot 3 \cdot 2 \cdot 1)(1)} = 1$

19. $\binom{6}{3} = \frac{6!}{(6-3)!3!} = \frac{6!}{3!3!} = \frac{6 \cdot 5 \cdot 4 \cdot 3 \cdot 2 \cdot 1}{(3 \cdot 2 \cdot 1)(3 \cdot 2 \cdot 1)} = 20$

21. $\binom{11}{1} = \frac{11!}{(11-1)!1!} = \frac{11!}{10!1!} = \frac{11 \cdot 10 \cdot 9 \cdot 8 \cdot 7 \cdot 6 \cdot 5 \cdot 4 \cdot 3 \cdot 2 \cdot 1}{(10 \cdot 9 \cdot 8 \cdot 7 \cdot 6 \cdot 5 \cdot 4 \cdot 3 \cdot 2 \cdot 1)(1)} = 11$

23. $\binom{4}{2} = \frac{4!}{(4-2)!2!} = \frac{4!}{2!2!} = \frac{4 \cdot 3 \cdot 2 \cdot 1}{(2 \cdot 1)(2 \cdot 1)} = 6$

25. $$\begin{aligned}(x+y)^4 &= \binom{4}{0}x^4 + \binom{4}{1}x^3y + \binom{4}{2}x^2y^2 + \binom{4}{3}xy^3 + \binom{4}{4}y^4 \\ &= x^4 + 4x^3y + 6x^2y^2 + 4xy^3 + y^4\end{aligned}$$

27. $$\begin{aligned}(x-y)^5 &= \binom{5}{0}x^5 + \binom{5}{1}x^4(-y) + \binom{5}{2}x^3(-y)^2 + \binom{5}{3}x^2(-y)^3 + \binom{5}{4}x(-y)^4 + \binom{5}{5}(-y)^5 \\ &= x^5 - 5x^4y + 10x^3y^2 - 10x^2y^3 + 5xy^4 - y^5\end{aligned}$$

29. $$\begin{aligned}(2m+1)^4 &= \binom{4}{0}(2m)^4 + \binom{4}{1}(2m)^3(1) + \binom{4}{2}(2m)^2(1)^2 + \binom{4}{3}(2m)(1)^3 + \binom{4}{4}(1)^4 \\ &= 1(16m^4) + 4(8m^3) + 6(4m^2) + 4(2m) + 1(1) \\ &= 16m^4 + 32m^3 + 24m^2 + 8m + 1\end{aligned}$$

31. $$\begin{aligned}(2r-3)^5 &= \binom{5}{0}(2r)^5 + \binom{5}{1}(2r)^4(-3) + \binom{5}{2}(2r)^3(-3)^2 + \binom{5}{3}(2r)^2(-3)^3 + \binom{5}{4}(2r)(-3)^4 + \binom{5}{5}(-3)^5 \\ &= 1(32r^5) + 5(16r^4)(-3) + 10(8r^3)(9) + 10(4r^2)(-27) + 5(2r)(81) + 1(-243) \\ &= 32r^5 - 240r^4 + 720r^3 - 1080r^2 + 810r - 243\end{aligned}$$

33. $$\begin{aligned}(a+b)^{10} &= \binom{10}{0}a^{10} + \binom{10}{1}a^9b + \binom{10}{2}a^8b^2 + \cdots \\ &= a^{10} + 10a^9b + 45a^8b^2 + \cdots\end{aligned}$$

35. $$\begin{aligned}(a-b)^{11} &= \binom{11}{0}a^{11} + \binom{11}{1}a^{10}(-b) + \binom{11}{2}a^9(-b)^2 + \cdots \\ &= (1)a^{11} + 11a^{10}(-b) + 55a^9b^2 + \cdots \\ &= a^{11} - 11a^{10}b + 55a^9b^2 + \cdots\end{aligned}$$

37. $$\begin{aligned}(2x+y)^8 &= \binom{8}{0}(2x)^8 + \binom{8}{1}(2x)^7y + \binom{8}{2}(2x)^6y^2 + \cdots \\ &= 1(256x^8) + 8(128x^7)y + 28(64x^6)y^2 + \cdots \\ &= 256x^8 + 1024x^7y + 1792x^6y^2 + \cdots\end{aligned}$$

39. $(4x-3y)^8 = \binom{8}{0}(4x)^8 + \binom{8}{1}(4x)^7(-3y) + \binom{8}{2}(4x)^6(-3y)^2 + \cdots$
$= 1(65{,}536x^8) + 8(16{,}384x^7)(-3y) + 28(4096x^6)(9y^2) + \cdots$
$= 65{,}536x^8 - 393{,}216x^7y + 1{,}032{,}192x^6y^2 + \cdots$

41. $\left(x+\frac{1}{x}\right)^7 = \binom{7}{0}x^7 + \binom{7}{1}x^6\left(\frac{1}{x}\right) + \binom{7}{2}x^5\left(\frac{1}{x}\right)^2 + \cdots$
$= 1(x^7) + 7x^6\left(\frac{1}{x}\right) + 21x^5\left(\frac{1}{x^2}\right) + \cdots$
$= x^7 + 7x^5 + 21x^3 + \cdots$

43. $n = 7$, $a = 2x$, $b = -1$, $r = 4$
$\binom{7}{4-1}(2x)^{7-4+1}(-1)^{4-1} = \binom{7}{3}(2x)^4(-1)^3 = 35(16x^4)(-1) = -560x^4$

45. $n = 6$, $a = x^2$, $b = (-y^2)$, $r = 2$
$\binom{6}{2-1}(x^2)^{6-2+1}(-y^2)^{2-1} = \binom{6}{1}(x^2)^5(-y^2) = 6x^{10}(-y^2) = -6x^{10}y^2$

47. $n = 9$, $a = y$, $b = -1$, $r = 5$
$\binom{9}{5-1}y^{9-5+1}(-1)^{5-1} = \binom{9}{4}y^5(-1)^4 = 126y^5(1) = 126y^5$

49. $n = 5$, $a = n$, $b = \frac{1}{n}$, $r = 2$
$\binom{5}{2-1}n^{5-2+1}\left(\frac{1}{n}\right)^{2-1} = \binom{5}{1}n^4\left(\frac{1}{n}\right) = 5n^4\left(\frac{1}{n}\right) = 5n^3$

Applying the Concepts

51. **a.** False; $0! \cdot 4! = 1 \cdot 4! = 24$

b. False; $\frac{4!}{0!} = \frac{4!}{1} = 24$

c. False.

d. False.

e. False.

f. True.

Chapter 12 Review Exercises

1. $\sum_{i=1}^{4} 2x^{i-1} = 2x^{1-1} + 2x^{2-1} + 2x^{3-1} + 2x^{4-1}$
$= 2 + 2x + 2x^2 + 2x^3$

2. $a_n = 3n - 2$
$a_{10} = 3(10) - 2 = 30 - 2 = 28$

The tenth term is 28.

3. $0.63\overline{3} = 0.6 + 0.03 + 0.003 + \cdots$

$$= \frac{6}{10} + \frac{3}{100} + \frac{3}{1000} + \cdots$$

$$S = \frac{a_1}{1-r} = \frac{\frac{3}{100}}{1-\frac{1}{10}} = \frac{\frac{3}{100}}{\frac{9}{10}} = \frac{3}{100} \cdot \frac{10}{9} = \frac{1}{30}$$

$$0.63\overline{3} = \frac{6}{10} + \frac{1}{30} = \frac{19}{30}$$

An equivalent fraction is $\frac{19}{30}$.

4. $r = \frac{a_2}{a_1} = \frac{\frac{3}{100}}{\frac{3}{10}} = \frac{1}{10}$

$$S = \frac{a_1}{1-r} = \frac{\frac{3}{10}}{1-\frac{1}{10}} = \frac{\frac{3}{10}}{\frac{9}{10}} = \frac{1}{3}$$

5. $\sum_{i=1}^{30} 4i - 1 = 3, 7, 11, \cdots, 119$

$$S_n = \frac{30}{2}(3 + 119) = 15(122) = 1830$$

6. $\sum_{n=1}^{5}(3n-2) = [3(1)-2] + [3(2)-2]$

$$+ [3(3)-2] + [3(4)-2] + [3(5)-2]$$
$$= 1 + 4 + 7 + 10 + 13$$
$$= 35$$

7. $0.2\overline{323} = 0.23 + 0.0023 + 0.000023 + \cdots$

$$= \frac{23}{100} + \frac{23}{10{,}000} + \frac{23}{1{,}000{,}000} + \cdots$$

$$S = \frac{a_1}{1-r} = \frac{\frac{23}{100}}{1-\frac{1}{100}} = \frac{\frac{23}{100}}{\frac{99}{100}} = \frac{23}{99}$$

An equivalent fraction is $\frac{23}{99}$.

8. $\binom{9}{3} = \frac{9!}{(9-3)!3!}$

$$= \frac{9!}{6!3!}$$
$$= \frac{9 \cdot 8 \cdot 7 \cdot 6 \cdot 5 \cdot 4 \cdot 3 \cdot 2 \cdot 1}{(6 \cdot 5 \cdot 4 \cdot 3 \cdot 2 \cdot 1)(3 \cdot 2 \cdot 1)} = 84$$

9. $\sum_{n=1}^{5} 2(3)^n = 6, 18, 54, \cdots$

$$r = \frac{a_2}{a_1} = \frac{18}{6} = 3$$

$$S_n = \frac{a_1(1-r^n)}{1-r}$$
$$= \frac{6(1-3^5)}{1-3}$$
$$= \frac{6(1-243)}{-2} = -3(-242) = 726$$

10.
$$d = a_2 - a_1 = 2 - 8 = -6$$
$$a_n = a_1 + (n-1)d$$
$$-118 = 8 + (n-1)(-6)$$
$$-126 = -6(n-1)$$
$$21 = n - 1$$
$$22 = n$$

There are 22 terms in the sequence.

11. $\sum_{n=1}^{8}\left(\frac{1}{2}\right)^n = \frac{1}{2}, \frac{1}{4}, \frac{1}{8}, \cdots$

$$r = \frac{a_2}{a_1} = \frac{\frac{1}{4}}{\frac{1}{2}} = \frac{1}{4} \cdot \frac{2}{1} = \frac{1}{2}$$

$$S_n = \frac{a_1(1-r^n)}{1-r}$$
$$= \frac{\frac{1}{2}\left(1-\left(\frac{1}{2}\right)^8\right)}{1-\frac{1}{2}}$$
$$= \frac{\frac{1}{2}\left(1-\frac{1}{256}\right)}{\frac{1}{2}}$$
$$= \frac{255}{256} \approx 0.996$$

12. $n = 8,\ a = 3x,\ b = -y,\ r = 5$

$$\binom{8}{5-1}(3x)^{8-5+1}(-y)^{5-1} = \binom{8}{4}(3x)^4(-y)^4$$
$$= 70(81x^4)y^4$$
$$= 5670x^4y^4$$

13. $a_n = \frac{(-1)^{2n-1}n}{n^2+2}$

$$a_5 = \frac{(-1)^{2(5)-1} \cdot 5}{(5)^2+2} = \frac{(-1)^9 \cdot 5}{25+2} = \frac{-5}{27}$$

The fifth term is $-\frac{5}{27}$.

14. $r = \frac{a_2}{a_1} = \frac{5\sqrt{5}}{5} = \sqrt{5}$

$$S_n = \frac{a_1(1-r^n)}{1-r}$$
$$S_7 = \frac{5(1-(\sqrt{5})^7)}{1-\sqrt{5}} = \frac{5(1-125\sqrt{5})}{1-\sqrt{5}}$$
$$= \frac{5-625\sqrt{5}}{1-\sqrt{5}}$$
$$= \frac{5-625\sqrt{5}}{1-\sqrt{5}} \cdot \frac{1+\sqrt{5}}{1+\sqrt{5}}$$
$$= \frac{5+5\sqrt{5}-625\sqrt{5}-3125}{1-5}$$
$$= \frac{-620\sqrt{5}-3120}{-4} \approx 1127$$

15. $d = a_2 - a_1 = -2 - (-7) = -2 + 7 = 5$
$a_n = a_1 + (n-1)d$
$a_n = -7 + (n-1)5 = -7 + 5n - 5 = 5n - 12$
$a_n = 5n - 12$

16. $n = 11, a = x, b = -2y, r = 8$

$$\binom{11}{8-1}(x)^{11-8+1}(-2y)^{8-1} = \binom{11}{7}x^4(-2y)^7 = 330x^4(-128y^7) = -42{,}240x^4y^7$$

17. $r = \frac{a_2}{a_1} = \frac{\sqrt{3}}{1} = \sqrt{3}$
$a_n = a_1 r^{n-1}$
$a_{12} = 1 \cdot (\sqrt{3})^{12-1} = (\sqrt{3})^{11} = 243\sqrt{3}$

18. $d = a_2 - a_1 = 13 - 11 = 2$
$a_n = a_1 + (n-1)d$
$a_{40} = 11 + 2(40-1) = 11 + 2(39) = 89$
$S_n = \frac{n}{2}(a_1 + a_n)$
$S_{40} = \frac{40}{2}(11 + 89) = 20(100) = 2000$

19.

$$\binom{12}{9} = \frac{12!}{(12-9)!9!} = \frac{12!}{3!9!} = \frac{12 \cdot 11 \cdot 10 \cdot 9 \cdot 8 \cdot 7 \cdot 6 \cdot 5 \cdot 4 \cdot 3 \cdot 2 \cdot 1}{(3 \cdot 2 \cdot 1)(9 \cdot 8 \cdot 7 \cdot 6 \cdot 5 \cdot 4 \cdot 3 \cdot 2 \cdot 1)} = 220$$

20.

$$\sum_{n=1}^{4} \frac{(-1)^{n-1}n}{n+1} = \frac{(-1)^{1-1} \cdot 1}{1+1} + \frac{(-1)^{2-1} \cdot 2}{2+1} + \frac{(-1)^{3-1} \cdot 3}{3+1} + \frac{(-1)^{4-1} \cdot 4}{4+1} = \left(\frac{1}{2}\right) + \left(-\frac{2}{3}\right) + \frac{3}{4} + \left(-\frac{4}{5}\right) = -\frac{13}{60}$$

21.

$$\sum_{i=1}^{5} \frac{(2x)^i}{i} = \frac{(2x)^1}{1} + \frac{(2x)^2}{2} + \frac{(2x)^3}{3} + \frac{(2x)^4}{4} + \frac{(2x)^5}{5} = 2x + \frac{4x^2}{2} + \frac{8x^3}{3} + \frac{16x^4}{4} + \frac{32x^5}{5} = 2x + 2x^2 + \frac{8x^3}{3} + 4x^4 + \frac{32x^5}{5}$$

22. $r = \frac{a_2}{a_1} = \frac{1}{3}$
$a_n = a_1 r^{n-1}$
$a_8 = 3\left(\frac{1}{3}\right)^7 = \frac{1}{729}$

23.

$$(x - 3y^2)^5 = \binom{5}{0}x^5 + \binom{5}{1}x^4(-3y^2) + \binom{5}{2}x^3(-3y^2)^2 + \binom{5}{3}x^2(-3y^2)^3 + \binom{5}{4}x(-3y^2)^4 + \binom{5}{5}(-3y^2)^5$$
$$= x^5 - 15x^4y^2 + 90x^3y^4 - 270x^2y^6 + 405xy^8 - 243y^{10}$$

24. $r = \frac{a_2}{a_1} = -\frac{1}{4}$

$$S = \frac{a_1}{1-r} = \frac{4}{1 - \left(-\frac{1}{4}\right)} = \frac{4}{\frac{5}{4}} = \frac{16}{5}$$

25. $r = \frac{a_2}{a_1} = \frac{\frac{4}{3}}{2} = \frac{2}{3}$

$$S = \frac{a_1}{1-r} = \frac{2}{1 - \frac{2}{3}} = \frac{2}{\frac{1}{3}} = 6$$

26.

$$\frac{12!}{5!8!} = \frac{12 \cdot 11 \cdot 10 \cdot 9 \cdot 8 \cdot 7 \cdot 6 \cdot 5 \cdot 4 \cdot 3 \cdot 2 \cdot 1}{(5 \cdot 4 \cdot 3 \cdot 2 \cdot 1)(8 \cdot 7 \cdot 6 \cdot 5 \cdot 4 \cdot 3 \cdot 2 \cdot 1)} = 99$$

27. $d = a_2 - a_1 = 5 - 1 = 4$
$a_n = a_1 + (n-1)d$
$a_{20} = 1 + (20-1)4 = 1 + 19(4) = 77$
$a_{20} = 77$

28.

$$\sum_{i=1}^{6} 4x^i = 4x + 4x^2 + 4x^3 + 4x^4 + 4x^5 + 4x^6$$

29. $d = a_2 - a_1 = -7 - (-3) = -7 + 3 = -4$
$a_n = a_1 + (n-1)d$
$-59 = -3 + (n-1)(-4)$
$-59 = -3 - 4n + 4$
$-59 = 1 - 4n$
$-60 = -4n$
$15 = n$

30. $r = \frac{a_2}{a_1} = \frac{5\sqrt{3}}{5} = \sqrt{3}$
$a_n = a_1 r^{n-1}$
$a_7 = 5(\sqrt{3})^{7-1} = 5(\sqrt{3})^6 = 135$

31. $r = \frac{a_2}{a_1} = \frac{2}{3}$
$s = \frac{a_1}{1-r} = \frac{3}{1-\frac{2}{3}} = \frac{3}{\frac{1}{3}} = 9$

32. $a_n = \frac{10}{n+3}$
$a_{17} = \frac{10}{17+3} = \frac{10}{20} = \frac{1}{2}$

33. $d = a_2 - a_1 = -4 - (-10) = 6$
$a_n = a_1 + (n-1)d$
$a_n = -10 + 9 \cdot 6 = -10 + 54 = 44$

34. $d = a_2 - a_1 = -12 - (-19) = -12 + 19 = 7$
$a_n = a_1 + (n-1)d$
$a_{18} = -19 + (18-1)7 = -19 + 119 = 100$
$S_n = \frac{n}{2}(a_1 + a_n)$
$S_{18} = \frac{18}{2}(-19 + 100) = 9(81) = 729$

35. $r = \frac{a_2}{a_1} = \frac{-16}{8} = -2$
$S_n = \frac{a_1(1-r^n)}{1-r}$
$S_6 = \frac{8(1-(-2)^6)}{1-(-2)}$
$= \frac{8(1-64)}{3}$
$= \frac{8(-63)}{3}$
$= 8(-21)$
$= -168$

36. $\frac{7!}{5!2!} = \frac{7 \cdot 6 \cdot 5 \cdot 4 \cdot 3 \cdot 2 \cdot 1}{(5 \cdot 4 \cdot 3 \cdot 2 \cdot 1)(2 \cdot 1)} = 21$

37. $n = 9, a = 3x, b = y, r = 7$
$\binom{9}{7-1}(3x)^{9-7+1}y^{7-1} = \binom{9}{6}(3x)^3y^6$
$= 84(27x^3)y^6$
$= 2268x^3y^6$

38. $\sum_{n=1}^{5}(4n-3) = 1 + 5 + 9 + 13 + 17 = 45$

39. $a_n = \frac{n+2}{2n}$
$a_8 = \frac{8+2}{2(8)} = \frac{10}{16} = \frac{5}{8}$

40. $d = a_2 - a_1 = 0 - (-4) = 4$
$a_n = a_1 + (n-1)d$
$a_n = -4 + (n-1)4$
$a_n = -4 + 4n - 4$
$a_n = 4n - 8$

41. $r = \frac{-2}{10} = -\frac{1}{5}$
$a_n = a_1 r^{n-1}$
$a_5 = 10 \cdot \left(-\frac{1}{5}\right)^{5-1} = 10 \cdot \left(-\frac{1}{5}\right)^4 = 10 \cdot \frac{1}{625}$
$= \frac{10}{625}$
$= \frac{2}{125}$

42. $0.36\overline{6} = 0.3 + 0.06 + 0.006 + \cdots$
$= \frac{3}{10} + \frac{6}{100} + \frac{6}{1000} + \cdots$
$S = \frac{\frac{6}{100}}{1-\frac{1}{10}} = \frac{\frac{6}{100}}{\frac{9}{10}} = \frac{6}{90} = \frac{2}{30}$
$0.36\overline{6} = \frac{3}{10} + \frac{2}{30} = \frac{11}{30}$

43. $d = a_2 - a_1 = -32 - (-37) = -32 + 37 = 5$
$a_n = a_1 + (n-1)d$
$a_{37} = -37 + (37-1)5 = -37 + 36 \cdot 5 = -37 + 180 = 143$

44. $r = \frac{a_2}{a_1} = \frac{4}{1} = 4$
$S_n = \frac{a_1(1-r^n)}{1-r}$
$S_5 = \frac{1(1-4^5)}{1-4} = \frac{-1023}{-3} = 341$

45. Strategy To find the number of weeks until a person is walking 60 min each day, use the formula for the *n*th term of an arithmetic sequence.
$a_1 = 15$, $d = 3$, $a_n = 60$. Find n.

Solution $a_n = a_1 + (n-1)d$
$60 = 15 + (n-1)3$
$60 = 15 + 3n - 3$
$60 = 12 + 3n$
$48 = 3n$
$16 = n$
The person will be walking 60 min each day in 16 weeks.

46. Strategy To find the temperature of the spa after 8 hours, use the Formula for the *n*th Term of a Geometric Sequence.

Solution $n = 8$, $a_1 = 0.95(102) = 96.9$, $r = 0.95$

$a_n = a_1 r^{n-1}$

$a_8 = 96.9(0.95)^7 = 67.7$

The temperature is 67.7°F.

47. Strategy To find the total salary for the nine-month period:

- Write the arithmetic sequence.
- Find the common difference of the arithmetic sequence.
- Use the Formula for the *n*th Term of an Arithmetic Sequence to find the 9th term.
- Use the Formula for the Sum of *n* Terms of an Arithmetic Sequence to find the sum of 9 terms of the sequence.

Solution \$2400, \$2480, \$2560, ...

$d = a_2 - a_1 = 2480 - 2400 = 80$

$a_n = a_1 + (n-1)d$

$a_9 = 2400 + (9-1)80 = 3040$

$S_n = \frac{n}{2}(a_1 + a_n)$

$S_9 = \frac{9}{2}(2400 + 3040) = 24{,}480$

The total salary for the nine-month period is \$24,480.

48. Strategy To find the amount of radioactive material at the beginning of the seventh hour, use the Formula for the *n*th Term of a Geometric Sequence.

Solution $n = 7$, $a_1 = 200$, $r = \frac{1}{2}$

$a_n = a_1 r^{n-1}$

$a = 200\left(\frac{1}{2}\right)^{7-1}$

$= 200\left(\frac{1}{2}\right)^6 = \frac{200}{64} = 3.125$

There will be a 3.125 mg of radioactive material in the sample at the beginning of the seventh hour.

Chapter 12 Test

1. $\sum_{n=1}^{4}(3n+1) = 4 + 7 + 10 + 13 = 34$

2. $\binom{9}{6} = \frac{9!}{(9-6)!6!}$

$= \frac{9!}{3!6!}$

$= \frac{9\cdot 8\cdot 7\cdot 6\cdot 5\cdot 4\cdot 3\cdot 2\cdot 1}{(3\cdot 2\cdot 1)(6\cdot 5\cdot 4\cdot 3\cdot 2\cdot 1)} = 84$

3. $r = \frac{a_2}{a_1} = \frac{4\sqrt{2}}{4} = \sqrt{2}$

$a_n = ar^{n-1}$

$a_7 = 4(\sqrt{2})^{7-1} = 4\cdot(\sqrt{2})^6 = 32$

4. $a_n = \frac{8}{n+2}$

$a_{14} = \frac{8}{14+2} = \frac{8}{16} = \frac{1}{2}$

The fourteenth term is $\frac{1}{2}$.

5. $\sum_{i=1}^{4} 3x^i = 3x^1 + 3x^2 + 3x^3 + 3x^4$

$= 3x + 3x^2 + 3x^3 + 3x^4$

6. $d = a_2 - a_1 = -19 - (-25) = -19 + 25 = 6$

$a_n = a_1 + (n-1)d$

$a_{18} = -25 + (18-1)6 = -25 + 17(6) = 77$

$S_n = \frac{n}{2}(a_1 + a_n)$

$S_{18} = \frac{18}{2}(-25 + 77) = 468$

7. $r = \frac{a_2}{a_1} = \frac{3}{4}$

$S = \frac{a_1}{1-r} = \frac{4}{1-\frac{3}{4}} = \frac{4}{\frac{1}{4}} = 16$

8. $d = a_2 - a_1 = -8 - (-5) = -8 + 5 = -3$

$a_n = a_1 + (n-1)d$

$-50 = -5 + (n-1)(-3)$

$-45 = -3(n-1)$

$15 = n - 1$

$16 = n$

9. $0.233\overline{3} = 0.2 + 0.03 + 0.003 + \cdots$

$= \frac{2}{10} + \frac{3}{100} + \frac{3}{1000} + \cdots$

$S = \frac{a_1}{1-r} = \frac{\frac{3}{100}}{1-\frac{1}{10}} = \frac{\frac{3}{100}}{\frac{9}{10}} = \frac{3}{90} = \frac{1}{30}$

$0.233\overline{3} = \frac{2}{10} + \frac{1}{30} = \frac{7}{30}$

An equivalent fraction is $\frac{7}{30}$.

10. $\frac{8!}{4!4!} = \frac{8 \cdot 7 \cdot 6 \cdot 5 \cdot 4 \cdot 3 \cdot 2 \cdot 1}{(4 \cdot 3 \cdot 2 \cdot 1)(4 \cdot 3 \cdot 2 \cdot 1)} = 70$

11. $r = \frac{a_2}{a_1} = \frac{2}{6} = \frac{1}{3}$

$a_n = a_1 r^{n-1}$

$a_5 = 6 \cdot \left(\frac{1}{3}\right)^{5-1} = 6 \cdot \left(\frac{1}{3}\right)^4 = 6 \cdot \frac{1}{81} = \frac{2}{27}$

12. $n = 7,\ a = x,\ b = -2y,\ r = 4$

$\binom{7}{4-1} x^{7-4+1}(-2y)^{4-1} = \binom{7}{3} x^4(-2y)^3$

$= 35x^4(-8y^3)$

$= -280x^4y^3$

13. $d = a_2 - a_1 = -16 - (-13) = -16 + 13 = -3$

$a_n = a_1 + (n-1)d$

$a_{35} = -13 + (35-1)(-3) = -13 + 34(-3) = -115$

14. $d = a_2 - a_1 = 9 - 12 = -3$

$a_n = a_1 + (n-1)d$

$a_n = 12 + (n-1)(-3)$

$= 12 - 3n + 3 = -3n + 15$

$a_n = -3n + 15$

15. $r = \frac{a_2}{a_1} = \frac{12}{-6} = -2$

$S_n = \frac{a_1(1-r^n)}{1-r}$

$S_5 = \frac{-6(1-(-2)^5)}{1-(-2)}$

$= \frac{-6(1-(-32))}{3}$

$= -2(33) = -66$

16. $a_n = \frac{n+1}{n}$

$a_6 = \frac{6+1}{6} = \frac{7}{6}$

$a_7 = \frac{7+1}{7} = \frac{8}{7}$

The sixth term is $\frac{7}{6}$.

The seventh term is $\frac{8}{7}$.

17. $r = \frac{a_2}{a_1} = \frac{\frac{3}{2}}{1} = \frac{3}{2}$

$S_n = \frac{a_1(1-r^n)}{1-r}$

$S_6 = \frac{1\left(1-\left(\frac{3}{2}\right)^6\right)}{1-\frac{3}{2}} = \frac{1-\frac{729}{64}}{-\frac{1}{2}} = \frac{-\frac{665}{64}}{-\frac{1}{2}} = \frac{665}{32}$

18. $d = a_2 - a_1 = 12 - 5 = 7$

$a_n = a_1 + (n-1)d$

$a_{21} = 5 + (21-1)7 = 5 + 20(7) = 145$

$S_n = \frac{n}{2}(a_1 + a_n)$

$S_{21} = \frac{21}{2}(5 + 145) = 1575$

19. **Strategy** To find how many skeins were in stock after the shipment on October 1:

- Write the arithmetic sequence.
- Find the common difference of the arithmetic sequence.
- Use the Formula for the *n*th term of an Arithmetic Sequence to find the tenth term.

Solution 7500, 6950, 6400,...

$d = a_2 - a_1 = 6950 - 7500 = -550$

$a_n = a_1 + (n-1)d$

$a_{10} = 7500 + (10-1)(-550) = 2550$

The inventory after the October 1 shipment was 2550 skeins.

20. **Strategy** To find the amount of radioactive material at the beginning of the fifth day, use the Formula for the *n*th Term of a Geometric Sequence.

Solution $n = 5,\ a_1 = 320,\ r = \frac{1}{2}$

$a_n = a_1 r^{n-1}$

$a_5 = 320\left(\frac{1}{2}\right)^{5-1} = 320\left(\frac{1}{2}\right)^4 = 20$

There will be 20 mg of radioactive material in the sample at the beginning of the fifth day.

Cumulative Review Exercises

1. $3x - 2y = -4$

$-2y = -3x - 4$

$y = \frac{3}{2}x + 2$

2. $2x^6 + 16 = 2(x^6 + 8)$

$= 2((x^2)^3 + 2^3)$

$= 2(x^2 + 2)(x^4 - 2x^2 + 4)$

3. $\dfrac{4x^2}{x^2+x-2} - \dfrac{3x-2}{x+2}$

$= \dfrac{4x^2}{(x+2)(x-1)} - \dfrac{3x-2}{x+2} \cdot \dfrac{x-1}{x-1}$

$= \dfrac{4x^2-(3x^2-5x+2)}{(x+2)(x-1)}$

$= \dfrac{x^2+5x-2}{(x+2)(x-1)}$

4. $f(x) = 2x^2 - 3x$
$f(-2) = 2(-2)^2 - 3(-2)$
$f(-2) = 2 \cdot 4 + 6$
$f(-2) = 14$

5. $\sqrt{2y}(\sqrt{8xy} - \sqrt{y}) = \sqrt{16xy^2} - \sqrt{2y^2}$
$= \sqrt{16y^2(x)} - \sqrt{y^2(2)}$
$= 4y\sqrt{x} - y\sqrt{2}$

6. $2x^2 - x + 7 = 0$
$a = 2, b = -1, c = 7$

$x = \dfrac{-b \pm \sqrt{b^2-4ac}}{2a}$

$= \dfrac{-(-1) \pm \sqrt{(-1)^2 - 4(2)(7)}}{2(2)}$

$= \dfrac{1 \pm \sqrt{1-56}}{4}$

$= \dfrac{1 \pm \sqrt{-55}}{4}$

$= \dfrac{1}{4} \pm \dfrac{\sqrt{55}}{4} i$

The solutions are $\dfrac{1}{4} + \dfrac{\sqrt{55}}{4} i$ and $\dfrac{1}{4} - \dfrac{\sqrt{55}}{4} i$.

7. $5 - \sqrt{x} = \sqrt{x+5}$
$(5-\sqrt{x})^2 = (\sqrt{x+5})^2$
$25 - 10\sqrt{x} + x = x + 5$
$-10\sqrt{x} = -20$
$\sqrt{x} = 2$
$(\sqrt{x})^2 = 2^2$
$x = 4$

Check:

$5-\sqrt{x}$	$=\sqrt{x+5}$
$5-\sqrt{4}$	$\sqrt{4+5}$
$5-2$	$\sqrt{9}$

$3 = 3$

The solution is 4.

8. $(x_1, y_2) = (4, 2), (x_2, y_2) = (-1, -1)$

$d = \sqrt{(x_1-x_2)^2 + (y_1-y_2)^2}$
$= \sqrt{(4-(-1))^2 + (2-(-1))^2}$
$= \sqrt{5^2+3^2}$
$= \sqrt{25+9}$
$= \sqrt{34}$

$(x-h)^2 + (y-k)^2 = r^2$
$(x-(-1))^2 + (y-(-1))^2 = (\sqrt{34})^2$
$(x+1)^2 + (y+1)^2 = 34$

9. (1) $3x - 3y = 2$
(2) $6x - 4y = 5$
Eliminate x.
Multiply Equation (1) by -2 and add Equation (2).
$-6x + 6y = -4$
$6x - 4y = 5$
$2y = 1$
$y = \dfrac{1}{2}$

Substitute $\dfrac{1}{2}$ for y in Equation (2).

$6x - 4\left(\dfrac{1}{2}\right) = 5$
$6x - 2 = 5$
$6x = 7$
$x = \dfrac{7}{6}$

The solution is $\left(\dfrac{7}{6}, \dfrac{1}{2}\right)$.

10. $\begin{vmatrix} -3 & 1 \\ 4 & 2 \end{vmatrix} = -3(2) - 4(1) = -6 - 4 = -10$

11. $2x - 1 > 3$ or $1 - 3x > 7$
$2x > 4$ $\quad -3x > 6$
$x > 2$ $\quad x < -2$

$\{x|x > 2\}$ or $\{x|x < -2\}$
$\{x|x < -2 \text{ or } x > 2\}$

12. $2x - 3y < 9$
$-3y < -2x + 9$
$y > \dfrac{2}{3}x - 3$

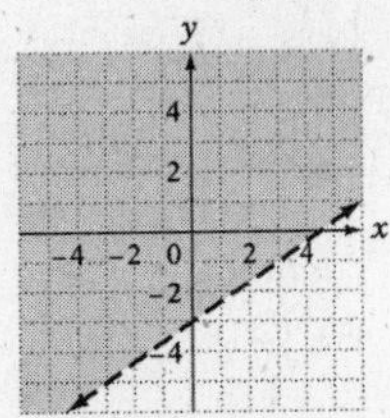

13. $\log_5\sqrt{\frac{x}{y}} = \frac{1}{2}[\log_5 x - \log_5 y]$
$= \frac{1}{2}\log_5 x - \frac{1}{2}\log_5 y$

14.
$4^x = 8^{x-1}$
$(2^2)^x = (2^3)^{x-1}$
$2^{2x} = 2^{3(x-1)}$
$2x = 3(x-1)$
$2x = 3x - 3$
$-x = -3$
$x = 3$
The solution is 3.

15. $a_n = n(n-1)$
$a_5 = 5(5-1) = 5(4) = 20$
$a_6 = 6(6-1) = 6(5) = 30$
The fifth term is 20.
The sixth term is 30.

16. $\sum_{n=1}^{7}(-1)^{n-1}(n+2) = (-1)^{1-1}(1+2) + (-1)^{2-1}(2+2) + (-1)^{3-1}(3+2) + (-1)^{4-1}(4+2) + (-1)^{5-1}(5+2)$
$+ (-1)^{6-1}(6+2) + (-1)^{7-1}(7+2)$
$= (-1)^0(3) + (-1)^1(4) + (-1)^2(5) + (-1)^3(6) + (-1)^4(7) + (-1)^5(8) + (-1)^6(9)$
$= 3 - 4 + 5 - 6 + 7 - 8 + 9 = 6$

17. $d = a_2 - a_1 = -10 - (-7) = -10 + 7 = -3$
$a_n = a_1 + (n-1)d$
$a_{33} = -7 + (33-1)(-3) = -7 + 32(-3) = -103$
The thirty-third term is -103.

18. $r = \frac{a_2}{a_1} = -\frac{2}{3}$

$S = \frac{a_1}{1-r} = \frac{3}{1-\left(-\frac{2}{3}\right)} = \frac{3}{\frac{5}{3}} = \frac{9}{5}$

19. $0.4\overline{6} = 0.4 + 0.06 + 0.006 + \cdots$
$= \frac{4}{10} + \frac{6}{100} + \frac{6}{1000} + \cdots$

$S = \frac{a_1}{1-r} = \frac{\frac{6}{100}}{1-\frac{1}{10}} = \frac{\frac{6}{100}}{\frac{9}{10}} = \frac{6}{100}\cdot\frac{10}{9} = \frac{1}{15}$

$0.4\overline{6} = \frac{4}{10} + \frac{1}{15} = \frac{14}{30} = \frac{7}{15}$

An equivalent fraction is $\frac{7}{15}$.

20. $n = 6,\ a = 2x,\ b = y,\ r = 6$

$\binom{6}{6-1}(2x)^{6-6+1}(y)^{6-1} = \binom{6}{5}(2x)^1 y^5$
$= 6 \cdot 2x \cdot y^5$
$= 12xy^5$

21. Strategy Amount of pure water:

	Amount	Percent	Value
Water	x	0	$0 \cdot x$
8%	200	0.08	200(0.08)
5%	$x+200$	0.05	$(x+200)(0.05)$

• The sum of the values before mixing equals the value after mixing.
$0 \cdot x + 200(0.08) = (x+200)(0.05)$

Solution
$$0 \cdot x + 16 = 0.05x + 10$$
$$16 = 0.05x + 10$$
$$6 = 0.05x$$
$$120 = x$$
The amount of water that must be added is 120 oz.

22. Strategy • Time required for the older computer: t
Time required for the new computer: $t - 16$

	Rate	Time	Part
Older Computer	$\frac{1}{t}$	15	$\frac{15}{t}$
New computer	$\frac{1}{t-16}$	15	$\frac{15}{t-16}$

• The sum of the part of the task completed by the older computer and the part of the task completed by the new computer is 1.

Solution
$$\frac{15}{t} + \frac{15}{t-16} = 1$$
$$t(t-16)\left(\frac{15}{t} + \frac{15}{t-16}\right) = 1(t)(t-16)$$
$$15(t-16) + 15t = t^2 - 16t$$
$$15t - 240 + 15t = t^2 - 16t$$
$$0 = t^2 - 46t + 240$$
$$0 = (t-40)(t-6)$$
$t - 40 = 0 \quad t - 6 = 0$
$t = 40 \quad t = 6$
$t = 6$ does not check as a solution since $t - 16 = 6 - 16 = -10$.
$t = 40$, $t - 16 = 24$

The new computer takes 24 min to complete the payroll. The older computer takes 40 min to complete the payroll.

23. Strategy • Rate of boat in calm water: x
Rate of current: y

	Rate	Time	Distance
With current	$x+y$	2	$2(x+y)$
Against current	$x-y$	3	$3(x-y)$

• The distance traveled with the current is 15 mi. The distance traveled against the current is 15 mi.
$2(x+y) = 15$
$3(x-y) = 15$

Solution
$$2(x+y) = 15 \quad \frac{1}{2} \cdot 2(x+y) = \frac{1}{2} \cdot 15$$
$$3(x-y) = 15 \quad \frac{1}{3} \cdot 3(x-y) = \frac{1}{3} \cdot 15$$
$$x + y = 7.5$$
$$x - y = 5$$
$$2x = 12.5$$
$$x = 6.25$$
$$x + y = 7.5$$
$$6.25 + y = 7.5$$
$$y = 1.25$$
The boat was traveling 6.25 mph and the current was traveling 1.25 mph.

24. Strategy To find the half-life, solve for k in the exponential decay equation.
Use $A_0 = 80$, $A = 55$, $t = 30$

Solution
$$A = A_0\left(\frac{1}{2}\right)^{t/k}$$
$$55 = 80\left(\frac{1}{2}\right)^{30/k}$$
$$0.6875 = (0.5)^{30/k}$$
$$\log 0.6875 = \frac{30}{k}\log 0.5$$
$$k = \frac{30\log 0.5}{\log 0.6875} \approx 55.49$$
The half-life is 55 days.

25. Strategy To find the total number of seats in the 12 rows of the theater:
- Write the arithmetic sequence.
- Use the Formula for the *n*th Term of an Arithmetic Sequence to find the 12th term.
- Use the Formula for the Sum of *n* Terms of an Arithmetic Sequence to find the sum of the 12 terms of the sequence.

Solution 62, 74, 86, ...

$d = a_2 - a_1 = 74 - 62 = 12$

$a_{12} = 62 + (12 - 1)12 = 62 + 132 = 194$

$S_n = \frac{n}{2}(a_1 + a_n)$

$S_{12} = \frac{12}{2}(62 + 194) = 1536$

There are 1536 seats in the theater.

26. Strategy To find the height of the ball on the fifth bounce, use the Formula for the *n*th Term of a Geometric Sequence.

Solution $n = 5, a_1 = 80\%$ of $8 = 6.4$,

$r = 80\% = 0.8$

$a_n = a_1 r^{n-1}$

$a_5 = 6.4(0.8)^{5-1}$

$= 6.4(0.4096) = 2.62144$

The height of the fifth bounce is 2.6 ft.

Final Exam

1. $12 - 8[3 - (-2)]^2 \div 5 - 3 = 12 - 8[5]^2 \div 5 - 3$
$$= 12 - 8(25) \div 5 - 3$$
$$= 12 - 200 \div 5 - 3$$
$$= 12 - 40 - 3$$
$$= -31$$

2. $\dfrac{3^2 - (-4)^2}{3 - (-4)} = \dfrac{9 - 16}{3 + 4} = \dfrac{-7}{7} = -1$

3. $5 - 2[3x - 7(2 - x) - 5x] = 5 - 2[3x - 14 + 7x - 5x]$
$$= 5 - 2[5x - 14]$$
$$= 5 - 10x + 28$$
$$= 33 - 10x$$

4. $\dfrac{3}{4}x - 2 = 4$
$$\frac{3}{4}x = 6$$
$$\frac{4}{3} \cdot \frac{3}{4}x = \frac{4}{3} \cdot 6$$
$$x = 8$$

5. $\dfrac{2 - 4x}{3} - \dfrac{x - 6}{12} = \dfrac{5x - 2}{6}$
$$12\left(\frac{2 - 4x}{3} - \frac{x - 6}{12}\right) = 12\left(\frac{5x - 2}{6}\right)$$
$$4(2 - 4x) - (x - 6) = 2(5x - 2)$$
$$8 - 16x - x + 6 = 10x - 4$$
$$14 - 17x = 10x - 4$$
$$-27x = -18$$
$$x = \frac{2}{3}$$

6. $8 - |5 - 3x| = 1$
$$-|5 - 3x| = -7$$
$$|5 - 3x| = 7$$

$5 - 3x = 7$ $\qquad$ $5 - 3x = -7$

$-3x = 2$ $\qquad$ $-3x = -12$

$x = -\dfrac{2}{3}$ $\qquad$ $x = 4$

The solutions are $-\dfrac{2}{3}$ and 4.

7. $2x - 3y = 9$ $\qquad$ $2x - 3y = 9$

$2x - 3(0) = 9$ $\qquad$ $2(0) - 3y = 9$

$2x = 9$ $\qquad$ $-3y = 9$

$x = \dfrac{9}{2}$ $\qquad$ $y = -3$

x-intercept: $\left(\dfrac{9}{2}, 0\right)$ $\quad$ y-intercept: $(0, -3)$

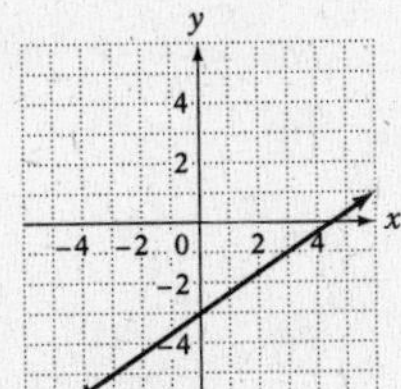

8. $(x_1, y_1) = (3, -2), (x_2, y_2) = (1, 4)$
$$m = \frac{y_2 - y_1}{x_2 - x_1} = \frac{4 - (-2)}{1 - 3} = \frac{6}{-2} = -3$$
$$y - y_1 = m(x - x_1)$$
$$y - (-2) = -3(x - 3)$$
$$y + 2 = -3x + 9$$
$$y = -3x + 7$$

The equation of the line is $y = -3x + 7$.

9. $3x - 2y = 6$
$$-2y = -3x + 6$$
$$y = \frac{3}{2}x - 3$$
$$m_1 = \frac{3}{2}$$
$$m_1 \cdot m_2 = -1$$
$$\frac{3}{2}m_2 = -1$$
$$m_2 = -\frac{2}{3} \qquad (x_1, y_1) = (-2, 1)$$
$$y - y_1 = m(x - x_1)$$
$$y - 1 = -\frac{2}{3}(x - (-2))$$
$$y - 1 = -\frac{2}{3}(x + 2)$$
$$y - 1 = -\frac{2}{3}x - \frac{4}{3}$$
$$y = -\frac{2}{3}x - \frac{1}{3}$$

The equation of the line is $y = -\dfrac{2}{3}x - \dfrac{1}{3}$.

10. $2a[5 - a(2 - 3a) - 2a] + 3a^2$
$$= 2a[5 - 2a + 3a^2 - 2a] + 3a^2$$
$$= 2a[5 - 4a + 3a^2] + 3a^2$$
$$= 10a - 8a^2 + 6a^3 + 3a^2$$
$$= 6a^3 - 5a^2 + 10a$$

11. $8 - x^3y^3 = 2^3 - (xy)^3$
$$= (2 - xy)(4 + 2xy + x^2y^2)$$

12. $x - y - x^3 + x^2y = x - y - x^2(x - y)$
$$= 1(x - y) - x^2(x - y)$$
$$= (x - y)(1 - x^2)$$
$$= (x - y)(1 - x)(1 + x)$$

13.
$$\begin{array}{r}
x^2 - 2x - 3 \\
2x - 3\overline{)2x^3 - 7x^2 + 0x + 4} \\
\underline{2x^3 - 3x^2} \\
-4x^2 + 0x \\
\underline{-4x^2 + 6x} \\
-6x + 4 \\
\underline{-6x + 9} \\
-5
\end{array}$$

$$\frac{2x^3 - 7x^2 + 4}{2x - 3} = x^2 - 2x - 3 - \frac{5}{2x - 3}$$

14. $\dfrac{x^2-3x}{2x^2-3x-5} \div \dfrac{4x-12}{4x^2-4}$

$= \dfrac{x^2-3x}{2x^2-3x-5} \cdot \dfrac{4x^2-4}{4x-12}$

$= \dfrac{x(x-3)}{(2x-5)(x+1)} \cdot \dfrac{4(x+1)(x-1)}{4(x-3)}$

$= \dfrac{x(x-3)(x+1)(x-1)}{(2x-5)(x+1)(x-3)}$

$= \dfrac{x(x-1)}{2x-5}$

15. $\dfrac{x-2}{x+2} - \dfrac{x+3}{x-3} = \dfrac{x-2}{x+2} \cdot \dfrac{x-3}{x-3} - \dfrac{x+3}{x-3} \cdot \dfrac{x+2}{x+2}$

$= \dfrac{x^2-5x+6-(x^2+5x+6)}{(x+2)(x-3)}$

$= \dfrac{x^2-5x+6-x^2-5x-6}{(x+2)(x-3)}$

$= \dfrac{-10x}{(x+2)(x-3)}$

16. $\dfrac{\frac{3}{x}+\frac{1}{x+4}}{\frac{1}{x}+\frac{3}{x+4}} = \dfrac{\frac{3}{x}+\frac{1}{x+4}}{\frac{1}{x}+\frac{3}{x+4}} \cdot \dfrac{x(x+4)}{x(x+4)}$

$= \dfrac{3(x+4)+x}{x+4+3x}$

$= \dfrac{3x+12+x}{4x+4}$

$= \dfrac{4x+12}{4x+4}$

$= \dfrac{4(x+3)}{4(x+1)} = \dfrac{x+3}{x+1}$

17.

$$\frac{5}{x-2} - \frac{5}{x^2-4} = \frac{1}{x+2}$$

$$(x+2)(x-2)\left(\frac{5}{x-2} - \frac{5}{(x+2)(x-2)}\right) = \frac{1}{x+2}(x+2)(x-2)$$

$$5(x+2) - 5 = x-2$$

$$5x+10-5 = x-2$$

$$5x+5 = x-2$$

$$4x = -7$$

$$x = -\frac{7}{4}$$

The solution is $-\dfrac{7}{4}$.

18.

$a_n = a_1 + (n-1)d$

$a_n - a_1 = (n-1)d$

$\dfrac{a_n - a_1}{n-1} = d$

$d = \dfrac{a_n - a_1}{n-1}$

19. $\left(\dfrac{4x^2y^{-1}}{3x^{-1}y}\right)^{-2}\left(\dfrac{2x^{-1}y^2}{9x^{-2}y^2}\right)^3$

$= \dfrac{4^{-2}x^{-4}y^2}{3^{-2}x^2y^{-2}} \cdot \dfrac{2^3x^{-3}y^6}{9^3x^{-6}y^6}$

$= 4^{-2} \cdot 3^{-(-2)} \cdot x^{-4-2}y^{2-(-2)} \cdot 2^3 \cdot 9^{-3}x^{-3-(-6)}y^{6-6}$

$= 4^{-2} \cdot 3^2x^{-6}y^4 \cdot 2^3 \cdot 9^{-3}x^3y^0$

$= \dfrac{9x^{-3}y^4 \cdot 8}{16 \cdot 729} = \dfrac{y^4}{162x^3}$

20. $\left(\dfrac{3x^{2/3}y^{1/2}}{6x^2y^{4/3}}\right)^6 = \dfrac{3^6x^4y^3}{6^6x^{12}y^8}$

$= \dfrac{729x^{4-12}y^{3-8}}{46656}$

$= \dfrac{1x^{-8}y^{-5}}{64}$

$= \dfrac{1}{64x^8y^5}$

21. $x\sqrt{18x^2y^3} - y\sqrt{50x^4y}$

$= x\sqrt{3^2x^2y^2(2y)} - y\sqrt{5^2x^4(2y)}$

$= 3x^2y\sqrt{2y} - 5x^2y\sqrt{2y}$

$= -2x^2y\sqrt{2y}$

22. $\dfrac{\sqrt{16x^5y^4}}{\sqrt{32xy^7}} = \sqrt{\dfrac{16x^5y^4}{32xy^7}}$

$= \sqrt{\dfrac{x^4}{2y^3}}$

$= \sqrt{\dfrac{x^4}{y^2(2y)}}$

$= \dfrac{x^2}{y}\sqrt{\dfrac{1}{2y}} \cdot \sqrt{\dfrac{2y}{2y}}$

$= \dfrac{x^2}{y}\sqrt{\dfrac{1 \cdot 2y}{(2y)^2}}$

$= \dfrac{x^2\sqrt{2y}}{2y^2}$

23. $\dfrac{3}{2+i} \cdot \dfrac{2-i}{2-i} = \dfrac{6-3i}{4+1} = \dfrac{6-3i}{5} = \dfrac{6}{5} - \dfrac{3}{5}i$

24.

$$(x-r_1)(x-r_2) = 0$$

$$\left(x-\left(-\frac{1}{2}\right)\right)(x-2) = 0$$

$$\left(x+\frac{1}{2}\right)(x-2) = 0$$

$$x^2 - \frac{3}{2}x - 1 = 0$$

$$2\left(x^2 - \frac{3}{2}x - 1\right) = 2(0)$$

$$2x^2 - 3x - 2 = 0$$

25. $2x^2 - 3x - 1 = 0$

$a = 2, b = -3, c = -1$

$$x = \frac{-b \pm \sqrt{b^2 - 4ac}}{2a}$$

$$= \frac{-(-3) \pm \sqrt{(-3)^2 - 4(2)(-1)}}{2(2)}$$

$$= \frac{3 \pm \sqrt{9+8}}{4} = \frac{3 \pm \sqrt{17}}{4}$$

The solutions are $\frac{3+\sqrt{17}}{4}$ and $\frac{3-\sqrt{17}}{4}$.

26. $x^{2/3} - x^{1/3} - 6 = 0$

$(x^{1/3})^2 - x^{1/3} - 6 = 0$

Let $u = x^{1/3}$.

$u^2 - u - 6 = 0$

$(u-3)(u+2) = 0$

$u - 3 = 0$	$u + 2 = 0$
$u = 3$	$u = -2$
$x^{1/3} = 3$	$x^{1/3} = -2$
$(x^{1/3})^3 = 3^3$	$(x^{1/3})^3 = (-2)^3$
$x = 27$	$x = -8$

The solutions are 27 and −8.

27.

$$\frac{2}{x} - \frac{2}{2x+3} = 1$$

$$x(2x+3)\left(\frac{2}{x} - \frac{2}{2x+3}\right) = 1[x(2x+3)]$$

$$2(2x+3) - 2x = 2x^2 + 3x$$

$$4x + 6 - 2x = 2x^2 + 3x$$

$$2x + 6 = 2x^2 + 3x$$

$$0 = 2x^2 + x - 6$$

$$0 = (2x-3)(x+2)$$

$2x - 3 = 0$	$x + 2 = 0$
$x = \frac{3}{2}$	$x = -2$

The solutions are $\frac{3}{2}$ and −2.

28. $f(x) = -x^2 + 4$

$-\frac{b}{2a} = -\frac{0}{2(-1)} = 0$

$f(x) = -0^2 + 4 = 4$

Vertex: (0, 4)

Axis of symmetry: ($x = 0$)

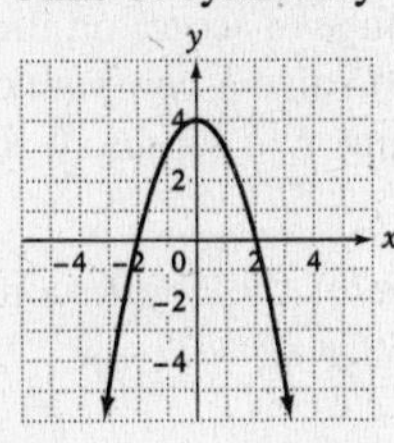

29. x-intercepts: (4, 0) and (−4, 0)

y-intercepts: (0, 2) and (0, −2)

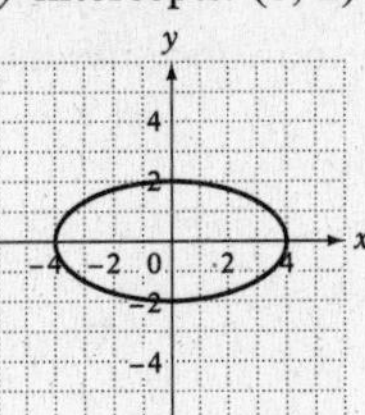

30.

$$f(x) = \frac{2}{3}x - 4$$

$$y = \frac{2}{3}x - 4$$

$$x = \frac{2}{3}y - 4$$

$$x + 4 = \frac{2}{3}y$$

$$\frac{3}{2}(x+4) = \frac{3}{2} \cdot \frac{2}{3}y$$

$$\frac{3}{2}x + 6 = y$$

The inverse function is $f^{-1}(x) = \frac{3}{2}x + 6$.

31. (1) $3x - 2y = 1$

(2) $5x - 3y = 3$

Eliminate y.

Multiply Equation (1) by −3 and Equation (2) by 2. Add the two new equations.

$-3(3x - 2y) = -3(1)$	$-9x + 6y = -3$
$2(5x - 3y) = 2(3)$	$10x - 6y = 6$
	$x = 3$

Substitute 3 for x in Equation (1).

$3x - 2y = 1$

$3(3) - 2y = 1$

$9 - 2y = 1$

$-2y = -8$

$y = 4$

The solution is (3, 4).

32. $\begin{vmatrix} 3 & 4 \\ -1 & 2 \end{vmatrix} = 3(2) - (-1)(4) = 6 + 4 = 10$

33. (1) $x^2 - y^2 = 4$
(2) $x + y = 1$
Solve Equation (2) for y and substitute into Equation (1).

$$\begin{aligned} x + y &= 1 \\ y &= -x + 1 \\ x^2 - y^2 &= 4 \\ x^2 - (-x+1)^2 &= 4 \\ x^2 - (x^2 - 2x + 1) &= 4 \\ x^2 - x^2 + 2x - 1 &= 4 \\ 2x &= 5 \\ x &= \frac{5}{2} \end{aligned}$$

Substitute $\frac{5}{2}$ for x in Equation (2).

$$\begin{aligned} x + y &= 1 \\ \frac{5}{2} + y &= 1 \\ y &= -\frac{3}{2} \end{aligned}$$

The solution is $\left(\frac{5}{2}, -\frac{3}{2}\right)$.

34. $2 - 3x < 6$ and $2x + 1 > 4$

$$\begin{aligned} -3x &< 4 & 2x &> 3 \\ x &> -\frac{4}{3} & x &> \frac{3}{2} \end{aligned}$$

The solution is $\left(\frac{3}{2}, \infty\right)$.

35.

$$\begin{aligned} |2x + 5| &< 3 \\ -3 < 2x &+ 5 < 3 \\ -3 - 5 < 2x &< 3 - 5 \\ -8 < 2x &< -2 \\ \frac{1}{2}(-8) < \frac{1}{2}(2x) &< \frac{1}{2}(-2) \\ -4 < x &< -1 \end{aligned}$$

$\{x \mid -4 < x < -1\}$

36.

$$\begin{aligned} 3x + 2y &> 6 \\ 2y &> -3x + 6 \\ y &> -\frac{3}{2}x + 3 \end{aligned}$$

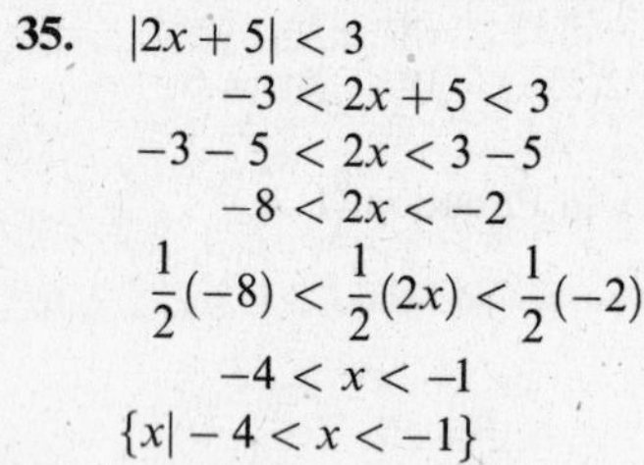

37.

$$\begin{aligned} f(x) &= \log_2(x+1) \\ y &= \log_2(x+1) \\ 2^y &= x + 1 \\ -1 + 2^y &= x \end{aligned}$$

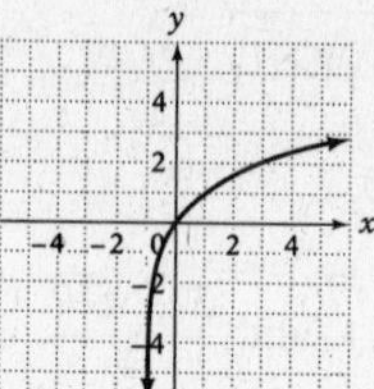

38. $2(\log_2 a - \log_2 b) = 2\log_2 \frac{a}{b} = \log_2 \frac{a^2}{b^2}$

39.

$$\begin{aligned} \log_3 x - \log_3 (x-3) &= \log_3 2 \\ \log_3 \left(\frac{x}{x-3}\right) &= \log_3 2 \end{aligned}$$

Use the fact that if $\log_b u = \log_b v$, then $u = v$.

$$\begin{aligned} \frac{x}{x-3} &= 2 \\ (x-3) \cdot \frac{x}{x-3} &= 2 \cdot (x-3) \\ x &= 2x - 6 \\ -x &= -6 \\ x &= 6 \end{aligned}$$

The solution is 6.

40.

$$\begin{aligned} \sum_{i=1}^{5} 2y^i &= 2y^1 + 2y^2 + 2y^3 + 2y^4 + 2y^5 \\ &= 2y + 2y^2 + 2y^3 + 2y^4 + 2y^5 \end{aligned}$$

41.

$$\begin{aligned} 0.5\overline{1} &= 0.5 + 0.01 + 0.001 + 0.0001 + \cdots \\ &= \frac{5}{10} + \frac{1}{100} + \frac{1}{1000} + \frac{1}{10{,}000} + \cdots \end{aligned}$$

$$S = \frac{a_1}{1-r} = \frac{\frac{1}{100}}{1 - \frac{1}{10}} = \frac{\frac{1}{100}}{\frac{9}{10}} = \frac{1}{100} = \frac{10}{9} = \frac{1}{90}$$

$$0.5\overline{1} = \frac{5}{10} + \frac{1}{90} = \frac{46}{90} = \frac{23}{45}$$

42. $n = 9$, $a = x$, $b = -2y$, $r = 3$

$$\begin{aligned} \binom{9}{3-1} x^{9-3+1}(-2y)^{3-1} &= \binom{9}{2} x^7(-2y)^2 \\ &= 36x^7 \cdot 4y^2 \\ &= 144x^7y^2 \end{aligned}$$

43. Strategy To find the range of scores on the fifth test, write and solve a compound inequality using x to represent the fifth test.

Solution $70 \leq$ average of the 5 test scores ≤ 79

$$\begin{aligned} 70 \leq \frac{64 + 58 + 82 + 77 + x}{5} &\leq 79 \\ 70 \leq \frac{281 + x}{5} &\leq 79 \\ 350 \leq 281 + x &\leq 395 \\ 69 \leq x &\leq 114 \end{aligned}$$

The range of scores is $69 \leq x \leq 100$.

44. Strategy
- Average speed of jogger: x
- Average speed of cyclist: $2.5x$

	Rate	Time	Distance
Jogger	x	2	$2x$
Cyclist	$2.5x$	2	$2(2.5x)$

- The distance traveled by the cyclist is 24 more miles than the distance traveled by the jogger.

Solution

$$2x + 24 = 2(2.5x)$$
$$2x + 24 = 5x$$
$$24 = 3x$$
$$8 = x$$
$$40 = 5x$$

The cyclist traveled 40 mi.

45. Strategy
- Amount invested at 8.5%: x
  Amount invested at 6.4%: $12000 - x$

	Principal	Rate	Interest
8.5%	x	0.085	$0.085x$
6.4%	$12000 - x$	0.064	$0.064(12000 - x)$

- The sum of the interest earned by the two investments is \$936.

Solution

$$0.085x + 0.064(12000 - x) = 936$$
$$0.085x + 768 - 0.064x = 936$$
$$0.021x + 768 = 936$$
$$0.021x = 168$$
$$x = 8000$$
$$12000 - x = 4000$$

The amount invested at 8.5% is \$8000.
The amount invested at 6.4% is \$4000.

46. Strategy
- The width of the rectangle: x
  The length of the rectangle: $3x - 1$
- Use the formula for the area of a rectangle ($A = lw$) if the area is 140 ft^2.

Solution

$$A = l \cdot w$$
$$140 = (3x - 1)x$$
$$140 = 3x^2 - x$$
$$0 = 3x^2 - x - 140$$
$$0 = (3x + 20)(x - 7)$$

$$3x + 20 = 0 \qquad x - 7 = 0$$
$$x = -\frac{20}{3} \qquad x = 7$$

The solution $-\frac{20}{3}$ does not check because the width cannot be negative.
$3x - 1 = 3(7) - 1 = 20$
The width is 7 ft. The length is 20 ft.

47. Strategy To find the number of additional shares, write and solve a proportion using x to represent the number of shares.

Solution

$$\frac{300}{486} = \frac{300 + x}{810}$$
$$(810 \cdot 486) \cdot \frac{300}{486} = \frac{300 + x}{810} \cdot (810 \cdot 486)$$
$$810(300) = (300 + x)486$$
$$243{,}000 = 145{,}800 + 486x$$
$$-486x = -97200$$
$$x = 200$$

200 additional shares would need to be purchased.

48. Strategy
- Rate of car: x
  Rate of plane: $7x$

	Distance	Rate	Time
Car	45	x	$\frac{45}{x}$
Plane	1050	$7x$	$\frac{1050}{7x}$

- The total time traveled is $3\frac{1}{4}$ h.

Solution

$$\frac{45}{x} + \frac{1050}{7x} = 3\frac{1}{4}$$
$$\frac{45}{x} + \frac{150}{x} = \frac{13}{4}$$
$$4x\left(\frac{45}{x} + \frac{150}{x}\right) = 4x\left(\frac{13}{4}\right)$$
$$180 + 600 = 13x$$
$$780 = 13x$$
$$60 = x$$
$$420 = 7x$$

The rate of the plane is 420 mph.

49. Strategy To find the distance the object has fallen, substitute 75 ft/s for v in the formula and solve for d.

Solution

$$v = \sqrt{64d}$$
$$75 = \sqrt{64d}$$
$$75^2 = (\sqrt{64d})^2$$
$$5625 = 64d$$
$$87.89 \approx d$$

The distance traveled is 88 ft.

50. Strategy • Rate traveled during the first 360 mi: x
Rate traveled during the next 300 mi: $x + 30$

	Distance	Rate	Time
First part of trip	360	x	$\frac{360}{x}$
Second part of trip	300	$x + 30$	$\frac{300}{x+30}$

• Total time traveled during the trip was 5 h.

Solution

$$\frac{360}{x} + \frac{300}{x+30} = 5$$
$$x(x+30)\left(\frac{360}{x} + \frac{300}{x+30}\right) = 5(x(x+30))$$
$$360(x+30) + 300x = 5(x^2 + 30x)$$
$$360x + 10800 + 300x = 5x^2 + 150x$$
$$660x + 10800 = 5x^2 + 150x$$
$$0 = 5x^2 - 510x - 10800$$
$$0 = 5(x^2 - 102x - 2160)$$
$$0 = (x+18)(x-120)$$
$$x + 18 = 0 \qquad x - 120 = 0$$
$$x = -18 \qquad x = 120$$

The solution −18 does not check because the rate cannot be negative. The rate of the plane for the first 360 mi is 120 mph.

51. Strategy To find the intensity:
• Write the basic inverse variation equation, replace the variable by the given values, and solve for k.
• Write the inverse variation equation, replace k by its value. Substitute 4 for d and solve for L.

Solution

$$L = \frac{k}{d^2}$$
$$8 = \frac{k}{(20)^2}$$
$$8 \cdot 400 = k$$
$$3200 = k$$
$$L = \frac{3200}{d^2}$$
$$L = \frac{3200}{4^2}$$
$$L = \frac{3200}{16} = 200$$

The intensity is 200 foot-candles.

52. Strategy • Rate of the boat in calm water: x
Rate of the current: y

	Rate	Time	Distance
With current	$x + y$	2	$2(x+y)$
Against current	$x - y$	3	$3(x-y)$

• The distance traveled with the current is 30 mi. The distance traveled against the current is 30 mi.
$2(x+y) = 30$
$3(x-y) = 30$

Solution

$$2(x+y) = 30 \quad \frac{1}{2} \cdot 2(x+y) = \frac{1}{2} \cdot 30$$
$$3(x-y) = 30 \quad \frac{1}{3} \cdot 3(x-y) = \frac{1}{3} \cdot 30$$
$$x + y = 15$$
$$x - y = 10$$
$$2x = 25$$
$$x = 12.5$$
$$x + y = 15$$
$$12.5 + y = 15$$
$$y = 2.5$$

The rate of the boat in calm water is 12.5 mph. The rate of the current is 2.5 mph.

53. Strategy To find the value of the investment after two years, solve the compound interest formula for P. Use $A = 4000$, $n = 24$, $i = \frac{9\%}{12} = \frac{0.09}{12} = 0.0075$.

Solution

$$P = A(1+i)^n$$
$$P = 4000(1 + 0.0075)^{24}$$
$$P = 4000(1.0075)^{24}$$
$$P \approx 4785.65$$

The value of the investment is \$4785.65.

54. Strategy To find the value of the house in 20 years, use the Formula for the nth Term of a Geometric Sequence.

Solution $n = 20$, $a_1 = 1.06(80{,}000) = 84{,}800$, $r = 1.06$

$$a_n = a_1(r)^{n-1}$$
$$a_{20} = 84{,}800(1.06)^{20-1}$$
$$= 84{,}800(1.06)^{19}$$
$$\approx 256{,}571$$

The value of the house will be \$256,571.